Library of Congress Cataloging-in-Publication Data

Electrical insulating liquids / R. Bartnikas, editor.
(Engineering dielectrics; v. 3)
Includes bibliographical references and index.
ISBN 0-8031-2055-9
1. Electric insulators and insulation—Liquids. 2. Dielectrics.
I. Bartnikas, R. II. Series.
TK3401.E53 vol. 3
[TK3441.L5]
621.319'37 s—dc20
[621.319'37] 93-27813
CIP

ASTM Publication Code Number (PCN)
31-002093-21

Printed in Chelsea, MI
January 1994

Monograph 2

Engineering Dielectrics Volume III Electrical Insulating Liquids

R. Bartnikas
Editor
Institut de Recherche d'Hydro-Québec
Varennes, Québec, Canada

ASTM
1916 Race Street
Philadelphia, PA
Printed in the U.S.A.

Foreword

The present publication, Volume III, concerns electrical insulating liquids and forms an integral part of a monograph series entitled *Engineering Dielectrics.* The subject matter of the series encompasses various practical and theoretical facets of dielectric materials and systems that are suitable for use in electrical apparatus, cables, and devices. The series is comprised of eight volumes, namely:

Volume I	Corona Measurement and Interpretation (ASTM STP 669)
Volume IIA	Electrical Properties of Solid Insulating Materials: Molecular Structure and Electrical Behavior (ASTM STP 783)
Volume IIB	Electrical Properties of Solid Insulating Materials: Measurement Techniques (ASTM STP 926)
Volume III	Electrical Insulating Liquids
Volume IV	Electrical Solid-Liquid Insulating Systems
Volume V	Cable Insulating Systems
Volume VI	Transformer and Rotating Machine Insulating Systems
Volume VII	Capacitor and Bushing Insulating Systems

The overall monograph series is sponsored by ASTM Committee D9 on Electrical and Electronic Insulating Materials, with the present volume being also cosponsored by ASTM Committee D27 on Electrical Insulating Liquids and Gases. As a result of the depth to which the subject matter of this series must be addressed in both the theoretical and practical aspects, the progress and preparation of the individual volumes has been proceeding at a substantially slower pace than one may have wished. It is, therefore, with some solicitude on my part that I have attempted to devote increasing amounts of my disponible time on efforts to accelerate the preparation of subsequent volumes. To all the users of the series, who have placed enquiries concerning the release of future volumes, I wish to express my gratitude for their patience and understanding.

Varennes, Sept. 18, 1991

R. Bartnikas
Editor

Contents

A glimpse of Brownian motion from antiquity

Do but observe:
Whenever beams enter and pour
The sunlight through the dark chambers of a house,
You will perceive many minute bodies mingling,
In a multiplicity of ways within those rays of light
Throughout the entire space, and as it were
In a never ending conflict of battle
Combating and contending troop with troop
Without pause, maintained in motion by perpetual
Encounters and separations; so that this
Should assist you to imagine what it signifies
When primordial particles of matter
Are always meandering in a great void.
To this extent a small thing may suggest
A picture of great things, and point the way
To new concepts. There is another reason
Why you should give attention to those bodies
Which are seen wavering confusedly
In the rays of the sun: such waverings indicate
That beneath appearance there must be
Motions of matter secret and unseen.

— — —

Contemplator enim, cum solis lumina cumque
inserti fundunt radii per opaca domorum:
multa minuta modis multis per inane videbis
corpora miscari radiorum lumine in ipso
et velut aeterno certamine proelia pugnas
edere turmatim certantia nec dare pausam,
conciliis et discidiis exercita crebis;
conicere ut possis ex hoc, primordia rerum
quale sit in magno iactari semper inani.
dumtaxat rerum magnarum parva potest res
exemplare dare et vestigia notitiai
hoc etiam magnis haec animum te advertare par est
corpora quae in solis radiis turbare videntur
guod tales turbae motus quoque materiai
significant clandestinos caecosque subesse.

Lucretius
in De Rerum Natura

Preface

The present volume constitutes an integral part of the monograph series on *Engineering Dielectrics* and concerns the electrical behavior and properties of insulating liquids. Some material on measurement techniques is also included in so far as it may be different from that, or is not described fully, in Vol. IIB, which is devoted to measurement techniques mainly as applied to solid dielectrics. Volume III is a precursor of Vol. IV, which deals primarily with the subject of electrical solid-liquid insulating systems.

A total of five chapters comprise Volume III. The electrical characteristics of insulating liquids are treated from the fundamental point of view in the first two chapters. Chapter 1 deals with the permittivity and dielectric loss and provides as well a description of the test methods and specimen cells particularly devised for the measurement of the dielectric properties of liquids over the entire frequency spectrum, extending into the optical frequency range. Since the vast majority of insulating liquids are utilized at power frequencies, a substantial amount of effort is expended in discussing the dielectric behavior at these frequencies. However, the dielectric frequency response of insulating liquids is also given considerable attention in that it permits to establish the nature of the dielectric loss at power frequencies and simultaneously reveals some pertinent data on the molecular structure of the liquids and on the manner in which that structure influences the dielectric loss and permittivity behavior. Chapter 2 considers the fundamental conduction processes, formation of space charges and prebreakdown and breakdown mechanisms from the quantitative point of view. The latter material prepares the reader for an introduction into Chapter 3, which presents a qualitative account of the phenomenological events that precipitate the electrical breakdown of liquids. Chapter 4 discusses the physical and chemical properties of mineral oils and provides a description of the various test methods for determining these properties. Finally, Chapter 5 considers the general molecular structural nature of insulating liquids and their composition.

Both SI and CGS-Gaussian units are employed in this volume as well as throughout the entire monograph series. The concurrent use of the two systems of measure is intentional, since it is desirable to remain *au fait* with both systems in order to sustain a fluently conversant link between the more recent and early parts of the literature on dielectrics. Useful conversion factors between the different systems of units are given in the appendices; a detailed treatment on the various systems of measure and their evolvement may be found in Vol. IIB.

I wish to take the opportunity to thank all the contributing authors of this volume as well as their respective institutions for the contribution of their valuable engineering time. My sincere thanks are also extended to Drs. H. Sharbough and W. Starr for their reading of, and constructive comments on, the manuscripts; in the same respect, I greatly appreciate the helpful comments received from Committee D27 on Electrical Insulating Liquids and Gases, who graciously joined D9 in cosponsoring Volume III. My thanks are equally due to the editing staff of ASTM for their assistance. I am grateful to my own institution for its encouragement, support, and permission for my continuing work on the monograph series. Since a substantial portion of my time and effort has been spent in my own premises, I am most grateful to my wife Margaret and my two children, Andréa and Thomas, for their patience.

Varennes, Sept. 18, 1991

R. Bartnikas

R. Bartnikas[1]

Chapter 1

Permittivity and Loss of Insulating Liquids

1.1 Introduction

Electrical insulating liquids are utilized either to insulate components of an electrical network from each other or ground, alone or in combination with solid insulating materials. Requirements for low capacitance systems entail the use of low dielectric constant or permittivity liquids, consistent with acceptable chemical and heat transfer properties. However, when the insulating liquids are employed as impregnants in capacitors, then a higher value of dielectric constant is sought so as to reduce the physical size of the capacitors. Frequently intermediate values of dielectric constant may be advantageous for attaining a more acceptable voltage distribution between contiguous liquid and solid insulations. In order to minimize the amount of energy dissipated, which manifests itself in the form of heat thereby leading to a temperature rise of the insulating system, it is important that the dielectric loss within the insulating system be small. That is, the insulating liquids should be characterized by low values of dissipation factor, which constitutes a measure of the energy loss occurring within the dielectric liquids.

Efficient utilization of electrical insulating liquids thus requires knowledge of the manner in which their dielectric constant and dissipation factor behave as functions of temperature, voltage, and frequency. Since the application of the vast majority of insulating liquids is confined to power apparatus and cables operating at power frequencies, one may be incipiently inclined to conclude that it would suffice to characterize their behavior only as a function of temperature and voltage within a very narrow range of power frequencies (say from 50 to 400 Hz). However, the dielectric behavior of a liquid is inextricably associated with its molecular structure and much of the information necessary to define the molecular structure can be derived from high frequency measurements. Consequently even though a given insulating liquid may be used only at a power frequency of 50 or 60 Hz, a proper insight into its electrical behavior can only be obtained by observing its dielectric response as a function of frequency with the temperature as a parameter. For this reason, in the present chapter, we intend to examine the dielectric behavior of liquids over a wide range of frequencies since such an approach will provide much of the required insight to comprehend more fully the intricacies of the dielectric behavior of liquids.

The interpretation of the dielectric behavior of liquids is rendered more difficult in that there is no

[1]Institut de Recherche d'Hydro-Québec, Varennes, Québec, Canada J3X 1S1.

satisfactory theory of the liquid state. For this reason liquids have often been considered either as approaching the state of a compressed gas or as a disordered solid. It is known that the molecules in a liquid are distributed randomly and that, therefore, there exists no long range order within a liquid. This is indicated by the fact that liquids have negligible shear strength so that the molecules in a liquid are therefore not constrained permanently with regard to their position [*1*]. However, that there is some degree of short range order in liquids has been recognized for sometime [2]: every molecule in relation to itself does establish a partially ordered arrangement of its neighbors. A radial distribution function can thus be used to characterize the local structure of the environment of any molecule; this distribution function is defined as the ratio of the local density at a given distance from an arbitrary molecule to the average density of the liquid in the bulk. The radial distribution function may be determined experimentally by Fourier analysis in terms of the angular distribution of the intensity of the X-rays scattered by a liquid [*3–5*]. Due to this limited degree of local order, it seems plausible to think that since the dipole molecules and ions would be faced by reproducible average values of energy barriers they would be required to surmount to yield an externally observed dielectric loss under an applied electrical field.

In dealing with the subject matter of dielectric loss in liquids, several important distinctions must be made from the approach normally used on solid dielectrics. First, because of the finite viscosity of liquids both the dipole orientation and ionic mobility will be much less hindered as compared to solids. Secondly, because both the dipole orientation and ionic mobility are influenced to a large extent by the macroscopic value of the viscosity, considerable attention will have to be paid to the manner in which the macroscopic viscosity is affected by temperature. Moreover, it must be remembered that in the historical context, the theory of dipole orientation as originally propounded by Debye [*6–11*], invoked the concept of a dipole sphere rotating in a continuum characterized by a macroscopic viscosity. In fact a substantial portion of the literature on liquid dielectrics is devoted to investigations concerned with the extent to which the dielectric behavior of liquids can or cannot be predicted by the macroscopic viscosity parameter.

The viscosity parameter of practical insulating liquids plays an important role in determining the dielectric behavior of these liquids, since these liquids, because of their large molecular size composition, exhibit relatively high values of viscosity at room temperature. Insulating liquids consist of covalently bonded molecules, and comprise a wide range of mineral oils, esters, synthetic hydrocarbons such as polyurethanes, and silicone fluids. These liquids are essentially nonpolar and may be regarded at most as dilute solutions of polar molecules; therefore, their dielectric losses are relatively low. Moreover, since the dielectric constants of these liquids are likewise relatively low, the dielectric loss due to electrolytic impurities (whose ionic dissociation probability is lessened due to the low dielectric constant of the liquid medium) is also small. Although the viscosities of the practical insulating liquids are relatively high at room temperature and thus tend to inhibit ionic movement, the losses due to dipole orientation are readily shifted into the power frequency régime when the temperature of the liquids falls considerably below room temperature.

1.2 Preliminary Considerations

The theory of dielectrics is concerned primarily with the relationship between the externally measured macroscopic quantities, such as the dielectric constant or real value of the permittivity, ϵ', and the molecular and atomic structures of the dielectric material. This relationship, as pertains to solids, has been dealt extensively within Vol. IIA of this series [*12*]. Although the approach on solids dielectrics is similar to that on liquids, there are nevertheless certain salient differences which we shall attempt to delineate in the present chapter.

1.2.1 Permittivity

If we consider a parallel plate capacitor *in vacuo* with its two plates having respective charge densities $+q$ and $-q$, then the insertion of a dielectric material between the plates will be manifested by a decrease of potential between the plates or, alternatively, by a rise in the capacitance between the plates. The description of the foregoing effect leads to the definition of the static dielectric constant or permittivity

$$\epsilon_s = \frac{C}{C_o} \tag{1.1}$$

where C_o is the capacitance *in vacuo* and C the actual capacitance with the dielectric material inserted. Since the insertion of a dielectric material between the parallel plates of the capacitor reduces the field strength E_o to E, this leads to a reduction of the surface charge density from

$$q = \frac{E_o}{4\pi} \tag{1.2}$$

to an effective value of

$$q' = \frac{E}{4\pi}$$

$$= \frac{E_o}{4\pi\epsilon_s} \tag{1.3}$$

In dielectric theory it is often expedient to use the Gaussian or electrostatic system of units, whereby $\epsilon_s = \epsilon_o = 1$; here ϵ_o is the permittivity *in vacuo* or free space and, accordingly, in this case the electric flux density vector $\overline{D}$ is given by $\overline{D} = \overline{E}_o$. This approach disposes with the introduction of the relative value of permittivity, ϵ_r, as is done when SI units are employed. The quantity ϵ_s is often referred to as the static value or low frequency value of permittivity or dielectric constant. At frequencies where the dielectric liquid exhibits negligible loss such that there is no phase shift between the $\overline{D}$ and $\overline{E}$ vectors, we have the well-known relationship

$$\overline{D} = \epsilon_s\overline{E} \tag{1.4}$$

1.2.2 Electric Polarization

The insertion of the dielectric material between the parallel-plates of a capacitor has been seen to result in a reduction of the voltage gradient between the plates, thereby inferring that the presence of a dielectric material tends to neutralize some of the charges on the plates. Accordingly, the dielectric material commonly is regarded as forming dipole chains that neutralize or bind the surface charges at the plate surface. To describe the observed effect, the term electric polarization, P, defined as the difference in the surface charge densities in accordance with Eqs 1.2 and 1.3, is introduced

$$P = q - q' \tag{1.5}$$

or

$$\overline{P} = (\epsilon_s - 1)\overline{E}/4\pi \tag{1.6}$$

Equations 1.5 and 1.6 constitute an important result, as they provide a direct relationship between the macroscopically determined values of $\overline{D}$, $\overline{E}$, and ϵ_s, and the polarization vector, $\overline{P}$, which can also be expressed in terms of the molecular properties of the dielectric liquid as shall become apparent subsequently.

1.2.3 Polarizability of Molecules

When in a simple neutral atom, such as hydrogen, in which the surrounding electron cloud of charge $(-)$ e is displaced a distance d from the positive proton charge e in the nucleus due to the action of an externally applied electrical field, a net induced dipole moment μ results

$$\mu = ed \tag{1.7}$$

In practice, the resultant charge distribution within the atoms of insulating liquids is rather complex, and it is thus common practice to express the induced dipole moment vector as

$$\overline{\mu} = \alpha_e \overline{E} \tag{1.8}$$

where α_e is the electronic polarizability of the atoms. Note that according to this expression the induced dipole moment is always proportional to the applied electric gradient, $\overline{E}$. If Eq 1.8 is written in dimensional form as

$$[\text{charge}]\,[\text{length}] = \alpha_e[\text{charge}]\,[\text{length}]^{-2} \tag{1.9}$$

then it is evident that α_e has the dimensions of volume. The electrons in the outer shells contribute more to the value of α_e, since they are less strongly bound to the nucleus. For this reason positively charged ions have lower values of α_e than the corresponding neutral atoms, whereas for negatively charged ions the converse is true. The value of α_e may be considered to be independent of frequency up to the ultraviolet range.

When dealing with whole molecules, the application of an external field will also cause the displacement of atoms and ions with respect to each other. This effect is described by the introduction of the atomic or ionic polarizability, α_a. For example, in the hydrochloric acid (HCl) molecule, the external field induces a net dipole moment because of a change in the interionic distances between the H^+ and Cl^- ions, whereas in a symmetrical molecule such as carbon tetrachloride (CCl_4), a net induced dipole moment results from a change in the bond angles between the C-Cl groups [*13*]. In analogy with the latter simple case, in the complex molecular structure of the hydrocarbon insulating liquids, the application of an external field would cause the bond angles to change between the C-H groups as well as the C-C chain itself. It should be emphasized that for nonpolar materials, the polarizability is determined primarily by α_e, since α_a is usually only approximately 10% of the value of α_e.

Some molecules, commonly referred to as polar molecules, have permanent dipole moments even in the absence of an external field. In these molecules the centroids of positive and negative charge distributions are separated permanently by fixed distances, thereby giving rise to permanent dipole moments. Accordingly, an additional type of polarizability, namely, the orientation polarizability, α_a, due to a permanent dipole moment is assigned to characterize polar molecules. In terms of the three polarizabilities, the polarization vector, $\overline{P}$, may be expressed as

$$\overline{P} = N(\alpha_e + \alpha_a + \alpha_d)\overline{E}_r \tag{1.10}$$

where N represents the number of molecules per unit volume and $\overline{E}_r$ is the directing field acting on the molecules and must be distinguished from the externally applied field, $\overline{E}$, because of molecular interaction effects.

In order to derive an expression for α_d in terms of the permanent dipole moment, μ, of the molecules, Debye [*6*] considered initially the polar molecules as being immersed in a low density gaseous medium so as to neglect molecular interaction effects. In the absence of an external field, the polar molecules will be oriented in random directions, and the gas as a whole will exhibit no resulting dipole moment. With a field applied, a torque will be exerted on the permanent dipoles, impelling them to orient in the direction of the field. The tendency for the dipoles to orient themselves at a given inclination, θ, to the field axis is given by the average value of $\cos\theta$. Using Boltzmann statistics, Debye [*loc. cit.*] demonstrated that in terms of the Langevin function [*14*], this average value may be expressed as

$$\overline{\cos\theta} = \frac{\mu E_r}{3\,kT} \tag{1.11}$$

where μ is the dipole moment of the permanent dipoles, k is the Boltzmann constant (1.38×10^{-16} ergs/k or 1.38×10^{-23} J/K) and T is the absolute temperature. Hence, the average value of the permanent dipole moment in the direction of the field is

$$\langle\mu\rangle = \mu\,\overline{\cos\theta} \tag{1.12a}$$

or

$$\overline{\mu} = \frac{\mu^2\overline{E}_r}{3\,kT} \tag{1.12b}$$

For very high electrical fields, the Langevin function tends to unity and here saturation effects are said to occur as all permanent dipoles become aligned in the direction of the field. In this régime, $\overline{\cos\theta}$ is no longer directly proportional to E_r and nonlinear behavior ensues. However, the values of the applied field necessary to produce the nonlinear behavior are well beyond those encountered in normal practice.

In view of the result expressed by Eq 1.12*b*, Eq 1.10 may be rewritten as

$$\overline{P} = N\left(\alpha_e + \alpha_a + \frac{\mu^2}{3\,kT}\right)\overline{E}_r \tag{1.13}$$

From Eq 1.6 it is apparent that the polarization vector has the units of charge per unit area, that is, coulombs per square centimeter. If the later is alternatively written as coulomb·centimeter per cubic centimeter, then $\overline{P}$ assumes the units of dipole moment per unit volume which now in fact corresponds to the units of $\overline{P}$ in Eq 1.13. This is entirely consistent with the physical situation represented in macroscopic terms for a parallel plate capacitor [*15,16*]. If the area of the capacitor plates is equal to A, then for a plate separation equal to d the macroscopic dipole moment vector, $\overline{M}$, between the plates is given by

$$\overline{M} = \overline{P}\,(Ad) \tag{1.14}$$

Equation 1.13 is not in a very useful form, since in practice the value of the directing field, $\overline{E}_r$, acting upon the atoms and molecules is not known. If we disregard this inherent difficulty, which arises with matter in the condensed state, and consider an idealized case of a pure gas at a sufficiently low pressure such that the number of molecules per unit volume is sufficiently small to neglect molecular interaction [*6,17,18*], then the directing field, $\overline{E}_r$, becomes equal to the externally applied field, $\overline{E}$. Thus, Eq 1.13 becomes

$$\overline{P} = N\left(\alpha_e + \alpha_a + \frac{\mu^2}{3\,kT}\right)\overline{E} \tag{1.15}$$

To derive the connection between the macroscopically determined static value of the dielectric constant, ϵ_s, and the atomic-molecular parameters of α_e, α_a, and μ, the value of $\overline{P}$ from Eq 1.6 is substituted into Eq 1.15 to yield

$$\epsilon_s - 1 = 4\pi N \left(\alpha_e + \alpha_a + \frac{\mu}{3\ kT}\right) \tag{1.16}$$

1.2.4 Field at a Molecular Dipole

Although the foregoing relationship of Eq 1.16 does provide a helpful mental concept of how the static value of the dielectric constant may be influenced by the atomic and molecular parameters, it is not applicable directly to dielectric liquids. The close proximity of atoms and molecules in liquids infers that the field at a molecular dipole is not only determined by the externally applied field but also by the resultant field created by other neighboring dipoles and particles. In order to comprehend more fully the situation in condensed matter, it is necessary to examine the relationship between the applied field, $\overline{E}$, the internal field, $\overline{E}_i$, and the directing field, $\overline{E}_r$. For this purpose, a spherical model along the postulates first propounded by Mosotti [*19*] and Clausius [*20*] is used. In their spherical model arrangement, the molecular dipole is situated at the center of the sphere where the internal field value is to be determined; the radius of the sphere is made sufficiently large so that the region outside the sphere can be represented as a *continuum* having a static dielectric constant, ϵ_s. Within the sphere the actual molecular structure of the dielectric material must be taken into account. On the premise of the foregoing assumptions, the internal field, $\overline{E}_i$, within the sphere at the dipole is determined by: the charge density contribution from the electrodes of the test cell given by the flux density vector $\overline{D}$ equal to $4\pi q$; the induced charges at the electrode-dielectric interface due to the polarization given by $-4\pi\,\overline{P}$; the charges induced at the spherical surface (field $\overline{E}_c$); and the contribution of the molecular fields within the spherical region (field E_d). The classical relationship due to Clausius-Mosotti for the internal field, $\overline{E}_i$, within the test sphere can thus be stated as

$$\overline{E}_i = \overline{D} - 4\pi\overline{P} + \overline{E}_c + \overline{E}_d \tag{1.17}$$

In terms of the selected spherical geometry, it can be demonstrated that $\overline{E} = (4\pi/3)\,\overline{P}$, so that Eq 1.17 reduces to

$$\overline{E}_i = \overline{E} + \frac{4\pi}{3}\overline{P} + \overline{E}_d \tag{1.18}$$

The calculation of $\overline{E}_d$ in most instances is not a trivial matter as it requires information on the geometrical disposition and polarizability of the atoms and molecules within the test sphere. Under certain conditions, the individual fields created at the dipole due to the effects of the surrounding molecules may cancel out at the center of the sphere with the consequence that $\overline{E}_d = 0$. This situation obtains if the surrounding molecules are electrically neutral without permanent dipoles, or if they are polar but are arranged in either highly symmetrical arrays or are in complete disorder. Hence, for $\overline{E}_d = 0$

$$\overline{E}_i = \overline{E} + \frac{4\pi}{3}\overline{P} \tag{1.19}$$

Substituting the value of $\overline{P}$ from Eq 1.6 yields

$$\overline{E}_i = \left[\frac{\epsilon_s + 2}{3}\right]\overline{E} \tag{1.20}$$

The foregoing value of the internal field commonly is referred to as the Clausius-Mosotti field, thereby being understood implicitly that it omits the contribution of the molecular fields within the spherical region.

It is well to emphasize here that the Clausius-Mosotti field also omits the field influence created by the molecular dipole in the center of the sphere upon the neighboring molecules. The inhomogeneous field of this dipole polarizes the adjacent molecules and the resultant nonuniform polarization of the adjacent molecules creates in turn at the central dipole a reaction field, $\overline{E}_R$, which is proportional and in the same direction as the permanent dipole moment, $\overline{\mu}$. Therefore

$$\overline{E}_R = B\overline{\mu} \tag{1.21}$$

where B is a proportionality constant. It can be shown that the appearance of the reaction field, $\overline{E}_R$, leads to a further modification of the internal field given by [17]

$$\overline{E}_i = \overline{E}_r\left[1 + \left(\frac{B}{1 - B\alpha}\right)\frac{\mu^2}{3\,kT}\right] \tag{1.22}$$

where α is the induced polarizability of the permanent dipole. The foregoing expression is valid only if the polarization induced by the discrete dipole molecule in its environment rotates without any phase lag with respect to the applied field. Materials that are characterized by atomic and dipole orientation polarization in addition to electronic polarization will exhibit a finite phase lag; in such circumstances, the difference in value between $\overline{E}_i$ and $\overline{E}_r$ will be somewhat less than that indicated by Eq 1.22.

1.2.5 Debye Equation

Translational or deformation effects due to the electronic and atomic polarizabilities are intermolecular phenomena and, consequently, are relatively insensitive to temperature. If the translational polarizability is denoted by a single term α, then

$$\alpha = \alpha_e + \alpha_a \tag{1.23}$$

and the resulting temperature independent polarization caused by the translational effects may be expressed as

$$\overline{P}_\alpha = \sum_{j=1}^{n} N_j\alpha_j(\overline{E}_i)_j \tag{1.24}$$

where N is the number of atoms or molecules per cubic centimeter, α is the total deformation polarizability due to the electronic polarizability, α_e, and the atomic polarizability, α_a, while j refers to the jth kind of atom or molecule.

The polarization due to dipole orientation is

$$\overline{P}_d = \sum_{j=1}^{n} N_j\overline{\mu}_j \tag{1.25}$$

where $\overline{\mu}_j$ is the average value of the jth permanent dipole moment. From Eq 1.25, it is apparent once more that the units of the polarization vector are those of dipole moment per unit volume, thus being consistent with the macroscopic definition of the dipole moment of a parallel plate capacitor, where the dipole moment is equal to the product of the polarization vector and the dielectric volume encompassed by the parallel plate system. The explicit temperature dependence of the $\overline{P}_d$ vector is obtained by substituting the expression of $\overline{\mu}$ for moderate values of the directing field, $\overline{E}_r$, such that

$$\overline{P}_d = \sum_{j=1}^{n} N_j \left[\frac{\mu_j^2}{3\,kT}\right](\overline{E}_r)_j \tag{1.26}$$

since the total polarization is

$$\overline{P} = \overline{P}_\alpha + \overline{P}_d \tag{1.27}$$

then substituting the appropriate values yields

$$\overline{P} = \sum_{j=1}^{n} N_j \left[\alpha_j(\overline{E}_i)_j + \frac{\mu_j^2}{3\,kT}(\overline{E}_r)_j\right] \tag{1.28}$$

Finally, the well known Debye equation is derived by substituting the value of the polarization vector, $\overline{P}$, from Eq 1.6

$$\left[\frac{\epsilon_s - 1}{4}\right]\overline{E} = \sum_{j=1}^{n} N_j \left[\alpha_j(\overline{E}_i)_j + \frac{\mu_j^2}{3\,kT}(\overline{E}_r)_j\right] \tag{1.29}$$

For low density gases, the separation between the molecules is sufficiently large to permit the approximation $\overline{E}_i \simeq \overline{E}_r$ to be made. In addition, making use of the Clausius-Mosotti field relationship gives

$$\left[\frac{\epsilon_s - 1}{\epsilon_s + 2}\right] = \frac{4\pi}{3}\sum_{j=1}^{n} N_j \left[\alpha_j + \frac{\mu_j^2}{3\,kT}\right] \tag{1.30}$$

It may be appreciated that for higher density materials such as insulating liquids, the application of this expression would lead to error. Nevertheless, it is helpful in providing a simplifying and illustrative relationship between the static value of the dielectric constant, ϵ_s, on one hand and the polarizability, α, and the dipole moment, μ, on the other.

1.2.6 Molar Polarization

The treatment of dielectric liquids deviates from that of solids in that it is convenient and advantageous to introduce the concept of a scalar quantity, namely, that of molar polarization, $[P]$. It is at this point of juncture that the theoretical treatment on dielectric liquids commences to diverge somewhat from that on solids. Molar polarization $[P]$ is defined by the quantity

$$[P] = \frac{\epsilon_s - 1}{\epsilon_s + 2}\left(\frac{M}{d}\right) \tag{1.31}$$

where d is the density and M the molecular weight. For an insulating liquid such as a mineral oil, which consists of a mixture of various molecular sizes, M represents the average molecular weight that may be expressed as

$$M = \sum_{j=1}^{n} x_j M_j \tag{1.32}$$

where x_j refers to the molar fractions of the components and M_j to their individual molecular weights. Since the number of molecules per cubic centimeter of the jth kind is given by [17]

$$N_j = \frac{d}{M}(x_j N_A) \tag{1.33}$$

where N_A is Avogadro's number, the Debye equation for a low density mixture such as a gas may be written as

$$\frac{\epsilon_s - 1}{\epsilon_s + 2} = \frac{4\pi}{3} \sum_{j=1}^{n} \left(\frac{d}{m}\right) x_j N_A \left[\alpha_j + \frac{\mu_j^2}{3\,kT}\right] \tag{1.34}$$

or upon rearranging

$$[P] = \frac{\epsilon_s - 1}{\epsilon_s + 2}\left(\frac{M}{d}\right)$$

$$= \frac{4\pi}{3} N_A \sum_{j=1}^{n} \left(\frac{d}{M}\right) x_j \left[\alpha_j + \frac{\mu_j^2}{3\,kT}\right] \tag{1.35}$$

Although the foregoing equation applies strictly speaking to gases, it has been often extended to liquids in the past and found to be valid within certain limits. For a pure liquid the molar polarization may be expressed as

$$[P] = \frac{4}{3} N_A \left[\alpha + \frac{\mu^2}{3\,kT}\right] \tag{1.36}$$

When the liquid contains no polar molecules, then the Debye equation reduces to the form

$$\overline{P} = \sum_{j=1}^{n} N_j \alpha_j (\overline{E}_i)_j \tag{1.37}$$

and is known as the Clausius-Mosotti equation. If the Clausius-Mosotti field is then substituted in Eq 1.37, we have that

$$\frac{\epsilon_s - 1}{\epsilon_s + 2} = \frac{4\pi}{3} \sum_{j=1}^{n} N_j \alpha_j \tag{1.38}$$

whence

$$[P] = \frac{4\pi}{3} N_A \sum_{j=1}^{n} x_j \alpha_j \tag{1.39}$$

and hence for a pure nonpolar liquid containing only a single type of molecule

$$[P] = \frac{4\pi}{3} N_A \alpha \tag{1.40}$$

The expression for molar polarization may be employed to obtain the dipole moment of polar molecules in insulating liquids. However, this approach is applicable only to insulating liquids that may be regarded as dilute solutions of polar molecules. If one considers a dilute solution of a single species of polar molecules dissolved in a nonpolar solvent, then the molar polarization may be expressed as [*17*]

$$\begin{aligned}[P] &= \frac{\epsilon_s - 1}{\epsilon_s + 2} \cdot \frac{M}{d} \\ &= \frac{4\pi}{3} N_A \left[(1 - x)\alpha_s + x \left(\alpha_d + \frac{\mu^2}{3\,kT} \right) \right]\end{aligned} \tag{1.41}$$

where α_s is the polarizability of the nonpolar solvent molecules, x is the molar fraction of the polar soluble molecules, and α_d and μ are, respectively, their polarizability and dipole moment. The molar polarization of the solution is essentially identical with that of the pure nonpolar solvent which is simply equal to

$$\begin{aligned}[P_s] &= \frac{\epsilon_s - 1}{\epsilon_s + 2} \cdot \frac{M_s}{d_s} \\ &= \frac{4\pi}{3} N_A \alpha_s\end{aligned} \tag{1.42}$$

where the subscript s refers to the solvent molecules; here, ϵ_s is the static value of the dielectric constant of the solvent, though in practical terms it is virtually indistinguishable from the value of the solution itself.

Similarly for the polar solute

$$[P_d] = \frac{4\pi}{3} N_A \left(\alpha_d + \frac{\mu^2}{3\,kT} \right) \tag{1.43}$$

If $[P_d]_o$ and $[P_d]_x$ are employed to denote the molar polarization of the dipole component at an extremely small concentration and at a finite polar molecule concentration, respectively, then the total molar polarization is given, respectively, by

$$[P] = (1 - x)[P_s] + x[P_d]_o \tag{1.44}$$

and

$$[P] = (1 - x)[P_s] + x[P_d]_x \tag{1.45}$$

Experimentally, $[P]$ and $[P_s]$ can be determined in terms of Eqs 1.41 and 1.42, respectively, as a function of the calculated x-value, using the measured values of ϵ_s, d, and M of the mixture and pure solvent, respectively [*21*]. The limiting value of $[P_d]_o$, corresponding to an infinitely low concentration of the polar molecules of the solute, is derived by extrapolation of the $[P_d]_x$ versus x plot to $x = 0$. This procedure is carried out for a number of temperatures and the values of $[P_d]_o$ are substituted for $[P_d]$ in Eq 1.43; then $[P_d]_o$ is plotted against $1/T$ to obtain the values of α_d and μ. Since Eq 1.43 is in the form

$$[P_d] = a + \frac{b}{T} \tag{1.46}$$

the value of α_d is obtained from the intercept of the resulting straight line with the $[P_d]_o$ axis at $1/T = 0$ (equal to $4\pi N_A\alpha_d/3$), while μ is calculated from the slope of the line equal to $4\pi N_A\mu^2/9k$. This approach has been verified by Smyth and Morgan [*10*] using their classical results obtained with chloroform ($CHCl_3$), monochlorobenzene (C_6H_5Cl), and ethyl bromide (C_2H_5Br) in the nonpolar solvent hexane (C_6H_{14}). The resulting $[P_d]_o$ versus $1/T$ curves are depicted in Fig. 1.1, in terms of which the dipole moments for $CHCl_3$, C_6H_5Cl, and C_2H_5Br in the liquid state were found to be 1.05, 1.52, and 1.86 D (Debye units).

The preceding approach for determining α_d and μ of polar molecules is hardly applicable in practice to electrical insulating liquids. Since, in general, insulating liquids constitute complex mixtures of polar and nonpolar molecules of different size and type, the aforedescribed technique is useful only to the extent that it does yield an indication of the average polarizability and polarity of the heterogeneous insulating liquids. Most dielectric liquids have low dielectric constants and, therefore, can be regarded as dilute solutions of polar molecules; hence, a plot of the total molar polarization $[P]$ versus $1/T$ may yield a straight line relationship whereby the intercept with the $[P]$ axis will be equal to $4\pi N_A \sum_{j=1}^{n} x_j\alpha_j/3$ while the slope of the line will be equal to $4\pi \sum_{j=1}^{n} N_A x_j\mu_j^2/9k$. It is apparent that should the polar molecules present in a given dielectric liquid be all of the same species, then the value of μ for the dipole molecules could be derived from the plot with the proviso that the molar fraction x of the polar soluble molecules were also known.

It is interesting to observe that quantum mechanical treatment of the dipole orientation process leads to results similar to those that would be obtained using Debye's classical approach by means of the general expression for the molar polarization as defined by Eq 1.36 [*22–27*]. Perhaps the most commonly referred to quantum mechanical correction is that derived by Van Vleck [*26*], which can be stated as

$$[P] = \frac{4\pi}{3}N_A\left[\alpha + \frac{\mu^2}{3kT}\right][1 - f(T)] \tag{1.47}$$

where

$$f(T) = \frac{h}{48\pi^2kT\mu^2}\left[\mu_x^2\left(\frac{1}{B}+\frac{1}{C}\right)\mu_y^2\left(\frac{1}{C}+\frac{1}{A}\right) + \mu_z^2\left(\frac{1}{A}+\frac{1}{B}\right)\right] \tag{1.48}$$

where μ_x, μ_y, and μ_z are dipole moments of the polar molecule along the three axes, h is Planck's constant and A, B, and C represent the moments of inertia about the x-, y-, and z-axes, respectively. Evidently in the classical theory, which corresponds essentially to h being reduced to zero, the quantum mechanical correction vanishes completely. As the moments of inertia of most polar molecules are relatively large, the numerical correction due to the $f(T)$ factor is only very slight. The significance of the quantum mechanical correction rests primarily with improving the correlation between the refractive index and the dielectric constant when higher accuracy is a requirement.

1.2.7 Onsager Equation

The Debye equation can be used effectively to calculate dipole moments of substances in a gaseous state, and, up to a certain degree of accuracy, the dipole moments of dilute solutions of

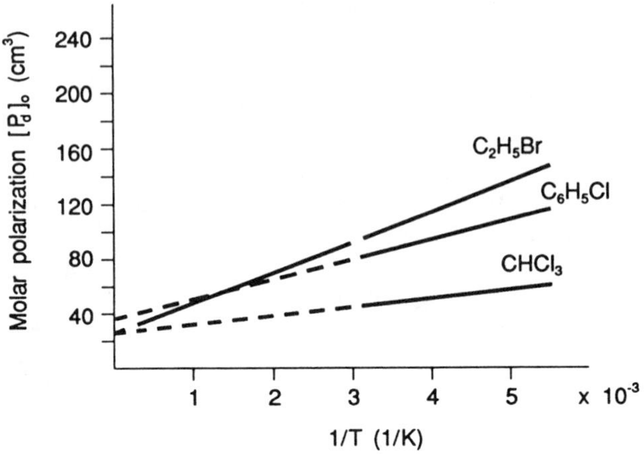

FIG. 1.1—*Molar polarization versus the inverse of the absolute temperature for dilute solutions of C_2H_5Br, C_6H_5Cl and $CHCl_3$ in a nonpolar solvent (hexane) (after Smyth and Morgan* [10]*).*

polar molecules. Müller [*28*] was perhaps the first to recognize some difficulties that may arise when it is attempted to determine dipole moments of polar molecule in dilute solutions. He observed that the calculated values of μ exhibited a definite tendency to decrease with the static value of the dielectric constant of the solvent, that is, the dielectric constant of the environment containing the dipoles. This behavior was attributed correctly to the molar polarization term $[(\epsilon_s - 1)/(\epsilon_s + 2)[M/d]$, which does not exceed unity even as ϵ_s approaches infinity. Deviations from the behavior predicted by the Debye equation become more pronounced as the concentration of polar molecules in the solution increases, until ultimately, for pure polar liquids the Debye equation fails entirely.

In an effort to remedy the difficulties arising with the Debye equation, which are associated inherently with the use of the Clausius-Mosotti field, Onsager [*29*] considered the effect of the reaction field E_R that is created in the spherical cavity containing the dipole as a consequence of the polarization which the dipole itself induces in its surroundings. Under these circumstances, the total field acting upon the dipole consists of the $\overline{E}_R$ plus the field $\overline{E}_G$, which would be produced in the cavity due to the external field in the absence of the dipole within the spherical cavity of Clausius-Mosotti that contains other molecules in addition to the dipole molecule. The spherical cavity postulated by Onsager is that of a single spherical polar molecule of radius, r. If the volume of the Onsager sphere is denoted as V_{sp}, then the induced dipole moment of the sphere may be expressed as

$$PV_{sp} = \alpha E \tag{1.49a}$$

Since the volume of the sphere is simply $4\pi r^3/3$ and the field within the dielectric sphere due to an external field, E, is equal to $3E/(\epsilon_s + 2)$, then substitution of the value P from Eq 1.6 gives [*29*]

$$\left[\frac{(\epsilon_s - 1)r^3}{3}\right]\left[\frac{3}{\epsilon_s + 2}\right] E = \alpha E \tag{1.49b}$$

from which

$$\alpha = \left[\frac{(\epsilon_s - 1)}{(\epsilon_s + 2)}\right] r^3 \tag{1.50a}$$

The preceding expression may be extended into the optical range in accordance with Lorenz [*30*] and Lorentz [*31*], using Maxwell's relationship [*32*] between the real value of the index of refraction, n', and the dielectric constant. Accordingly

$$\alpha = \left[\frac{(n')^2 - 1}{(n')^2 + 2}\right] r^3 \tag{1.50b}$$

Onsager [*loc. cit.*] employed this expression to define the polarizability of the spherical dipole in his model. Hence $(n')^2 = \epsilon_\infty$, since Onsager's model considers polar molecule liquids. Note that for nonpolar dielectrics $(n')^2 = \epsilon_s$.

In an electric field, E_i, the vector sum of the permanent and induced dipole moment is [*29*]

$$\overline{m} = \mu\overline{u} + \alpha\overline{E}_i \tag{1.51}$$

where μ represents the permanent dipole moment of the spherical molecule *in vacuo* and $\overline{u}$ designates the unit vector. In the mathematical treatment of his model, Onsager appears to have adopted the calculations described by Bell [*17,33*] which were also carried out in reference to a point dipole within a spherical cavity in solution. In solving Laplace's equation with the dipole point singularity of the electric field, he derives the effective external dipole moment due to the spherical dipole as

$$\overline{m}_* = \left(\frac{3\epsilon_s}{2\epsilon_s + 1}\right)\overline{m} \tag{1.52}$$

and the reaction field as

$$\overline{E}_R = \left[\frac{2(\epsilon_s - 1)}{2\epsilon_s - 1}\right]\frac{\overline{m}}{r^3} \tag{1.53}$$

In a similar solution, but now with the spherical dipole removed, it is demonstrated that due to an externally applied field $\overline{E}$ the electric moment of spherical region assumes the form

$$\overline{m}_1 = \left(\frac{\epsilon_s - 1}{2\epsilon_s + 1}\right) r^3\overline{E} \tag{1.54}$$

and the field within the empty cavity is then

$$\overline{E}_G = \left(\frac{3\epsilon_s}{2\epsilon_s + 1}\right)\overline{E} \tag{1.55}$$

Combining Eqs 1.52 to 1.55 yields the total field E_i acting upon the spherical dipole within a polarized dielectric

$$\begin{aligned}\overline{E}_i &= \overline{E}_G + \overline{E}_R\\ &= \left[\frac{3\epsilon_s}{2\epsilon_s + 1}\right]\overline{E} + \left[\frac{2(\epsilon_s - 1)}{(2\epsilon_s + 1)\, r^3}\right]\overline{m}\end{aligned} \tag{1.56}$$

whereas Eq 1.53 states the internal equilibrium of the spherical dipole of the Onsager model, Eq 1.56 defines the conditions for the equilibrium in the proximity of the molecular dipole sphere. Combining Eqs 1.52, 1.53, and 1.56 leads to [*29*]

$$\overline{m} = \left[\frac{[(n')^2 + 2]\,[2\epsilon_s + 1]}{3\,[2\epsilon_s + (n')^2]}\right] \mu\overline{u} + \left[\frac{\epsilon_s[(n')^2 - 1]}{[2\epsilon_s + (n')^2]}\right] r^3\overline{E} \tag{1.57}$$

and using Eq 1.51 gives

$$\overline{m} = \mu_1\overline{u} + \left[\frac{\epsilon_s[(n')^2 - 2]}{[2\epsilon_s + (n')^2]}\right] \alpha\overline{E} \tag{1.58}$$

here μ_1 denotes the effective dipole moment of the spherical dipole and is given by [*29*]

$$\mu_1 = \left[\frac{[(n')^2 + 2]\,[2\epsilon_s + 1]}{3\,[2\epsilon_s + (n')^2]}\right] \tag{1.59}$$

Onsager observed that the value of $\overline{E}_i$ is contingent upon the orientation. Therefore, he proceeded to compute the orienting force couple for every individual direction of $\overline{u}$, noting that the reaction field vector, $\overline{E}_R$, acts parallel to the dipole vector $\overline{m}$ and, hence, does not contribute to the orienting force couple that equals the vector product of $\overline{E}_i$ and $\overline{m}$, and $\overline{E}_i \times \overline{m}$. Consequently, in terms of Eq 1.56, this force couple is given by

$$\begin{aligned} \overline{M} &= \overline{E}_i \times \overline{m} \\ &= \overline{E}_G \times \overline{m} \end{aligned} \tag{1.60}$$

Since $\overline{E}_G = \overline{E}\,[3\epsilon_s/(2\epsilon_s - 1)]$ and $r^3\overline{E}_R = \overline{m}\,[2\,(\epsilon_s - 1)/(2\epsilon_s + 1)]$, then in terms of Eq 1.59

$$\overline{M} = \left[\frac{3\epsilon_s}{2\epsilon_s + 1}\right] \overline{E} \times \overline{m} \tag{1.61}$$

and using Eq 1.58 and considering only the portion in parallel to $\overline{u}$ such that $\overline{E}_G \times \overline{E}$ vanishes, yields

$$\begin{aligned} \overline{M} &= \mu_1\overline{E} \times \overline{u} \\ &= \mu^*\overline{E} \times \overline{u} \end{aligned} \tag{1.62}$$

where μ^* is designated by Onsager [*loc. cit.*] as the external characteristic moment of the molecule and is equal to

$$\mu^* = \frac{3\epsilon_s\mu_1}{(2\epsilon_s + 1)} \tag{1.63}$$

In nonvector notation, Eq 1.62 may be expressed as

$$M = \mu^* E \sin\theta \tag{1.64}$$

where θ is the angle which the spherical dipole makes with the axis of the field. The work of orientation, W, may be found by integration of $M = \partial W/\partial\theta$ to yield [*29*]

$$W = -\mu^* E \cos\theta \tag{1.65}$$

As in the case of the Debye model, the mean orientation of the molecules in the field in the Onsager model is also obtained in terms of Boltzmann statistics as

$$\overline{\cos\theta} = \frac{\int \cos\theta \exp[-W/kT] \sin\theta \, d\theta \, d\phi}{\int \exp[-W/kT] \sin\theta \, d\theta \, d\phi} \tag{1.66}$$

where ϕ is the solid angle within which the dipole is subtended. Making use of the Langevin function [*29*] and considering moderate values of the field strengths, E, it is found that

$$\overline{\cos\theta} = \mu^* E/3\,kT \tag{1.67}$$

In Eq 1.58, the unit vector $\overline{u}$ may thus be replaced by $\overline{\cos\theta}$, yielding

$$\overline{m} = \mu_1\mu^* (\overline{E}/3\,kT) + \left[\frac{\epsilon_s[(n')^2 - 1]}{[2\epsilon_s + (n')^2]}\right] \alpha\overline{E} \tag{1.68}$$

If N is defined as the number of dipole molecules per unit volume, then the polarization per unit volume, $\overline{P}$, is given by

$$\overline{P} = N\overline{m}$$

$$= N\left[\frac{\mu_1\mu^*}{3\,kT} + \frac{\epsilon_s[(n')^2 - 2]}{[2\epsilon_s + (n')^2]}\alpha\right]\overline{E} \tag{1.69a}$$

or in terms of Eq 1.6

$$(\epsilon_s - 1) = 4\pi N\left[\frac{\mu_1\mu^*}{3\,kT} + \frac{\epsilon_s[(n')^2 - 2]}{[2\epsilon_s + (n')^2]}\alpha\right] \tag{1.69b}$$

Onsager assumes that the volume of the dielectric liquid equals the algebraic sum of the volumes of all the molecules, that is

$$N(4\pi r^3/3) = 1 \tag{1.70}$$

Then considering Eqs 1.50*b* and 1.70, we can write that

$$4\pi N[(n')^2 + 2]\alpha = 3[(n')^2 - 1] \tag{1.71}$$

Therefore, elimination of α from Eq 1.69*b*, leads to [*29*]

$$(\epsilon_s - 1) = 4\pi N\frac{\mu_1\mu^*}{3\,kT} + \frac{3\epsilon_s[(n')^2 - 1]}{[2\epsilon_s + (n')^2]} \tag{1.72}$$

Further simplification and rearrangement gives

$$\frac{[2\epsilon_s + 1][\epsilon_s - (n')^2]}{[2\epsilon_s + (n')^2]} = 4\pi N\frac{\mu_1\mu^*}{3\,kT} \tag{1.73}$$

Also expressing μ_1 in terms of μ^* according to Eq 1.63 yields

$$\frac{[2\epsilon_s + 1][\epsilon_s - (n')^2]}{[2\epsilon_s + (n')^2]} = 4\pi N \frac{(\mu^*)^2}{3\,kT} \cdot \frac{(2\epsilon_s + 1)}{3\epsilon_s} \tag{1.74}$$

or noting the relationship between μ_1 and μ in Eq 1.59, gives an alternate expression

$$\frac{[\epsilon_s - (n')^2][2\epsilon_s + (n')^2]}{\epsilon_s[(n')^2 + 2]^2} = \frac{4\pi N\mu^2}{9\,kT} \tag{1.75}$$

The dipole moment μ in Eq 1.75 represents essentially the same quantity as defined by μ in the Debye equation. It is demonstrated by Onsager [*29*] that Eq 1.75 may be expressed in a more explicit manner as

$$\frac{\epsilon_s - 1}{\epsilon_s + 2} = \frac{[(n')^2 - 1]}{[(n')^2 + 2]} + \frac{3\epsilon_s[(n')^2 + 2]}{[2\epsilon_s + (n')^2](\epsilon_s + 2)} \cdot \frac{4\pi N\mu^2}{9\,kT} \tag{1.76}$$

If the foregoing form of the equation is compared with that of the Debye equation, it will be immediately recognized that the value of μ^2 in the Debye equation is modified by the factor $3\epsilon_s[(n')^2 + 2]/[2\epsilon_s + (n')^2](\epsilon_s + 2)$. Furthermore, since the Avogadro's number $N_A = N\,M/d$ then in analogy with Eq 1.35, the total molar polarization $[P]$ in terms of the Onsager equation is given by

$$\begin{aligned}[P] &= \frac{\epsilon_s - 1}{\epsilon_s + 2} \cdot \frac{M}{d} \\ &= \frac{(n')^2 - 1}{(n')^2 + 2} \cdot \frac{M}{d} + \frac{3\epsilon[(n')^2 + 2]}{[2\epsilon_s + (n')^2](\epsilon_s + 2)} \cdot \frac{4\pi N_A \mu^2}{9\,kT}\end{aligned} \tag{1.77}$$

Even though both the Debye and Onsager theories on the dielectric constant neglect local molecular forces and thus are less applicable to dielectric solids, their validity appears to hold reasonably well for liquids. Since the polarization per unit volume in the Onsager case is equal to $(\epsilon_s - 1)(2\epsilon_s + 1)/9\epsilon_s$, it differs substantially from that of Debye's case, which is based on the Clausius-Mosotti term $(\epsilon_s - 1)/(\epsilon_s + 2)$. That the results obtained with the two models would be anticipated to display an agreement with liquids of low dielectric constant is substantiated in Fig. 1.2, which portrays the polarization per unit volume as a function of the static dielectric constant, ϵ_s, for the two cases [*34*]. A close agreement is discernible for dielectric constant values less than two, a region corresponding to dielectric constants of extremely dilute solutions of polar molecules in nonpolar solvents. This is indeed close to the case for most mineral insulating oils, whose dielectric constants fall within the range of 2.0 to 2.2.

1.2.8 Kirkwood Equation

Although the Onsager model represents an important advance over the Debye model, the treatment of the spherical dipole by Onsager as a cavity in a statistical *continuum* having a uniform dielectric constant equal to that of the liquid precludes it from being an exact model precisely because of this *ad hoc* assumption. Moreover, the Onsager model fails to take into account entirely any hindrances to the dipole orientation process by the adjacent molecules of the liquid. The first attempt to examine hindered rotation appears to have been made by Van Vleck [*35*], but his method

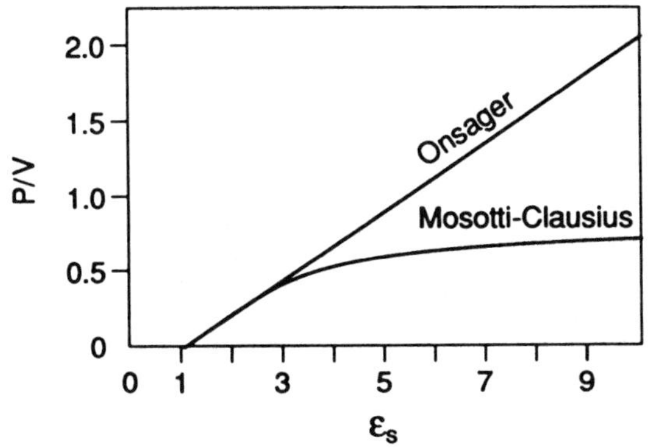

FIG. 1.2—*Polarization per unit volume as a function of the dielectric constant according to the Clausius-Mosotti and Onsager fields (after Oster* [34]*).*

of calculation proved to be too difficult for practical application to dielectric liquids. However, Kirkwood [*36,37*] was successful in generalizing the Onsager theory within the limits of applicability of classical statistical mechanics and by disposing of the necessity of equating the local dielectric constant to the macroscopic value of the dielectric constant of the medium. In this sense, Kirkwood's theory must be regarded as rigorously statistical in all its aspects.

In Kirkwood's approach the dielectric constant is related to the moment μ_k of the single molecular dipole and the moment μ'' representing the sum of the molecular dipole moment and the average moment induced by the polar molecule in its environment by impeding the rotation of its neighbors relative to itself. Hindrance to rotation was considered by Kirkwood to result solely from electrostatic interactions and short range molecular forces. In this model should a molecule be assumed to orient its neighbors due to its inherent dipole field and further should they be treated as part of a statistical continuum, then $\mu_k = \mu''$ and the Onsager model obtains. Kirkwood deals with the problem of hindered rotation by relating the dipole moment, μ'', to the potential of the torque restricting the relative rotation of the neighboring molecules. In Kirkwood's model, the local dielectric constant of the statistical continuum surrounding the spherical dipole differs from the macroscopic dielectric constant, ϵ_s, of the liquid, because of the influence of the molecular-spherical dipole upon the average distribution of orientation of the neighboring molecules. Kirkwood derives the relationship between ϵ_s and $\mu_k\mu''$ in terms of the molar polarizability $[P]$ as

$$[P] = \frac{(\epsilon_s - 1)(2\epsilon_s + 1)}{9\epsilon_s} \cdot \frac{M}{d}$$

$$= \frac{4\pi N_A}{3}\left(\alpha + \frac{\mu_k\mu''}{3\,kT}\right) \tag{1.78}$$

Here, it is once more appropriate to emphasize the distinction between μ_k and μ''. Whereas μ represents simply the dipole moment of the permanent dipoles in the liquid dielectric, the quantity μ'' refers to the quantity μ_k plus the added induced moment arising from the hindered rotation of the adjacent molecules contained within the spherical cavity region. The manner in which the product $\mu_k\mu''$ is related to μ^2 in the Debye equation depends upon a correlation factor, g, which has the form

$$g = \left[1 + \frac{N_A}{V}\int_o^{v_o}\int_o^{\phi} \cos\phi \exp\left[-W/kT\right] d\phi\, dv\right] \tag{1.79}$$

where

$$\int_{o}^{v_o} \exp\left[-W/kT\right] dv = 1 \tag{1.80}$$

here V represents the molar volume, v_o is the volume of the sphere containing the molecular dipole, ϕ is the angle between the dipole moment of an arbitrary pair of molecules; the integration is carried out over all relative orientations and positions of centers of gravity of the molecular pair within the sphere of volume v_o outside whose exterior the local value of the dielectric constant is given by the macroscopic dielectric constant of the liquid medium. Due to the difficulties posed by the calculations in statistical mechanics for obtaining W in the absence of exact data on the structure of liquids, Kirkwood proposed the approximation

$$g = [1 + n\,\overline{\cos\phi}] \tag{1.81}$$

The preceding approximation is based on the assumption that the local dielectric constant may be represented by the macroscopic dielectric constant outside a region containing the dipole and its n nearest neighbors. Here $\overline{\cos\phi}$ is the average of the cosine of the angle ϕ that the test dipole makes with one of its neighbors; the value $\overline{\cos\phi}$ denotes, therefore, the average over all possible orientations of the dipole and its neighboring molecules such that

$$\cos\phi = \int\int \cos\phi \exp\left[W_o/kT\right] d\psi_1 d\psi_2 \tag{1.82}$$

where W_o is now the potential of the average torque acting on a pair of the nearest molecular neighbors and may arise both due to dipole-dipole interaction or from short range molecularly forces; alternatively, it may simply be viewed as the value of the potential barrier restricting orientation. The quantity $\overline{\cos\phi}$ is thus a measure of the correlation that exists between the orientations of neighboring molecules under the effects of hindered relative rotation in the liquids. In terms of the correlation function, g

$$\mu_k \mu'' = g\mu_k^2 \tag{1.83}$$

So that Eq 1.78 may be expressed as

$$\frac{(\epsilon_s - 1)(2\epsilon_s + 1)}{9\epsilon_s} \cdot \frac{M}{d} = \frac{4\pi N_A}{3}\left(\alpha + \frac{g\mu_k^2}{3\,kT}\right) \tag{1.84}$$

The extent of hindered rotation is indicated by the amount by which the value of g departs from unity. It was shown by Oster and Kirkwood [*38*] that some associated (hydrogen bonded) liquids, such as alcohols, may exhibit g values as high as 2.57, indicating a preference of the neighboring molecular dipoles for parallel orientation. In extending Kirkwood's theory to dilute solutions of polar molecules, where dipole-dipole interactions may be neglected, Oster [*34*] found that for low molar fractions of polar constituents the values of g were less than unity. The negative deviations would be expected to result when the hindering forces are more predisposed to give rise to antiparallel orientation of the neighboring molecular dipoles. Since adequate structural information is lacking on most complex insulating liquids, the g parameter cannot be calculated. Kirkwood's approach is useful, however, in that it provides a model that assists one at least mentally in visualizing hindered dipole rotation. In practice, g represents a disponible parameter which in most cases can be only estimated empirically.[2]

[2]Recent data by Jorat et al. (*IEEE Transactions on Electrical Insulation*, Vol. 26, 1991, pp. 763–769) has shown a dependence of g upon temperature in the case of propylene carbonate toluene mixtures.

1.2.9 Fröhlich Equation

Fröhlich [39] employed a somewhat more simplified approach to derive a more general expression for the Kirkwood equation. As did Kirkwood, he too applied classical statistical mechanics to derive a molecular model that takes into account both the short range interactions between the molecules themselves and the resulting deformation polarization. He followed the classical approach by considering a sphere within a dielectric continuum of static dielectric constant, ϵ_s, containing N dipole units per unit volume and demonstrated that

$$\epsilon_s - 1 = \frac{4\pi N}{3}\left(\frac{3\epsilon_s}{2\epsilon_s + 1}\right)\frac{\langle \overline{m}(x)\overline{m}_*(x)\rangle}{kT} \tag{1.85}$$

where $\overline{m}(x)$ is the dipole moment of a unit volume for a given set of x displacements of its elementary charges and $\overline{m}_*(x)$ is the average dipole moment of a spherical region of volume V within the dielectric specimen, with the proviso that the set of x displacements of a unit volume at the center persists for an extended period of time. Thereby, the quantity $\langle \overline{m}(x)\, \overline{m}_*(x)\rangle$ represents the average value of $\overline{m}(x)\, \overline{m}_*(x)$ over all possible displacements, x, within the unit volume.

In Fröhlich's model the spherical region consists of a set of charges, e_j, embedded in a continuum of dielectric constant, $(n')^2$, with the remainder of the specimen represented by the static dielectric constant, ϵ_s. Upon the application of an external field, assuming that the charge configuration within the sphere is not altered, the field is increased to [40]

$$\overline{E}'_G = \frac{3\epsilon_s}{2\epsilon_s + (n')^2}\overline{E} \tag{1.86}$$

and the corresponding average moment, M_e, of the charges e_j is now given by

$$M_e = \frac{\epsilon_s - (n')^2}{4\pi} VE \tag{1.87}$$

Accordingly, due to the elastic displacement of the charges, Eq 1.85 must be rearranged as

$$\epsilon_s - (n')^2 = \frac{4\pi N}{3\,kT}\left[\frac{3\epsilon_s}{2\epsilon_s + (n')^2}\right][\langle \overline{m}(x)\, \overline{m}_*(x)\rangle] \tag{1.88}$$

If it is further assumed that the unit volume contains one molecule having a permanent dipole moment of μ whose energy is independent of its direction, then x denotes the angles of rotation of the test dipole such that under these circumstances $m(x) = \mu_x$ and $m_*(x) = \mu_*$. Hence, Eq 1.88 becomes

$$\epsilon_s - (n')^2 = \frac{4\pi N}{3\,kT}\left[\frac{3\epsilon_s}{2\epsilon_s + (n')^2}\right]\mu_x\mu_* \tag{1.89}$$

Fröhlich [*loc. cit.*] then notes that in accordance with the Kirkwood relation

$$\mu_x\mu_* = (1 + n\cos\phi)\mu_k^2 \tag{1.90a}$$

or in terms of Kirkwood's correlation function, g

$$\mu_x\mu_* = g\mu_k^2 \tag{1.90b}$$

Alternatively, Eq 1.89 may be expressed as

$$\epsilon_s - (n')^2 = \left[\frac{3\epsilon_s}{2\epsilon + (n')^2}\right]\frac{4\pi N}{3\,kT}g\mu_k^2 \tag{1.91}$$

Fröhlich [*41*] observes that if the relation between the dipole moment μ_k and that of a spherical dipole molecule *in vacuo,* μ, is taken as

$$\mu = \frac{3}{(n')^2 + 2}\mu_k \tag{1.92}$$

Then Eq 1.91 becomes

$$\frac{[\epsilon_s - (n')^2][2\epsilon_s + (n')^2]}{\epsilon_s[(n')^2 + 2]^2} = \frac{4\pi N}{9\,kT}g\mu^2 \tag{1.93}$$

Evidently, for $n' = 1$, Eq 1.93 reduces to the Onsager equation. It is further demonstrated by Smyth [*42*] that the rearrangement of terms and the introduction of Avogadros number, N_A, yield the molar polarizability expression

$$\frac{\epsilon_s - 1}{\epsilon_s + 2}\cdot\frac{M}{d} = \frac{(n')^2 - 1}{(n')^2 + 2}\cdot\frac{M}{d}$$

$$= \frac{3\epsilon_s[(n')^2 + 2]}{[2\epsilon_s + (n')^2](\epsilon_s + 2)}\frac{4\pi N_A}{9\,kT}g\mu^2 \tag{1.94}$$

Fröhlich's equation is somewhat more general than that of Kirkwood's; however, as pointed out by Brown [*43*], the two equations lead to different results only when the term due to the distortion polarization is included. It must be said that the models proposed by Debye, Onsager, and Kirkwood have essentially withstood the test of time in the sense that, although numerous changes and modifications have been suggested to ameliorate these models, no great improvement has resulted. In summary, it is appropriate that we make a few cursory comments on some of these attempts. Cole [*44*] generalized his approach to consider alternating voltages using a specialized treatment of the distortion polarization. Buckingham [*45*] derived an equation which took into account the moment induced in the neighboring solvent molecules adjacent to nonspherical soluble molecules. Wilson [*46*] showed that the Onsager model may lead to more representative values of the calculated dipole moments if the molecular anistropy and molecular association effects are not neglected. Other modifications in the expression for the dipole moment were obtained by Scholte [*47*] and Ross and Sack [*48*], who considered the dipole molecules as ellipsoidal in shape, and by Abbott and Bolton [*49*] who approximated the form of the polar molecules by that of prolate spheroids.

Very recently, an alternate equation to that of Onsager has been derived by Omini [*50–54*]; it is somewhat more complex in form than the Onsager equation, but does lead to a slight improvement in the agreement between the calculated values of μ and those measured in the vapor state. For a more theoretical treatment on the molecular statistical interpretation as regards the suitability of the microscopic models chosen to describe the dielectric constant in terms of the dipole moment of polar molecules, the reader is referred to the treatise by Kielich [*55*]. It is also worthy of note that more recently considerable effort has gone into computer assisted dipole modeling studies to test the accuracy of the existing dipole theories. One such study is that of Adams [*56*], who applied computer simulation to the behavior of polar liquids, using as its basis the Gaussian model potentials for molecular interaction as proposed by Berne and Pechukas [57].

The inherent difficulties associated with the various models that are used to provide a connection between the external measurement of the dielectric constant and the microscopic quantities of the liquid dielectric can perhaps be best delineated by means of some rather pertinent and astute observations made by Van Vleck [58]. He noted that for a dilute solution of polar molecules in a nonpolar solvent, one can write the general expression

$$\frac{3(\epsilon_s - 1)}{(\epsilon_s + 2)} = 4\pi\left[N\alpha + N_s\alpha_s + \frac{N_s\mu^2}{3kT}f(\alpha_s, \epsilon_s)\right] \tag{1.95}$$

where N and N_s are the number of solvent and soluble molecules per unit volume respectively, μ is the dipole moment of the soluble molecules and α and α_s are, respectively, the polarizability of the solvent and soluble molecules. He points out that to determine the partition function $f(\alpha_s, \epsilon_s)$ requires a rigorous and accurate statistical-mechanical treatment, with the usual attending insurmountable difficulties. Evidently, if $f(\alpha_s, \epsilon_s) = 1$, then Eq 1.95 reduces simply to the Debye expression. Van Vleck [58] makes the further observation that the intrinsic problem with the various models and their numerous modifications (that were propounded in order to achieve a suitable form of Eq 1.95), is that they are essentially semiphenomenological models that assume the partition function $f(\alpha_s, \epsilon_s)$ to be expressible in terms of α_s and ϵ_s without dependence upon the molecular geometry or of how the nonpolar solvent molecules are arranged or how the polar molecules are placed. Van Vleck was able to demonstrate that for a given density of the dielectric liquid, different formulae are obtained for the partition function $f(\alpha_s, \epsilon_s)$ when certain mean molecular arrangements in the liquid are postulated. It is thus apparent that the primordial difficulty with any theory on the dielectric constant of liquids is associated inextricably with our lack of detailed knowledge of the structure of dielectric liquids.

1.3 Dielectric Constant and Molecular Structure

From considerations in the previous sections, it is apparent that in the case of nonpolar liquids, the dielectric constant of the liquid is determined by its electronic and atomic polarizability, molecular weight, and density. Thus, the dielectric constant depends on the electronic structure of the atoms; in addition, if the liquid is comprised of more than one type of atom, an asymmetrical sharing of electrons may arise as the stronger binding atoms will cause the electron clouds to be displaced eccentrically towards them. Atoms will thereby acquire charges of opposite polarity, whereby the application of an external field will impel a change in the equilibrium positions of the atoms leading to atomic polarization. Atomic polarization need not be uniquely due to atomic or ionic displacements within the molecule, it may also result from changes in the bond angles between the atoms or ions. With polar liquids which exhibit a permanent asymmetrical charge distribution between dissimilar atoms, a permanent dipole moment exists even in the absence of an external field. As a consequence of their permanent dipole moment, polar molecules are characterized by a nonsymmetrical molecular structure. Evidently as a result of their additional permanent dipole contribution, polar liquids are characterized by higher dielectric constants. Since all liquids exhibit some atomic polarizability, the use of the (n') term in the development of the Onsager, Kirkwood, and Fröhlich equations takes into account only the electronic polarizability contribution; therefore, it is customary to replace the (n') term by ϵ_∞, which is the dielectric constant at optical frequencies and includes also the atomic polarization contribution.

A rudimentary estimate of the atomic polarization places it at about 10% of the electronic polarization [59]. However, this estimate may be in considerable error when compared to some actual values available. For example, McAlpine and Smyth [60] found that for the monosubstituted benzenes of fluorobenzene (C_6H_5F), chlorobenzene (C_6H_5), and nitrobenzene ($C_6H_5NO_2$), the values of the electronic and atomic polarization, P_e and P_a, were, respectively, 24.8 and 1.0 cm^3, 29.9 and 1.8 cm^3, and 31.6 and 6.1 cm^3, exhibiting appreciable difference in the P_e to P_a ratio. These values were obtained in the vapor state by subtracting the P_e value from the total induced polarization value to yield the P_a value [*61*]. The measurements had to be performed in the gaseous state rather than in the liquid state in order to eliminate the solvent effect [*62*], which becomes more pronounced when the soluble molecules deviate appreciably from a spherical shape.

In so far as the atomic polarization, P_a is always considerably less than the electronic polarization, P_e, and with polar molecules the dipole orientation polarization is substantially greater than the same $P_a + P_e$, it is convenient for illustrative purposes to represent the polarization as a function of frequency, f, as depicted in Fig. 1.3. Evidently, the variation of the dielectric constant with frequency is similar to that of Fig. 1.3.

In neutral atoms or molecules where the electric centroids of gravity of positive and negative charges coincide, no permanent polarity exists such that $P_d = 0$ and the total polarization is equal to the induced polarization ($P_a + P_e$). Hence, for example, liquid helium because of its low density and zero dipole moment has an unusually low dielectric constant, ϵ', of 1.055; likewise for liquid argon, $\epsilon' = 1.533$. Similarly, other cryogenic liquids of the diatomic gases fall into the same category: H_2 ($\epsilon' = 1.231$), O_2 ($\epsilon' = 1.507$) and N_2 ($\epsilon' = 1.435$). The lower molecular weight diatomic cryogenic liquids may be more easily compressible, resulting in correspondingly higher increases in their density. This effect is quite perceptible when the dielectric constant of liquid nitrogen is compared to that of liquid hydrogen as a function of pressure as depicted in Fig. 1.4 [*63,64*].

In contrast to the diatomic cryogenic liquids, acids that have ionic bonds are characterized by permanent dipole moments and have, therefore, substantially higher dielectric constants. For instance, hydrochloric acid (HCl) has a relatively high dielectric constant of 6.35 which reflects its permanent dipole moment value of $\mu = 1.08$ D (Debye units); one Debye unit $= 10^{-18}$ electrostatic units (ESU) or 333×10^{-30} C m. The value of one Debye unit has been fixed conveniently at 10^{-18} ESU, because the interatomic distances are of the order of angstroms or 10^{-8} cm while the electronic charge in ESU is equal to 4.8025×10^{-10}. Likewise another acid, HBr, has a dipole moment of 0.79 D with a corresponding value of $\epsilon' = 7.00$ at room temperature. In comparing the ϵ' values of HCl and HBr, it must be emphasized that a higher permanent dipole moment value does not necessarily lead to a correspondingly larger permittivity value since the density of the

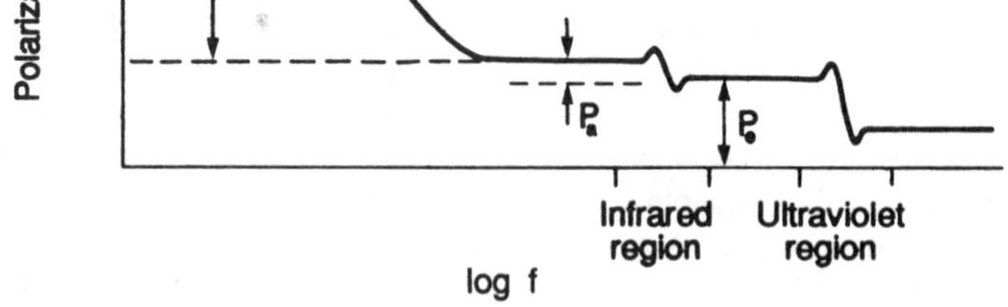

FIG. 1.3—*A schematic plot of polarization as a function of frequency.*

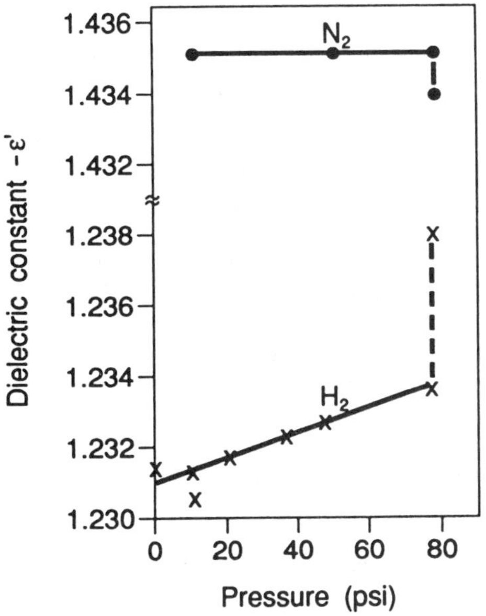

FIG. 1.4—*Variation of dielectric constant, ε', with pressure for liquid H_2 and liquid N_2 (after Jeffries and Mathes* [64]*).*

liquids as well as their respective electronic and atomic polarizability values also contribute to the final value of the dielectric constant. Simple ionic compounds are unique in the sense that their experimentally determined permanent dipole moments, either in the liquid or gaseous phases, may be compared readily with the easily calculated values obtained in terms of the ionic charge and the interionic distances of their relatively simple molecular structure. For example, the salt potassium iodide (KI) is characterized by a permanent dipole moment of 7.8 D [*10*], which is found to be appreciably less than the calculated value of 12.7 D using the interionic distances. This discrepancy arises from the mutual polarization effects of the K and I ions, whereby the mutually induced dipole moments oppose the direction of the permanent dipole moment leading to a lower measured value of μ. With some ionic bonds, the resulting dipole moments may be further reduced as the proton charge penetrates a significant distance into the larger negative ion, thereby effectively decreasing the interionic distance. It is apparent that if the dipole vectors are added, between the atoms comprising the molecule, the overall permanent dipole moment for the entire molecule should approximate the resultant of the component moments along the individual chemical bonds. However, because of the individual polarization effects of the dipole component moments, errors in the calculated values would be certain to occur. In practice, a procedure is often followed in the reverse direction, that is, measurement of the permanent dipole moment of a liquid reveals some intrinsic details of the molecular structure. A case in point is the nonpolar liquid carbon tetrachloride (CCl_4) for which μ is found to be zero, indicating that the molecular structure of CCl_4 is symmetrical.

In the instance of nonionic or covalent bonds, a permanent dipole moment of the molecule arises due to the tendency of the electron clouds to shift preferentially to the atomic nuclei containing the largest positive charge. Most insulating liquids used in practice are covalently bonded hydrocarbons, having varying amounts of paraffinic, naphthenic, and aromatic constituents. The simplest paraffin molecule is that of methane (CH_4), which is symmetrical and has a dipole moment of zero. It may be appropriate here to interject an historical note in that it was precisely the zero dipole measurement data that finally resolved the question of the CH_4 molecular symmetry. If any of the hydrogen atoms in the CH_4 molecule are substituted by a halogen atom, say chlorine, then the charge distribution in the resulting CH_3Cl molecule becomes asymmetrical with a permanent dipole moment of 1.87 D [*42*]. As the chain length of the paraffin molecules is increased, the charge distribution in the overall molecule of the saturated aliphatic hydrocarbon still retains its symmetry. Thus zero dipole moment remains a characteristic of the larger paraffins such as ethane (C_2H_6), propane (C_3H_8), pentane (C_5H_{12}), dodecane ($C_{12}H_{26}$), and hexadecane ($C_{16}H_{34}$) [*42*]. Branching in the paraffins does not appear to lead to any significant polarity of the overall molecule; for instance, branched chain heptanes (C_7H_6) and 2,2,4-trimethylpentane [$(CH_3)_3$ CH_2—$CH(CH_3)_2$] are found to have dipole moments indistinguishable from zero. The behavior of the unsaturated aliphatic hydrocarbons is similar to the saturated ones. For example, ethylene ($CH_2 = CH_2$) trans-2-butene ($CH_3CH = CHCH_3$) and 1,3-butadiene ($CH_2 = CHCH = CH_2$) have zero dipole moments. However, depending upon the position of the unsaturated bond some charge asymmetry may be introduced, yielding a small net permanent dipole moment. Such is the case with propylene ($CH_3CH = CH_2$) for which $\mu = 0.35$ D. Also replacement of any hydrogen by a halogen atom in unsaturated aliphatics leads to a marked charge imbalance in the molecule and, hence, a sizeable dipole moment. If one considers ethylene, then several possibilities may arise as has been shown by the early work of Errera [*65*]. When one of the hydrogens is substituted by a chlorine, the structure of vinyl chloride ($CH_2 = CHCl$) results with a net dipole moment of 1.44 D. If now two of the hydrogen atoms are substituted with chlorine, then two different molecular structures become possible, namely, the *cis* and *trans* $CHCl = CHCl$ forms as depicted in Fig. 1.5. As anticipated, the balanced *trans* structure has a zero dipole moment as contrasted with the *cis* structure for which $\mu = 1.89$ D; this demonstrates that the molecular structure of vinyl chloride is planar. Also that the molecule is nonspherical is further supported by a perceptible solvent effect as the μ value of 1.89 D (the *cis* form) in the gaseous state falls to 1.80 D when the measurements are carried out in solution [*66*]. In terms of electron diffraction measurements the angle subtended by the $C = C -$ Cl bonds is found to be 123° [*67*]. An approximate size of the overall molecule may be estimated by noting that the intermolecular distance between the carbon and chlorine atoms is 1.69 *Å* [*68*] and that between carbon and hydrogen is 1.04 *Å* [*67*].

Nitromethane (CH_3NO_2), is a rather interesting liquid, which has been effectively used by Smyth and Walls [*69*] to illustrate the effect of permanent dipole orientation upon the dielectric constant. The substitution of a hydrogen by NO_2 in methane leads to a highly imbalanced molecule with a dipole moment of 3.50 D when measured in the gaseous state and 3.15 D when measured in solution, indicating considerable deviation in molecular shape from that of a perfect sphere. The dielectric constant versus temperature plot for nitromethane as obtained by Smyth and Walls [*69*] at 70 kHz is depicted in Fig. 1.6. Presumably the authors used the frequency of 70 kHz, because with the earlier measuring instruments, higher frequencies were more convenient for obtaining more reproducible results. The curve demonstrates the striking transition behavior as nitromethane passes from the solid to the liquid phase with increasing temperature. From the curve it can be perceived that in the solid phase, where the orientation of the polar molecules is restricted by the lattice

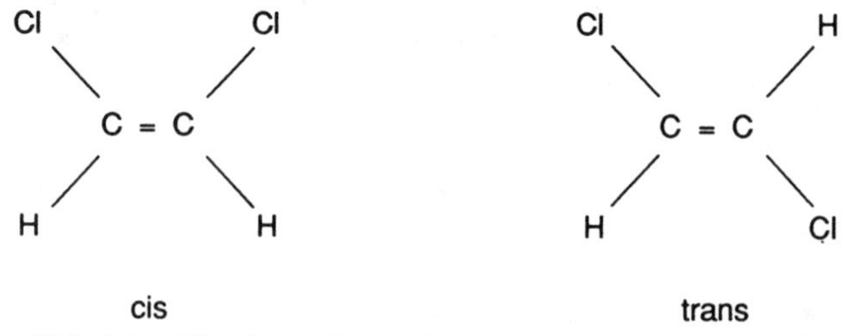

FIG. 1.5—*The cis- and transforms of halogenated ethylene.*

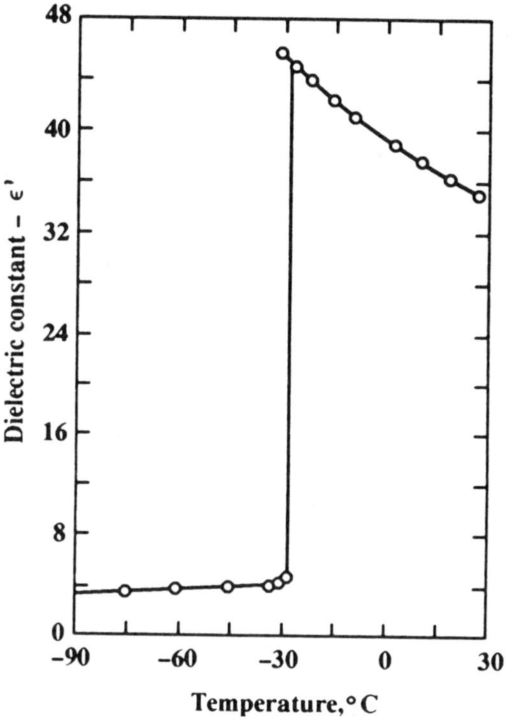

FIG. 1.6—*Effect of temperature upon the dielectric constant of nitromethane at 70 kHz (after Smyth and Walls* [69]*).*

forces, the dielectric constant remains nearly constant with temperature. As the substance changes from the solid to the liquid phase with increasing temperature, there is an abrupt increase in the dielectric constant as the permanent dipoles become suddenly free to execute orientation motion with the alternating electric field. As the temperature is further increased past the solid-liquid transition point, the dielectric constant commences to decrease gradually as a consequence of the $\mu^2/3\ kT$ term in Eq 1.34; in addition, there occurs a minuscule decrease in the dielectric constant due to a small decrease in the density of the fluid with temperature. However, the extremely small but definite increase in dielectric constant with temperature discernible in the solid phase portion of the curve is likely due to an increasing but very limited number of discrete dipoles gaining sufficient energy with rising temperature to surmount their individual energy barriers.

The aromatic mineral oils, employed as insulting liquids, are characterized by a molecular structure that generally contains a number of benzene rings. That the benzene rings themselves are of an essentially planehexagonal structure with zero dipole moment has been established by dielectric measurements [*61,70*] and later confirmed by X-ray diffraction measurements [*67*]. However, it usually is found that attachment of the benzene rings to the remainder of the oil molecule tends to lead to a charge unbalance in the overall molecule, which is manifest by a somewhat higher dielectric constant of the so-called aromatic oils. Since the concentration of the aromatic molecules in the mineral oils is carefully controlled, the resulting increase in the dielectric constant of the aromatic oils is not very pronounced. For example in Table 1.1, mineral oils A, B, and C having respective aromatic contents of 21.5, 15.0, and 0% within approximately the same viscosity range at 37°C have dielectric constants of 2.282, 2.203, and 2.093, respectively, all

TABLE 1.1—*Properties of some typical mineral oils and silicone fluids* [71].

Liquid	Oil A	Oil B	Oil C	Oil D	Oil E	S1	S2
Aromatic carbon atoms, %	21.5	15.0	0	12.0	11.0	—	—
Naphthenic carbon atoms, %	36.5	40.0	45.0	38.5	36.0	—	—
Paraffinic carbon atoms, %	42.0	45.0	55.0	49.5	53.0	—	—
Average molecular weight	335	350	410	535	212	1200	25 000
Viscosity at 27°C, centipoises	87	80	63	1260	9.2	10	970
Index of refraction	1.5119	1.5015	1.4784	1.5077	1.4837	1.399	1.4035
Density at 20°C	0.9230	0.9100	0.8730	0.9236	0.8790	0.940	0.972
Dielectric constant at 27°C and 1000 Hz	2.282	2.203	2.093	2.224	2.143	2.535	2.632

measured at 1 kHz [*71*]. Some of this difference must also be attributed to a difference in the density of these oils. The density effect can be more definitely defined if one compares two oils, D and E, having respective densities of 0.9236 and 0.8790 with highly different viscosities but approximately equal aromatic contents of 12.0 and 11.0%, respectively: their respective dielectric constants are found to be 2.224 and 2.143 indicating a perceptible density effect.

Whilst the benzene ring structure itself is nonpolar, the substitution of a hydrogen atom by an equivalent CH_3 group would lead to a formation of an alkylbenzene having an asymmetrical molecular structure of the type shown in Fig. 1.7. It is further pointed out by Smyth [*42*] that due to hyperconjugation [*72*] a total of nine polar structures are possible, thereby explaining the relatively low permanent dipole moment of 0.4 D for the toluene molecule. In practice larger molecular weight alkylbenzenes have been proposed as possible substitutes for mineral oils used in cables. The molecular structure of these alkylbenzenes is depicted in Fig. 1.8, for which the value of x may range from two to approximately six [*73*]. These synthetic insulating fluids have a density of 0.87 and a dielectric constant of approximately 2.1 as compared to benzene itself, which has density of 0.879 and a dielectric constant of 2.29 at 20°C [*74*]. Evidently, although the alkylbenzene molecule in Fig. 1.8 is asymmetrical, it is probably too large to undergo rotation as a whole in an electrical field with perhaps only some sidelink orientation taking place that leads to a small permanent dipole moment. Consequently due to its slightly lower density than that of benzene, the alkylbenzene

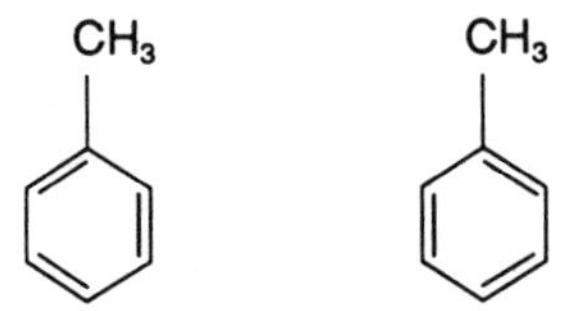

FIG. 1.7—*Two possible Kekulé structures of a simple alkylbenzene.*

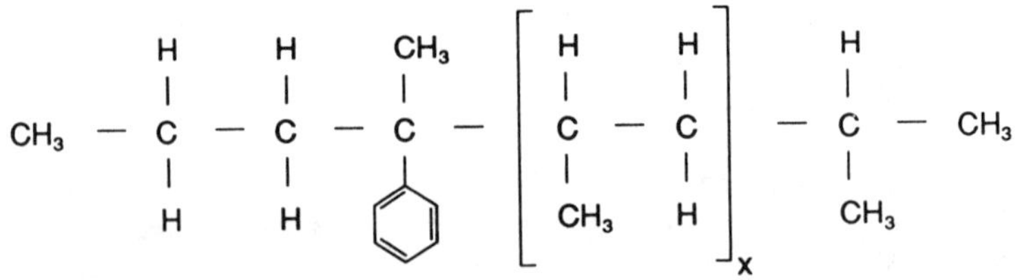

FIG. 1.8—*Typical molecular structure of an alkyl benzene, used as a dielectric impregnating liquid* [73].

insulating fluid is characterized by a dielectric constant that does not substantially differ much from that of benzene having a nonpolar plane-hexagonal molecular structure.

The molecular structure of mineral oils is rather complex in the sense that the molecular weights obtained on these oils are only average values and do not necessarily represent mean values. If we consider once again the oils A, B, C, D, and E alluded to earlier in our discussion, for which some of the properties are tabulated in Table 1.1, then for example the stated average molecular weight of 212 for oil E must be reinterpreted in the light of some additional information available on the molecular size distribution in this oil. It is known that for this particular oil, which is a standard oil used in self-contained cables, the bulk of the molecules contain between 15 to 20 carbon atoms, although small numbers of molecules, having anywhere between 4 and 40 carbon atoms, may also be present. The amounts of the paraffinic, naphthenic, and aromatic constituents in the oil are expressed as percentages of the number of carbon atoms associated with these structures in accordance with the method developed by Martin et al. [*75–78*], which for oil E from Table 1.1 are found to be 53.0, 36.0, and 11.0%, respectively. Again, these values do not infer that every molecule has these definite proportions of paraffinics, naphthenics, and aromatics in its structure, but that these are the average amounts characterizing the insulating oil as a whole. Accordingly, we can stipulate that one possible type of molecule present in this oil could have the structure portrayed in Fig. 1.9. The exact shape of the oil molecules is not known, but because of the ring structures, both aromatic and naphthenic present in most insulating oils, the molecules would tend to assume a rather compact quasi-spherical shape with paraffinic chain structures protruding from the main body. It is most important to emphasize here that in the terminology used with mineral oils, an unsaturated benzene ring signifies an aromatic structure whereas a saturated benzene ring along with any five carbon rings designate naphthenic structures. This type of designation [*75–81*] is contrary to that commonly accepted in practice of naming any six-carbon atom ring structure as aromatic. The same type of designation is also employed in ASTM Standard on Tentative Method for Test for Carbon Type Composition of Insulating Oils of Petroleum Origin (D 2140).

In any given insulating oil containing a certain percentage of paraffinic, aromatic, and naphthenic constituents, it is quite possible that there may be some molecules that may be only paraffinic and whereas others may be comprised of only paraffinic and naphthenic components, while others may have only aromatic and paraffinic components. However, the majority of molecules would be expected to be composed of all three constituents as depicted in Fig. 1.9. Using the interatomic distances provided in the references by Hückel [*82,83*] as 1.54 Å for the C—C bond in the linear

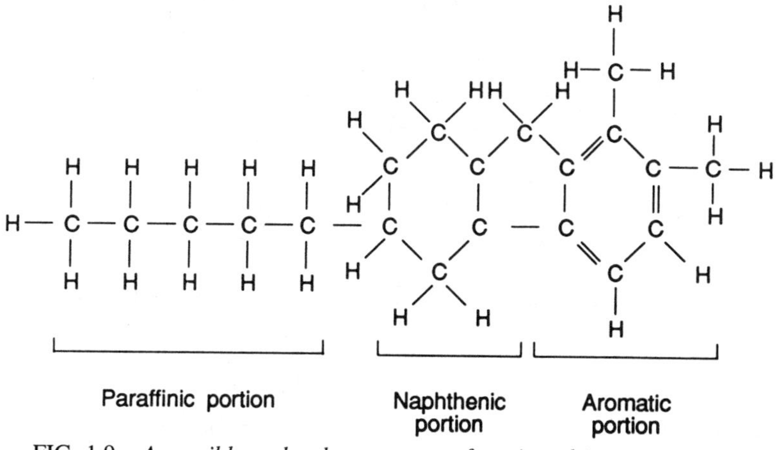

FIG. 1.9—*A possible molecular structure of a mineral impregnating oil.*

chain, 1.40 Å for the C—C bond in the saturated benzene ring, 1.54 Å for the C—C bond in the five carbon atom naphthene ring, 1.34 Å for the C═C bond and 1.09 Å for the C—H bond, one can speculate as to the overall size of the oil molecules. Evidently when such rudimentary estimates are carried out it must be remembered that in the molecular structure representation, the entire molecule cannot be considered as lying in a single plane. Assuming a relatively compact spherical arrangement of the molecules, it is doubtful whether the average molecular diameter of the low viscosity, low molecular weight oil E would be much in excess of 15 Å.

We have already examined the polarity of paraffin and benzene molecules, but nothing was said on the naphthenic type molecules due to the paucity of experimental data available. As can be seen from Fig. 1.9, naphthenic molecules may consist of combinations of six and five carbon ring structures, with a possible propensity towards some molecular charge asymmetry. To what degree this would induce the molecule to assume a net dipole moment is difficult to say. It is known, however, that oils which are mainly naphthenic do not exhibit any significant polarity, so long as the aromatic content in these oils is negligible [*71*].

A great deal of information on the molecular structure of mineral oils may be derived from their infrared spectra. Figures 1.10 and 1.11 depict the infrared spectra of the oils A, B, C, D, and E, respectively (see Table 1.1). The amount of aromatics present in the oils is proportional to the intensity of the band at 1600 cm^{-1}, which is characterized by the stretching vibrations of the C═C bonds within the benzene ring structures. The bending vibrations of the aromatic C—H groups give rise to the numerous bands between 900 to 700 cm^{-1}, with the exception of the band at 720^{-1}; however, the complexity of these bands prevents their use in quantitative analysis. The 720 cm^{-1} band results from the rocking vibrations of the chains containing the methylene groups (—CH_2—),

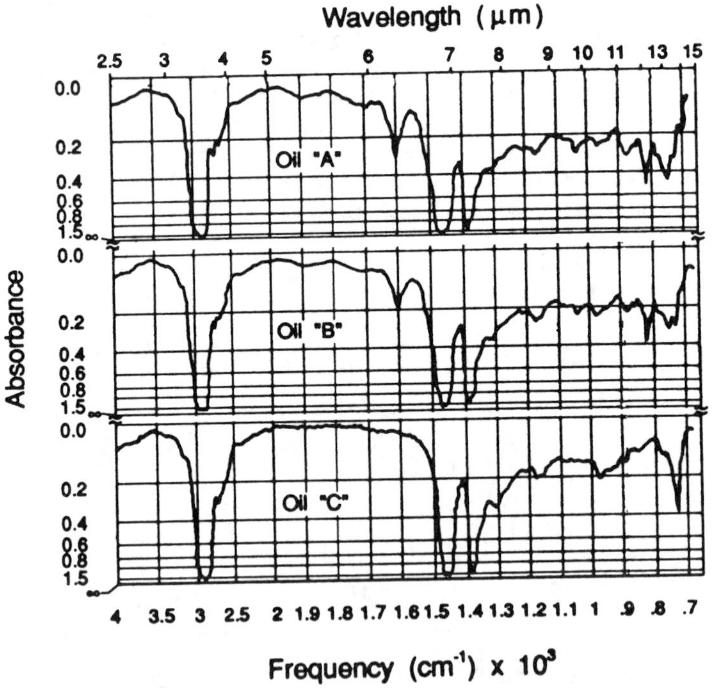

FIG. 1.10—*Infrared spectra of three mineral oils with three different aromatic contents (after Bartnikas* [71]*).*

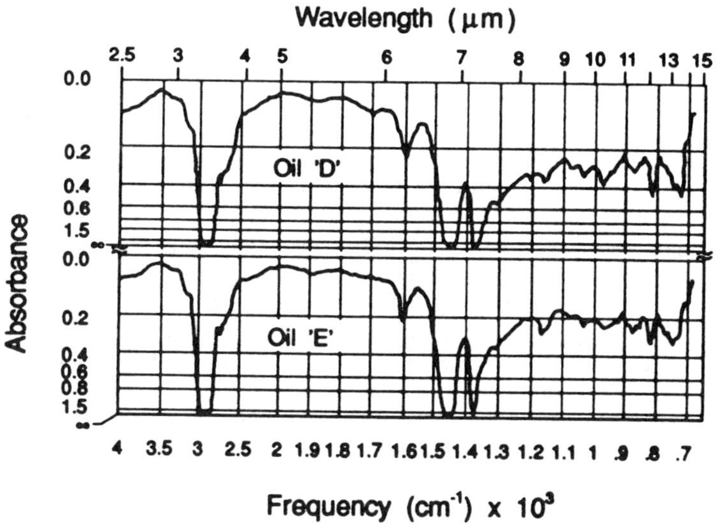

FIG. 1.11—*Infrared spectra of two oils with approximately equal aromatic contents but highly different viscosities (after Bartnikas* [71]*).*

consisting of four or more base units; consequently, its intensity represents an effective measure of the paraffinic content. This band is perceived to be most pronounced in oil C, which has the highest paraffinic content of the three given mineral oils and which is entirely free of aromatics as is indicated by the absence of the 1600 cm^{-1} band and the aromatic bands between 900 and 700 cm^{-1}. It is noteworthy that in oils containing aromatic fractions, the latter bands tend to obscure the paraffinic band. The rather strong bands between 3200 and 2500 cm^{-1} result from the stretching vibrations of the —CH_3, —CH_2, and —CH groups, while the bending vibrations of the same groups are responsible for the pronounced bands in the 1500 to 1425 and 1400 to 1350 cm^{-1} regions of the infrared spectrum. The naphthenes do not give rise to any specific bands. The two cable oils D (high viscosity, solid cable-type impregnating oil) and E (low-viscosity, self contained hollow core cable oil) exhibit a striking similarity in their infrared spectra, notwithstanding their highly different viscosities, molecular weights, and the fact that they originate from different petroleum sources.

One synthetic fluid that has been already used extensively in transformers is silicone. Silicone fluids form the family of organosiloxane liquids of which the polydimethylsiloxanes, that are used for electrical applications, have the well defined chemical structure represented in Fig. 1.12 [*84*]. Increasing values of x (that is, the number of repeat units in the molecular chain) result in silicone oils with higher viscosities. Although at room temperature over the lower frequency regime (<100 MHz) all electrical silicone insulating oils behave essentially as nonpolar fluids [*71*], the polydimethylsiloxane molecule has been demonstrated to have a weak dipole moment that increases with

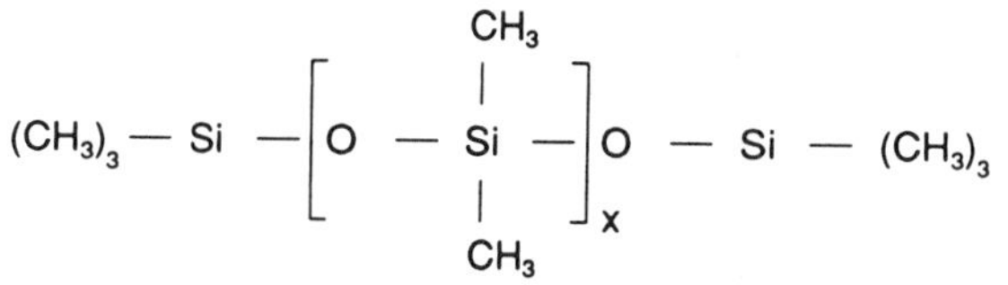

FIG. 1.12—*Molecular structure of polydimethylsiloxane (silicone oil).*

the chain length [*85–87*]. The existence of a permanent dipole moment is evident from Fig. 1.13, which represents schematically the bond angles in a polydimethylsiloxane molecule. Compared to the varied molecular structure of mineral oils, the helical molecular structure of silicone fluids is relatively simple and well defined, thus being more amenable to theoretical investigation. Thorough experimental and theoretical studies on the configuration of the polydimethylsiloxane chains reported by Flory et al. [*88–90*], indicate the angles θ' and θ'' in Fig. 1.13 to be 37 and 70°, respectively. The interatomic distances for the Si—O and Si—C bonds are taken as 1.64 and 1.90 Å, respectively [*91*].

Component dipole moment calculations [*87*] on the polydimethylsiloxane molecule have employed the method described by Eyring [*92*], which is based on the assumption that the individual bond moments are constant and that a given molecule may be idealized as a framework of lines along the valence bonds connecting the individual atoms. This type of calculation of the average resultant moment can be readily carried out provided the potential energy associated with the distortion of the molecule is known. The calculated dipole moments obtained by Dasgupta and Smyth [*87*], indicate the dipole moment to be a linear function of x, that is, the length of the polydimethylsiloxane molecule. This explains the marked increase of dielectric constant with the viscosity of silicone fluids, which is a direct function of the molecular length. These authors suggest that dipole orientation occurs by molecular rotation and by intramolecular motion, that includes group rotation around the Si—O bonds as well as probable bending of the Si—O—Si bonds.

The Eyring procedure [*92*] yields the mean square sum of all the individual dipole moments in the molecule, taking account of their direction and magnitude; this equals also the square of the dipole moment calculated from the measured value of the dielectric constant with the proviso that the molecules are free to orient as a whole in the applied electrical field. Mark [*93*] and Sutton and Mark [*94*] consider the freely rotating chain model of the polydimethylsiloxane molecule employed by Dasgupta and Smyth as untenable, because there is evidence that the abundance of interaction between the methyl groups not only causes the various conformations or rotational states to differ in energy but also introduces interdependence or cooperativeness between such states [*90*].

While Sutton and Mark use the actual value of 143° for the Si—O—Si bond angle, Dasgupta and Smyth use an approximation of 150°. The more rigorously calculated dipole moments obtained at 20°C by Sutton and Mark [*94*] are presented in Table 1.2 as a function of the repeat unit, x. Other pertinent data such as the index of refraction, n', optical and static values of the dielectric constant as well as the density are also tabulated. The static value of the dielectric constant, ϵ_s, has been determined at a frequency of 2 MHz, this being a sufficiently low frequency of measurement at room temperature, since the dipole loss in the silicone fluids is manifest only over much higher frequencies. The values of μ in Table 1.2 do not differ significantly from those found by Dasgupta and Smyth [*87*], assuming unhindered molecular rotation.

Dasgupta and Smyth [*87*] have also found that polydimethylsiloxane is characterized by a rather sizeable atomic polarization of the order of 13.46, 19.11, 24.77, and 36.61 cm as compared to the electronic polarization values of 65.86, 83.64, 101.90, and 138.22 cm for the respective repeat units of x = 1, 2, 3, and 5 at 25°C. They attributed this relatively large value of the atomic polarization

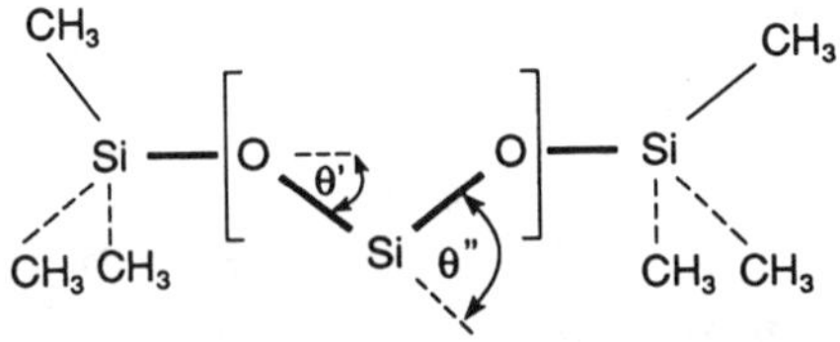

FIG. 1.13—*Schematic representation of bond angles in the polydimethylsiloxane molecule, with the accented solid lines representing the component dipole moments (after Flory et al.* [88–90]*).*

TABLE 1.2—*Dipole moment, μ as a function of* x *for polydimethylsiloxane at 20°C (after Sutton and Mark* [89]).

x	Density-d, g/cm^3	n'	ϵ_∞	ϵ_s	μ, D
3	0.8748	1.3807	2.215	2.476	0.94
5	0.9014	1.3840	2.262	2.589	1.21
7	0.9182	1.3870	2.286	2.645	1.41
9	0.9265	1.3880	2.299	2.669	1.57
300	0.970	1.3917	2.359	2.785	8.44

principally to the bending and twisting vibrations of the Si—O and Si—O—Si bonds, whose absorption frequencies extend well into the far infrared region. Figure 1.14 depicts the infrared spectra obtained on two silicone fluids. The S1 fluid is of a low viscosity comparable to that of a hollow core cable oil (see oil E in Table 1.1) and the S2 fluid is of a high viscosity comparable to that of a solid type impregnating oil (see oil D in Table 1.1) [*71*]. It is appropriate to emphasize at this point that it is difficult to draw an adequate comparison between mineral oils and silicone fluids solely in terms of viscosity, because the latter exhibit a considerably less pronounced change in viscosity with temperature than do the mineral oils.

The similarity of the infrared spectra of the S1 and S2 fluids reflects the identical chemical structure of the two silicone fluids. The main differences apparent in the two spectra are the intensity, which is attributable to a difference in the respective molecular weights, and the band at 755 cm^{-1} in the case of the S1 fluid which is probably caused by the —$Si(CH_3)_2$ terminating groups—these being relatively more copious in the low molecular weight fluid. The bands observed in the two liquids between 3200 and 2700 cm^{-1} and 1475 to 1375 cm^{-1}, respectively, are caused by the stretching and bending vibrations of the C—H groups; in fact these bands are analogous to those exhibited by the mineral oils in the same regions of the spectrum. The $Si(CH_3)_2$ groups are responsible for the intense bands between 1300 to 1225 and 900 to 775 cm^{-1}. The

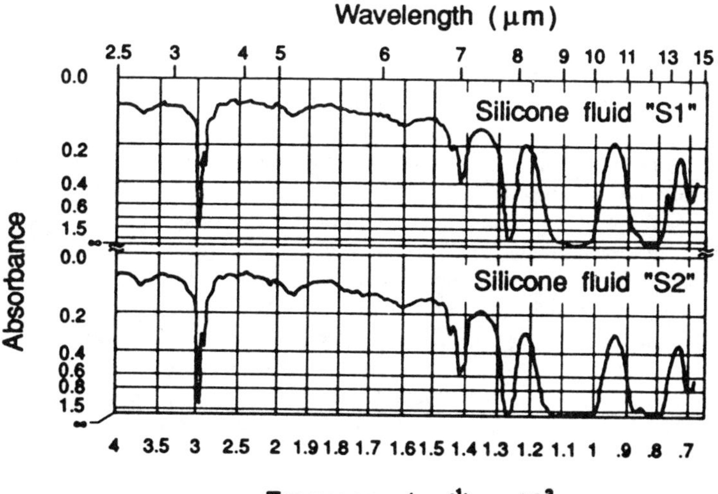

FIG. 1.14—*Infrared spectra of a low viscosity (S_1) and high viscosity (S_2) silicone fluid (after Bartnikas* [71]).

Si—O—Si groups give rise to the broad and strong band in the region extending from 1200 to 950 cm^{-1}.

The inherent polarity of the polydimethylsiloxane molecules causes the silicone liquids to behave somewhat differently from mineral oils, which essentially act as very dilute solutions of polar molecules. This is strikingly evident from Table 1.1, if one compares oils D and E with silicone fluids S1 and S2. Although oil E and silicone fluid S1 have identical viscosities at the temperature of comparison, S1 has a dielectric constant substantially higher than oil E. Likewise as the viscosity of the liquid is augmented, the dielectric constant of the silicone fluid tends to increase more rapidly. It is conceded that these exhibited trends are also influenced by the density changes of these liquids; yet, the density change alone cannot account for the entire variation observed in the dielectric constant, since it is apparent from Table 1.2 that the polarity of the polydimethylsiloxane molecule increases markedly with the molecular length as is indicated by the difference in the values of ϵ_s and (n').

Figure 1.15 demonstrates the difference in polarity between the silicone fluids and the mineral oils. The plots of the molar polarization versus the inverse of the absolute temperature for the mineral oils are virtually straight lines parallel to the $1/T$ axis, thereby indicating that even the most polar of aromatic oils constitute at best only extremely dilute solutions of polar molecules. The intercepts at $1/T = 0$ obtained from the projections of these lines show that the value of the electronic and atomic molar polarizations ranges from approximately 65 to 165 cm^3 for the least viscous to the most viscous of the mineral oils. In contradistinction to the behavior of the mineral oils, the $[P]$ versus $1/T$ characteristics of the silicone fluids exhibit a definite inclination to the $1/T$ axis, indicial of their polar nature. If the S1 fluid characteristic is approximated by a straight line and the same is done with the S2 fluid characteristic, but only over the lower values of $1/T$ where the S2 curve appears more linear, then the intercepts of those straight lines with the $[P]$/-axis at $1/T = 0$ yield the respective values of 400 and 8100 cm^3. These values representing the sum of the electronic and atomic molar polarizations are substantially in excess of these determined for the mineral oils, reflecting to a large extent the much higher molecular weights of the silicone fluids.

The permanent dipole moments, determined in terms of the slopes of the approximating straight lines for the $[P]$ versus $1/T$ plots of the S1 and S2 fluids, are found to be 1.33 and 22.18 D, respectively. Since the molecular structure of the polydimethylsiloxane fluids is well defined, the repeat unit x for both S1 and S2 may be estimated from Eq 1.96, which expresses the molecular weight, M, as

$$M = 74x + 162 \tag{1.96}$$

from which it is established that for S1, $x = 14$ and for S2, $x = 336$ repeat units. Comparing the values of μ with those of Sutton and Mark [*94*] presented in Table 1.2, the value for S1 appears somewhat too low and that for S2 is very much too high. The apparent disagreement between the two sets of values may be attributed to two reasons: firstly, S1 and S2 are industrial grade insulating fluids thereby being mixtures of different molecular size fluids with the result that M used in Eq 1.97 represents only an average value, thus yielding a corresponding average value of x; secondly, the graphical determination of μ is based on the Debye equation which does not apply to wholly polar liquids.

The inherent polarity of the silicone fluids necessitates the use of the Onsager equation which takes into account the reaction field and is, thus, applicable to polar liquids. The Onsager equation given by Eq 1.73 may be simplified in form to

$$\mu^2 = \frac{[\epsilon_s - (n')^2]}{\epsilon_s[(n')^2 + 2]^2} \frac{9\ kT}{4\pi N_A} \frac{M}{d} \tag{1.97}$$

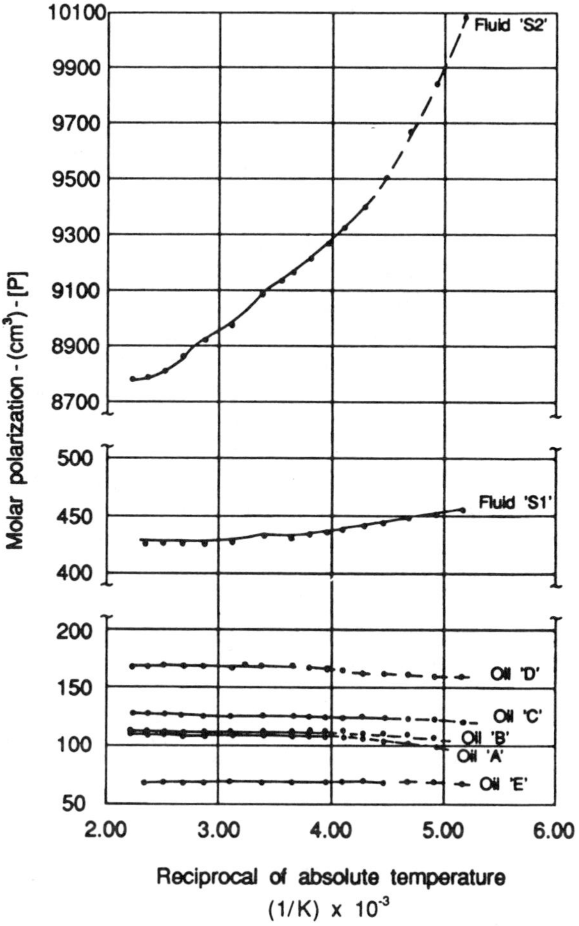

FIG. 1.15—*Molar polarization versus reciprocal absolute temperature for a number of mineral oils and silicone fluids at 1 kHz (after Bartnikas* [71]*).*

At a temperature of 27°C, substitution of the appropriate values yields a μ equal to 0.957 D and 4.51 D for the S1 and S2 fluids, respectively. These values are appreciably smaller than those indicated by the calculations of Sutton and Mark [*loc. cit.*]. The discrepancies in these particular calculated values are believed to be entirely the result of the average molecular weight value used in Eqs 1.96 and 1.97. It should be also emphasized that the Onsager model postulates a spherical molecule, which is not at all the case for the polydimethylsiloxane molecule which is of helical form.

Although in the past polychlorinated biphenyls (PCBs) have been extensively used as insulating liquids in transformers and capacitors, their manufacture has been terminated in North America as a consequence of their environmental unacceptability. Due to their lingering presence in some of the older electrical apparatus that may still be in operation, a few cursory remarks on their dielectric behavior appear to be in order. The molecular structure of the PCBs consists of two attached benzene rings, with varying numbers of chlorine atoms substituted for the normal hydrogen atoms of the benzene ring. One common PCB used in the earlier capacitors is that of pentachlor diphenyl,

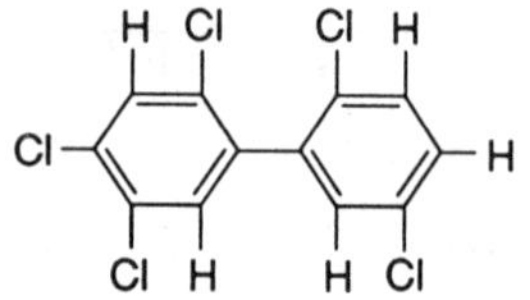

FIG. 1.16—*Molecular structure of pentachlor diphenyl.*

consisting of three and two substituted chlorine atoms in the first and second benzene rings, respectively, as depicted in Fig. 1.16.

The introduction of the substituting chlorine atoms into the otherwise nonpolar biphenyl molecule results in a substantial charge asymmetry, causing the molecule to become polar. This accounts for the relatively high dielectric constant of 4.25 at 100°C at the power frequency. PCBs are relatively highly viscous fluids, and it has been customary practice to lower their viscosity by suitable diluents such as trichlor benzene [95]. The latter due to its three substituting chlorine atoms in the benzene molecule, exhibits the usual striking behavior of intrinsic polar compounds at the solid-liquid transition temperature as evinced in Fig. 1.17. The behavior of the dielectric constant within the liquid phase is governed by the $\mu^2/3\ kT$ term; the somewhat gradual fall in the ϵ' values with decreasing temperature within the solid phase region suggest some limited but persistent orientation of the permanent dipoles within the solid state when the temperature is not too low to allow sufficient energy for some of the permanent dipoles to surmount their discrete potential barriers.

1.4 Dielectric Loss

It would be indeed a grave misconception to construe that the theory of dielectric loss originated in its entirety with the Debye model. In fact the theory evolved very gradually over many decades, following the somewhat now primeval observations of Siemens in 1864 that capacitors subjected to

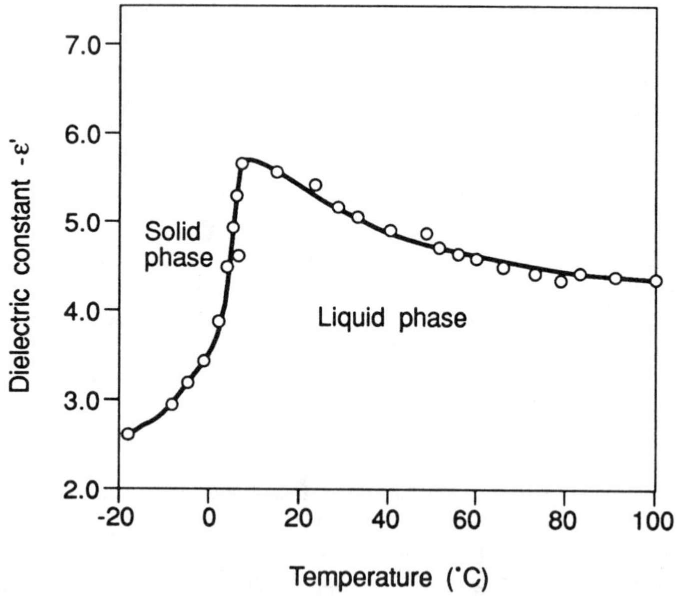

FIG. 1.17—*Effect of temperature upon the dielectric constant of trichlor benzene at 1 kHz (after Clark* [95]*).*

an alternating voltage tend to undergo appreciable heating [*96*]. The heating in dielectrics was attributed to the so-called absorption current effect, which was extensively studied by Hess [*97*], Rowland [*98*], Hopkinson [*99*], and Tank [*100*]. The results of these studies led to the conclusion that the observed dielectric loss is a direct consequence of an energy loss current component in the dielectric which arises from a lag in the charging current vector behind the applied voltage vector. Wagner [*101*] extended the Maxwell theory of absorption further by deriving expressions of the imaginary and real permittivities as functions of frequency for a double layer capacitor.

Independently Pellat [*102,103*] derived equivalent expressions, which were based on the assumption that a sudden application of an electric field to a dielectric will cause the dielectric displacement to assume instantaneously a finite value and, thereafter, rise logarithmically to a limiting value [*104*]. This concept agrees remarkably well with the current approach in dielectric theory where the atomic and electronic polarization are taken to reach the final values instantaneously following the application of the field, and where only the polarization due to the permanent dipoles lags behind the applied field. Pellat's equations are essentially of the same form as the Debye's equations for the a-c case; for this reason, the equations for the complex permittivity as a function of frequency are sometimes also referred to as the Pellat-Debye equations.

Under sufficiently low values of an alternating electric field the behavior of a dielectric liquid will be contingent upon the frequency of the applied field, the temperature and its molecular structure. If the liquid represents a nearly perfect dielectric, then no detectable phase difference will occur between the voltage gradient and displacement vectors, $\overline{D}$ and $\overline{E}$, and, consequently, the ratio $\overline{D}/\overline{E}$ will be defined by a constant equal to the real value of the dielectric constant, ϵ'. Alternatively, a finite dielectric loss in the liquid dielectric will lead to a phase angle difference, δ, between the $\overline{E}$ and $\overline{D}$ vectors and as a consequence the permittivity will become a complex quantity. If E_o and D_o represent the amplitudes of $\overline{E}$ and $\overline{D}$, then the relationship between $\overline{E}$ and $\overline{D}$ may be stated as

$$D_o \exp[j(\omega t - \delta)] = \epsilon^* E_o \exp[j(\omega t)] \tag{1.98}$$

where ϵ^* is the complex value of the permittivity or dielectric constant defined by

$$\epsilon^* = \epsilon' - j\epsilon'' \tag{1.99}$$

where ϵ'' is the imaginary part of the dielectric constant and is also commonly referred to as the loss factor or loss index of the dielectric liquid. Since $\exp[j\delta] = \cos\omega t + j\sin\omega t$, it follows that

$$\epsilon' = \frac{D_o}{E_o}\cos\delta \tag{1.100}$$

and

$$\epsilon'' = \frac{D_o}{E_o}\sin\delta \tag{1.101}$$

from which the definition of the dissipation factor, $\tan\delta$, follows

$$\tan\delta = \frac{\epsilon''}{\epsilon'} \tag{1.102}$$

The energy loss, W, in the dielectric material may be expressed as

$$W = \frac{\omega}{2\pi}\int_o^{2\pi/\omega} \overline{J}\cdot\overline{E}\,dt \tag{1.103}$$

where $\overline{J}$ is the current density vector given by $(1/4\ \pi)\ \partial\overline{D}/\partial t$. Since for a periodically varying field $\overline{E} = \overline{E}_o \cos\omega t$ and $\overline{D} = \overline{D}_o \cos(\omega t - \delta)$ or

$$\overline{D} = \overline{D}_o \cos\omega t \cos\delta + \overline{D}_o \sin\omega t \sin\delta \tag{1.104}$$

then substituting the $\overline{E}$ and $\overline{D}$ vectors in Eq 1.103 yields

$$W = \frac{\omega}{8\pi}\overline{D}_o\overline{E}_o \sin\delta \tag{1.105}$$

Evidently, if the phase or loss angle δ is zero, then the energy loss, W, in the dielectric is also reduced to zero. It is further apparent that the energy loss may be expressed alternatively in terms of the imaginary portion of the permittivity or the loss factor ϵ'' as (see Eq 1.101)

$$W = \frac{\omega E_o^2}{8\pi}\epsilon'' \tag{1.106}$$

In common practice in the field of electrical engineering, one is concerned frequently with the capacitance and dissipation factor terms of insulating systems used in capacitor, cable, or transformer applications. In such applications, it may be expedient to deal with dielectrics as equivalent circuits. In such circumstances a resistive element is selected to represent the loss part of the dielectric, while an ideal capacitance is employed to represent the dielectric displacement. Various combinations of resistances and capacitances may be arranged to represent correctly the response of a dielectric material over a wide frequency regime [*12,15*]. However, for convenience in the most general case, a simple *RC*-parallel equivalent circuit is utilized with the tacit assumption that the resistance and capacitive elements of the circuit are appropriate functions of temperature and frequency so as to yield the observed loss and dielectric constant variations.

In the parallel equivalent circuit representation, one considers the dielectric as consisting of a conductance, G, shunted in parallel by a capacitance, C, as portrayed in Fig. 1.18*a*. With an alternating voltage $\overline{V}$ applied across the parallel circuit, the insulation leakage or loss current, $\overline{I}_l$, which is in phase with the voltage vector $\overline{V}$ is given by

$$\overline{I}_l = G\overline{V} \tag{1.107}$$

and the charging or displacement current, $\overline{I}_c$, by

$$\overline{I}_c = j\omega C\overline{V} \tag{1.108}$$

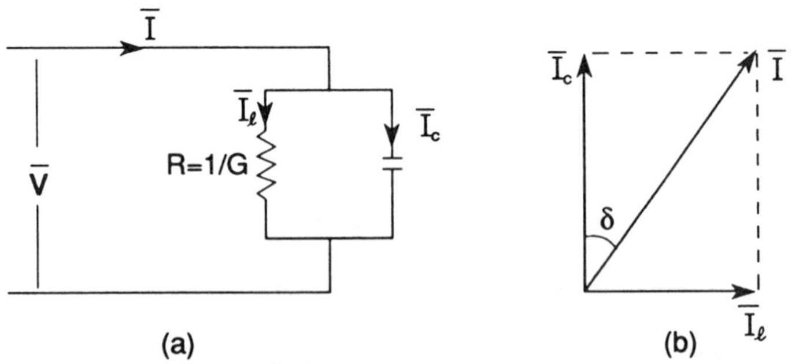

FIG. 1.18—*Parallel equivalent circuit of a dielectric* (a) *with corresponding phasor diagram* (b).

The total current, $\bar{I}$, is thus the vector sum of these two components

$$\bar{I} = (G + j\omega C)\bar{V} \tag{1.109}$$

If C_o represents the capacitance *in vacuo* of the parallel plate capacitor, then the currents may be expressed alternatively as

$$\bar{I}_c = j\omega\epsilon' C_o\bar{V} \tag{1.110}$$

and

$$\bar{I}_l = \omega\epsilon'' C_o\bar{V} \tag{1.111}$$

whence the total current becomes

$$\bar{I} = (\omega\epsilon'' + j\omega\epsilon')C_o\bar{V} \tag{1.112}$$

If we farther assume that the dielectric is contained between two parallel plates of area A separated by a distance d, such that the electric field $\bar{E} = \bar{V}/d$ and the capacitance *in vacuo* $C_o = A\epsilon_o/d$ (where ϵ_o is the permittivity in free space and is equal to unity in ESU units), then the current density vector is given by

$$\begin{aligned}\bar{J} &= \bar{J}_l + \bar{J}_c \\ &= (\omega\epsilon'' + j\omega\epsilon')\bar{E}\end{aligned} \tag{1.113}$$

with its loss component, $\bar{J}_l$, and charging component, $\bar{J}_c$, having the same phase relationships as $\bar{I}_l$ and $\bar{I}_c$ shown in the vector representation of Fig. 1.18*b*.

It is apparent from the vector diagram that the dissipation factor

$$\tan\delta = \frac{\bar{J}_l}{\bar{J}_c} = \frac{\epsilon''}{\epsilon'}$$

$$= \frac{1}{\omega RC} \tag{1.114}$$

where ω is the radial frequency term in radians, and R and C are, respectively, the parallel resistance and capacitance of the dielectric. When using Eq 1.114 in practice, it should be borne in mind that R is not simply equal to the insulation resistance measured under d-c conditions. Here R is actually the a-c insulation resistance at a given radial frequency ω or $2\pi f$ and is normally determined from the measurement of tan δ and C at the frequency $f = 1/2\pi\omega$, where f is the frequency term in hertz.

The conductivity σ of a dielectric is defined as the ratio of the leakage current density $\bar{J}_l$ to the electric field gradient $\bar{E}$,

$$\sigma = \frac{\bar{J}_l}{\bar{E}} \tag{1.115}$$

Hence, in terms of Eq 1.113

$$\sigma = \omega\epsilon'' \tag{1.116}$$

and consequently

$$\tan\delta = \frac{\sigma}{\omega\epsilon'} \tag{1.117}$$

Here σ is the a-c conductivity and need not only arise solely from dipole orientation loses but may also be caused by a migration of charge carriers (electrons and ions) under the action of the electric field $\overline{E}$.

1.4.1 Phenomenological Approach

In order to describe the manner in which the energy loss of a dielectric material changes with frequency, one must determine the dependence of the complex permittivity upon the frequency of the applied alternating field. When a d-c field, $\overline{E}$, is suddenly applied across a dielectric, the dielectric almost instantaneously, or in a very short time interval, Δt, will attain a finite polarization value. The polarization value is reached almost instantaneously, because it is governed essentially by the electronic and atomic polarizability mechanisms. Here the limiting value of the real dielectric constant, ϵ', is defined by the high frequency value of the dielectric constant, ϵ_∞, which is dependent upon the atomic and electronic polarizabilities. Hence, the resulting high frequency dielectric displacement is $\epsilon_\infty\overline{E} = \overline{D}$. In accordance with Eq 1.6, we can proceed and define the resultant instantaneous or high frequency polarization, $\overline{P}_\infty$, as

$$\overline{P}_\infty = \left[\frac{\epsilon_\infty - 1}{4\pi}\right]\overline{E} \tag{1.118}$$

The slower processes, involving permanent dipole orientation and free charge carrier migration, will give rise to a polarization phenomenon that will attain its saturation value appreciably more gradually because of such effects as the inertia of the dipoles and the finite transit times of the mobile charge carriers involved. The dielectric displacement vector, $\overline{D}$, affected by these processes will also be influenced by the atomic and electronic polarization contributions and will be given by $\epsilon_s\overline{E}$; here ϵ_s is evidently the static value of the real dielectric constant, ϵ'. The total polarization of the dielectric as determined by the permanent dipole and any mobile charge contributions plus the atomic and electronic polarization components will be that given by Eq 1.6, that is, $\overline{P} = (\epsilon_s - 1)\overline{E}/4\pi$.

The idealized polarization-time curve is depicted in Fig. 1.19, in which P_s is the achieved saturation value of polarization as may result from the orientation of permanent dipoles or the movement of free charge carriers. Depending upon the temperature and the chemical and physical structure of the dielectric, the final saturation value of P_s can be reached in a time that may easily vary anywhere from seconds to several days or even longer. If for Fig. 1.19, $P(t)$ is used to denote the time dependent portion of P_s, then the equation of the curve may be represented by a form typically characteristic of a charging capacitor

$$P(t) = P_s\left[1 - \exp\left(-\frac{t}{\tau}\right)\right] \tag{1.119}$$

where τ is the time constant of the charging process. The time constant, τ, is a measure of the time lag and is, thus, referred to as the relaxation time of the polarization process. Solving for $\exp(-t/\tau)$ in Eq 1.119 gives

$$\exp(-t/\tau) = \frac{P_s - P(t)}{P_s} \tag{1.120}$$

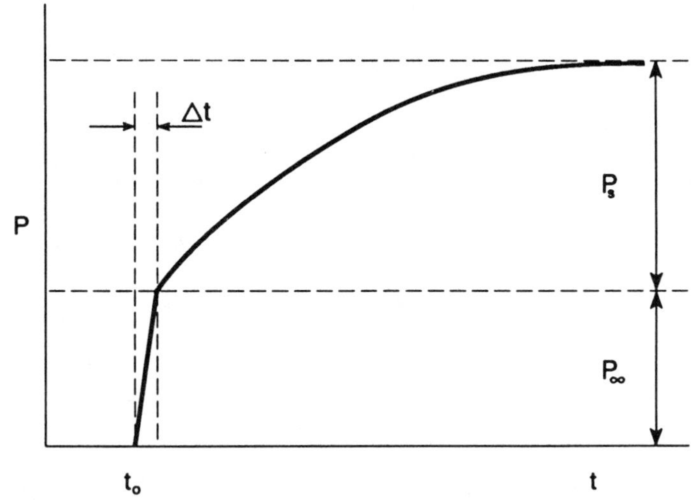

FIG. 1.19—*Polarization versus time characteristic of a dielectric material.*

Differentiation of Eq 1.119 and substitution of the result in Eq 1.120 yields

$$\frac{d\,P(t)}{dt} = \frac{1}{\tau}[P_s(t) - P(t)] \tag{1.121}$$

In the foregoing equation, P_s is replaced by $P_s(t)$, which designates the saturation value of P_s that would be obtained in an alternating field, $E(t)$, whose instantaneous value equals that of the static field giving rise to P_s [*17*].

Since the total polarization of the dielectric $\overline{P}$, as defined by Eq 1.6, is equal to the same value of $\overline{P}_\infty$, as defined by Eq 1.118, we have that

$$\overline{P}_s = \left[\frac{\epsilon_s - \epsilon_\infty}{4\pi}\right]\overline{E} \tag{1.122}$$

and alternatively in complex notation

$$P_s^* = \left(\frac{\epsilon_s - \epsilon_\infty}{4}\right) E_o \exp(\omega t) \tag{1.123}$$

Substitution of the latter into Eq 1.121 gives the differential equation

$$\frac{d\,P^*(t)}{dt} = \frac{1}{\tau}\left[\frac{\epsilon_s - \epsilon_\infty}{4\pi} E_o \exp(j\omega t) - P^*(t)\right] \tag{1.124}$$

whose solution is of the form

$$P^*(t) = \frac{1}{4\pi}\left[\frac{\epsilon_s - \epsilon_\infty}{1 + j\omega\tau}\right] E_o \exp(j\omega t) \tag{1.125}$$

Since in complex notation we can write that

$$D^* = E^* + 4\pi[P_\infty^* + P^*(t)] \tag{1.126}$$

Substitution of the respective values leads to the Pellat-Debye equations, which express the real and imaginary values of the dielectric constant as a function of the radial frequency, ω, and the relaxation time, τ, of the dielectric loss process

$$\epsilon' = \epsilon_\infty + \frac{\epsilon_s - \epsilon_\infty}{1 + \omega^2\tau^2} \tag{1.127}$$

and

$$\epsilon'' = \frac{(\epsilon_s - \epsilon_\infty)\omega\tau}{1 + \omega^2\tau^2} \tag{1.128}$$

These equations are similar in form to the antecedent Maxwell-Wagner equations, which were developed by Wagner using Maxwell's theory and applied to a multilayered capacitor system [*101*].

Examination of Eq 1.128 shows that when ϵ'' is plotted against the radial frequency term, ω, it attains a maximum at $\omega = 1/\tau$ or $f = 1/2\pi\omega$, where f is the frequency of the field in Hz. At this frequency, the maximum value of ϵ'' becomes equal to

$$\epsilon'' = \frac{\epsilon_s - \epsilon_\infty}{2} \tag{1.129}$$

and ϵ' assumes the value

$$\epsilon' = \frac{\epsilon_s + \epsilon_\infty}{2} \tag{1.130}$$

The curves of ϵ'' and ϵ' versus $\log\omega$ are symmetrical and are depicted in Fig. 1.20. In practice, the dielectric loss of specimens is measured in terms of the $\tan\delta$ value rather than the loss factor, ϵ''. From Eqs 1.127 and 1.128, it follows that

$$\tan\delta = \frac{\epsilon''}{\epsilon'} = \frac{(\epsilon_s - \epsilon_\infty)\omega\tau}{\epsilon_s + \epsilon_\infty\omega^2\tau^2} \tag{1.131}$$

with a maximum of $\tan\delta$, when $\partial(\tan\delta)/\partial\omega = 0$, occurring at a frequency

$$\omega = \frac{1}{\tau}\left(\frac{\epsilon_s}{\epsilon_\infty}\right)^{1/2} \tag{1.132}$$

and having a value equal to

$$\tan\delta = \frac{\epsilon_s - \epsilon_\infty}{2(\epsilon_s\epsilon_\infty)^{1/2}} \tag{1.133}$$

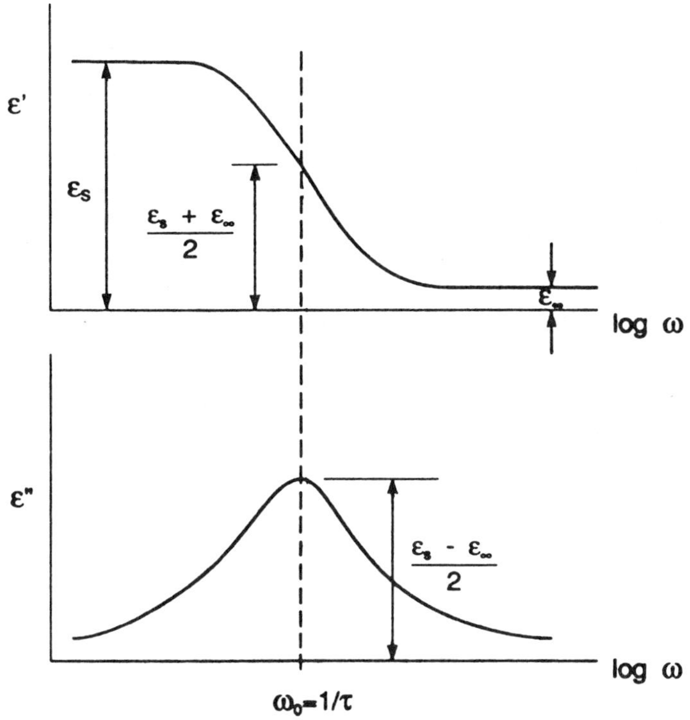

FIG. 1.20—*Curves of ϵ' and ϵ'' versus ω on a logarithmic scale in accordance with the Pellat-Debye equations.*

It is apparent that the maximum of tanδ does not coincide with that of ϵ''. The amount of shift or displacement of the tanδ peak toward the higher frequencies depends upon the degree of polarity of the dielectric material, that is, the difference between the ϵ_s and ϵ_∞ values. With most practical insulating materials this difference is relatively small, with the consequence that the resulting shift in the tanδ peak is also correspondingly small.

An alternative method for representing ϵ'' and ϵ', is that of each plotted point being representative of the corresponding values of ϵ'' and ϵ' measured at one particular frequency, ω. The resulting plots approximate semicircular arcs and are commonly referred to as Cole-Cole plots [*105*]. In an idealized case, as the frequency, ω, is increased from zero to infinity, a perfect semicircular plot on the complex plane (with ϵ' plotted on the real axis and ϵ'' on the j or imaginary axis) results as is portrayed in Fig. 1.21. At the boundary limit of $\omega = 0$, $\epsilon' = \epsilon_s$, and $\epsilon'' = 0$; at the other boundary extreme of $\omega = \infty$, $\epsilon' = \epsilon_\infty$, and, once again, $\epsilon'' = 0$. Here the center of the semicircle is located at $(\epsilon_s + \epsilon_\infty)/2$ and the vector drawn from the origin to any point on the semicircular locus defines the complex permittivity vector $\epsilon^* = \epsilon' - j\epsilon''$.

1.4.2 Molecular Theories

In the previous section, the dielectric response due to an externally applied alternating electric field was described phenomenologically in terms of the externally determined values of the real and imaginary dielectric constant. A much more fundamental approach is to define the dielectric

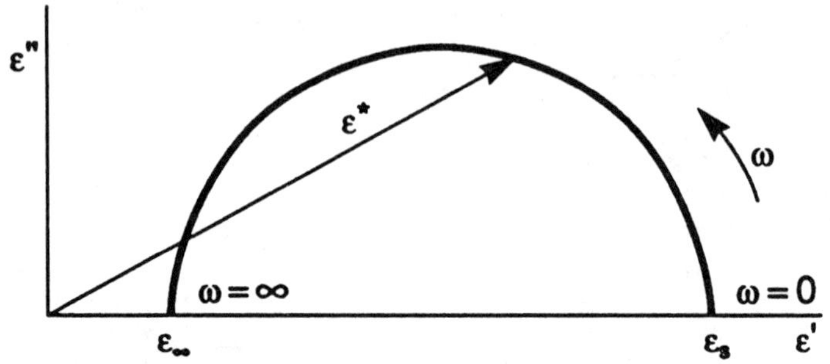

FIG. 1.21—*Idealized Cole-Cole plot of* ϵ'' *versus* ϵ'.

behavior in terms of the molecular structure of the material. Debye [*10*] has accomplished this task by the use of the molar polarization expression modified for the a-c case and the introduction of the intrinsic relaxation time τ^* and its consequent relationship to the macroscopic viscosity of the dielectric liquid.

According to Eq 1.31, one can introduce the complex molar polarization quantity that would arise under a-c conditions as

$$[P^*] = \frac{\epsilon^* - 1}{\epsilon^* + 2}\left(\frac{M}{d}\right) \tag{1.134}$$

Since in complex rotation

$$P^* = P^*_\infty + P^*(t) \tag{1.135}$$

and further noting the accordingly revised form of Eq 1.126, Eq 1.135 becomes

$$P^* = \frac{(\epsilon_s - 1)}{4\pi}E^* + \frac{\epsilon_s - \epsilon_\infty}{4\pi(1 + j\omega\tau)}E^* \tag{1.136}$$

Equation 1.135, therefore, may be rewritten in its molar polarization form as

$$[P^*] = [P_\infty] + \frac{[P] - [P_\infty]}{1 + j\omega\tau^*} \tag{1.137}$$

or alternatively as

$$\frac{\epsilon^* - 1}{\epsilon^* + 2} = \frac{\epsilon_\infty - 1}{\epsilon_\infty + 2} + \left[\frac{\epsilon_s - 1}{\epsilon_s + 2} - \frac{\epsilon_\infty - 1}{\epsilon_s + 2}\right]\left(\frac{1}{1 + j\omega\tau^*}\right) \tag{1.138}$$

where the relaxation time, τ^*, now refers to the intrinsic value. Solving for the complex permittivity ϵ^*, we find

$$\epsilon^* = \epsilon_\infty + \frac{(\epsilon_s - \epsilon_\infty)}{1 + j\omega\tau^*\left(\frac{\epsilon_s + 2}{\epsilon_\infty + 2}\right)} \tag{1.139}$$

Since the complex permittivity is given by $\epsilon^* = \epsilon' + j\epsilon''$, then in terms of Eqs 1.127 and 1.128

$$\epsilon^* = \epsilon_\infty + \frac{\epsilon_s - \epsilon_\infty}{1 + j\omega\tau} \tag{1.140}$$

Comparison of Eqs 1.139 and 1.140 indicates that the intrinsic relaxation time, τ^*, is related to the relaxation time τ by the expression

$$\tau = \tau^* \left(\frac{\epsilon_s + 2}{\epsilon_\infty + 2}\right) \tag{1.141}$$

It must be emphasized that the concept of an intrinsic relaxation time is principally useful where the molar polarization can be expressed in terms of the molecular quantities of the atomic and electronic polarizability and the permanent dipole moment. Such a relationship is in fact implicit in Eq 1.137, namely, that

$$[P^*] = \frac{4\pi}{3} N_A \left[\alpha + \frac{\mu^2}{3\,kT} \cdot \frac{1}{1 + j\omega\tau^*}\right] \tag{1.142}$$

In the Debye molecular model for dipole relaxation in liquids, Eq 1.142 constitutes the underlying basis of a relationship between the intrinsic relaxation time, τ^*, and the viscosity of the liquid medium. Since Debye considered spherical molecules of radius r rotating in a liquid continuum of viscosity η, the intrinsic relaxation time, τ^*, was found by establishing the manner in which the distribution function, $G(\tau^*)$ of a spherical dipole ensemble changes due to the abrupt application or removal of an electric field.

If a system of N dipoles is considered, then within the solid angle θ and $(\theta + d\theta)$ at any time in the direction of a directing field E_r, the number of dipoles, dN, is

$$dN = \frac{G(\tau^*)N \sin\theta\, d\theta}{2} \tag{1.143}$$

where the distribution function $G(\tau^*)$, defined in accordance with Boltzmann statistics, is given by [*10*]

$$G(\tau^*) = A \exp\left[\mu E_r \cos\theta / kT\right] \tag{1.144}$$

where A is a constant and k is the Boltzmann constant (1.38×10^{-16} ergs/K or 1.38×10^{-23} J/K). Expansion of Eq 1.144 and neglection of the saturation effects [*10*] yields

$$G(\tau^*) = A\left[1 + \frac{\mu E_r}{kT} \cos\theta\right] \tag{1.145}$$

Note that with the directing field E_r reduced to zero, $G(\tau^*) = A$. It is further observed that the derivative $d\theta/dt$ represents the angular velocity of the orienting dipole due to the exertion of a torque, Y, by the imposed electric field, such that

$$T = \xi \frac{d\theta}{dt} \tag{1.146}$$

where ξ is a friction proportionality constant, whose magnitude is contingent upon the shape of the molecular dipole and its interaction with the neighboring molecules of the liquid. For a liquid medium with a viscosity of η, the friction constant, ξ, for a rotating molecular sphere of radius r is given by Stokes law as

$$\xi = 8\pi\eta r^3 \tag{1.147}$$

Debye calculated the distribution function $G(\tau^*)$, at a time t after the removal of the field and found that

$$G(\tau^*) = A\left[1 + \frac{\mu E_r}{kT}\exp\left(-2\,kT\,t/\xi\right)\right] \tag{1.148}$$

In terms of Boltzmann statistics the average component of the dipole vector in the original direction of the field at a time t is thus given by

$$\langle\mu_t\rangle = \int_o^{\pi} G(\tau^*)\,\mu\cos\theta\left(\frac{1}{2}N\sin\theta\right)d\theta \Big/ \int_o^{\pi} G(\tau^*)\left(\frac{1}{2}N\sin\theta\right)d\theta \tag{1.149}$$

Substitution of Eq 1.148 into Eq 1.149 yields

$$\langle\mu_t\rangle = \frac{\mu^2 E_r}{3\,kT}\exp\left[-2\,kTt/\xi\right] \tag{1.150}$$

From the exponential term of Eq 1.150, it is apparent that the relaxation process of a spherical dipole rotating in a viscous continuum, may be described by a time constant which is given by the intrinsic value of the relaxation time

$$\tau^* = \frac{\xi}{2\,kT} \tag{1.151}$$

alternatively, in terms of Eq 1.147

$$\tau^* = \frac{4\pi\eta r^3}{kT} \tag{1.152}$$

The sudden removal of the directing field, $\overline{E}_r$, causes an exponential decay of the ordered state of the aligned spherical dipoles as governed by Brownian movement due to thermal agitation. Equation 1.152 demonstrates that larger molecular dipoles will be characterized by longer relaxation times, while any increase in the temperature will lead to a reduction in the relaxation time. Hence, the viscosity parameter must necessarily play a dominant role in determining the behavior of insulating liquids: higher viscosity insulating oils will tend to exhibit dipole losses at considerably lower frequencies.

The Debye model represents an oversimplification of the state of affairs in the sense that it is not truly a molecular model because of the macroscopic viscosity, η, term. Thus, it is pointed out correctly by Daniel [*1*] that while the shear viscosity, η, is determined macroscopically, the friction occurring within the liquid itself requires a specific viscosity term that would be a direct measure of the friction between the rotating spherical dipole and its neighboring liquid molecules. Evidently, in practice this type of measurement is not attainable. For ellipsoidally shaped molecules, Eq 1.152 has been modified by Fischer [*106*] to the form

$$\tau^* = \frac{4\pi\eta abcf^*}{kT} \tag{1.153}$$

where f^* is a function of the axial ratios and a, b, and c which represent the axes of the ellipsoidal dipole. Although Eq 1.153 has been found to yield reasonable results, it requires knowledge on the parametric values of the dipole ellipsoids. However, in practice, it is often desirable to proceed in the reverse direction in order to estimate the molecular dimensions of the dipoles responsible for the observed relaxation process.

Since the viscosity of the soluble or dipole molecules is normally different from that of the solvent solution, several attempts have been made to take this into account. Hill [*107*] defined a mutual viscosity η_{12} between two miscible liquids having respective macroscopic viscosities η_1 and η_2 of the polar and solvent molecules, in terms of which the derived relaxation time was found to be

$$\tau = \left[\frac{I_{21}I_1}{I_{21} + I_1}\right]\left[\frac{m_1m_2}{m_1 + m_2}\right]\frac{3\eta_{12}d_{12}}{kT} \tag{1.154}$$

The mutual viscosity, η_{12}, is a function of the molecular masses of the respective liquids, their eigenvalue frequencies of vibration, ν_1 and ν_2, the average distance d_{12} separating the polar and solvent molecules and a proportionality constant K which in turn is dependent upon the momentum transfer characteristics between the polar and solvent molecules; I_1 and I_{21} represent moment of inertia quantities such that I_{21} is the moment of inertia of the soluble molecules about the polar molecules whose moment of inertia about themselves is I_1. An alternate expression for Hill's relaxation time is obtained by eliminating the mutual viscosity η_{12} term from Eq 1.154

$$\tau = \left[\frac{I_{21}I_1}{I_{21} + I_1}\right]\frac{K(\nu_1 + \nu_2)}{kT} \tag{1.155}$$

A similar but somewhat more empirical approach for the correction of the macroscopic viscosity term has been followed by Gierer et al. [*108*]. They apply a multiplication factor of $\frac{1}{6}(r_1/r_2) + \{1 + (r_1/r_2)\}^{-3}$ to the value of η appearing in Eq 1.152, such that

$$\tau = \frac{4\pi r_1^3}{kT}\eta\left[\tfrac{1}{6}(r_1/r_2) + \{1 + (r_1/r_2)\}^{-3}\right] \tag{1.156}$$

where r_1 is the radius of the polar molecules (assumed to be spherical in shape) and r_2 is the equivalent radius of a spherical layer of solvent molecules enveloping the polar molecule entities. While the treatments of Fischer, Hill, and Gierer et al. elucidate quite effectively the inadmissibility of the macroscopic viscosity concept to dipole orientation at the microscopic or molecular level, it is also evident at the same time that they themselves can be applied, at best, only to simple dielectric liquids having well-known and defined molecular structures. Their application to practical insulating liquids, consisting of a large melange of molecular sizes and shapes (both polar and solvent) would meet with almost insurmountable difficulties. Hence, notwithstanding the theoretical advances made in the various refinements of Debye's original model consisting of a dipole sphere rotating in a continuum of macroscopic viscosity, η, Debye's simplistic model still remains perhaps the only viable one that is capable of yielding some practical information when applied to insulating liquids having complex and poorly defined molecular structures.

An alternative quasi-molecular model that has been propounded to explain the dependence of the relaxation time upon temperature, without the introduction of the macroscopic viscosity, η, is that of Eyring [*109*] and Bauer [*110*]. According to this model, the molecular dipole may be considered as having two or more equilibrium positions; thus, to surmount the potential barrier between two possible equilibrium positions, the dipole must acquire sufficient energy. The marked time dependence of the relaxation time upon temperature obtains because the molecular dipoles vibrate around their equilibrium positions until they acquire by thermal fluctuation the activation energy necessary to pass over the energy barriers facing them into another equilibrium position. The rate at which such discrete jumps or rotations take place is given by

$$\kappa_o = \frac{kT}{h}\exp\left[-\Delta F/RT\right] \tag{1.157}$$

κ_o is known as the rate constant for the activation of dipoles and is numerically equal to the mean value of the number of jumps made by the dipole per unit time; h is the Planck constant, and R the gas constant, while ΔF is defined as the free energy of dipole relaxation. The intrinsic relaxation time of the dipole orientation process is then expressed as $\tau^* = 1/\kappa_o$ or in terms of Eq 1.157 as

$$\tau^* = \frac{h}{kT} \exp [\Delta F/RT] \tag{1.158}$$

Note that since

$$\Delta F = \Delta H - T\Delta S \tag{1.159}$$

then Eq 1.158 may be expressed as

$$\tau^* = \frac{h}{kT} \exp [\Delta H/RT] \exp [-\Delta S/R] \tag{1.160}$$

or more commonly as

$$\tau^* = \frac{h}{kT} \exp [\Delta H/kT] \exp [-\Delta S/k] \tag{1.161}$$

where ΔH is the heat of activation of dipole orientation or simply the activation energy of dipole orientation and ΔS is the entropy of activation. Since, in practice, most insulating liquids can be considered as dilute solutions of polar molecules, the intrinsic relaxation time τ^* does not differ substantially from the macroscopic value τ, and, therefore, the latter is often substituted for τ^* in dielectric measurements. Moreover, as the measurements of ϵ' and $\tan\delta$ are recorded as functions of frequency, it is more expedient to use the value of the absorption frequency, f_m, at the $\tan\delta$ maximum in lieu of τ. Hence, because $f_m = 1/2\pi\tau$, Eq 1.161 becomes

$$f_m = f_o \exp [-\Delta H/kT] \tag{1.162}$$

where

$$f_o = \frac{kT}{2\pi h} \exp [\Delta S/k] \tag{1.163}$$

The activation energy of dipole orientation may be determined either in terms of Eq 1.160 or Eq 1.162. Taking natural logarithms of both sides of Eq 1.160, yields

$$\ln \tau T = \ln \frac{h}{k} - \frac{\Delta S}{k} + \frac{\Delta H}{k} \left[\frac{1}{T} \right] \tag{1.164}$$

From the nature of the foregoing equation, it is apparent that a dipole loss process described by a single value of ΔH will manifest itself in the form of a straight-line relationship between $\ln \tau T$ and $1/T$ in terms of whose slope ΔH can be determined. Alternatively the same information may be obtained by taking the natural logarithms of both sides of Eq 1.162, whence

$$\ln\left[\frac{f_m}{T}\right] = \ln\frac{k}{h} - \ln 2\pi + \frac{\Delta S}{k} - \frac{\Delta H}{k}\left[\frac{1}{T}\right] \tag{1.165}$$

Again ΔH is obtained from the slope of the ln $[f_m/T]$ versus $1/T$ characteristic.

The application of Eq 1.165 may be illustrated by considering a number of mineral oils whose properties are presented in Table 1.1. Figure 1.22 portrays the ln $[f_m/T]$ versus $1/T$ characteristics obtained by Bartnikas [*111*] with four different insulating oils. The form of the four characteristics is similar in shape and their displacement from each other is indicative of the difference in their viscosity characteristics. Thus the near coincidence of the characteristics for oils A and B results directly from their nearly identical viscosity values. The perceptible curvature exhibited by the characteristics of the four oils, demonstrates that the activation energy of dipole orientation cannot be described by a single activation energy, ΔH; in fact, the behavior infers both the activation energy, $\Delta H(T)$, and entropy, $\Delta S(T)$ of dipole orientation to be functions of temperature. A number of tangential values of $\Delta H(T)$ were determined graphically and the corresponding values of $\Delta S(T)$ were estimated from the intercepts as $(1/T) \rightarrow 0$; Eq 1.159 was employed to verify these values with respect to the ΔF calculated, using the relation

$$f_m = \frac{kT}{2\pi h}\exp\left[-\Delta F/kT\right] \tag{1.166}$$

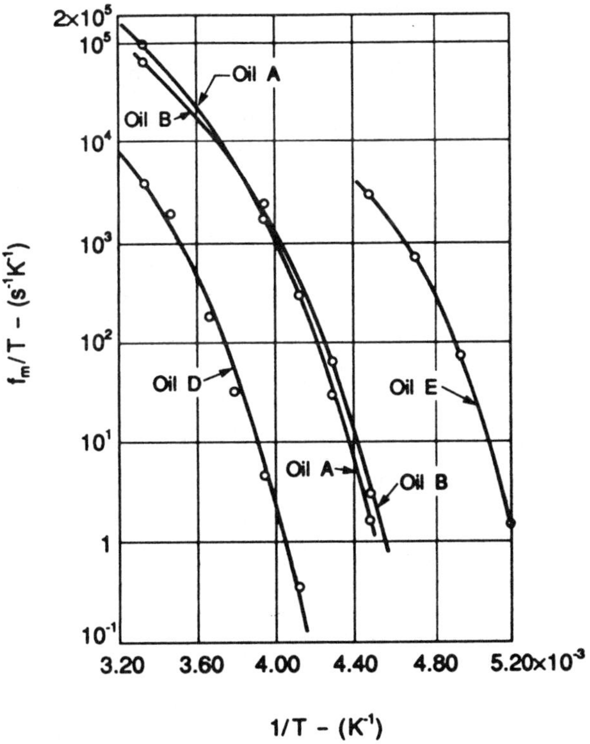

FIG. 1.22—*Variation of* f_m/T *with* $1/T$ *for a number of cable insulating oils with moderate aromatic contents (after Bartnikas* [111]*).*

Figure 1.23 depicts a typical plot of ΔF as a function of the temperature T. The monotonic decrease of ΔF with temperature is thus suggestive of changes in the dipole orientation mechanism with temperature.

In Table 1.3, a comparison is presented of the calculated values of ΔF and those determined in terms of $\Delta H(T)$ and $\Delta S(T)$ as found graphically from Fig. 1.22; a reasonably good agreement is apparent between the experimentally determined and calculated values of ΔF. The values of $\Delta H(T)$ and $\Delta S(T)$ provide some useful information concerning the nature of molecular dipole orientation in the tested electrical oils. The finite values of the apparent entropies of activation, $\Delta S(T)$, evident from Table 1.3, suggest that molecules in the proximity of the dipole molecule must undergo some rearrangement before dipole rotation can take place. This is in agreement with the observations made by Jackson [*112*] and Frank [*113*] on chlorinated diphenyls, where substantially larger entropies were found to characterize the dipole orientation process in the more polar fluids. Relatively high entropies were found with glasses by Baker and Smyth [*114*], which in the limit may be regarded as fluids of excessively high viscosity. In the case of the mineral oils the fact that $\Delta S(T)$ increases rapidly with diminishing temperature indicates that increasingly more of the adjacent molecules must rearrange themselves prior to dipole orientation at the lower temperature, that is, at higher densities where greater packing of the molecules occurs. The increase of the entropy of activation is accompanied also by pronounced increases in the apparent activation energy, $\Delta H(T)$ with decreasing temperature.

The variation of the activation energy, $\Delta H(T)$, as a function of the temperature, T, for the four different mineral oils is depicted in Fig. 1.24. The observed decrease of $\Delta H(T)$ with T for all the oils, strongly infers a gradual change (with the exception of a single discontinuity in each curve) in the orientation mechanism with temperature. This type of behavior is to be anticipated because molecular interaction in the oils decreases as their viscosity falls with rising temperature. Consequently, an increasingly smaller number of molecules adjacent to the rotating dipole need to readjust prior or during the orientation process, thus in effect creating a lower potential energy barrier. The single discontinuity in each curve suggests an abrupt change in the orientation mechanism in the vicinity of the discontinuity, perhaps due to some glass transition point.

It is conceivable that a relationship may exist between the ln τT versus $1/T$ and the ln $[f_m/T]$ versus $1/T$ characteristics. Some empirical data on mechanical shear relaxation have been obtained [*115*], which show that for hydrogen bonded liquids a nonlinear plot of ln η versus $1/T$ is common

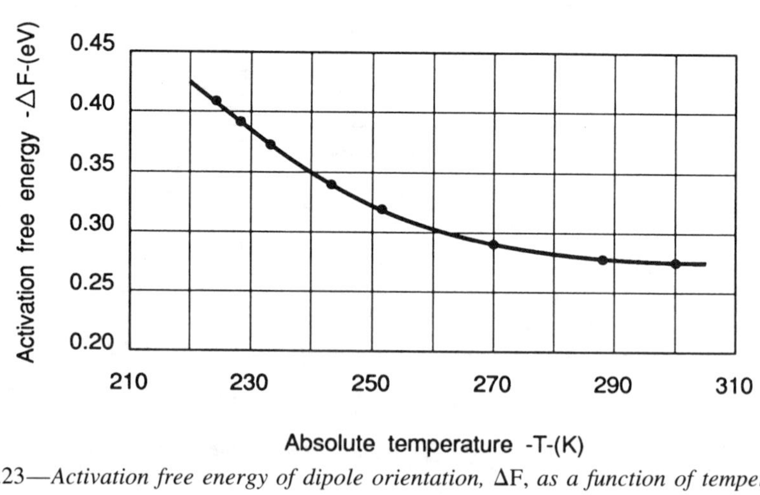

FIG. 1.23—*Activation free energy of dipole orientation,* ΔF, *as a function of temperature of a high aromatic content mineral oil A (after Bartnikas* [111]*).*

TABLE 1.3—ΔF, ΔH(T) *and* ΔS(T) *as a function of temperature for different cable insulating oils (after Bartnikas* [111]*).*

Oil	T, °K	Calculated ΔF, eV	$[\Delta H(T) - T\Delta S(T)]$, eV	$\Delta H(T)$, eV	$\Delta S(T)$, eV/°K
	223	0.41	0.41	1.66	0.0056
	227	0.39	0.39	1.42	0.0045
	233	0.37	0.37	1.17	0.0034
A	243	0.34	0.33	0.99	0.0027
	253	0.31	0.31	0.79	0.0019
	270	0.29	0.29	0.58	0.0011
	286	0.28	0.27	0.46	0.00066
	300	0.27	0.27	0.41	0.00046
	223	0.40	0.40	1.53	0.0051
	227	0.38	0.37	1.32	0.0042
	233	0.36	0.36	1.10	0.0032
B	243	0.33	0.32	0.86	0.0022
	253	0.31	0.31	0.66	0.0014
	270	0.29	0.29	0.51	0.00081
	286	0.283	0.29	0.39	0.00035
	300	0.279	0.28	0.36	0.00025
	243	0.48	0.49	1.59	0.0045
	253	0.44	0.44	1.32	0.0035
D	263	0.42	0.35	1.10	0.0029
	273	0.39	0.38	0.95	0.00208
	288	0.356	0.354	0.66	0.00107
	300	0.352	0.351	0.51	0.00053
	193	0.36	0.35	1.59	0.0064
	198	0.33	0.30	1.37	0.0054
	203	0.31	0.31	0.99	0.0034
E	208	0.29	0.29	0.85	0.0027
	213	0.28	0.28	0.70	0.0019
	217	0.272	0.268	0.61	0.0016
	223	0.267	0.265	0.43	0.00075

and that this generally leads in turn to a variation in the average value of the dielectric relaxation times and, hence, to a change in activation energy. Implicit in Eq 1.152 is the assumption that a relationship exists between the ln η versus $1/T$ and the ln $[f_m/T]$ versus $1/T$ characteristics. Indeed a plot of ln η versus $1/T$ is similar in form to that of ln $[f_m/T]$ versus $1/T$, but the activation energies derived from these two plots are not necessarily equal. The viscosity η of an insulating oil may be expressed by

$$\eta = A(\eta) \exp [\Delta H(\eta)/kT] \tag{1.167}$$

where $\Delta H(\eta)$ is the activation energy for viscosity and $A(\eta)$ is a preexponential factor which is equal to (Nk/V) exp $[-\Delta S(\eta)]$, N being the number of molecules and V the molar volume of the liquid. In analogy with Eqs 1.162 and 1.163, one would anticipate τT to be proportional to ηV if the enthalpies of dipole orientation and viscous flow were identical. The activation energy of dipole orientation has been compared with that of viscosity of oils by Hakim [*116*] and the values were found to differ considerably, so that the suggested proportionality is not valid. Figures 1.25 and 1.26 show the dependence of the frequency, f_m at the maximum of the dipole loss with the viscosity, η, as a function of the inverse absolute temperature, respectively. It can be seen that the

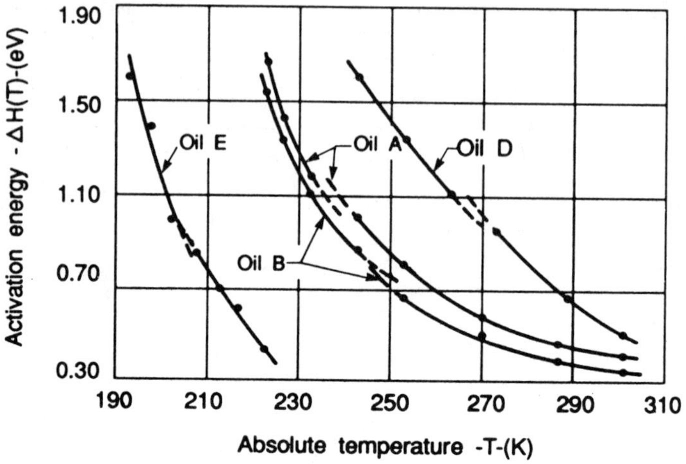

FIG. 1.24—*Activation energy of dipole orientation as a function of temperature for a number of mineral oils (after Bartnikas* [111]*).*

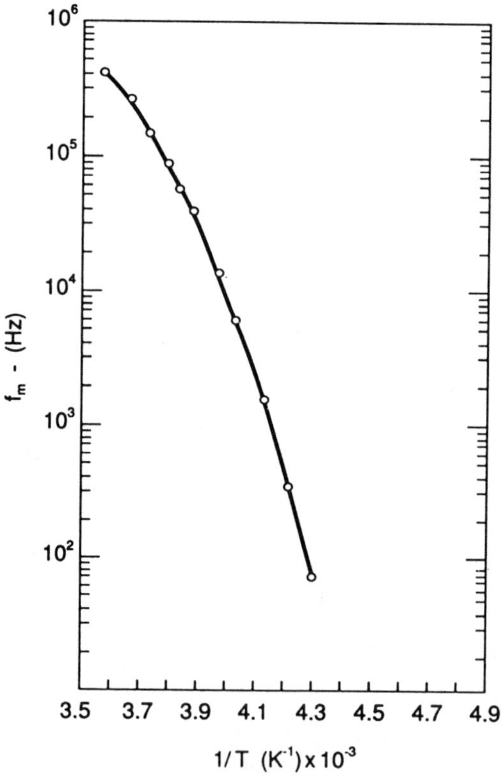

FIG. 1.25—*Dependence of frequency at maximum of dielectric loss on the inverse absolute temperature (after Hakim* [116]*).*

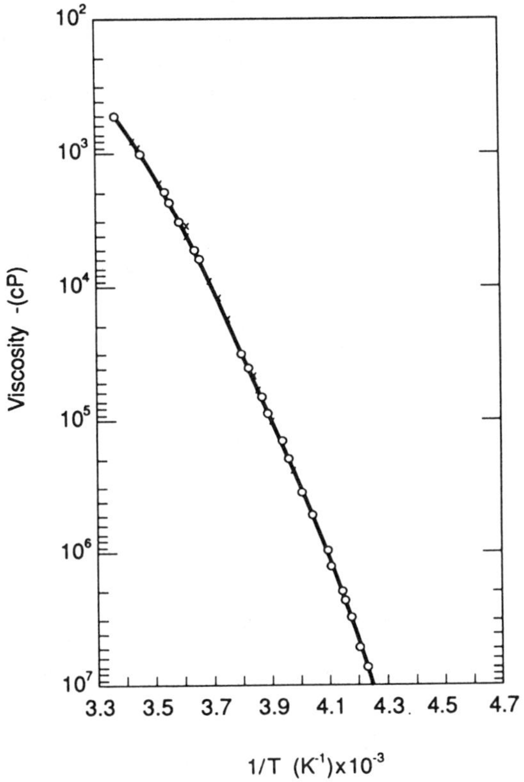

FIG. 1.26—*Dependence of viscosity on the inverse basolute temperature; different point notations refer to different spindles (after Hakim* [116]*).*

TABLE 1.4—*Dependence of activation energy of dipole orientation,* ΔH, *and viscous flow,* $\Delta H(\eta)$, *upon temperature (after Hakim* [116]*).*

Temperature, K	$\Delta H(T)$, kcal/mol	$\Delta H(\eta)$, kcal/mol	$\frac{\Delta H(\eta)}{\Delta H(T)}$
282.0	8.7	19.0	2.19
273.0	11.8	18.6	1.58
268.0	18.4	23.0	1.25
263.0	21.9	25.3	1.15
258.5	22.4	25.6	1.15
252.0	25.2	27.9	1.10
248.5	27.7	27.2	0.98
242.5	31.6	28.8	0.91
237.0	40.5	28.1	0.70

curvature of the two curves differs perceptibly; the calculated $\Delta H(T)$ and $\Delta H(\eta)$ values are given in Table 1.4 as functions of temperatures. As anticipated the two activation energies differ appreciably, because the viscosity is dependent upon frictional forces arising mainly as a result of collisions between the solvent molecules while the dipole relaxation mechanism involves primarily collisions between the polar molecules or dipoles and the surrounding solvent molecules. Moreover dipole orientation involves rotational motion only, whereas in the case of viscous flow both translational and rotational motions of the molecules occur.

Although all mineral oils generally exhibit nonlinear ln η versus $1/T$ characteristics, it is difficult to relate this behavior directly to the nonlinearity of the ln $[f_m/T]$ versus $1/T$ curves. If ΔH of the tested oils in Fig. 1.24 were plotted against the viscosity, η, then evidently ΔH would be found to increase with viscosity very much in the same manner as it does with decreasing temperature. It can be observed from Fig. 1.24 that the apparent activation energy of dipole orientation for the oils at the lowest test temperature is relatively high and falls between 1.50 and 1.70 eV. This proximity in the values of ΔH is not surprising when it is observed that for the four oils tested, the viscosities lie within a relatively narrow range of 10^6 to 10^7 cP, and the dipole diameters extend from a few angstroms to a maximum of 5.5 Å in proportion to the decreasing viscosity of the oil. At the highest test temperature, the ΔH values are found to lie between 0.30 and 0.5 eV; however, now the spread in the viscosity is considerably greater, extending from 10 to 10^3 cP. Evidently, if we attempt to construct an empirical relationship between the frictional force restraining dipole orientation and the potential energy barrier to orientation, then we can write that

$$\Delta H \simeq [8\pi\eta r^3]f(x) \tag{1.168}$$

where $f(x)$ is a function of the molecular dipole configuration and the fluctuating interaction of the molecular environment, the latter being in turn a function of space and time within the insulating liquid. Equation 1.168 indicates that ΔH would be expected to increase with the viscosity, η, as well as the dipole radius, r—both dependences being implied by the results of Fig. 1.24. The exact manner in which ΔH varies with viscosity; hence, temperature is determined by the modifying function, $f(x)$. The calculated dipole diameters, $2r$, do not exhibit any systematic increase with decreasing temperature, so that no perceptible viscosity drag effects are manifested [*111*]. Perceptible drag effects would only be anticipated with dipoles having excessively large diameters. One must conclude, therefore, that the probability of a significant number of dipoles having such macroscopic dimensions in insulating oils is low.

In solids, the dipole molecules are considered to undergo orientation by the execution of discrete jumps between two or more equilibrium positions as opposed to a more gradual orientation of the dipoles in a liquid medium characterized by a macroscopic viscosity, η, in accordance with the Debye model. Since in liquids the average distance between neighboring molecules is approximately of the same order of magnitude as in solids, one would surmise approximately the same degree of interaction to occur between the molecules in liquids. If in a liquid a molecular dipole is displaced from its equilibrium position, then the adjacent molecules will tend to rearrange themselves to accommodate this new position—that is, the dipole may execute a jump to a new equilibrium position provided the adjacent molecules would rearrange their positions. Alternatively, the dipole may jump to another equilibrium position, provided one is available, even without rearrangement of the neighboring molecules. Thus, a dipole in a liquid could conceivably undergo orientation by executing discrete jumps of the type that may characterize dipole orientation in a solid phase.

On the basis of the foregoing considerations, Fröhlich [*41*] developed a barrier type model to describe dipole orientation in both liquids and solids. Fröhlich's barrier model has two equilibrium directions, namely 1 and 2 with the dipoles situated in (+) and (−) sites, where Γ_{12}^{+} represents the probability per second for a transition of a dipole on a (+) site from direction 1 into the direction 2.

Using the same notation, N_1^+ and N_1^- denote the number of dipoles pointing in the direction 1 at the (+) and (−) sites, respectively. Hence, if N is the total number of dipoles, then

$$\begin{aligned}\frac{N}{2} &= N_1^+ + N_2^+ \\ &= N_1^- + N_2^-\end{aligned} \tag{1.169}$$

and at equilibrium with no external field applied $N_1^+ = N_2^-$ and $N_1^- = N_2^+$; this implies an equality of the transition probabilities also. That is the transition probability from the direction 1 to the direction 2 at the (+) site, Γ_{12}^+, will equal that from the direction 2 to direction 1 at the (−) site, Γ_{21}^-; likewise, $\Gamma_{12}^- = \Gamma_{21}^+$.

If an external field is applied and suddenly removed, then the rates of change at a (+) site are given by [*41*]

$$\begin{aligned}\frac{dN_1^+}{dt} &= -\frac{dN_2^+}{dt} \\ &= -N_1^+\Gamma_{12}^+ + N_2^+\Gamma_{21}^+\end{aligned} \tag{1.170}$$

or in terms of Eq 1.169 by

$$\frac{d}{dt}(N_1^+ - N_2^+) = -(\Gamma_{12}^+ + \Gamma_{21}^+)(N_1^+ - N_2^+) + (\Gamma_{21}^+ - \Gamma_{12}^+)\frac{N}{2} \tag{1.171a}$$

Similarly, for the (−) site

$$\frac{d}{dt}(N_1^- - N_2^-) = -(\Gamma_{12}^+ + \Gamma_{21}^+)(N_1^- - N_2^-) - (\Gamma_{21}^+ - \Gamma_{12}^+)\frac{N}{2} \tag{1.171b}$$

Since

$$\Delta N = (N_1^+ - N_2^+) + (N_1^- - N_2^-) \tag{1.172}$$

then addition of Eqs 1.171*a* and 1.171*b* yields

$$\frac{d(\Delta N)}{dt} = -(\Gamma_{12}^+ + \Gamma_{21}^+)\,\Delta N \tag{1.173}$$

but, by definition, the polarization P is equal to the moment per unit volume, that is, $P = \mu(\Delta N)$ or $\Delta N = P/\mu$. Hence, $d(\Delta N)/dt$ is proportional to the rate of change of polarization as the dielectric discharges after the field is removed. Consequently [*41*]

$$\Delta N \propto \exp\left[-(\Gamma_{12}^+ + \Gamma_{21}^+)\,t\right] \tag{1.174}$$

and the relaxation time, τ, is given by (see Eq 1.119)

$$\tau = \frac{1}{\Gamma_{12}^+ + \Gamma_{21}^+} \tag{1.175}$$

The Fröhlich's barrier model in general will approximate the Debye behavior, though as pointed out by Hill [*117*] some deviations may occur since the field acting on the dipoles is contingent upon both the dielectric constant and the external field. The nature of the equilibrium positions, into which the dipoles may orient, is not fully understood. But in view of the fluctuating interaction of the molecular environment surrounding the discrete dipoles, it has been suggested by Glarum [*118*] that dipoles may orient themselves into lattice defect sites. Anderson and Ullmann [*119*] considered these sites as free volume sites created by a fluctuating environment; Johari and Dannhauser [*120*] were able to explain the relaxation process in a number of alcohols in terms of the fluctuating free volume model.

1.4.3 Mobile Charge Carriers

In the foregoing section, we have seen that dipole loss originates with liquids containing polar molecules and results from a temporal phase difference which occurs between the orienting dipoles and the externally applied field. The dipole loss mechanism is influenced greatly by the frequency of the field and temperature, but is independent of the field strength below the saturation region which usually lies well beyond the field intensities encountered in high voltage apparatus. However, the leakage current flowing through an insulating liquid subjected to an alternating electric field does not necessarily arise from only the dipole orientation losses; in particular, at power frequencies under elevated voltages, the leakage current may also be caused by mobile charges, such as ions or electrons. The question of conductivity due to mobile charge carriers constitutes the subject matter of Chapter 2 of this volume, so that in this section conduction by ions and electrons will be considered only from the point of view of dielectric loss occurring under a-c conditions.

1.4.4 Losses Due to Ionic Conduction

Mobile ions are generated either intrinsically by the dissociating molecules of the liquid itself or extrinsically by the electrolytic impurities contained within the liquid. It has been recognized for some time [*121*] that in complex molecular structure insulating liquids, such as mineral oils, the ions are not only present as simply charged constituents but may also exist in the form of molecular aggregates or micelles in which the individual ions are enveloped by an atmosphere of neutral molecules with the induced dipole axes pointing towards the ions. The application of alternating field induces the ions to undergo oscillation, thereby causing energy to be dissipated in an internal frictional process. Since the amplitude of this oscillation is contingent upon the electric field strength, the ionic loss is necessarily voltage dependent [*122,123*]. Furthermore, temperature changes exert a marked influence upon the magnitude of the ionic a-c loss as a result of the viscosity variation with temperature. That is, a rise in temperature is accompanied by a decrease in viscosity and, hence, an increase in the mobility of the ions resulting in a higher conductivity or loss. Generally it is found that with insulating liquids of low dielectric constant having low ion concentrations, the a-c conductivity will tend to remain relatively invariant with frequency over the lower frequency regime ($<10^4$ Hz) [*71*]; however, a frequency dependence is manifested when the product of the Langevin relaxation time (which is the time required for the establishment of equilibrium between the charge distribution and the external electrical field) and frequency is of the order of unity [*124*].

A rise in temperature causes simultaneously a reduction in viscosity or resistance to the movement of ions and an increase in the number of ions as a consequence of an enhanced dissociation rate. It is intuitively apparent that insofar as the conduction loss within liquids is a consequence of the movement of the mobile ionic charge carriers, the observed conductivity must necessarily be directly proportional to the concentration of mobile ions and inversely proportional to the viscosity of the insulating liquid. If the observed loss results solely from ionic conduction then the product of the macroscopic viscosity and conductivity may be taken as a meaningful measure of the ionic content with the liquid. It has been found by Whitehead [*121*] some time ago that the product of viscosity and conductivity is nearly constant for mineral oils in the temperature range from 30 to 60°C; although with oils that undergo deterioration while in service, the value of this product exhibits some variation between 20 and 80°C [*125*]. It is a well established empirical relationship (Walden's rule) that for electrolytic solutions, the product of the absolute viscosity, η, and the equivalent conductance, Λ, varies between one half and unity [*126*]. If we assume along with Dunkley and Sillars [*127*] that the lower limit is more appropriate for insulating liquids, then

$$\eta\Lambda = \frac{1}{2} \tag{1.176}$$

Since the equivalent conductance is by definition equal to the ratio of the conductivity to the ion concentration (expressed in gram-equivalent weights per cm^3), Eq 1.176 may be utilized to derive a rudimentary estimate of the ionic charge carrier concentration in insulating liquids. This has been done by Bartnikas [*71*], using the a-c conductivity values of a number of insulating oils and silicone fluids. Figure 1.27 compares the ion concentration of three different oils having approximately the same viscosities but with different aromatic contents. It can be seen that the ion content is higher for the more aromatic oils; however, the ion concentration is not a function of the aromatic content *per se,* rather it is more a function of the degree of oxidation of the respective oils since it is known that some aromatic molecules are particularly susceptible to oxidation. The ionic content tends to increase with temperature for all the oils with the exception of the paraffinic oil C, where an initial decrease is discernible. Since the ion concentration is determined in terms of the a-c conductivity, the results may be influenced to some extent by a remnant dipole loss; however, this dipole loss over the range of test temperature is negligibly small in comparison to the ionic losses. There are several explanations that may be propounded to account for the increase of ionic content with the aromaticity of the oils. The larger dielectric constants of the more polar oils enhance the ionic dissociation constant because of the inverse relationship between the coulomb force of attraction

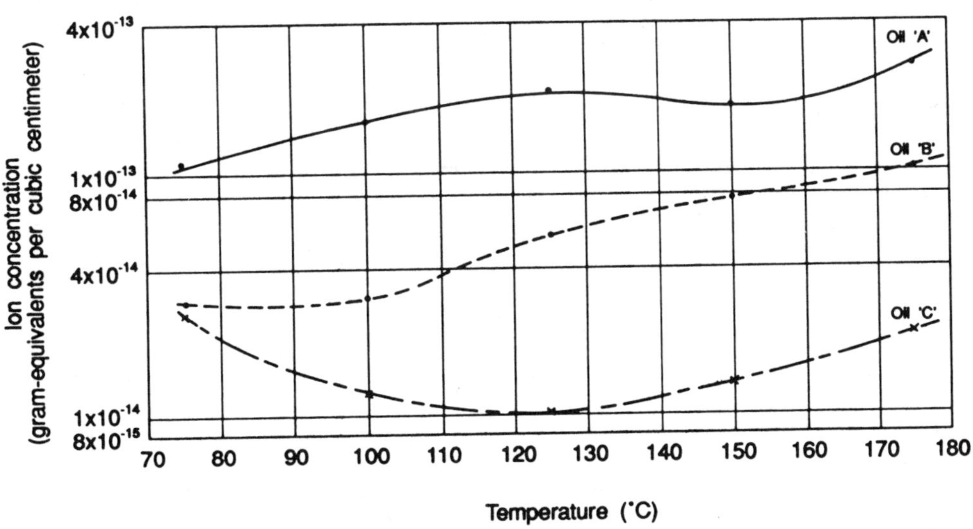

FIG. 1.27—*The influence of temperature on the ion concentration at 1 kHz in mineral oils of different aromatic content but of approximately equivalent viscosity (after Bartnikas* [71]*).*

between the oppositely charged ions and the macroscopic dielectric constant of the continuum separating the ions. Yet this dependency would be expected to be relatively weak due to the small disparity in the dielectric constants of the investigated oils: oil A (2.282), oil B (2.203), and oil C (2.093) at room temperature. However, the ion concentration may be further augmented in the more polar oils by the formation of dipole (Debye-Hückel) atmospheres which envelope the individual ions. The formation of such micelles greatly reduces the attractive forces between the discrete ions, thereby decreasing their recombination tendency into undissociated molecules or ion pairs and thus enhancing the conductivity of the oils. The foregoing hypothesis is supported by the fact that identical treatment of oils A and B with Fuller's earth produced at 85°C a reduction in the tanδ value of some 26% higher for the more aromatic oil, even though spectrographical analysis indicated equal removal of oxidation products in both oils [*71*]. Since no measurable change was detected in the aromatic content, the result suggests that the conduction ions originate with the oxidation products whose dissociation is enhanced in the presence of aromatic or polar molecules.

Electrical grade mineral oils contain aromatic constituents in order to provide the oils with an ability to absorb any gas that may be evolved under intense field conditions [*128,129*]. Some other early investigations have already indicated that the aromatic constituents thus retained may influence other electrical properties of the oil such as the breakdown strength [*130*] and conductivity [*131*]. Although a portion of the aromatic constituents are susceptible to oxidation, certain polynuclear aromatic constituents have been found on the contrary to enhance oxidation stability [*132*]. The oxidation reaction of mineral oils has been examined extensively by a number of workers [*133–136*], and while it generally is agreed that oxidation of certain aromatic fractions gives rise to sludge formation and increases the viscosity of the oils, it is the naphthenic and paraffinic constituents that yield significant amounts of acids upon oxidation; apart from possible contamination effects, these acids would appear to be the most probable sources of dielectric loss in oils due to ionic conduction. The dissociation of the acids should provide a copious supply of high mobility hydrogen ions or protons. Another source of protons conceivably could be from other impurities such as small traces of water. Although the dissociation of water has been found to account for losses in highly pure liquids with well defined molecular structure such as cyclohexane [*137*] and perhaps low loss silicone fluids, its effects on the considerably higher loss mineral oils conceivably is considerably less pronounced. This would follow from the fact that the traces of water in mineral oils are usually very small (<10 ppm) and the low dielectric constant of the essentially nonpolar oils does not provide an effective medium for the dissociation of a highly polar substance, such as water. This is also borne out in practice, as has been observed by Sillars [*138*] sometime ago, that exposure of mineral oils to moist conditions does not result in any perceptible increase of the tanδ value.

The a-c conductivity arising from the movement of ionic charge carriers is a strong function of temperature as a consequence of the direct dependence of the macroscopic viscosity upon temperature. Thus, at room temperature, a lower viscosity oil will be characterized by a substantially higher a-c conductivity than a high viscosity oil, notwithstanding the fact that the higher viscosity oil may have a higher ion concentration [*71*] as is evident from Fig. 1.28 which compares oils D and E with their characteristics given in Table 1.1. At lower temperatures the higher viscosity oils tend to exhibit lower losses, because their higher viscosity impedes free ionic movement; it is precisely for this reason that low viscosity oils are highly susceptible to contamination with ionic impurities. The higher intrinsic ion content in the more viscous oils is probably a result of a higher ion dissociation as affected by the greater dielectric constant of the more viscous oils. The foregoing behavior is not confined to mineral oils as the same predisposition to increasing ionic content with viscosity is also in evidence with silicone fluids as depicted in Fig. 1.29. Note that for comparative purposes, fluids S1 and S2 have been selected specifically to match the viscosities of the oils E and D, respectively at room temperature (see Table 1.1), though with rising temperature, the reduction in viscosity for silicone liquids is considerably less than that with mineral oils.

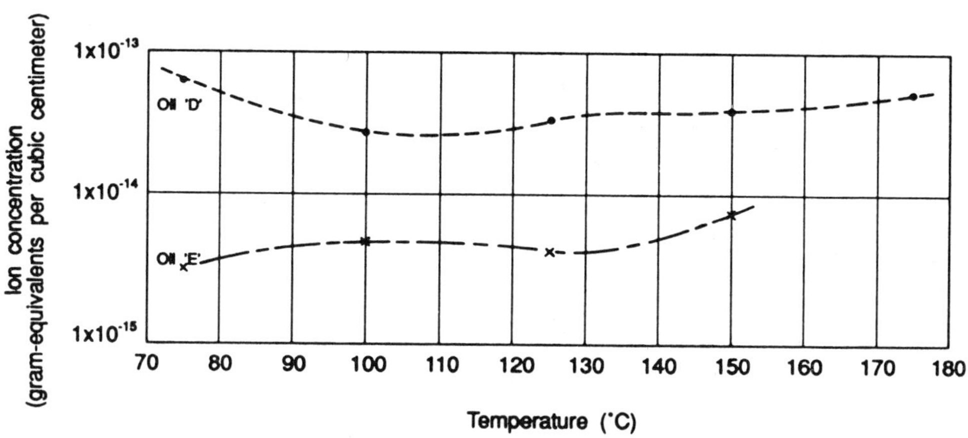

FIG. 1.28—*The influence of temperature on the ion concentration at 1 kHz in two mineral oils of equal aromatic content but different viscosity (after Bartnikas* [71]*).*

Perhaps one of the most peculiar characteristics of ionic conduction in insulating liquids is that it tends to be invariant with frequency. With E′ also constant (see Eq. 1.117), this behavior is most strikingly manifested in the tanδ value's monotonic decrease with frequency over the lower frequency regime at elevated temperatures where the ionic loss becomes the preponderant loss mechanism. This rather common intrinsic property of insulating liquids is illustrated in Figs. 1.30 and 1.31, which were obtained at 90°C so as to minimize any possible dipole contribution effects. The behavior of tanδ with frequency over the lower frequency range, suggests that the observed loss differs considerably from what one would usually attribute as being typical of a normal type of relaxation mechanism. This type of behavior would infer a relaxation time distribution that is independent of temperature. The calculated a-c conductivity, σ, values in terms of Eq 1.117 are given in Table 1.5, from which the approximate constancy of σ exhibits a slight increase due to a finite increase in the contribution to the overall dielectric loss magnitude by the dipole mechanism.

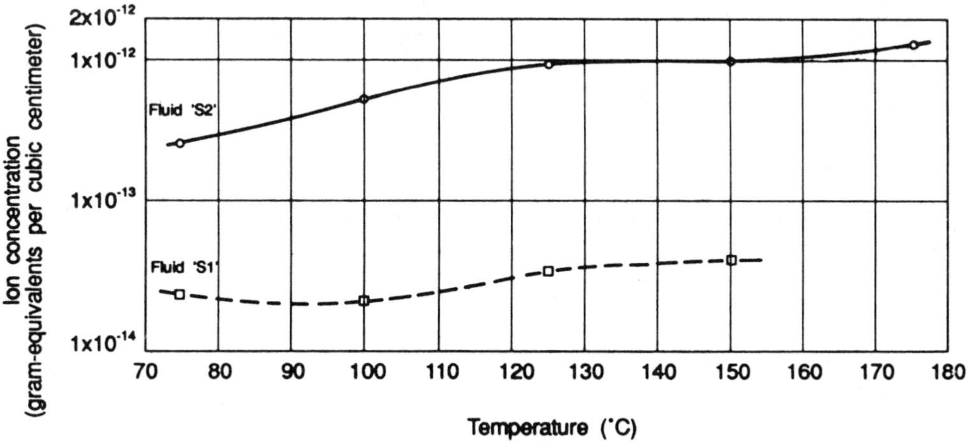

FIG. 1.29—*The influence of temperature on the ion concentration at 1 kHz in a low (S_1) and high (S_2) viscosity silicone fluid (after Bartnikas* [71]*).*

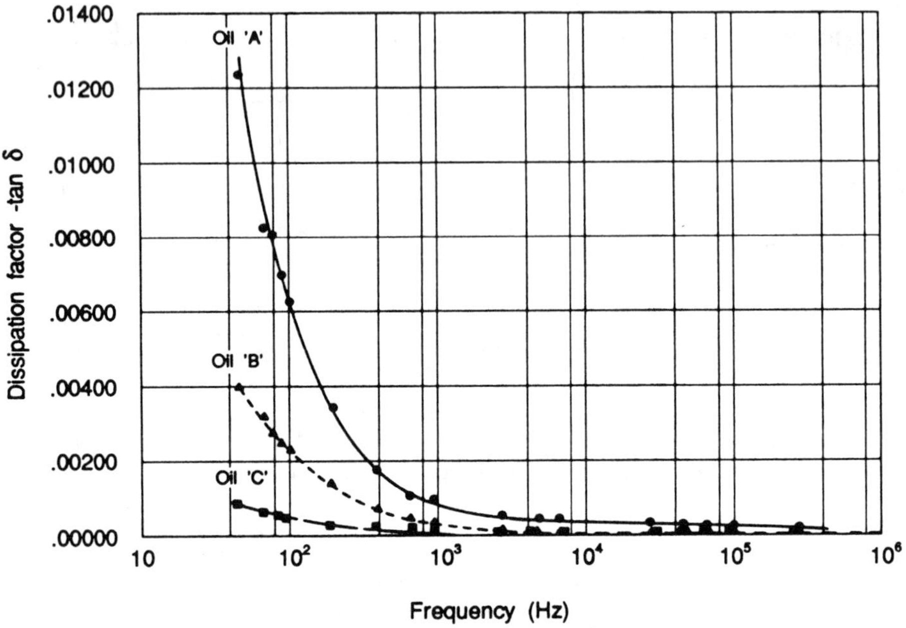

FIG. 1.30—*Dissipation factor versus frequency characteristics at 90°C of three mineral oils with different aromatic contents but of approximately equal viscosities (after Bartnikas* [71]*).*

Invoking Stoke's law, the ionic mobility, μ_m, of insulating liquids may be approximately expressed as

$$\mu_m = q/6\pi\eta r_o \tag{1.177}$$

with the tacit assumption that the ions may be treated as spheres of radii r_o; here q represents the charge of the ions. Equation 1.177 indicates that the mobility will be lower in high viscosity fluids and in fluids in which the ionic conduction species will possess larger radii. It can be appreciated that Eq 1.177 represents at best only a first approximation since it contains the macroscopic viscosity term, η, which does not take into account adequately the microscopic molecular environment impeding the ionic movement. Since the conductivity, σ, may also be written as

$$\sigma = qN\mu_m \tag{1.178}$$

where N in the concentration of the charge carriers involved with each one transporting a charge, q. In terms of Eqs 1.177 and 1.178, then

$$\sigma = q^2N/6\pi\eta r_o \tag{1.179}$$

The conductivity term, σ, generally is composed of the conductivity contributions from both positive and negative ions, such that

$$\sigma = q[N_+\mu_{m+} + N_-\mu_{m-}] \tag{1.180}$$

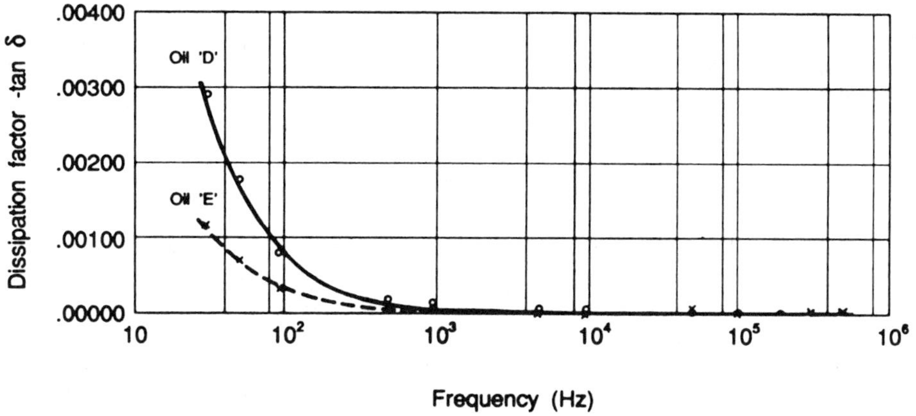

FIG. 1.31—*Dissipation factor versus frequency characteristics at 90°C of two mineral oils with approximately equal aromatic contents but different viscosities (after Bartnikas* [71]*).*

It should be emphasized that in Eq 1.180 the concentrations of the positive and negative ions, N_+ and N_-, are not necessarily equal, while their respective mobilities, μ_{m+} and μ_{m-}, are a characteristic of each of the ionic species involved and increase with temperature. The ionic concentrations, N_+ and N_-, are augmented markedly with temperature as the latter does induce dissociation of the impurity or ionic complexes. In contrast, decreasing temperature facilitates the ions to reassociate themselves into complexes. When an a-c field is applied across a dielectric material, the oppositely charged ions will tend to become increasingly segregated and thereby be prevented from recom-

TABLE 1.5—*Conductivity and dielectric constant of oils A, B, and C as a function of frequency (after Bartnikas* [71]*).*

	Oil A at 90°C		Oil B at 90°C		Oil C at 90°C	
Frequency, Hz	Dielectric Constant	Conductivity, $(\text{ohm-cm})^{-1}10^{-13}$	Dielectric Constant	Conductivity, $(\text{ohm-cm})^{-1}10^{-13}$	Dielectric Constant	Conductivity, $(\text{ohm-cm})^{-1}10^{-13}$
47	2.190	7.08	2.135	2.20	2.022	0.419
70	2.187	7.00	2.133	2.59	2.019	0.458
80	2.187	7.76	2.133	2.60	2.024	0.826
100	2.187	7.52	2.131	2.65	2.022	0.550
200	2.187	8.12	2.131	3.10	2.022	0.605
400	2.187	8.66	2.131	3.04	2.022	1.10
700	2.187	8.49	2.131	3.53	2.024	1.75
1×10^3	2.187	10.8	2.133	3.47	2.026	2.70
3×10^3	2.187	17.5	2.131	4.54	2.020	5.19
5×10^3	2.185	25.7	2.133	7.56	2.022	8.94
7×10^3	2.187	35.8	2.131	10.5	2.024	9.45
1×10^4	2.187	46.0	2.131	8.02	2.022	8.21
3×10^4	2.187	102	2.131	7.20	2.022	—
5×10^4	2.187	149	2.131	26.7	2.020	—
7×10^4	2.185	186	2.131	26.2	2.020	—
1×10^5	2.185	244	2.135	—	2.020	—
2×10^5	2.183	433	2.135	—	2.022	—
3×10^5	2.185	731	2.133	—	2.022	—

bining into complexes. The enhanced segregation of the ions with the field will cause a proportional rise in the ion concentrations N_+ and N_-, with the field, with a subsequent rise in the conductivity and tanδ. Since the value of the dissipation factor, tanδ, is given by $\sigma/\omega\epsilon''$, where σ is now the a-c conductivity term, then

$$\tan\delta = q\left[\frac{N_+\mu_{m+} + N_-\mu_{m-}}{\omega\epsilon'}\right] \tag{1.181}$$

The foregoing model considers only two ionic species of opposite charge; however, in heterogeneous insulating liquids such as mineral oils more than one pair of oppositely charged species may occur with the consequence that the conductivity expression would be comprised of a corresponding number of concentration-mobility product terms.

The ionic losses in insulating liquids may arise from an ionic mobility that is activated thermally having a general Arrhenius form of $\mu_m = \mu_{om} \exp[\Delta H/kT]$. To explain this type of conduction process a simple model has been propounded by Fröhlich [*13,41,139*] which in its modified form had also a direct application to the dipole relaxation process as described previously. The Fröhlich model considers the ions as having two equilibrium positions at A and B separated by a distance 2a as delineated in Fig. 1.32. In the absence of an electric field, the energy of the ions situated in Sites A and B is equal, so that the same number of ions is contained in each well. In order for the ions to cross from one well into the other, they must gain sufficient thermal energy from their environment to surmount the potential energy barrier, ΔH, which separates the two respective well sites. In accordance with Boltzmann statistics, the probability per second that an ion will jump from Site A to B or vice versa may be expressed as

$$\Gamma = \nu \exp[-\Delta H/kT] \tag{1.182}$$

where ν is the vibration frequency of the ions within their respective wells. Alternatively, it may be regarded as the number of attempted jumps per second by the ion (usually of the order of 10^{12} per second). Since the thermal agitation term is equal to kT, the units of the activation energy, ΔH, are in electron volts; the units of ΔH may also be expressed in kcal/mol, in which case the thermal agitation term RT is used, where R is the gas constant. Note that 1 eV = 23.06 kcal/mol.

Under equilibrium in the absence of an electric field, ions jump continuously between Sites A and B, so that if one considers N as the total number of ions per cubic centimeter, $N/2$ ions would be expected to be found in each well at any one instant. When an external electrical field, E, is applied in the direction indicated in Fig. 1.32, the ions in Site A will face a potential barrier ($\Delta H - eaE$), while those in Site B will be faced with a larger potential barrier equal to ($\Delta H + eaE$), assuming that the charge borne by all the ions is equal to the electronic charge e. The site preferred by the

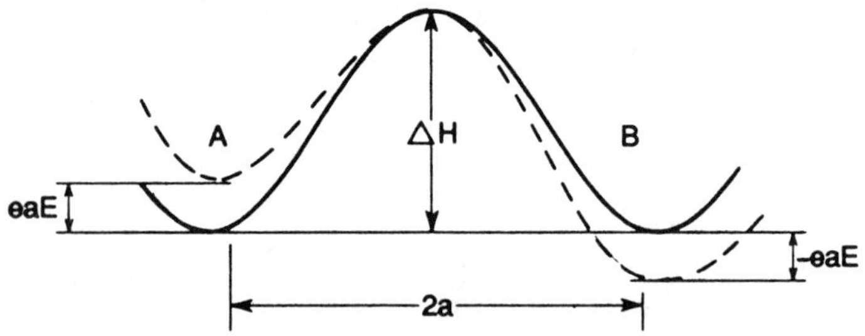

FIG. 1.32—*Two potential well model.*

ions under these circumstances, will thus be at Site B and the probability of the ions in moving a distance a from Site A to B will be

$$\Gamma_{AB} = \nu \exp\left[-(\Delta H - eaE)/kT\right] \tag{1.183}$$

or in terms of Eq 1.182

$$\Gamma_{AB} = \Gamma \exp\left[eaE/kT\right] \tag{1.184}$$

Similarly, a transition from Site B to A, would entail a probability equal to

$$\Gamma_{BA} = \Gamma \exp\left[-eaE/kT\right] \tag{1.185}$$

If N_A denotes the number of ions in Site A and N_B denotes the number of ions in Site B at a given time t, then the polarization, $P(t)$, resulting from the charge imbalance between the two ionic sites is

$$P(t) = (N_B - N_A)ea \tag{1.186}$$

Evidently, in the absence of an electric field $N_A = N_B = N/2$ and $P(t) = 0$. Following the application of an electric field at a time, t, the change in the number of ions at Site A becomes

$$\frac{dN_A}{dt} = -N_A\Gamma_{AB} + N_B\Gamma_{BA} \tag{1.187}$$

Similarly at Site B

$$\frac{dN_B}{dt} = N_A\Gamma_{AB} - N_B\Gamma_{BA} \tag{1.188}$$

Subtracting Eq 1.187 from Eq 1.188 and observing that under moderate fields $eaE \ll kT$, yields

$$\frac{d}{dt}(N_B - N_A) = -2\Gamma(N_B - N_A) + 2\Gamma N\frac{eaE}{kT} \tag{1.189}$$

The solution of this differential equation is [*139*]

$$-(N_B - N_A) = \frac{N\,eaE}{2kT}\left[1 - \exp(-2\Gamma t)\right] \tag{1.190}$$

Utilizing the expression for the polarization, $P(t)$, defined by Eq 1.186, we obtain

$$P(t) = \frac{Ne^2a^2E}{kT}\left[1 - \exp(-t/\tau)\right] \tag{1.191}$$

where the relaxation time, τ, is related to the ionic jump probability by

$$\tau = 1/2\Gamma \tag{1.192}$$

The foregoing model predicts a decrease in the relaxation time with absolute temperature as is observed also in practice. Furthermore, Eq 1.191 is of a form similar to that of Eq 1.119 which commonly is utilized in the development of the phenomenological theory of dielectrics; here the saturation value of the polarization, P_s is seen to be equivalent to (Ne^2a^2E/kT). Substitution of the relaxation time, τ, from Eq 1.192 into the Pellat-Debye equations leads to

$$\epsilon' = \epsilon_\infty + \frac{\epsilon_s - \epsilon_\infty}{1 + \omega^2/4\Gamma^2} \tag{1.193}$$

and

$$\epsilon'' = \frac{(\epsilon_s - \epsilon_\infty)\omega}{2\Gamma(1 + \omega^2/4\Gamma^2)} \tag{1.194}$$

With the foregoing model, the maximum loss will occur at a frequency of $f = \Gamma/\pi$; much below this value the ions will surmount their respective potential barriers almost instantaneously, thus contributing very little to the loss factor, ϵ''. At the other extreme for frequencies exceeding appreciably the absorption frequency value of Γ/π, the ions will remain almost stationary due to their inertia and once again contribute very little to the loss factor value.

In dealing with dielectric losses under alternating voltage conditions, one is inevitably concerned with the a-c value of the conductivity. Although, at times the measured d-c and a-c values of conductivity may exhibit an agreement, this agreement is rather fortuitious and has little physical significance particularly *vis-à-vis* the variations in the methods used for arriving at the d-c value, that is, the arbitrary time of voltage application prior to measurement. It is also pointed out by Stevels [*140*] and Taylor [*141*] that in terms of physical considerations *per se,* the d-c and a-c values of the conductivity should exhibit a difference. The difference results from the fact that under a-c conditions, the ions will tend to oscillate over potential barriers separated by relatively short distances. Whereas under d-c conditions, a smaller number of conduction ions will be involved, since only those that can surmount all the potential barriers over the relatively large intervening distance between the electrodes will be the prime contributors to the d-c conduction current.

It has been mentioned previously that the presence of oxidation products in hydrocarbon insulating liquids leads to a formation of organic acids. The dissociation of these acids results in the availability of high mobility protons or hydrogen ions that conceivably may become the major contributors to an ionic loss mechanism at power frequencies. The manner in which ions migrate in a mineral oil is not understood fully, but one of the more obvious possibilities is that they may jump from one dipole extremity of opposite charge to another, assuming that the liquid possesses a sufficient number of dipoles to support such a mechanism. The latter model has also been invoked by Hakim [*142*] to explain the ionic losses in mineral oils, using an approach that has been applied antecedently on amorphous materials. He considered the potential barrier, ΔH, facing the ions as comprised of two components, namely, a binding energy component ΔW_b, which is the electrostatic energy required to move the ion from one equilibrium site to another, plus an elastic strain energy ΔW_s necessary to deform the molecular structure existing at that instant in order to generate a free volume sufficiently large to allow passage of the ion.

The quantity ΔW_b is equal by definition to the difference between the electrostatic energy of the ion when separated from an oppositely charged ion by a distance equal to the sum of the two ionic radii and the electrostatic energy term corresponding to a separation distance of half the jump distance of the ion (one half the jump distance plus the sum of the ionic radii). Since the attractive interaction energy between a nonpolarizable dipole of moment μ and ions of charge e is $e\mu/4\pi\epsilon_o r^2$, then the binding energy, ΔW_b, is given by [*loc. cit.*]

$$\Delta W_b = \frac{e\mu}{4\pi\epsilon_o r_1^2} - \frac{e\mu}{4\pi\epsilon'(x + r_1)^2} \tag{1.195}$$

where ϵ_o is the permittivity *in vacuo*, ϵ' is the real value of the permittivity of the oil, r_1 is the sum of the radii, and x is one half the jump distance of the ion. In terms of experimental data, Hakim [*142*] finds $\mu = 3.5$ D and $r_1 = 4.0$ Å based on Stoke's law; to determine the separation, x, when the ion is halfway across its jump distance (equal to one half the distance between the two respective dipoles plus r_1), he assumes a dipole concentration of 1% ($N = 1.3 \times 10^{27}$ dipoles/cm) which yields an average separation value of 23 Å and hence $x = 16$ Å. Substituting the appropriate values in Eq 1.195 leads to a value of 14 kcal/mol for ΔW_b.

The strain energy is obtained using the Frenkel equation

$$\Delta W_s = 8\pi G r_o (r - r_o)^2 \tag{1.196}$$

where G is the shear modulus, r is the radius of the ion, and r_o is the radius of the free volume hole. In lieu of G, Hakim uses the compressional modulus value of 3.2×10^{10} dynes/cm^2 applicable for oils of varying viscosity [*143*]. The value of r_o is determined from

$$r_o = \frac{1}{8\pi}\left(\frac{kt}{\zeta_s}\right)^{1/2} \tag{1.197}$$

where ζ_s is the surface tension. Taking $r_o = 0.4$ Å at room temperature and $r = 3r_o$ (= 1.2 Å), yields $\Delta W_s = 2.9$ kcal/mol, and, hence, $\Delta H = 16.9$ kcal/mol thus giving a reasonably good agreement with the experimental data obtained by Hakim [*142*]. Hakim's ionic conduction model, results in a temperature dependent activation energy, ΔH, due to the dependence of ΔW_s upon temperature via the value of the modulus for compression and r_o (see Eq 1.197). Increasing temperature will cause a reduction in the compression modulus [*144*] and an increase in r_o, resulting in a reduction of the strain energy term, ΔW_s. This tendency is in agreement with the experimental results on oils, for which ΔH exhibits a decrease with rising temperature.

The foregoing model conceivably could be also applicable to explain ionic conduction losses in highly purified silicone fluids, in which the hydrogen conduction ions may result from the dissociation of water absorbed by the silicone fluid following its exposure to air [*145*]. Although more recent evidence has indicated that silanol may be responsible for some of the observed ionic losses in silicone fluids [*146,147*]. Silanol, characterized by the terminal unit ≡ SiOH, arises directly from the incomplete termination of the silicone molecule or polymer.

Whereas the dipole loss is invariant with voltage over the usual voltage stresses encountered in practice, the dielectric loss due to ionic conduction may change very appreciably with voltage. The conductivity of a dielectric is said to be ohmic when it remains constant with the electric field strength. At the more elevated field strengths, the conductivity of dielectric liquids begins to increase with the field, thereby inferring that the number of charge carriers becomes larger or that perhaps new charge carriers of another kind are injected into the dielectric at the electrodes.

The behavior of conductivity with field strength in weak electrolytes has been studied extensively by Wien [*148,149*], and the increase of conductivity with field strength is now commonly referred to as the Wien effect. The early investigations on the Wien effect have been well documented by Nikuradse [*150*] in 1933. There have been two theories propounded to explain the Wien effect: one by Plumley [*151*] and another one by Onsager [*152*]. Plumley's model appears to be the lesser known of the two and is based on a field enhanced ionization process involving the lowering of the energy of the hydrogen bond by the field. If N is the concentration of the undissociated molecules in molecules/cubic centimeter, then the dissociation constant $K(E)$ at the field E is defined by

$$NK(E) = -\frac{dN}{dt} \tag{1.198}$$

Using statistical mechanics Plumley, shows that

$$\frac{d}{dt}[\ln K(E)] = \frac{W_E}{kT^2} \tag{1.199}$$

where W_E is the energy required for the removal of an ion from a molecule in a liquid phase under the action of an applied field E. Integration of both sides leads to

$$K(E) = A \exp[-W_E/kT] \tag{1.200}$$

where A is an integration constant. Since in the presence of a field E, the ionization energy of a favorably oriented ionic bond is lowered by

$$\Delta W_E = -e^2/\epsilon' a - Eae \tag{1.201}$$

where e is the electronic charge, ϵ' is the real value of the dielectric constant or permittivity of the dielectric liquid, and a is the separation of the ions in the two respective potential wells given by

$$a = (e/\epsilon' E)^{1/2} \tag{1.202}$$

If W_o is the normal ionization energy, then it is shown by Plumley [*loc. cit.*] that

$$\begin{aligned} W_x &= W_o + \Delta W_E \\ &= W_o - 2e^{3/2}(E/\epsilon')^{1/2} \end{aligned} \tag{1.203}$$

from which

$$\frac{K(E)}{K(O)} = \exp\left[\frac{2e^{3/2}(E/\epsilon')^{1/2}}{kT}\right] \tag{1.204}$$

where $K(O)$ is the dissociation constant at zero field.

In the alternative model developed for weak electrolytes by Onsager [*152*], the dissociation constant $K(E)$ was computed in terms of the equations of Brownian motion of ions in a combined Coulomb and external field. He used the procedure of von Smoluchowski [*153*] in applying the equations of Brownian motion to the problem of two particles. In his derivation Onsager adopted Bjerrums [*154*] convention in that he considered the ions as bound to each other as long as the distance between the ion pair, r, was such that $r < Q$, where Q is equal to $e^2/2\epsilon' kT$. He further assumed that the concentration of the free ions, N, was sufficiently small so that the Debye-Hückel radius $1/K$ of the ionic atmosphere was much greater than the effective range Q of the ions, such that

$$KQ = -\kappa e^2/2\epsilon' kT \ll 1 \tag{1.205}$$

and where

$$\kappa^2 = 4\pi N/\epsilon' kT \tag{1.206}$$

Based on the foregoing considerations, Onsager [*loc. cit.*] was able to demonstrate that

$$\frac{K(E)}{K(O)} = \left\{1 + \frac{e^3E}{2\epsilon' k^2T^2} + \frac{1}{3}\left(\frac{e^3E}{2\epsilon' k^2T^2}\right)^2 + \frac{1}{18}\left(\frac{e^3E}{2\epsilon' k^2T^2}\right)^3 + \cdots\right\} \tag{1.207}$$

tending to a limiting value at high fields of

$$\frac{K(E)}{K(O)} = (2\pi)^{1/2} 8 \left(\frac{e^3E}{2\epsilon' k^2T^2}\right)^{-3/4} \exp\left[\frac{8e^3E}{2\epsilon' k^2T^2}\right]^{1/2} \quad (1.208)$$

or

$$\frac{K(E)}{K(O)} = \frac{J_1[(-e^3E/\pi\epsilon' k^2T^2)^{1/2}]}{(-4e^3E/\pi\epsilon' k^2T^2)^{1/2}} \quad (1.209)$$

where J_1 is the first order Bessel function.

At high field strengths both the Onsager and Plumley equations lead to the same result, but at low field strengths the departure is very substantial as has indeed been confirmed experimentally by Brière [*155*], using polar liquids. In addition, Brière found that for pure polar liquids the Plumley equation is more valid. However, with weak, nonpolar electrolytes the Onsager equation leads to satisfactory results and its validity has been verified by a number of other workers [*156–158*]. In applying the Onsager theory to weakly polar or nonpolar dielectric liquids subjected to high voltages, Denat, Gosse and Gosse and co-workers [*159–162*] have shed considerable light on the manner in which intrinsic ionic-type charge carriers are generated in a liquid. Nonpolar liquids are characterized by dielectric constants in the order of about two, such that association between ions at zero or weak electric fields is favored as the Coulomb forces of attraction are able to overcome the dissociating forces of thermal agitation for distances up to approximately 28 nm, thereby exceeding considerably the molecular dimensions. Denat et al. [*loc. cit.*] have propounded a two-step process for the formation of the ionic charge carriers in the bulk of the liquid in the presence of an external field: the initial step involves the formation of ionic-dipoles, which is subsequently followed by the separation of these ion-pairs under the action of the applied field. In order to account for the tanδ increase with field strength at the more elevated fields another mechanism is invoked that concerns injection of charge carriers at the electrodes; once more in a two step process: first the ions are created at the electrodes and then extracted out of the image-force region. The latter effect becomes only active for applied fields of 200 kV/cm or greater and has, therefore, little practical relevance as this is considerably beyond the average voltage stresses to which practical insulating liquids are subjected. It should be emphasized here that the concept of charge carrier creation in the liquid bulk and at the electrodes is an old one, as has been well documented in the classical treatise by Sharbaugh and Watson [*163*].

Figure 1.33*a* compares the behavior of oils A, B, and C (see Table 1.1) at room temperature and 85°C, having approximately identical viscosity characteristics but possessing differing amounts of aromatics. All three mineral oils exhibit a typical Wien effect, that is, the tanδ value increases with field strength because the application of a field causes ions of opposite charge to be segregated, thereby reducing their recombination tendency. Since the rate of creation of new ions remains unaltered, as it is determined by thermal agitation, which is constant at any given test temperature, the ion concentration continues to increase with the field, thus causing a proportional increase in the a-c conductivity or tanδ. Whereas the increase of conductivity with the field results from an increase in the ion concentration, the difference in conductivity for any of the three oils at room temperature and 85°C must be ascribed to differences in both the rate of increase in the ion concentration as well as mobility of the ions at the higher temperatures. Insofar as the ion concentration is concerned, at the base stress and room temperature the ion concentration of oils A, B, and C is, respectively, 8.10×10^{-14}, 2.21×10^{-14} and 1.55×10^{-15} gram equivalents per cubic centimeter, while at 85°C these values are augmented to 8.95×10^{-14}, 2.95×10^{-14} and 2.86×10^{-15}, respectively. The more aromatic oils, which are more susceptible to ionic substances arising from oxidation products, are clearly seen to exhibit higher ionic losses over the entire high electric field region.

The effect of ion mobility becomes most striking if a comparison is carried out between two

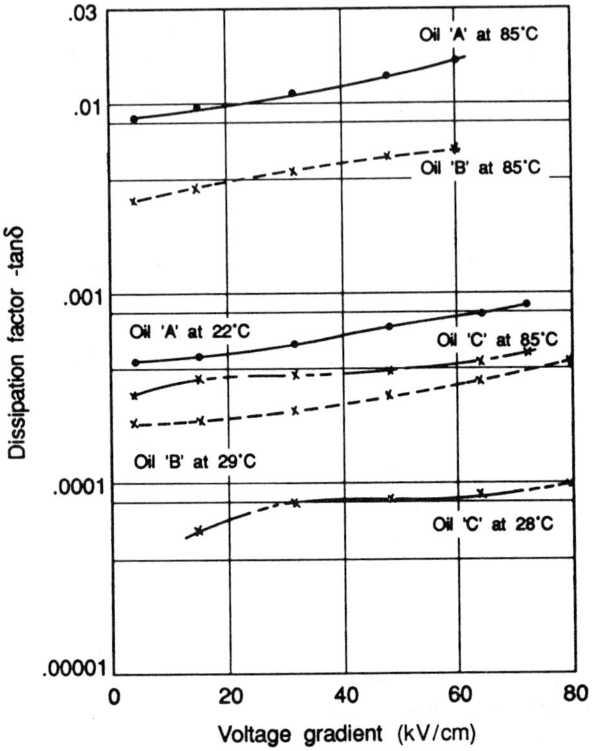

FIG. 1.33*a*—*Dissipation factor versus voltage gradient characteristics at 60 Hz of three mineral oils with different aromatic contents but equivalent viscosities, with temperature as a parameter (after Bartnikas* [71]*).*

mineral oils having widely different molecular weights and viscosities, but approximately equal aromatic contents as is the case with oils D and E whose properties are given in Table 1.1. The tanδ versus average voltage stress characteristics for these oils are portrayed in Fig. 1.33*b*. The dielectric losses at room temperature for the low viscosity oil E are seen to exceed those of the high viscosity oil D; however, at 85°C, the converse is seen to hold. The results demonstrate lucidly that, although the ion content in the higher viscosity oil D is greater, the losses in that oil become only pronounced at temperatures sufficiently elevated to permit a significant increase in the ion mobility. Hence, one would anticipate from the observed behavior that with two mineral oils of highly different viscosity but equal ion content, the losses in the lower viscosity oil would be higher. This explains why in practice, low viscosity oils are found to be highly susceptible to contamination. Not all insulating liquids are characterized by the rapid fall in viscosity with temperature that is typical of mineral oils. In this regard silicone fluids are a noteworthy exception in that their viscosity exhibits a relatively low rate of decrease with temperature. As a consequence, the losses in silicone fluids that have a high viscosity at room temperature do not increase at the same rate as for mineral oils, having viscosities in the same range at room temperature. Again considering silicone fluids S1 and S2 (see Table 1.1), whose viscosities match approximately those of oils E and D, respectively, it is apparent from Fig. 1.34 that notwithstanding the higher ion content of the higher viscosity fluid S2 as compared to S1 (see Fig. 1.29), the reduction of viscosity of the S2 fluid at 85°C is not sufficient to cause its losses to exceed those of the S1 fluid at 85°C.

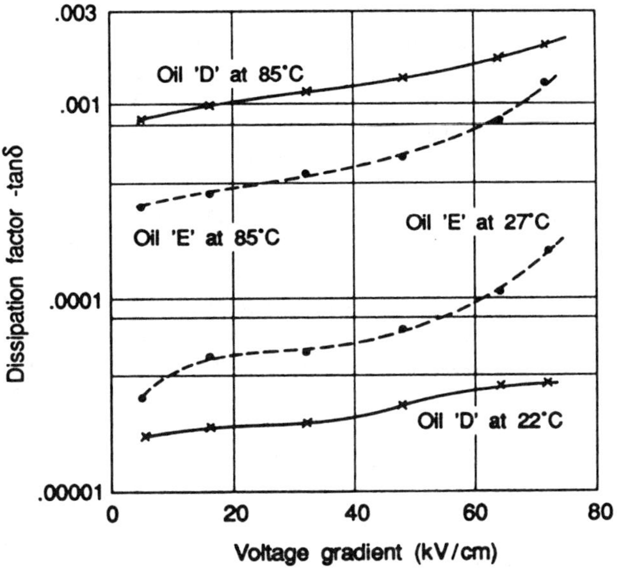

FIG. 1.33*b*—*Dissipation factor versus voltage gradient characteristics at 60 Hz of two mineral oils with approximately equal aromatic contents but different viscosities, with temperature as a parameter (after Bartnikas* [71]*).*

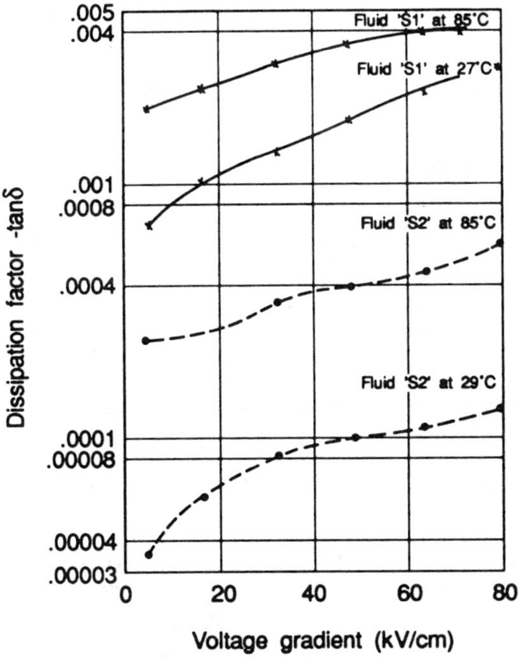

FIG. 1.34—*Dissipation factor versus voltage gradient characteristics at 60 Hz of two silicone fluids with different viscosities, with temperature as a parameter (after Bartnikas* [71]*).*

The propensity to contamination of the mineral oils by ionic impurities can be ascertained most effectively by means of intentional contamination. For illustrative purposes, Fig. 1.35 shows the effect of an introduction of one part in 5000 parts per volume of calcium naphthanate (a soluble impurity) into oils D and E. Comparison of Figs. 1.33*b* and 1.35 reveals a most pronounced contaminating effect; only in the high viscosity oil D at room temperature, the losses are at the upper level of acceptability. The losses in the contaminated low viscosity oil E at room temperature are seen to exceed those in the high viscosity oil even at 85°C. It is difficult to account for the saturation effect apparent in the 85°C characteristic of the contaminated oil E as this behavior can neither be attributed to the first Wien effect, which occurs normally at much higher fields in strong electrolytes nor can one invoke the possibility of hindered ionic motion in view of the relatively large oil gap used. Perhaps some form of a space charge mechanism may be active at this point.

The use of calcium naphthanate presents a convenient means for obtaining a rudimentary estimate of the ion mobilities that may be encountered in oils under a-c conditions. Particularly with the low viscosity oil E, the addition of calcium naphthanate is seen to enhance the ionic loss to an extent where any contribution of the ions originally present in the virgin oil may be neglected. Assuming that the calcium naphthanate dissociates completely into one calcium and two naphthanate ions, and that the mobility of both ionic species is approximately equal, yields mobilities of 2.3×10^{-5} and 13.0×10^{-5} $cm^2\ V^{-1}s^{-1}$ at room temperature and 85°C, respectively. Even though the ion mobilities of the calcium and naphthanate ions may differ appreciably, the obtained values are still expected to provide a first order approximation of the ion mobilities. These ion mobility values agree approximately with those found in both oils and silicones by other workers [*164–168*]. It is seen that the mobility increases by about an order of magnitude at 85°C as compared to room temperature, being independent of the field strength. While, the ion concentra-

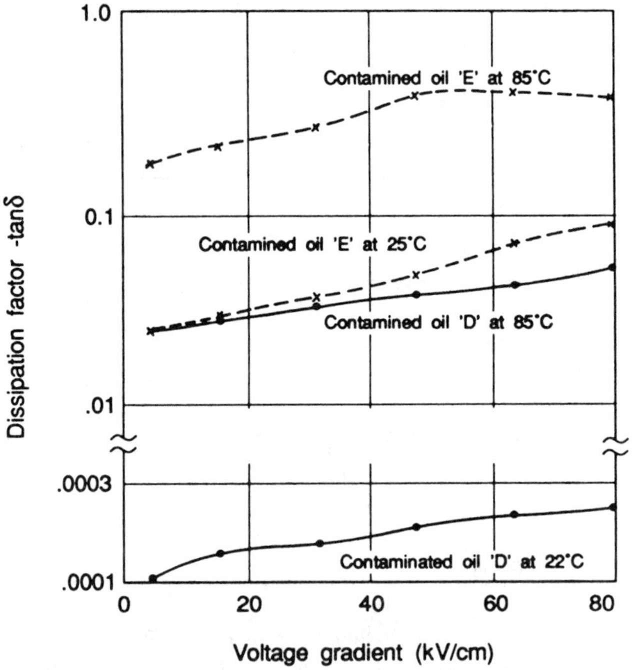

FIG. 1.35—*Effect of calcium naphthanate contamination upon the tanδ-voltage gradient characteristics of a low (oil E) and high viscosity (oil D) mineral oils (after Bartnikas* [71]*)*.

tion is dependent upon the field strength and is found to increase from a value of 2.10×10^{-13} at the base stress to 7.58×10^{-13} gram equivalents per cubic centimeter at the maximum testing stress. No calculation of the ion mobility in the contaminated oil D was feasible at room temperature, because losses resulting from other than the calcium naphthanate ions could not be neglected. Nevertheless, at 85°C such omissions were justified and the ion mobility was determined to be equal to 8.4×10^{-6} cm^2 V^{-1}s^{-1}. Comparing this value with that of the contaminated oil E at 85°C elucidates the profound effect of the oil viscosity upon the ion mobility. The lower ion mobility of the higher viscosity oil D explains the larger a-c losses inherent with the low viscosity contaminated oil E despite the greater ion concentration in the contaminated oil D: at the base stress and at a temperature of 85°C, the ion concentrations in contaminated oils D and E are 11.5×10^{-13} and 5.0×10^{-13} gram equivalents per cubic centimeter, respectively.

Under intense field conditions, the a-c loss behavior of thin insulating films is usually quite distinct from that which may occur in the liquid bulk. This is primarily due to the fact that the transit times of the ions in thin films at power frequencies are of the same order as the period of the sinusoidal wave of the applied voltage. The phenomenon is most strikingly manifest if one examines the current wave shapes in the dielectric as has been done initially by Leibscher [*169,170*] in 1943 as depicted schematically in the idealized waveform in Fig. 1.36. With increasing magnitude of the alternating electric field, the oscillatory movement of the ions becomes limited at the outer boundaries of the thin film dielectric resulting in a current waveform characterized by a peak which is followed by a lower and approximately constant plateau. The losses due to ionic motion under a-c field conditions in thin films have been considered by Schumann [*171*] and two early models specifically related to limited ion motion within the pores of oil impregnated papers were proposed by Böning [*172*] and Garton [*173*]. Garton's model is applicable to the case where the number of free ions is insufficient to give rise to a space charge field comparable in magnitude to the applied external field. Böning's model is more general in the sense that it considers both the ions trapped in the structure of the paper and the free mobile ions in the oil film.

For thin oil films in the range of 1 to *ca.* 10 μm, the dielectric losses are governed by space-charge polarization mechanisms. There have been a number of generalized space-charge polarization theories proposed to explain ionic motion that is blocked or limited at the electrodes. In order to circumvent nonlinear effects at the more elevated fields, a number of such theories have been developed over the low field region [*174–178*]; one of these, developed by Friauf [*177*], has been described in Vol. IIA of this series [*12*]. More recently, there has been an increased tendency to deal with the nonlinear behavior of the space charge field by means of numerical methods. In this respect Stern and Weaver [*179*] report an interesting approach, which we shall adopt here in order

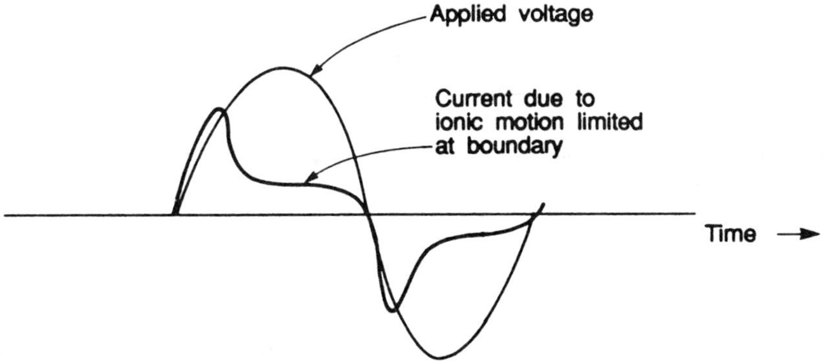

FIG. 1.36—*Current waveform due to limited ionic motion in a thin oil film (after Liebscher* [169,170]*).*

to illustrate space charge phenomena. A complete description of their approach to the space charge problem is deemed to be of particular interest since it has been shown to agree closely with the results obtained using the Garton model as well.

Stern and Weaver [*loc. cit.*] consider a simple plane-parallel capacitor, having electrode areas A separated by a distance $x = d$. The movement of free charge carriers in the direction of an electric field E then results in a current, I, of the form

$$I = A\left(J_l + \epsilon_\infty \frac{\partial E}{\partial t}\right) \tag{1.210}$$

where J_l represents the leakage current density, $\epsilon_\infty\ \partial E/\partial t$ the displacement or capacitive current density component and ϵ_∞ the high frequency value of the real permittivity at a distance x between the two parallel place electrodes. Since Eq 1.210 must be integrated from $x = o$ to $x = d$, then

$$Id = A\int_o^d \left(J_l + \epsilon_\infty \frac{\partial E}{\partial t}\right) dx \tag{1.211}$$

The integral involving the leakage current density, J_l, may be rewritten as [*179*]

$$A\int_o^d J_l\, dx = A\int_o^d \rho \upsilon\, dx = (Ad)\frac{dP}{dt} \tag{1.212}$$

where P is the mean polarization resulting from the movement of charge carriers at a velocity υ and ρ is the charge density. By definition

$$P = \frac{1}{d}\int_o^d \rho x\, dx \tag{1.213}$$

and hence the total current can now be expressed as

$$I = A\frac{dP}{dt} + A\,\epsilon_\infty \frac{dV}{dt} \tag{1.214}$$

or since the capacitance is $C = \epsilon_\infty A/d$, then

$$I = C\left[\left(\frac{d}{\epsilon_\infty}\right)\frac{dP}{dt} + \frac{dV}{dt}\right] \tag{1.215}$$

For a sinusoidal voltage of the form, $V = V_o \exp(j\omega t)$, the resulting polarization may be represented as a periodic function, having the Fourier series [*179*]

$$P = \sum_r P_r \exp\left[j(r\omega t - \theta_r)\right] \tag{1.216}$$

For simplification purposes Stern and Weaver use only the first term P_1, in Eq 1.216 for $r = 1$. Since the general expression for a current flowing through a dielectric of complex permittivity $\epsilon^* = (\epsilon' - j\epsilon'')$ is given by

$$I = (\epsilon' - j\epsilon'')j\omega C_o V_o \exp(j\omega t) \tag{1.217}$$

then combining Eqs 1.215 and 1.217 yields

$$(\epsilon' - \epsilon_\infty) - j\epsilon'' = \left(\frac{d}{V_o}\right) P_1 \exp(-j\theta_1) \tag{1.218}$$

For the voltage, V_1, of sinusoidal form V_o sin ωt, it follows that

$$\epsilon' - \epsilon_\infty = \left(\frac{d}{V_o}\right) \frac{\omega}{\pi} \int_o^{2\pi/\omega} P \sin\omega t \, dt \tag{1.219}$$

and

$$\epsilon'' = \left(\frac{J}{V_o}\right) \frac{\omega}{\pi} \int_o^{2\pi/\omega} P \cos\omega t \, dt \tag{1.220}$$

In order to determine ϵ' and ϵ'', the polarization P must be found. For simplicity Stern and Weaver [*179*] consider the charge carriers as having a charge, e, with n_o representing a constant density value of the negative charge carriers and $n(x)$ that of the positive charge carriers, such that

$$\frac{\partial E}{\partial x} = \frac{\rho}{\epsilon_\infty} = \frac{-en_o + en(x)}{\epsilon_\infty} \tag{1.221}$$

Integration of Eq 1.221 with respect to x

$$\int_o^x \frac{\partial E}{\partial x} dx = \int_o^x \left[\frac{-en_o + en(x)}{\epsilon_\infty}\right] dx \tag{1.222}$$

yields

$$E(x) = E(0) - \frac{en_o x}{\epsilon_\infty} + \frac{e}{\epsilon_\infty} f(x) \tag{1.223}$$

where

$$f(x) = \int_o^x n(x)\, dx \tag{1.224}$$

and the field at $x = 0$ is defined by

$$E(0) = \frac{V}{d} + \frac{1}{2}\frac{en_o d}{\epsilon_\infty} - \frac{e}{d\epsilon_\infty} \int_o^d f(x)\, dx \tag{1.225}$$

In order to determine the mobile charge density $n(x)$ as a function of time in Eq 1.224, the parallel plate capacitor is divided into N layers, each having a width equal to d/N in analogy to the multilayered Wagner capacitor. The electric field and the current density across the boundaries of the layers at $x = 0$, d/N, $2d/N$ . . . d, are then denoted as E_o, E_1, E_2 . . . E_n and J_o, J_1, J_2 . . . J_n, respectively. The density $n(x)$ of the positive charge carriers within each of the N layers is different and is denoted by n, n_2, n_3, . . . n_n. Accordingly

$$\frac{dn_i}{dt} = \frac{N}{d}(J_{i-1} - J_i) \tag{1.226}$$

for $i = 1, 2, 3 \ldots N$. The boundary conditions at the two blocking electrodes entail that

$$J_o = J_n = 0 \qquad (1.227)$$

as opposed to the interlayer boundaries, where

$$J_i = \mu_+ n_i E_i \qquad (1.228)$$

and where μ_+ represents the mobility of the positive charge carriers. The space charge model of Stern and Weaver is simplistic in the sense that it neglects diffusion effects. For $x = id/N$, Eq 1.224 may be expressed in terms of the multilayer capacitor notation as

$$f\left(\frac{id}{N}\right) = f_i = \frac{d}{N}\sum_{r=1}^{i} n_r \qquad (1.229)$$

for large values of N

$$\int_o^d f(x)\,dx = \frac{d}{N}\left[\sum_{i=1}^{N-1} f_i + \frac{1}{2} f_N\right] \qquad (1.230)$$

Equations 1.223 and 1.230 thus define the positive mobile ion concentration $n(x)$ at all points of the dielectric. For computational purposes Stern and Weaver [*loc. cit.*], specify the variation of $n(x)$ with respect to time at successive instants of equal time intervals of T/m, where the period $T = 2\pi/\omega$ and m is an integer. Considering the change in the positive mobile ion density, n_i between $t = jT/m$ and $t = (j + 1)T/m$ for $j = 0, 1, 2 \ldots m - 1$, Eq 1.226 assumes an alternative form of

$$\Delta n_i = \frac{N}{d}(J_{i-1} - J_i)\left(\frac{T}{m}\right) \qquad (1.231)$$

since the mean polarization of the dielectric is

$$P = -\frac{n_o ed}{2} + \frac{e}{d}\int_o^x x\,n(x)\,dx \qquad (1.232)$$

then

$$P = -\frac{n_o ed}{2} + \frac{ed}{N^2}\sum_{j=1}^{N} i\,n_i - \frac{e}{2N} f_N \qquad (1.233)$$

Equations 1.219 and 1.220 may then be rewritten as [*179*]

$$\epsilon' - \epsilon_\infty = \frac{2d}{V_o m}\sum_{j=1}^{m-1} P_j \sin\left(\frac{2\pi j}{m}\right) \qquad (1.234)$$

and

$$\epsilon'' = -\frac{2d}{V_o m}\sum_{j=0}^{m-1} P_j \cos\left(\frac{2\pi j}{m}\right) \qquad (1.235)$$

The presence of the voltage amplitude term, V_o, in Eqs 1.234 and 1.235 indicates that both the real and imaginary values of the permittivity, ϵ' and ϵ'', are voltage dependent. To examine more fully the voltage dependence of ϵ' and ϵ'', Stern and Weaver introduce the dimensionless parameters $A = n_o ed/\epsilon_\infty V_o$ and $\omega_c = 2\mu_+ V_o/d^2$, thereby modifying Eqs 1.234 and 1.235 to

$$\frac{\epsilon' - \epsilon_\infty}{\epsilon_\infty} = \frac{A}{m}\sum_{j=1}^{m-1} P_j \sin\left(\frac{2\pi j}{m}\right) \tag{1.236}$$

and

$$\frac{\epsilon''}{\epsilon_\infty} = -\frac{A}{m}\sum_{j=0}^{m-1} P_j \cos\left(\frac{2\pi j}{m}\right) \tag{1.237}$$

In plotting $\tan\delta$ or ϵ''/ϵ' against ω/ω_c, they were able to demonstrate a shift in the $\tan\delta$ peak frequency with voltage, which became much more pronounced with small specimem thicknesses or small values of d. In fact for $A < 1$, that is, for specimens of small thickness and large values of applied voltage V_o, their model yielded good agreement with the theory of Garton [*173*] which deals specifically with thin liquid films where the ionic motion is restricted by the electrode or adjacent solid dielectric boundaries and the ion density is insufficient to create a space charge gradient comparable to the magnitude of the applied electrical field. The simplified expressions for $\tan\delta$ under these conditions are obtained by Stern and Weaver as

$$\tan\delta = \frac{A\omega_c}{2\pi\omega}\left[\omega t_c - \frac{1}{2}\sin\omega t_c\right] \tag{1.238}$$

for $\omega < \omega_c$, and

$$\tan\delta = \frac{A\omega_c}{2\omega} \tag{1.239}$$

for $\omega > \omega_c$, where t_c is defined by the equation

$$\cos\omega t_c = 1 - \frac{2\omega}{\omega_c} \tag{1.240}$$

Garton's model is more simplistic and easier to visualize physically than that of Stern's and Weaver's, though it does not result in explicit expressions for ϵ'' and ϵ' as a function of frequency. Garton [*173*] considers the motion of ions in oil containing pores of a kraft type insulating paper subjected to a sinusoidal field, $E \sin\omega t$. If the diameter of the pores is taken as d, then at a sufficiently high amplitude of the field E_o, the oscillatory motion of the ions following the changes in the polarity of the field will become limited by the boundaries of the pores. Any further increase in the amplitude of the applied field, E, for $E > E_o$, will only increase the stationary period of the ions blocked at the pore boundaries. This stationary period of the ions will result in no energy loss, with the consequence that for $E > E_o$ the value of $\tan\delta$ will exhibit a decrease with the applied voltage; a $\tan\delta$ peak will thus be manifested at $E = E_o$, while for field $E < E_o$, $\tan\delta$ will display the usual increase observed with the electric field up to the value $E = E_o$.

In the development of his model, Garton [*loc. cit.*] postulates a number of simplifying assumptions, namely, that positive and negative ions have the same mobility and that their motion is controlled solely by the applied field and the macroscopic viscosity of the liquid. Furthermore, he neglects interionic attraction forces and the contact potential and image forces at the boundaries. Thus under the influence of an applied field E, the ions are assumed to be congregated into a layer of thickness $d(1 - m)$, where $m = E/E_o$ and d is the thickness of the oil film. Under an applied field the two ionic layers of opposite sign will thus execute oscillations from one side of the oil film to the other. At $E = E_o$, the ionic layers will become compressed to a negligible thickness at the boundaries of the oil films.

In his derivation of the tanδ value of a thin oil film, Garton postulates that a recombination of the positive and negative ions occurs only during the time when the ions interpenetrate as their two respective layers move past each other. Although the electric field segregates the ions of opposite sign into two separate layers, it is not assumed to have any effect on the rate of ionic dissociation as the latter process is governed primarily by temperature. The equation for tanδ derived by Garton [*173*] is of the form

$$\tan\delta = \frac{-Q + [Q^2 + 4(s - 1)sQ]^{1/2}}{2\pi(s - 1)}\{(\theta - \sin\theta\cos\theta)\tan\delta_0\} \qquad (1.241)$$

where for $0 < m < 0.5$

$$Q = \frac{(1 - m)^2}{1 - 1.636\,m} \qquad (1.242)$$

and for $0.5 < m < 1$

$$Q = \frac{\pi(1 - m)\sin\phi}{2(\phi\sin\phi + \cos\phi - 1)} \qquad (1.243)$$

where $\sin\phi = (1 - m)/m$. The value of s in Eq 1.241 refers the ratio of the number of molecules that are capable of dissociation to the number of dissociated ions of one sign. The term $\tan\delta_o$ represents the maximum of the tanδ value and θ is defined by the expression

$$\cos\theta = 1 - 6\pi r\eta(\omega d)/Ee \qquad (1.244)$$

where r is the ionic radius, η the macroscopic viscosity of the liquid, ω the radial frequency term, and e the charge of the ions. Garton [*loc. cit.*] has furthermore demonstrated that Eq 1.241 may be utilized to derive some pertinent data on the number, mobility, and radius of the ions. However, it is evident from the nature of Eq 1.241 that in view of the uncertainty of the magnitude of the m and s parameters, its implementation in practice is difficult or perhaps even impossible.

Perhaps one of the most comprehensive models proposed to expand the space charge behavior is that of Held and Wenzel [*180*]. The basis for this model appears to have its origin in the early observations of Liebscher [*169,170*] as concerns the waveform of the current in oil film layered strata within oil impregnated paper insulating systems (see Fig. 1.36). Their derived expression for the dissipation factor, tanδ, which predicts a peak in tanδ as a function of voltage and frequency, is expressed as

$$\tan\delta = \frac{n(Ne)^2}{3\pi\epsilon'\eta\omega r}[y(\alpha)] \qquad (1.245)$$

where N represents the total number of ions of charge e and n their density. Since Stoke's law is invoked in the derivation of this expression, the macroscopic viscosity, η, together with the ionic radius, r appear as explicit terms in Eq 1.245. The function $y(\alpha)$ is given by

$$y(\alpha) = \frac{1}{2\pi}[4 \arcsin (1/\alpha)^{1/2} - \sin \{4 \arcsin (1/\alpha)^{1/2}\}] \tag{1.246}$$

here the parameter α is directly proportional to the rms value of the electric field, $\hat{E}$, and is defined by

$$\alpha = \frac{Ne\,\hat{E}}{3\pi\eta r\omega d} \tag{1.247}$$

Implicit in the real value of the permittivity, ϵ', in Eq 1.245 is its frequency dependence over the space charge absorption region.

Shimokawa et al. [*165*] have carried out an experimental study of the space charge losses that occur in silicone fluid films contaminated with a water solution of sodium chloride; for this purpose a silicone fluid, having a viscosity of 0.85 cP at 25°C was employed. Their results illustrate qualitatively very well the space charge losses that are caused by ionic motion blocked at the electrodes. A Garton/Böning effect is clearly discernible in Fig. 1.37, which portrays the variation of tanδ as a function of the electric field with temperature and frequency as parameters. Moreover, the experimental results agree qualitatively with the models Held and Wenzel [*180*] and Stern and Weaver [*179*] both of which infer that the position of the tanδ peak on the electrical field scale is contingent upon the frequency of the applied field. The temperature parameter effect is also in accordance with the relaxation behavior of a space charge polarization process. Figure 1.38 shows the relaxation spectra obtained on a 4 μm silicone film with the electrical field and temperature as parameters. It is interesting to observe that an increase of the electrical field from 7.5 to 23.0 kV/cm, causes the absorption peak due to the limited motion of charge carriers to shift into the vicinity of the power frequency range at 42°C.

In their studies of the current waveforms across thin liquid oil films under a-c conditions, Liebscher [*169,170*] and Held and Wenzel [*180*] found that at constant frequency, the peak of the current due mobile charge limited motion (see Fig. 1.36) increases with temperature and field strength; also the current peak occurs earlier in the power frequency cycle as the field or the temperature is increased. By assuming that the position of the current peak with respect to the phase scale of the power frequency wave is a direct measure of the transit time of the ions across the fluid film, Shimokawa et al. [*165*] were able to estimate the mobility of the ionic charge carriers in their silicone fluid specimen. They assumed that the ion concentration is sufficiently low so that the resulting space charge field will be too small to influence the movement of the ions and, under space charge conditions, they approximated the thickness of the silicone fluid film, d, by the expression

$$d = \int_o^{t_t} \mu_\pm \frac{V \sin\omega t}{d} dt \tag{1.248}$$

where $\mu_\pm$ is the mobility of the positive or negative ions, V is the amplitude of the applied voltage, and t_t is the transit time of the ions determined in terms of the positive current peak. Integration of Eq 1.248 yields the value of the mobility, $\mu_\pm$, as [*165*]

$$\mu_\pm = \frac{\omega d^2}{2V} \frac{1}{[\sin(\omega t_t/2)]^2} \tag{1.249}$$

Their estimated mobility value, $\mu_\mp$, under space charge limited conditions is found to be 2.7×10^{-5} cm^2 V^{-1}s^{-1}, being presumably a composite value for the sodium and chlorine ions in the intentionally contaminated silicone fluid specimen.

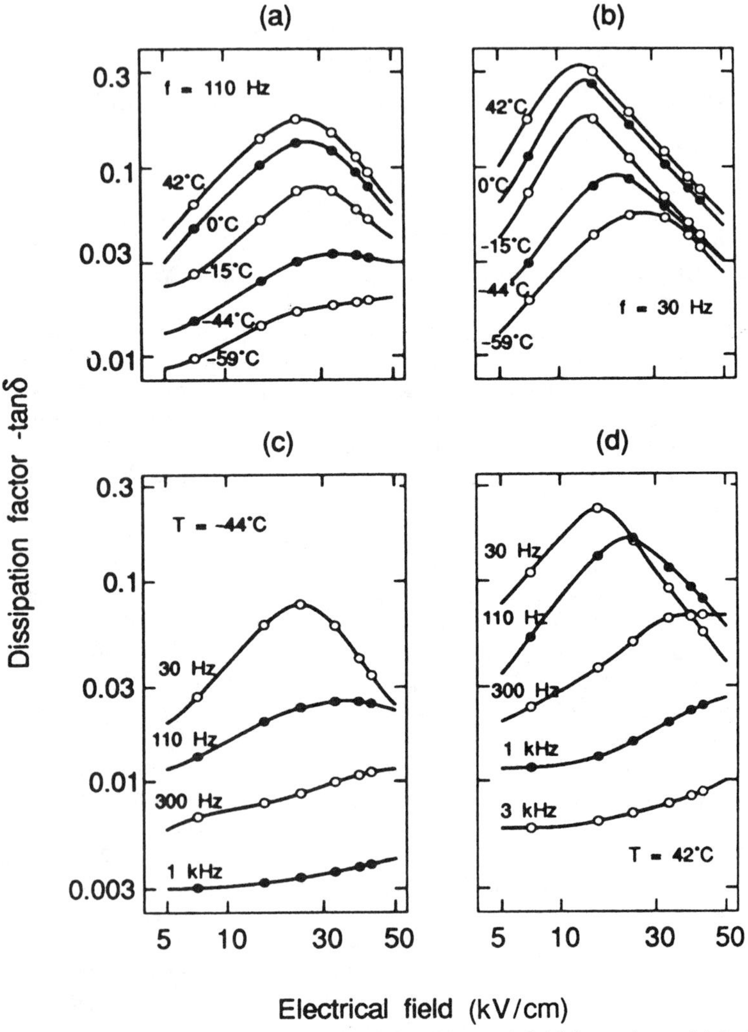

FIG. 1.37—*Dissipation factor as a function of the electrical field for a 4 μm thick silicone fluid film contaminated with NaCl at:* (a) *110 Hz and* (b) *30 Hz with temperature as a parameter and* (c) *−44°C and* (d) *42°C with frequency as a parameter (after Shimokawa et al.* [65]*).*

1.4.5 Losses Due to Electronic Conduction

In dealing with the a-c dielectric loss resulting from the conduction and space charge polarization losses, the mobile charge carriers involved were assumed to be ions. However, it is well known that electrons or holes may also give rise to similar phenomena as that observed with ions. Although conduction by electrons can occur readily under a-c conditions, it rarely constitutes a major portion of the overall dielectric loss magnitude in practical insulating liquids. In highly purified liquids, the electronic losses may predominate, but in practice their contribution is obscured by the very much higher ionic losses that arise either as for example a result of oxidation or the presence of other extraneous electrolytic substances in the insulating liquids. Since the subject of electronic conduc-

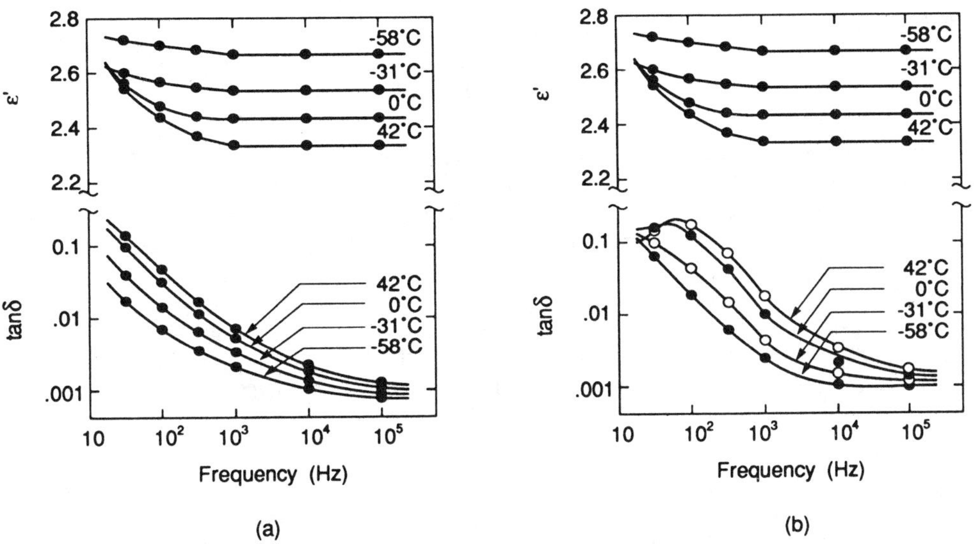

FIG. 1.38—*Relaxation spectra for a 4 μm thick silicone fluid film contaminated with NaCl and subjected to an electrical field of* (a) *7.5 kV/cm and* (b) *23 kV/cm (after Shimokawa et al.* [65]*).*

tivity is dealt with very extensively in Chapter 2 of this volume, it behooves us to allude to it only in a cursory manner here.

The subject of electronic conductivity in insulating liquids has been extensively examined by Schmidt et al. [*181,182*]. In their work they invoked the electron conduction model propounded by Bagley [*183*], which is based on the rate theory since it assumes the electron motion to be a thermally activated process with an equilibrium existing between the free and trapped electrons at zero applied field. If the exchange frequency is taken as

$$f_o = \nu \exp[-\Delta H_a/kT] \tag{1.250}$$

where ν is the attempted jump frequency of the electrons and ΔH_a is the average height of the multiple potential barriers. Under the application of an external field, E, the net jump exchange frequency is increased to [*183*]

$$\begin{aligned} f &= f_o\left[\exp\left(\frac{aeE}{2kT}\right) - \exp\left(-\frac{aeE}{2kT}\right)\right] \\ &= 2f_o \sinh[aeE/2kT] \end{aligned} \tag{1.251}$$

where a is the distance separating the potential barriers and e is the electronic charge. Since the electronic drift mobility is of the form $\mu_- = af/E$, then substituting the value for f gives the mobility $\mu_-(E)$ as a function of the electric field

$$\mu_-(E) = \frac{a}{E}[2\nu \exp(-\Delta H_a/kT) \sinh(aeE/2kT)] \tag{1.252}$$

for small fields this value reduces to

$$\mu_-(0) = \frac{a^2 e}{kT}[\nu \exp(-Ea/kT)] \tag{1.253}$$

and therefore [*182,183*]

$$\frac{\mu_-(E)}{\mu_-(0)} = \frac{2kT}{aeE}[\sinh(aeE/2kT)] \tag{1.254}$$

The positive temperature coefficient together with the low electron mobilities observed in liquid hexane by Schmidt et al. [*182*], indicates that the trapped electron model is applicable in a highly purified liquid where the electron motion takes place through thermally activated jumps executed by the electrons between the different traps. The activation energy for this process is relatively small, being equal to 0.08 eV.

Under some circumstances, permanent and induced dipoles within the liquid may form bridges or chains between the opposite polarity electrodes along which free electrons may hop with relative ease from dipole to dipole. Kok [*183a*] has demonstrated that the equilibrium condition between the voltage gradient force and the diffusion force of thermal agitation is given by

$$\mu E = 0.866\, kT \tag{1.255}$$

where μ is the permanent dipole moment and E is the minimum value of the voltage gradient necessary to cause permanent dipole alignment. Considering oils A and B (see Table 1.1), dipole alignment at 27°C would be anticipated at E = 445 kV/cm and at 85°C a field strength of 531 kV/cm would be required. Evidently, the possibility of this type of mechanism frequently arising in practice must be discarded since the field strengths are well beyond those encountered with normal operating stresses. Nevertheless, it is conceivable that field enhancement at asperities may under certain conditions lead to such inordinately high field values. However, the formation of bridges by induced dipoles of macroscopic size is readily realizable and, in fact, often such dipole chains are observed to be a prelude to breakdown of actual oil gaps undergoing testing.

1.5 Distribution of Relaxation Times

When the values of ϵ'' and ϵ' are calculated, using the Pellat-Debye equations, and plotted against $\log\omega$, the resulting curves are generally less dispersed in form than those measured experimentally. This behavior is readily apparent from Fig. 1.39, which compares the theoretically derived values of ϵ'' with the experimentally obtained ones for an oil filled cable oil at temperatures of -80, -70, and -60°C. The disparity between the two sets of curves is a direct consequence of the tacit assumption in the application of the Pellat-Debye equations that the relaxation process may be described by a single relation time. The increased dispersion or broadening of the experimental anomalous absorption curves is attributable to a distribution in the relaxation times, which arises from every ionic or molecular dipole having in its particular environment its own intrinsic relaxation time because the molecular interactions and thermal agitation forces as well as the directing field are functions of both space and time. Consequently, a measurement of ϵ'' as a function of frequency represents necessarily an average measure over a given set of spacial conditions prevailing in the vicinity of every molecular or ionic dipole and charge carrier at a particular point in time. This leads to a spread in relaxation times with a distribution centered around the most probable value.

The concept of a distribution function to describe the broadening in the absorption curves is an old one and has in fact been used both by von Schweidler [*184*] and Wagner [*101*] even before the formulation of the Debye relaxation time model. If one considers a distribution function $G(\tau)$, such that $G(\tau)d\tau$ defines the fraction of ionic or molecular dipoles associated at a particular instant with relaxation times within the interval τ to $(\tau + d\tau)$, then [*184*]

$$\int_o^\infty G(\tau)d\tau = 1 \tag{1.256}$$

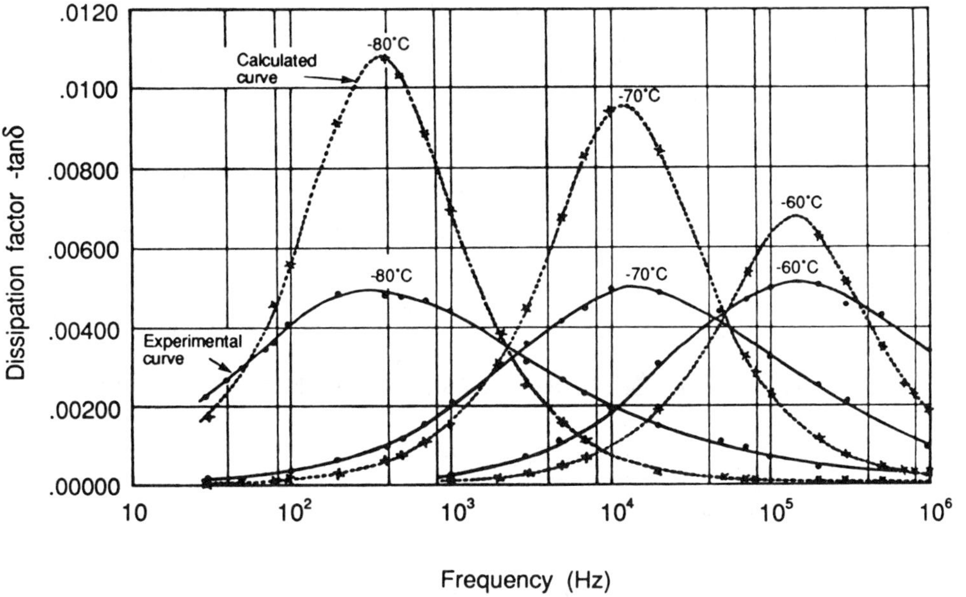

FIG. 1.39—*Experimental and calculated absorption curves for oil E, a low viscosity cable oil (after Bartnikas* [71]*).*

in terms of which Eqs 1.127 and 1.128 for the real and imaginary permittivity respectively become modified to

$$\epsilon' = \epsilon_\infty + (\epsilon_s - \epsilon_\infty) \int_o^\infty \frac{G(\tau)d\tau}{(1 + \omega^2\tau^2)} \tag{1.257}$$

and

$$\epsilon'' = (\epsilon_s - \epsilon_\infty) \int_o^\infty \frac{\omega\tau\, G(\tau)d\tau}{(1 + \omega^2\tau^2)} \tag{1.258}$$

The precise magnitudes of the dipole interaction forces as well as the thermal agitation forces surrounding the permanent dipoles are generally not known, nor is the value of the directing field with the result that the form of $G(\tau)$ cannot be derived theoretically but must be obtained by fitting the experimental data. Occasionally a relaxation process may be described by an exact mathematical distribution function. One such function, the log Gaussian distribution function, first utilized by Wagner [*101*], appears to be capable of approximating adequately a number of relaxation processes. It is defined by the equation

$$G(\tau)d\tau = \frac{\zeta}{\sqrt{\pi}} \exp\left[-\zeta^2 z^2\right] dz \tag{1.259}$$

where ζ is a measure of the width of the distribution curve and $z = \ln \tau/\tau_o$; in the application of this function, Wagner assumed that there is an infinite number of causes disturbing the original relaxation time, τ_o. It has been demonstrated by Hakim [*116*] that for parameter values of ζ ranging from 0.375 to 0.458 a good fit will result between the experimental and calculated curves as is evident from Fig. 1.40 obtained at -24.5°C on a high viscosity cable oil with ($\eta = 5 \times 10^2$ cP at $T =$ 20°C).

The value of $\zeta = 0.384$ also yields a good fit between the distribution function, $G(\tau)$ derived experimentally and the log Gaussian approximation portrayed in Fig. 1.41. The experimentally derived value of $G(\tau)$ is obtained by a deconvolution of the experimental data [*185*]. That is, Eq 1.258 is first expressed as a summation of n terms

$$\epsilon'' = (\epsilon_s - \epsilon_\infty) \sum_{j=i}^{n} \frac{G(\tau_j)\omega_i \tau_j \Delta\tau_j}{(1 + \omega_i^2 \tau_j^2)} \tag{1.260}$$

If the summation is arranged in the form of a matrix M, where the elements of the matrix are given by

$$A_{ij} = (\epsilon_s - \epsilon_\infty) \left[\frac{\omega_i \tau_j \Delta\tau_j}{(1 + \omega_i^2 \tau_j^2)} \right] \tag{1.261}$$

then $\epsilon'' = M\,G(\tau)$, and hence the distribution function may be obtained in terms of the inverse matrix

$$G(\tau) = M^{-1}\epsilon'' \tag{1.262}$$

Since the log Gaussian distribution tends to provide the best fit for many insulating liquid systems, it may be somewhat redundant to consider in great detail other mathematical distributions such as for example the Poisson and skewed-types. However, some remarks must be made as concerns the effect of a distribution of relaxation times upon the Cole-Cole or ϵ'' versus ϵ' plots. Cole and Cole [*105*] demonstrated that under the occurrence of a distribution of relaxation times, the complex permittivity given by Eq 1.140 may be expressed in the form

$$\epsilon^* = \epsilon_\infty + \frac{\epsilon_s - \epsilon_\infty}{1 + (j\omega\tau_o)^{1-\zeta}} \tag{1.263}$$

where ζ is a constant such that for $\zeta = 0$, Eq 1.263 reduces to Eq 1.144 for the standard case of a single relaxation time. When the value of ζ falls within the limits of $0 \leq \zeta \leq 1$, the relaxation

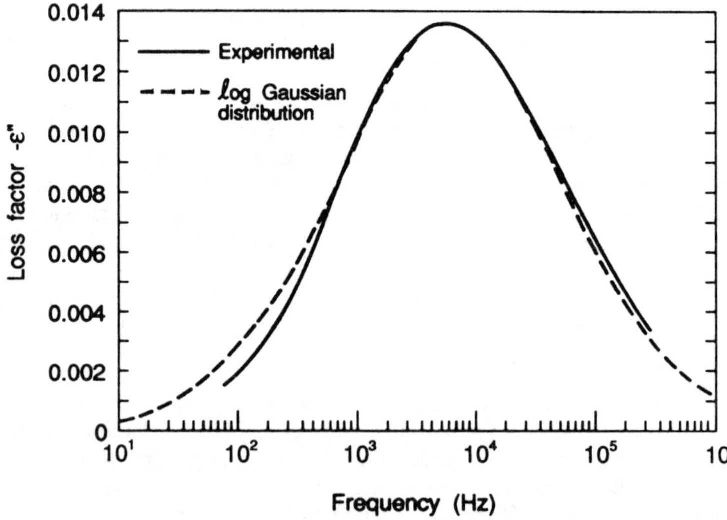

FIG. 1.40—*Experimental and calculated loss factor, ϵ'', characteristics of a cable oil at 24.5°C, using the log-Gaussian approximation with $\zeta = 0.384$ (after Hakim* [116]*).*

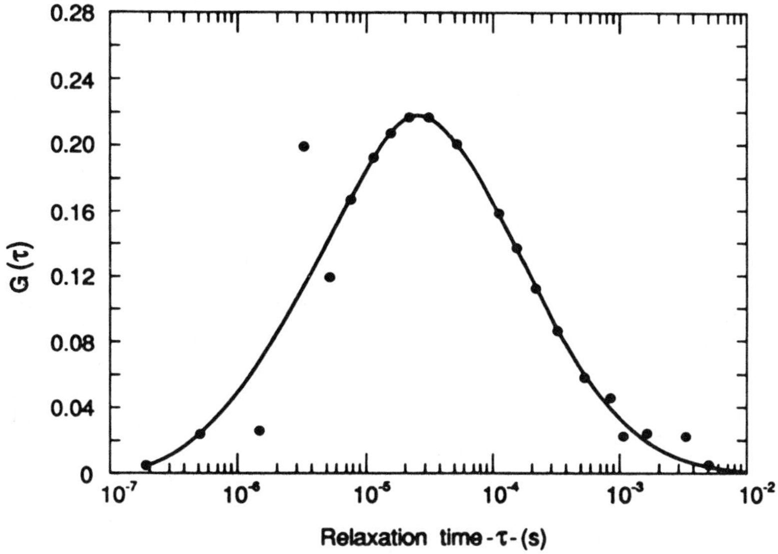

FIG. 1.41—*Log-Gaussian distribution function,* G(τ), *of a cable oil with* ζ = *0.384* (solid line) *as compared to the experimentally derived distribution* (points) *(after Hakim* [116]*).*

process is then governed by more than one relaxation time. Under the occurrence of a distribution of relaxation times, the center of the semicircular arc of Fig. 1.21 will be depressed to a point given by $(\epsilon_s + \epsilon_\infty)/2$, $[-(\epsilon_s - \epsilon_\infty)/2] \cot [(\zeta - 1)\pi/2]$ and will assume a radius equal to $[(\zeta - 1)\pi/2]$. This is illustrated in Fig. 1.42. When measurements are carried on actual electrical insulating liquids, an ideal circular arc rarely is approached. For example, Fig. 1.43 obtained on a cable insulating oil reveals a slight distortion in the circular arc. When plotting ϵ'' versus ϵ' at every frequency point of measurement, it is apparent that the values of ϵ_∞ and ϵ_s must be obtained by extrapolation to yield the ϵ_∞ and ϵ_s values at $\omega = \infty$ and $\omega = 0$, respectively. The displacement of the center of the circle from the ϵ'-axis, or alternatively the magnitude of the factor, ζ, is a measure of the deviation from the single-relaxation-time-determined process. There have been other emperical formulae derived to approximate the Cole-Cole plots; one particular approximation is that of Cole-Davidson relating to the skewness of the ϵ'' versus ϵ' plots. However, skewness rarely is observed with the relaxation process in dielectric liquids, and the reader is, therefore, referred elsewhere [*186*] for its treatment. Indeed, as pointed out by Hakim [*116*], the fact that skewness is not observed in the ϵ'' versus ϵ' plots for mineral oils would tend to negate the defect diffusion and dipole-dipole interaction models proposed, respectively, by Glarum [*118*] and Nee and Zwanzig [*187*] which are found to result in skewed type plots.

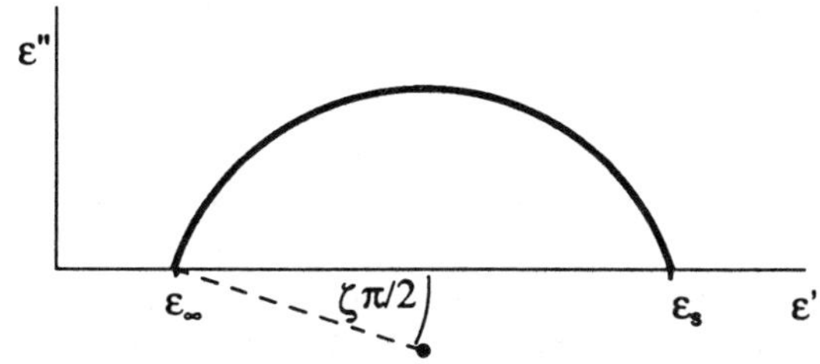

FIG. 1.42—*Cole-Cole plot in accordance with Eq 1.263.*

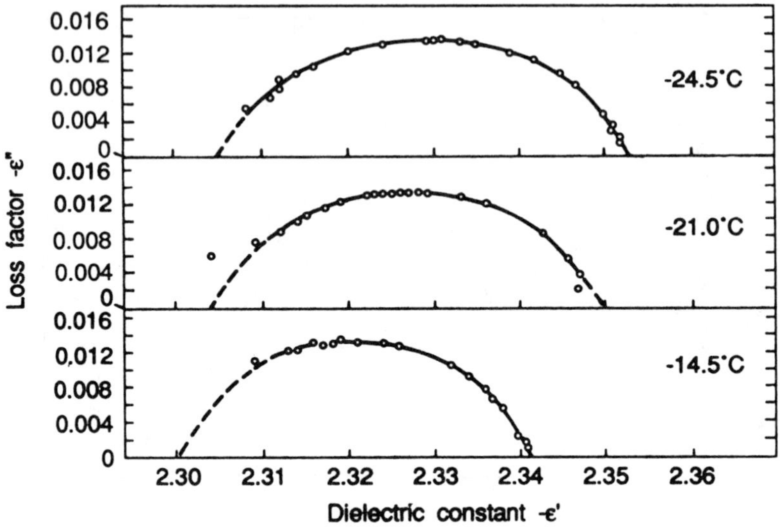

FIG. 1.43—*Cole-Cole plots of a 500 cP cable oil, with temperature as a parameter (after Hakim* [116]*).*

Irrespective of the relaxation process involved, a distribution of relaxation times is accompanied necessarily by a distribution of the activation energies. Since the intrinsic relaxation time of a given relaxation process, whether dipole or ionic, is defined by Eq 1.160, then the macroscopically determined relaxation time, τ, may be expressed as

$$\tau = A(\tau) \exp \left[\frac{\Delta H}{kT} \right] \tag{1.264}$$

where the preexponential term $A(\tau)$ includes the usual entropy, ΔS, contribution. Assuming that $A(\tau)$ is a constant common to all the relaxation spectrum, Tanke et al. [*188*] note that the fractional number of mechanisms with a relaxation time τ_i between $[\tau_i - \frac{1}{2}\, d(\ln \tau)]$ and $[\tau_i + \frac{1}{2}\, d(\ln \tau)]$ is

$$G(\tau_i)\, d(\ln \tau) = G(\Delta H_i)\, d(\Delta H) \tag{1.265}$$

where $G(\Delta H_i) d(\Delta H)$ represents the fractional number of mechanisms with an activation energy between $[\Delta H_i - \frac{1}{2}\, d(\Delta H)]$ and $[\Delta H_i + \frac{1}{2}\, d(\Delta H)]$. Also

$$\Delta H_i = RT[\ln \tau_i - \ln A(\tau)] \tag{1.266}$$

The average and most probable values of the activation energy ΔH_i coincide, because $G(\Delta H_i)$ is Gaussian in ΔH and not in $\ln(\Delta H)$. Using the approach of Tanke et al. [*188*], Hakim [*116*] calculated the activation energy distribution of a high viscosity cable oil as shown in Fig. 1.44. Here it is appropriate to remark that a distribution in the activation energy implies a change in the activation energy with temperature. This question will receive further attention in the subsequent section.

Figure 1.39 obtained by Bartnikas [*71*] indicates that with increasing temperature, the dispersion in the tanδ versus frequency curves decreases as the absorption curves calculated on the basis of a

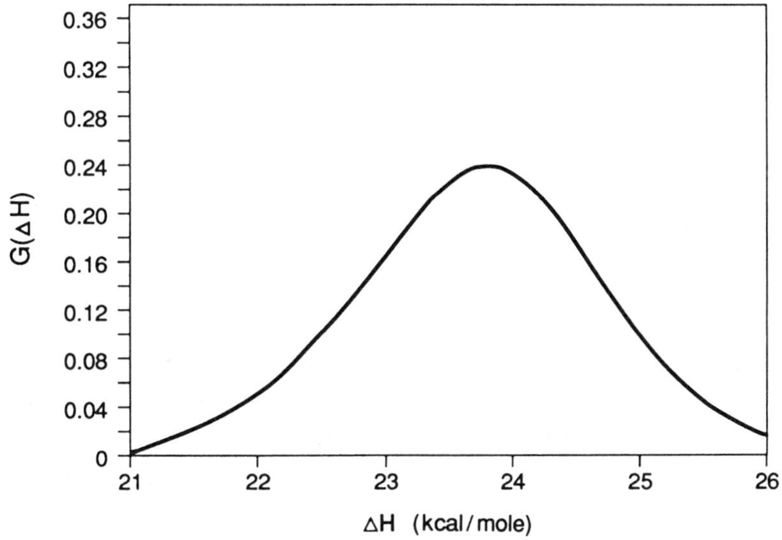

FIG. 1.44—*Calculated distribution of activation energies at −24.5°C for a 500 cP cable oil (after Hakim* [116]*).*

single relaxation time become increasingly better approximations for the experimentally observed curves. Furthermore, the experimental characteristics evince a slight but readily perceptible diminution in the magnitude of the absorption peak with falling temperature. This behavior is less discernible in the relatively high viscosity oils (see oil D in Table 1.1) over the lower temperature régime as may be seen by examining Fig. 1.45. Hakim [*116*], who carried out tests on a mineral oil over a more restricted frequency range but within the same temperature régime, found the change in the intensity of the loss peaks negligible and concluded that the loss peak magnitude remains unaltered with temperature. However, Bartnikas [*71*] also finds that lower viscosity oils with different aromatic contents (see oils A and B in Table 1.1) exhibit a readily noticeable decrease in the $\tan\delta$ maxima with reduced temperature as is apparent from Figs. 1.46 and 1.47. A decreasing loss peak magnitude would infer that the number of dipoles or ions capable of executing jumps over the thermal barriers is decreasing with falling temperature; this would seem a plausible situation since the thermal energy available for the dipoles to surmount the barriers is certainly decreasing with falling temperature.

1.6 Frequency Response of Insulating Liquids

The frequency response of liquids is defined essentially by the manner in which ϵ' and ϵ'' vary with frequency, with the temperature and applied electric field as parameters. The behavior of ϵ' and ϵ'' with frequency permits to establish the character of the loss mechanism prevailing under a given set of conditions; also the data allow some important deductions to be made as regards the chemical and physical structure of the liquid. In order to ascertain the predominant loss mechanism within the various frequency regimes of interest, the measurement of ϵ'' and ϵ' must be performed over as wide a frequency range as practically feasible. Since the application of electrical insulating liquids is confined primarily within power frequencies, one is concerned principally with the magnitude and mechanism of the dielectric loss at the power frequencies. In order to determine the nature of these losses at the power frequencies, it is helpful to examine the frequency spectrum over

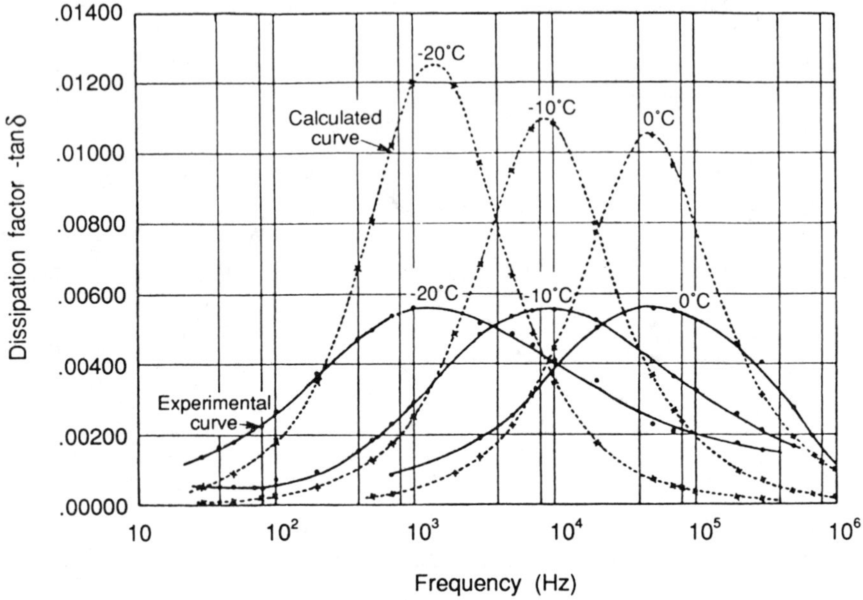

FIG. 1.45—*Experimental and calculated absorption curves for oil D, a high viscosity cable oil (after Bartnikas* [71]*).*

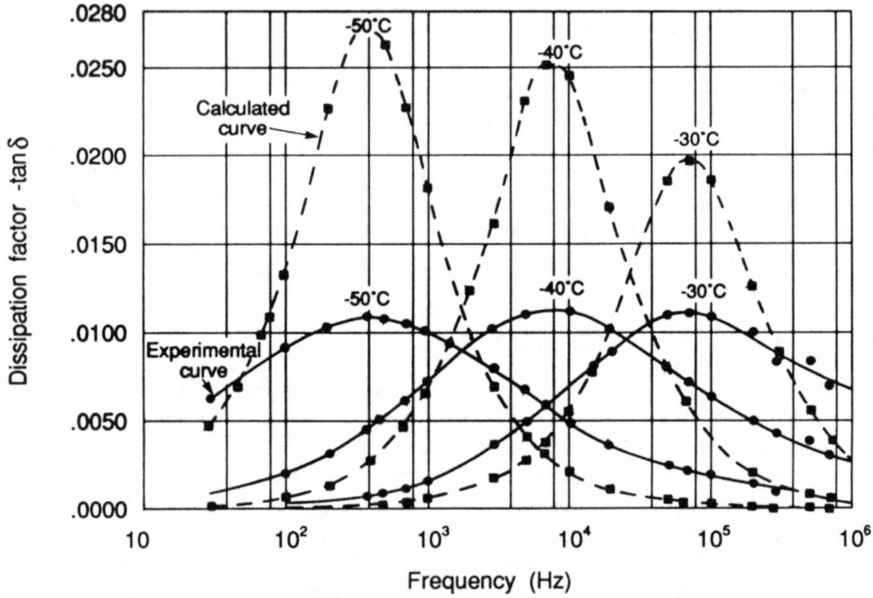

FIG. 1.46—*Experimental and calculated absorption curves for high aromatic oil A, an 87 cP mineral oil (after Bartnikas* [71]*).*

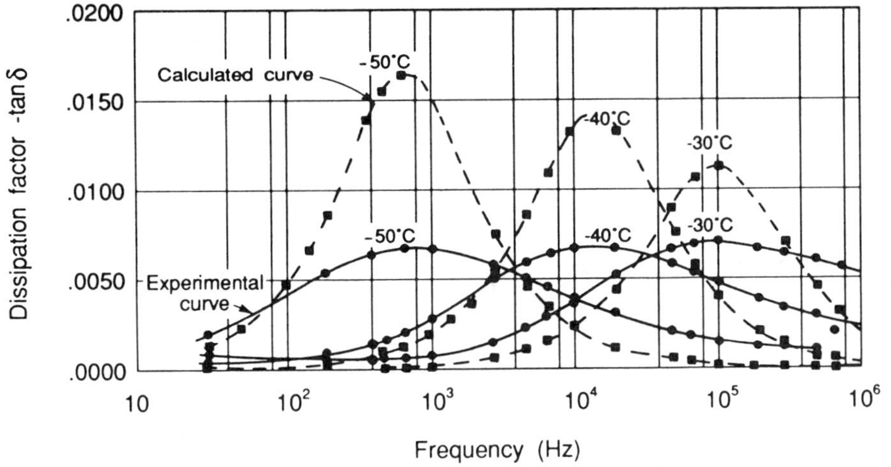

FIG. 1.47—*Experimental and calculated absorption curves for medium aromatic oil B, an 80 cP mineral oil (after Bartnikas* [71]*).*

several decades above and below the power frequency regime. However, if information is desired on the molecular structure of the insulating liquids, then the measurement range must be extended into the microwave frequency region.

Whereas at room temperature and above, dipole losses at 60 Hz constitute only a small fraction of the overall loss in typical mineral oils, they nevertheless can lead to prodigiously high losses at lower temperatures. Figure 1.48 depicts the relaxation spectra of a high viscosity cable insulating oil (see oil D in Table 1.1), which evinces a dipole absorption peak at -30°C in the vicinity of the power frequency. As the viscosity, η, of the oil is reduced, increasingly lower temperatures are required to shift the dipole absorption peak towards 60 Hz as is apparent from Fig. 1.49, which shows the relaxation spectra of a low viscosity hollow-core type cable filling oil (see oil E in Table 1.1). The general tendency in the shift of the absorption peaks with temperature and, hence, viscosity is predictable from Eq 1.152, that is, $\tau = 4\pi r^3/kT$. Making the rather rudimentary assumption that the dipole shape may be approximated in the form of a sphere leads to dipole diameters of 1 and 5.5 Å for oils D and E, respectively. The former values suggest the involvement of side or end link orientation mechanisms within the molecule, while the latter value would appear to imply the rotation of the entire molecule in view of the low molecular weight of oil E.

Returning to Fig. 1.48, it can be perceived that at -80°C, the losses in the high viscosity oil D are virtually reduced to an insignificant level over the frequency range form 30 to 10^6 Hz; a space charge loss is discernible over the power frequencies at -10, 0, and 15°C. Due to the low viscosity of oil E, the dipole absorption peaks are manifested only at the very low test temperatures over the measuring frequency range; the dipole loss becomes only perceptible as tanδ commences to increase significantly beyond 10^7 Hz. The increases of tanδ with decreasing frequency at -20, -40, -50, and -60°C over the power and audio frequencies is indicative of a space charge loss. The ionic losses in both the low and high viscosity oils tend to predominate at the more elevated temperatures as has been already observed in Fig. 1.31.

Perhaps the most expedient manner in which the extent or predominance of the three losses (dipole, space charge, and conduction by charge carriers) may be demonstrated over the different temperature regions is in terms of a plot of ϵ' and tanδ against temperature at a fixed frequency. The most convenient frequency of measurement for this purpose is 1 kHz, since it is removed sufficiently from the power frequency and its harmonics to eliminate interference difficulties. Such a

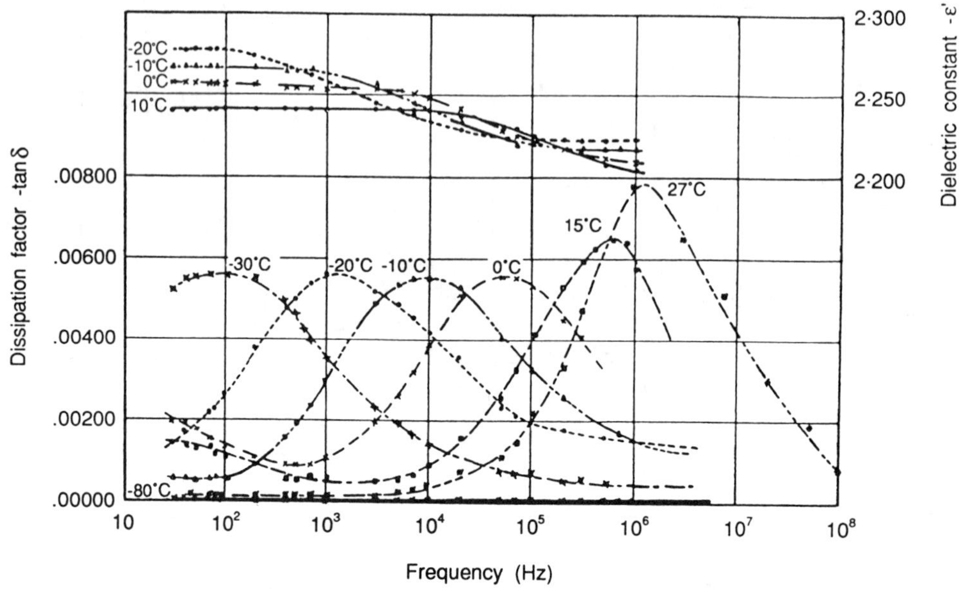

FIG. 1.48—*Relaxation spectra of oil D, a high viscosity cable oil (after Bartnikas* [71]*).*

a plot is given for oils D and E in Fig. 1.50. The usual dipole peaks for oils E and D are evinced at −78 and −20°C, respectively, while the space charge peaks that are appreciably more evasive on the frequency plots are clearly discernible at −8°C for oil E and at 15°C for oil D. Since the electrode separation of the measuring cell would be too large to attribute the space charge loss to

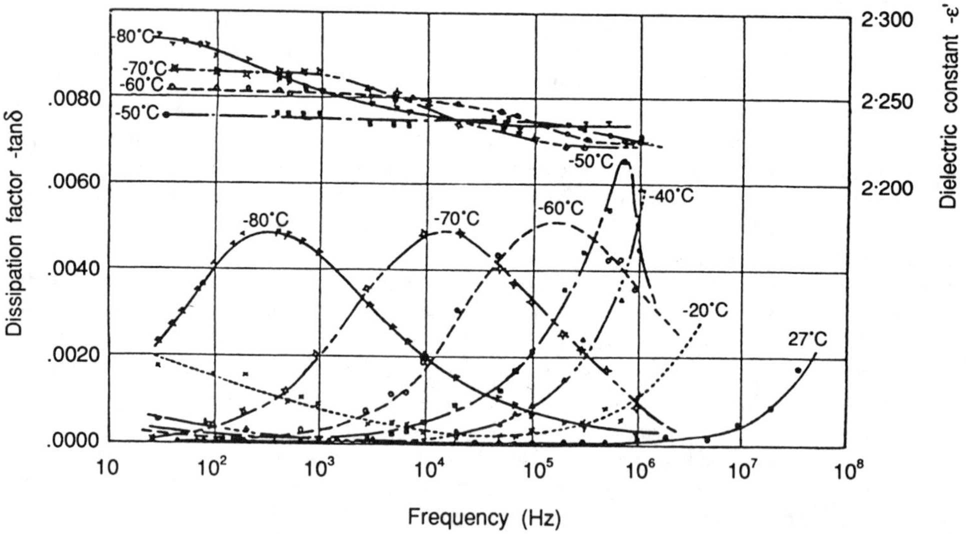

FIG. 1.49—*Relaxation spectra of oil E, a low viscosity cable oil (after Bartnikas* [71]*).*

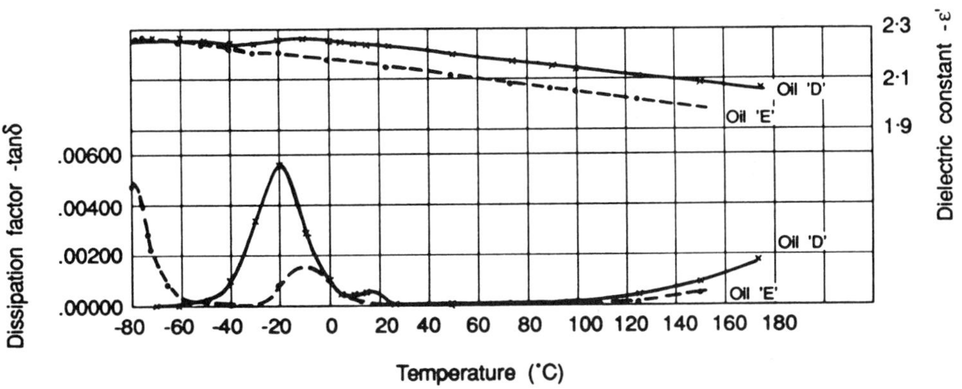

FIG. 1.50—*Dissipation factor and dielectric constant versus temperature characteristics of oils D and E at 1 kHz (after Bartnikas* [71]).

blocking electrode effects, it is difficult to speculate on the exact cause for these space charge peaks.

The ionic conduction losses are clearly manifest over the higher temperature range. Note that the monotonic decrease in ϵ' for both the low and high viscosity oils with temperature over the elevated temperature régime is attributable to a decrease in their density. Figure 1.51 shows the tanδ and ϵ' versus frequency characteristics obtained on a medium viscosity oil with a viscosity of $\eta = 500$ cP at 20°C, obtained over the frequency range of 10^{-2} to 10^{2} Hz for temperatures between 25 and 85°C. Although the test results refer to a thin oil specimen of thickness 0.15 mm, no space charge peaks in tanδ are apparent at the lower frequencies; in fact, tanδ is seen to decrease linearly with falling frequency on the log-log scale, while ϵ' remains unchanged. The behavior indicates that no electrode limiting effects are present with a film thickness even as low as 0.15 mm and that the losses are purely ionic in nature, that is, the a-c conductivity remains virtually independent of and constant with frequency. If desired, under a-c conditions, the ionic conductivity of the oil may be calculated using Eq 1.122 and subsequently be equated to the term of $qN\mu_m$ of Eq 1.181, where N is the concentration of the ions, q their charge, and μ_m their mobility in cm^2 V^{-1}s^{-1}. On the assumption that only a single ionic species with a single electronic charge are involved, then either the ion concentration, N, or their mobility, μ_m, may be estimated provided either one is known. To improve matters, the use of the Nernst-Einstein equation

$$\frac{D}{\mu_m} = \frac{kT}{q} \tag{1.267}$$

is made to determine first the mobility of μ_m and then the concentration N (as has been done by Reed [*189*] for example). Reed correctly points out that Eq 1.267 must apply equally well for weak electrolytes as for strong electrolytes, since the forces restraining the movement of ions under the action of an electric field are necessarily the same as those that also oppose the process of thermal diffusion of the ions. However, the diffusion coefficient D (cm^2/s) of the ions within the insulating liquid under test must be known if μ_m is to be determined; also the nature of the charge carriers must be known in order to define the value of q.

The presence of aromatics in mineral oils has been implicated as the prime source of the observed dipole loss [*71*]. However, it has not been established whether the aromatics *per se* are responsible for the observed dipole loss or whether due to the oxidation susceptibility of the

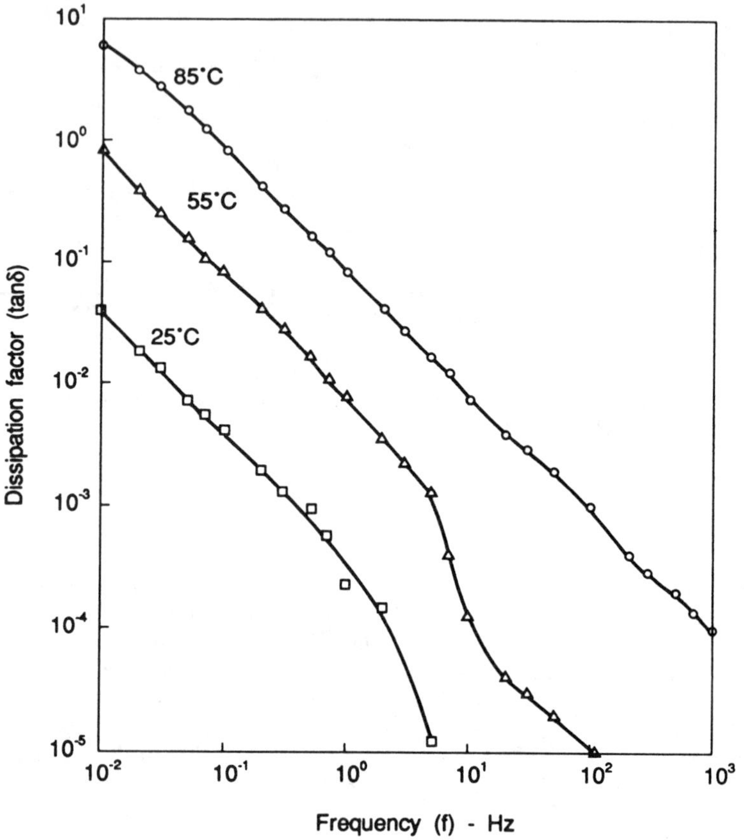

FIG. 1.51—*Dissipation factor versus frequency of a 1.5 mm thickness film of a medium viscosity pipe-filling cable oil, with temperature as a parameter (after Bartnikas, unpublished work* ca. *1972).*

aromatics it is the resulting oxidized aromatic molecules themselves that lead to the dipole loss. Figures 1.52, 1.53, and 1.54 depict the relaxation spectra obtained on a highly aromatic, moderately aromatic and zero aromatic content oil, respectively, with the oils being especially selected to have nearly equivalent viscosity characteristics (see Table 1.1). The two aromatic oils (Figs. 1.52 and 1.53) are characterized by strong absorption peaks, whose intensity appears to be a function of the aromatic content of the oils. Furthermore, both oils A and B exhibit tanδ maxima that diminish with decreasing frequency; also the oils indicate the presence of space charge losses over the lower frequencies at temperatures of 0, −10, and −20°C. In contradistinction, the losses of the nonaromatic oil C, portrayed in Fig. 1.54, exhibit only extremely small dipole losses below −30°C. For temperatures less than room temperature down to −40°C, the losses over the lower frequency band are predominantly those of space charge polarization. Space charges losses in the three oils (A, B, and C) appear only at the higher viscosities and, consequently, are completely absent at higher test temperatures. For example, the peak of the space charge loss of oil C at 0°C in Fig. 1.54 does not shift to higher frequencies as the temperature is increased, but rather vanishes entirely as is apparent from the room temperature curve. The minuscule absorption peaks discernible at −60, −50, and

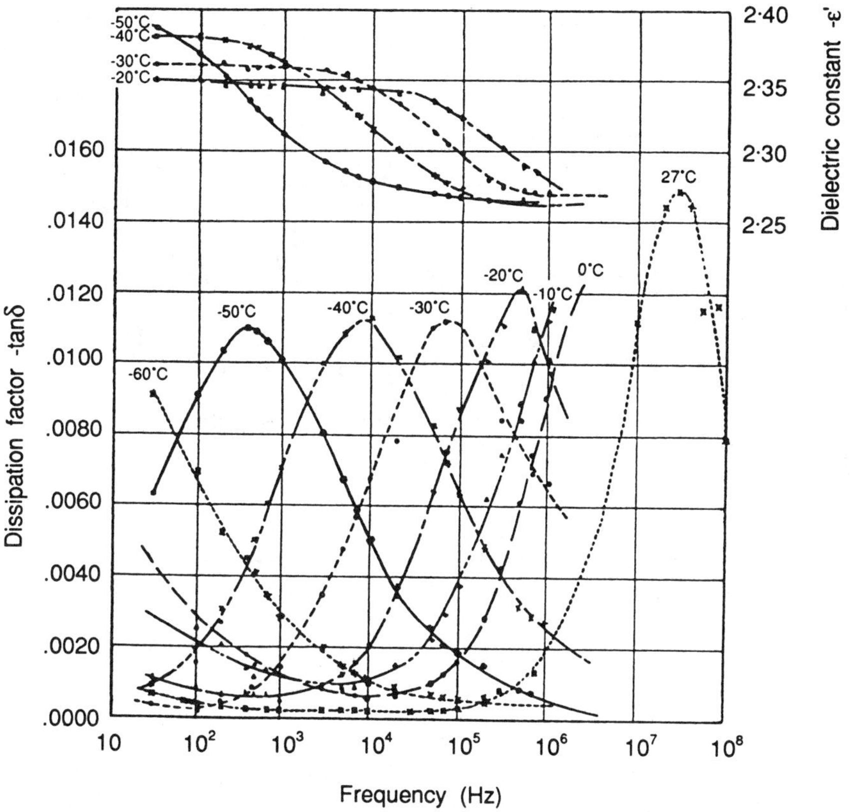

FIG. 1.52—*Relaxation spectra of oil A, a high aromatic content mineral oil (after Bartnikas* [71]*).*

−40°C in oil C may be due to the presence of some residual aromatics in the oil; in particular, it is difficult to speculate as to the nature of the dipole peaks that appear to be inferred by the increases in the tanδ value with frequency at all test temperatures over the higher frequency régime. On the other hand, there would seem to be little uncertainty that the dipole losses in the two aromatic oils A and B must originate with the aromatic constituents. Yet, since both oils A and B behave essentially as dilute solutions of polar molecules notwithstanding their relatively high aromatic content, it must as a consequence be concluded that not all of these aromatics constitute polar molecules. It had been conjectured initially by Hakim [*142*] that perhaps the dipoles in the oil consist of the residuals of various nitrogen, oxygen, and sulfur compounds that are soluble in the aromatics and, thus, are removed with the aromatics in any purification process. However, it was demonstrated subsequently by the same author that neither sulfur nor nitrogen content has any relation to the dipole loss.

The oxidation of mineral oil has a rather peculiar effect on the dipole loss [*142*]. Although oxidation appears to have no measurable effect upon the activation energy of the dipole orientation loss, it has a perceptible influence on the frequency of the absorption peak as illustrated in Fig. 1.55. Each of the indicated oxidation steps corresponds to a heating of the previously degassed and dried oil rapidly to 115°C and maintaining it for a period of 68 h at this temperature. The oxidation

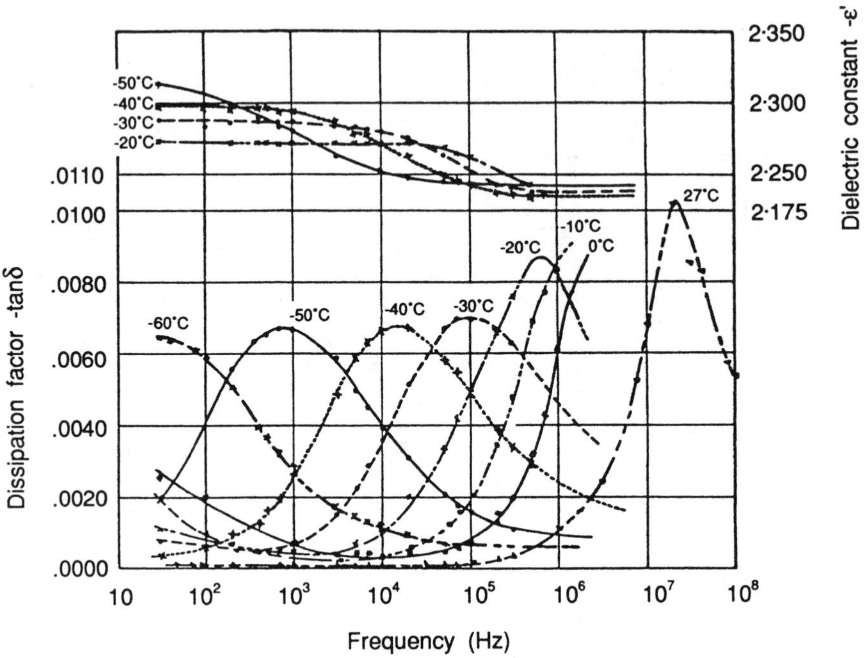

FIG. 1.53—*Relaxation spectra of oil B, a medium aromatic content mineral oil (after Bartnikas* [71]*).*

procedure followed was intentionally mild, with no detectable sludge formation occurring; the degree of oxidation did not progress any further after the completion of the second run. The study carried out by Hakim [*142*] shows that although oxidation tends to increase the dipole losses only marginally, its effect on the a-c ionic conductivity is rather pronounced with the ionic loss being augmented by about one order of magnitude. This agrees with the practical experience on mineral oils, where great care is taken to prevent their oxidation. In manufacturing processes where the mineral oils may be reused a number of times in several impregnation sequences, the oils tend to assume an increasingly darker color which is accompanied generally by an increase in their a-c conductivity or the 60 Hz tanδ value. Typical relaxation spectra of a used high viscosity oil (see Table 1.1, oil D) are portrayed in Fig. 1.56. Comparing these relaxation spectra on the used oil with those of its virgin state in Fig. 1.48, it becomes apparent that the used oil just as the intentionally oxidized oil in Fig. 1.55 exhibits a slight decrease in the frequency of the loss maximum and tanδ maxima that are somewhat higher than those of the original oil D. The effect of oxidation upon the ionic conductivity is, however, much more striking when the 90°C characteristic is compared to that of the virgin oil in Fig. 1.31. The difference infrared spectrum of the used versus the virgin oil D reveals a weak band at 1692 cm^{-1}, resulting from the presence of the unsaturated aromatic $C = C$ groups. The intensity of the carbonyl band fixes the amount of oxidation products in the region from 0.05 to 0.1%. The slightly stronger aromatic $C = C$ absorption of the used oxidized oil specimen suggests the formation of small amounts of oxidized and condensed polynuclear aromatics, which accounts for the darker color of the oil. Further analysis of the used oil D indicated that the oxidation products in the oil consist of a mixture of aliphatic and aromatic carboxylic acids. A possible sequence of reactions which may arise in the production of the aliphatic and aromatic carboxylic acids are the following

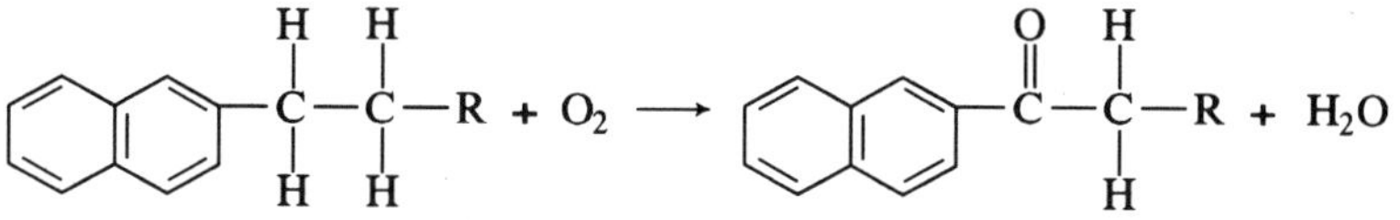

(1.268)

(unstable ketone)

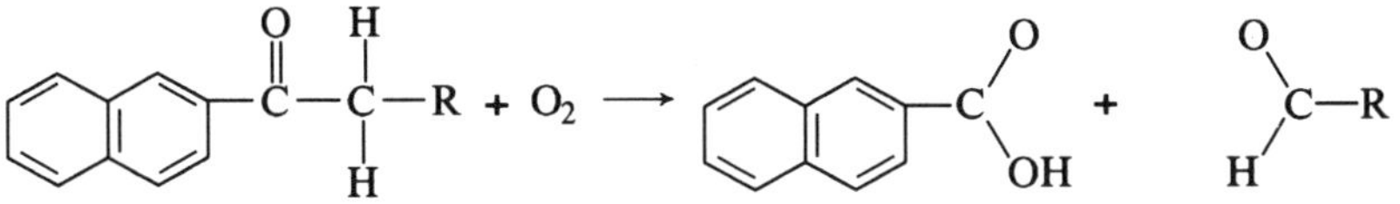

(1.269)

(aromatic acid) (unstable aldehide)

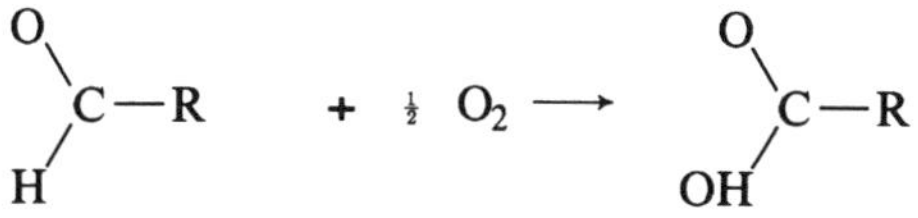

(1.270)

(unstable aldehyde) (aliphatic acid)

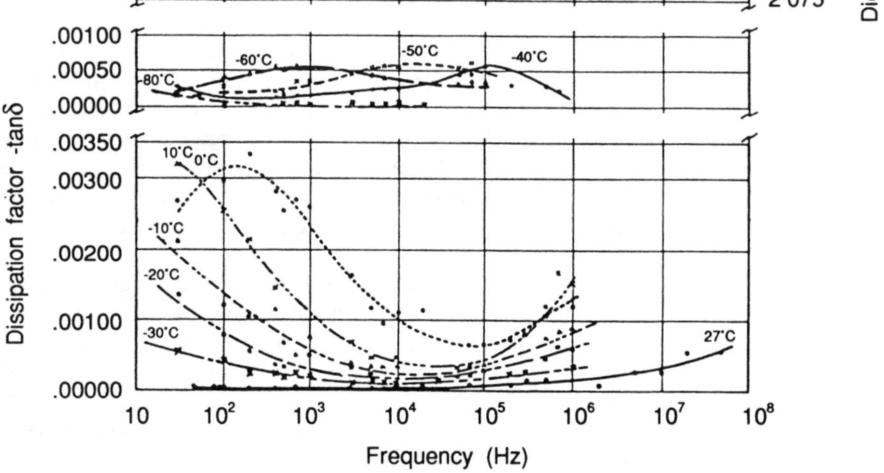

FIG. 1.54—*Relaxation spectra of oil C, a mainly paraffinic type mineral oil (after Bartnikas [71]).*

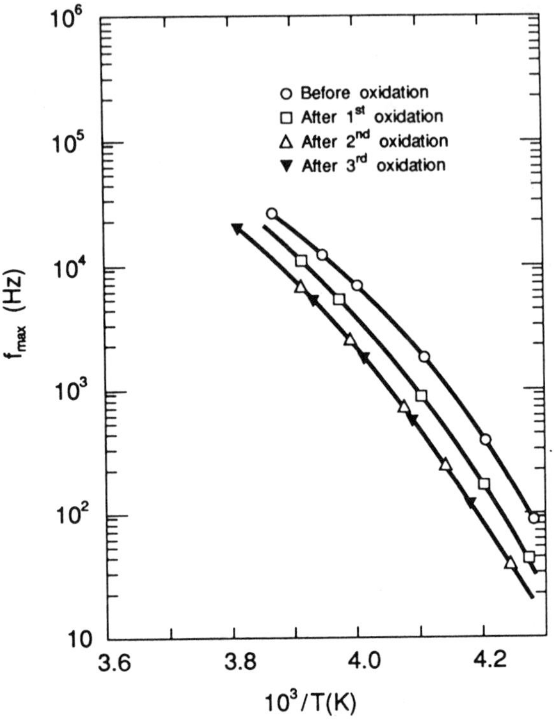

FIG. 1.55—*Effect of oxidation on the frequency of the tanδ maximum as a function of the inverse absolute temperature of a 500 cP cable oil (after Hakim* [143]*).*

The formation of the two organic acids and water as shown in Eqs 1.268 to 1.270 may account for the considerably enhanced ionic conductivity of the used oil at 90°C.

Although thus far we have confined our discussion on the dielectric losses in mineral oils to specifically those used in cables, mineral oils used in other electrical apparatus exhibit essentially a similar dielectric behavior. Nor, apart from a reduction in their loss magnitude, does there appear to be any great changes made to these oils over the past number of decades as concerns their dielectric response. This is illustrated in Figs. 1.57 and 1.58 that were obtained in 1935 by Rieche [*190*] on a rather lossy type of transformer oil. Figure 1.57 indicates a slight increase in the magnitude of the loss peak with temperature as a parameter. The dispersion is also seen to be significantly higher at lower temperatures, with the ionic losses very much predominating at the higher temperatures over the power frequency range. The corresponding dielectric constant versus frequency characteristics are portrayed in Fig. 1.58.

In an effort to replace PCBs, a number of substitute oils have been made commercially available. For application to capacitors, it is desirable to have oils with moderately high dielectric constants and yet low losses. Only a relatively small number of the substitute liquids have been fully characterized as concerns their dielectric response behavior. One liquid, which has been reasonably well examined, is castor oil—an oil of vegetable origin having the structure $CH_3(CH_2)_5$ $CHOHCH_2CH = CH(CH_2)_7$ COOH [*191*]. The OH group is mainly responsible for the slight polarity of the oil and hence, its elevated value of the dielectric constant. The high frequency experimental data obtained by Ramu [*191*] extend up to 10 GHz; however, it does not extend sufficiently low to provide information on the ionic conduction losses that normally predominate

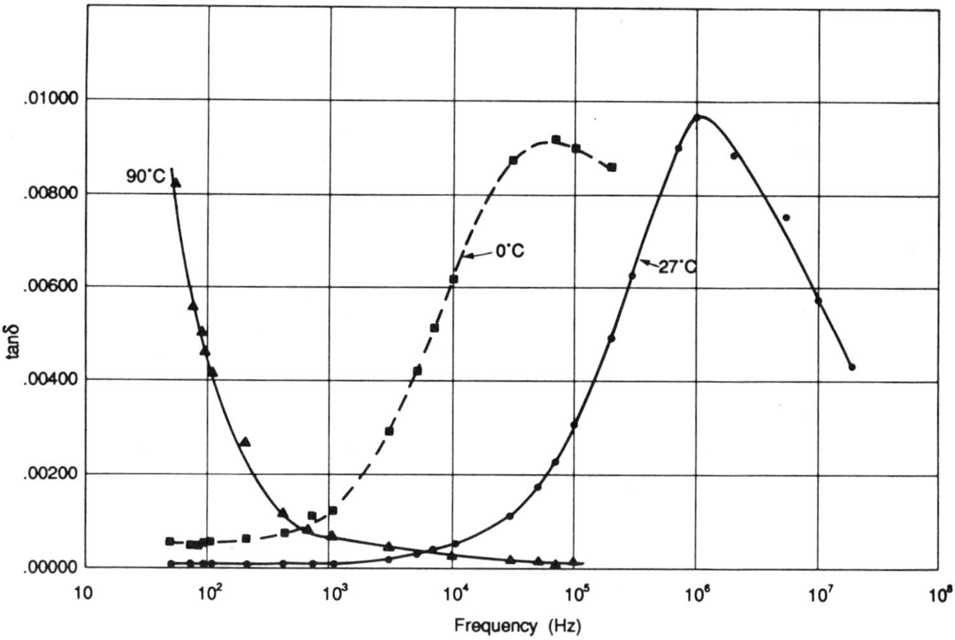

FIG. 1.56—*Relaxation spectra of oxidized oil D, a high viscosity cable oil (after Bartnikas, unpublished work* ca. *1963).*

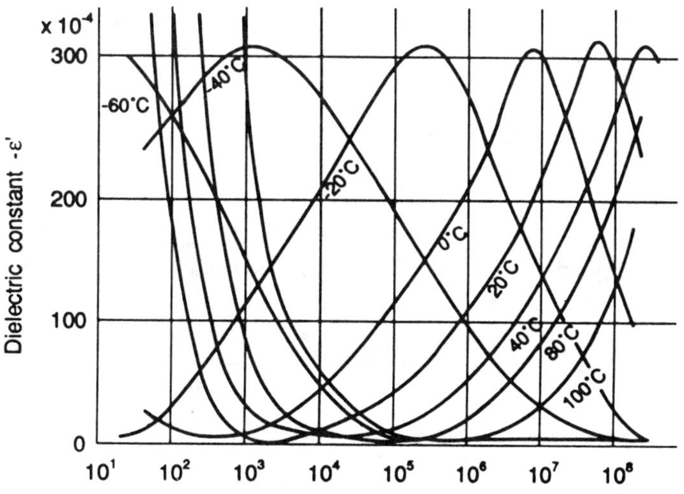

FIG. 1.57—*Absorption curves of a transformer oil (after Rieche* [190]*).*

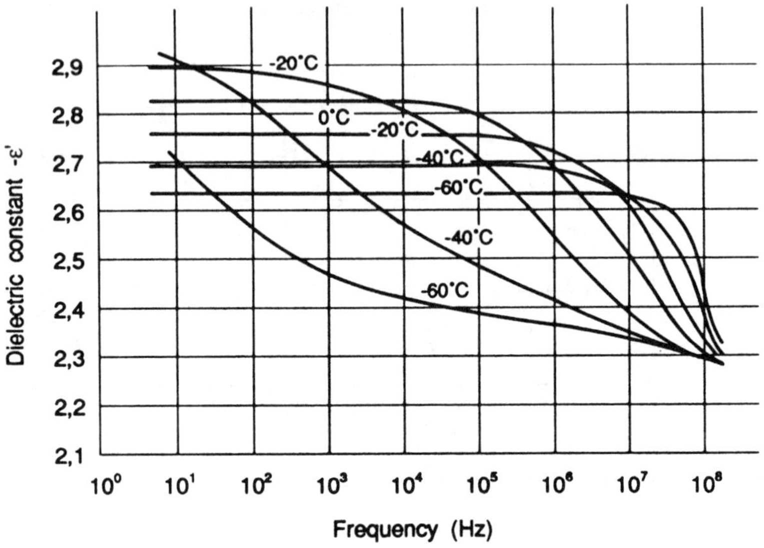

FIG. 1.58—*Dielectric constant as a function of frequency of a transformer oil (after Rieche* [190]*).*

over the power frequencies at or above room temperature. The relaxation spectrum at room temperature is depicted in Fig. 1.59. The peak dipole loss is seen to occur in the proximity of 100 MHz, with a tanδ value corresponding to approximately 0.30, which is appreciably above the ones obtained with either used or virgin state mineral cable oils that are intrinsically dilute solutions of polar molecules; however, the losses in the vicinity of the power frequencies are acceptably low. The loss peak of the electrical grade castor oil exhibits considerable dispersion indicating a distribution of relaxation times, a fact which is substantiated by its Cole-Cole plot with its center depressed below the ϵ'-axis, as portrayed in Fig. 1.60.

The value of the empirical distribution parameter ζ (see Eq 1.163) as determined by Ramu [*192*] for castor oil is 0.089; this value is used in turn to determine the relaxation times and the associated dipole diameters. It is assumed that the portion of the semicircle above the ϵ'-axis can be approximated by an ellipse of the form [*192*]

$$\frac{a}{b} = \frac{2\epsilon''_{max}}{(\epsilon_s - \epsilon_\infty)} \tag{1.271}$$

where a and b are, respectively, the semimajor and semiminor axes of the ellipse. Equation 1.271 may be expressed alternatively as

$$\frac{2}{A}\{\tan^{-1}[\tanh(A/4)]\} = \frac{2\epsilon''_{max}}{(\epsilon_s - \epsilon_\infty)} \tag{1.272}$$

Ramu [*192*] further invokes the empirical approximation that

$$A^2 = 37.6\,\zeta \tag{1.273}$$

in terms of which two distinct relaxation times τ_1 and τ_2 are defined as

$$\tau_1 = \tau_o \exp[A/2] \tag{1.274}$$

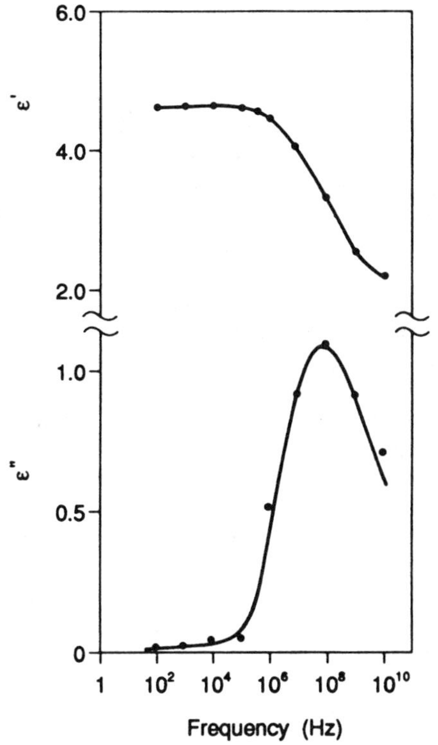

FIG. 1.59—*Relaxation spectrum of castor oil at 27°C (after Ramu* [192]*).*

and

$$\tau_2 = \tau_o \exp[-A/2] \tag{1.275}$$

yielding the corresponding values of the dipole diameters 1.54 and 2.97 Å, respectively, in accordance with Eq 1.156. Ramu [*192*] suggests that the size of the dipoles is indicative of the major portion of the molecule being involved in the orientation mechanism rather than the possibility of the C—H and O—H groups alone. Yet, since the interatomic distances are generally of the order of 1 Å, the value of 1.54 Å could conceivably still imply the involvement of the C—H and O—H groups in the orientation mechanism. The validity of the two discrete relaxation times characterizing the electrical grade castor oil rests on how well the semielliptical plot approximates the

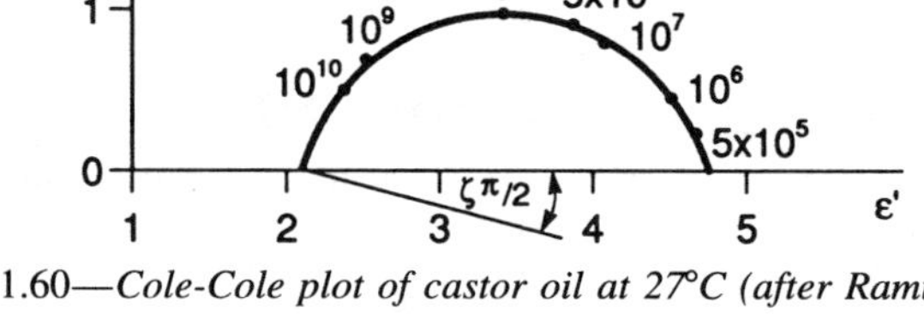

FIG. 1.60—*Cole-Cole plot of castor oil at 27°C (after Ramu* [192]*).*

semicircle on the ϵ''-ϵ' plane. The Cole-Cole trace in Fig. 1.60 deviates appreciably from an ellipsoidal fit, and it would appear that a Gaussian approximation is more appropriate for insulating liquids.

Silicone dimethylsiloxane fluids constitute a group of synthetic liquids that have become viable substitutes for PCBs; they also represent a group of liquids that has been studied extensively electrically. One of the earliest studies carried out on silicone fluids was that of Bartnikas [*71*], who reported data on silicone liquids of high and low viscosity whose properties are given in Table 1.1. Figure 1.61 shows the dielectric response of a 10 cP silicone fluid with temperature as a parameter. The dielectric losses for temperatures below −50°C are virtually indistinguishable from zero. The ionic losses are seen to predominate over the lower frequencies at room temperature and above, while the influence of a dipole loss becomes perceptible in the vicinity of 100 MHz for the room temperature curve. A low intensity space charge loss that is too weak to affect the dielectric constant values is in evidence between room temperature and −40°C. The interelectrode separation is too large to attribute the observed space charge loss to an electrode blocking mechanism, and, consequently, residual moisture in the silicone fluid may be a more tangible cause.

The relaxation spectra of a high viscosity (970 cP) silicone fluid are portrayed in Fig. 1.62. The dimethylsiloxane fluid S2 (see Table 1.1) exhibits space charge loss behavior over the region from room temperature down to −50°C, while ionic conduction losses predominate for temperatures above room temperature. A rather peculiar behavior is evinced by the tanδ peaks observed at −60, −70, and −80°C; the absorption peak intensity is augmented with falling temperature, but the peaks do not shift to lower frequencies. Thus the losses cannot be ascribed to the normal type of dipole relaxation mechanism, particularly as this peak is observed to vanish completely above −60°C; only traces of a very small shifting type peak are discernible at −50°C. Bartnikas [*71*] had

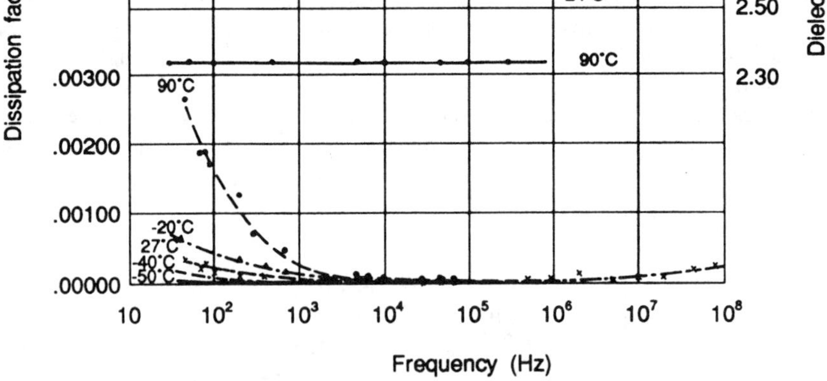

FIG. 1.61—*Dissipation factor and dielectric constant as a function of frequency of a low viscosity silicone fluid S_1 (after Bartnikas* [71]*).*

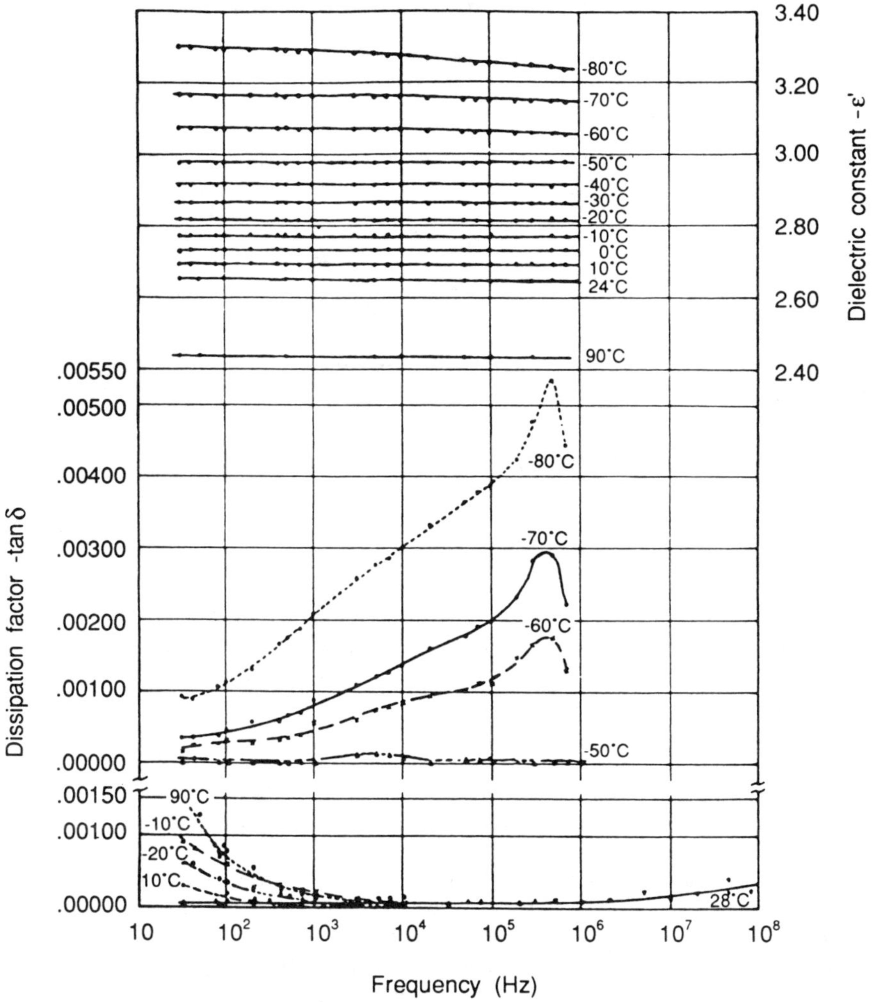

FIG. 1.62—*Dissipation factor and dielectric constant as a function of frequency of a high viscosity silicone fluid S_2 (after Bartnikas* [71]*).*

suggested initially that perhaps these losses may be associated with some vibrational mechanisms in the crystal lattices, since the silicone fluid S2 commences to solidify at approximately −50°C. Later the same anomaly in the loss peaks was confirmed by Liu et al. [*193*], but their differential scanning calorimetric data were not able to substantiate a clear relationship between the anomalous loss peaks and the crystallization state. The exact mechanism underlying these anomalous peaks thus remains somewhat of an enigma, even though their existence has been further substantiated by means of the thermally stimulated current technique by Mizatani et al. [*194*]. It is interesting to note that the dielectric constant undergoes the usual decrease with frequency in the vicinity of the absorption frequency of the anomalous loss peaks.

Although both the low and high viscosity silicone fluids exhibit the presence of space charge

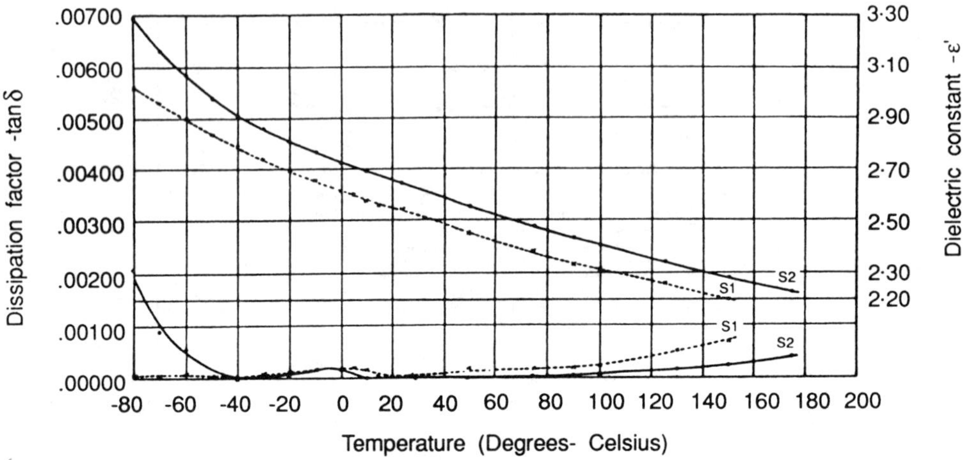

FIG. 1.63—*Dissipation factor and dielectric constant as a function of temperature of silicone fluids S_1 and S_2 at 1 kHz (after Bartnikas* [71]*).*

polarization losses over the lower frequencies at the lower temperature range, the frequencies traversed are not sufficiently low to evince the absorption peaks themselves. The existences of these loss peaks, however, is evident from Fig. 1.63, which portrays the dielectric response as a function of temperature at a convenient measurement frequency of 1 kHz. A slight temperature dependence is apparent with the high viscosity fluid peak centered at approximately −5°C and the low viscosity fluid peak at 5°C. The suggestion by Bartnikas [*71*] in his early studies of the silicone fluids that moisture may be responsible for the occurrence of the space charge loss has been confirmed more recently by Liu et al. [*193*], who observed that increased exposure to humidity results in higher space charge losses.

Neither the investigations of Bartnikas [*71*] nor Liu et al. [*193*] detected any discontinuities in the dielectric response characteristics of the type that were reported by Hakim et al. [*145*]. High resolution of measurement used by Hakim et al. is not sufficient *per se* to account for this disparity, since the observed discontinuities extend to temperature spans as high as 10°C. It should be emphasized, however, that while Hakim et al. [*loc. cit.*] studied a fluid with a viscosity of 48 cP, the two fluids examined by Bartnikas [*71*] differed very appreciably in viscosity being respectively 10 and 970 cP. Figures 1.64 and 1.65 depict the tanδ and permittivity versus temperature characteristics respectively obtained on a 48 cP polymethylsiloxane fluid at 1 kHz, using a cooling rate of 8°C/h [*145*]. A typical dipole relaxation peak is seen to occur at −107°C, but what is most striking are the abrupt discontinuities in the two characteristics at −72°C and in the loss curve only at −44°C. The rapid fall in the dielectric constant with decreasing temperature beyond −125°C is attributable to the solidification of the fluid which prevents dipole orientation. The abrupt falls in both tanδ and ϵ' for temperatures in excess of −107°C is indicative of structural changes occurring in the liquid. Lee et al. [*195*] have observed multiple crystallization transitions in polydimethylsiloxanes, which they have attributed to a formation of metastable crystalline phases which upon further cooling recrystallize into a more stable form. A range of crystallization temperatures was detected by Hakim et al. [*145*] on the 48 cP silicone fluid using low-temperature wide-angle X-ray diffraction. They explain the change of ϵ' in the region between −70 and 120°C solely in terms of its temperature dependence. They assume that in the vicinity of −70°C and higher, the real value of

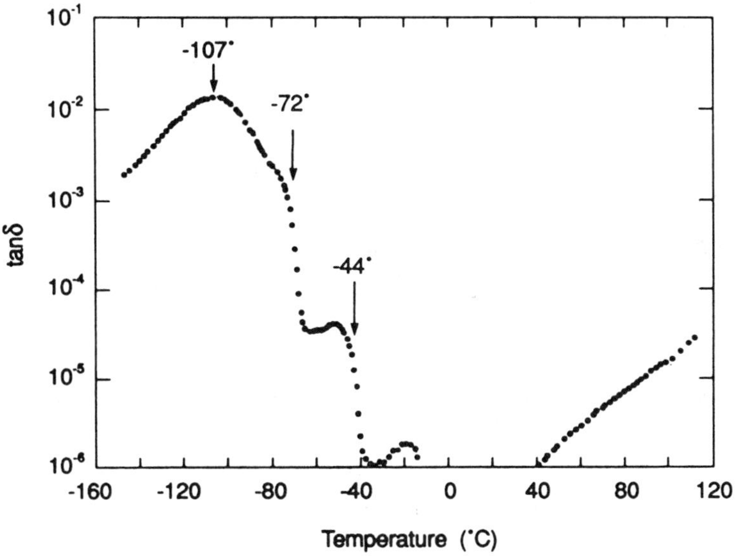

FIG. 1.64—*Dissipation factor versus temperature of a 48 cP silicone fluid at 1 kHz (after Hakim et al.* [145]*).*

the permittivity, ϵ', is essentially equal to the value of the static dielectric constant, ϵ_s, which in turn can be estimated from the Onsager equation in terms of the permanent dipole moment μ

$$\mu^2 = \frac{(\epsilon_s - \epsilon_\infty)(2\epsilon_s + \epsilon_\infty)}{\epsilon_s(\epsilon_\infty + 2)^2}\frac{9kTU}{4\pi N_A} \tag{1.276}$$

where U is the molar volume. If the small temperature dependence of ϵ_∞ is neglected, then differentiation of Eq 1.276 with respect to the absolute temperature T yields [*145*]

$$A\frac{\partial(\ln \epsilon_s)}{\partial T} = \frac{\partial(\ln \mu^2)}{\partial T} - \frac{\partial(\ln U)}{\partial T} - \frac{1}{T} \tag{1.277}$$

where

$$A = \frac{2\epsilon_s^2 + \epsilon_\infty^2}{(\epsilon_s - \epsilon_\infty)(2\epsilon_s + \epsilon_\infty)} \tag{1.278}$$

Using typical values of 1×10^{-3} K^{-1} for the expansion coefficient $\partial(\ln U)/\partial T$ and 0.65×10^{-3} K^{-1} for the value of $\partial(\ln \mu^2)/\partial T$ corresponding to 50 monomer units of the silicone molecule [*93*], they note that the permanent dipole contribution in Eq 1.277 is negated very nearly by the density term. Hence, as the thermal agitation of the dipoles increases with temperature, the $1/T$ contribution in Eq 1.277 predominates.

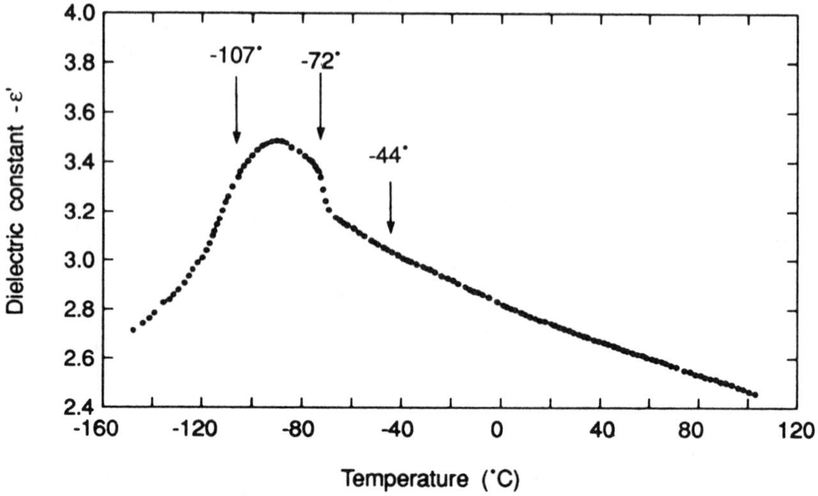

FIG. 1.65—*Dielectric constant versus temperature of a 48 cP silicone fluid at 1 kHz (after Hakim et al.* [145]*).*

1.7 Measurements at Optical Frequencies

The foregoing section dealt with dispersion mechanisms that in their historical context have been often referred to as giving rise to anomalous dispersion. The normal type of dispersion was considered to be the one that occurs in the region encompassing the frequencies extending from the infrared to the ultraviolet, where the index of refraction exhibits an increase as the absorption frequency is approached in contradiction to the decrease in dielectric constant observed near the vicinity of a dielectric loss peak. If one considers the resonance absorption resulting from a vibration of electrons in the ultraviolet region, then for an electron vibrating at an eigenfrequency of ν_o (the free vibration frequency of the electron in the absence of an external field) the real and imaginary indices of refraction, n' and n'', are given, respectively by

$$n' - 1 = \frac{Ne^2}{2\pi m_e} \cdot \left[\frac{\nu_o^2 - \nu^2}{(\nu_o^2 - \nu^2) + \dfrac{\gamma^2\nu^2}{4\pi^2}} \right] \tag{1.279}$$

and

$$n'' = \frac{Ne^2}{4\pi^2 m_e} \left[\frac{\gamma\nu}{(\nu_o^2 - \nu^2) + \dfrac{\gamma^2\nu^2}{4\pi^2}} \right] \tag{1.280}$$

where m_e is the mass of the electron and e its charge, γ is a damping factor, and N is the number of atoms per cubic centimeter and is equal to $N_A d/M$; here, N_A is Avogadro's number, d is the density of the liquid, and M its molecular weight. The damping factor γ must be introduced to take into account the emission of radiation by the vibrating electrons. Since the electrons are forced to vibrate at the frequency of the external electromagnetic field, the resulting damped vibrations give rise to energy absorption at and in the vicinity of the eigenfrequency, ν_o. In the infrared region the resonance absorption behavior due to the vibration of atoms and molecules is essentially similar in form to that of resonance absorption due to the electrons shown in Fig. 1.66, except that now the damping factor describes the collisions and interactions between the various atoms and molecules.

From the practical viewpoint, there appears to be very little interest in studying the absorption losses of insulating liquids in the infrared and ultraviolet regions. However, there is considerable value in having information in the manner in which the real part of the index of refraction, n', varies with frequency, since then the values of the electronic and atomic polarizations may be established. This in turn would indicate what percentage of the dielectric constant is due to atomic and electronic polarization and what remainder, if any, is due to the presence of polar molecules in the insulating liquid.

For the electronic molar polarizability, $[P_e]$, one can write that

$$[P_e] = \frac{n_e^2 - 1}{n_e^2 + 2}\frac{M}{d} \tag{1.281}$$

where n_e represents the real value of the index of refraction in the visible light region. Here the frequency of the electromagnetic radiation is too slow to excite the electrons and too rapid to induce vibration of the atoms or molecules. Thus, Eq 1.181 can be used to calculate the electronic molar polarizability, $[P_e]$. A measurement of the index of refraction in the infrared region can then be employed to determine $[P_e] + [P_a]$, where $[P_a]$ is the atomic molar polarizability [*196*]. If n_{ir} is the index of refraction in the infrared region, then the atomic molar polarizability may be obtained from [*197*]

$$\begin{aligned}[P_a] &= ([P_e] + [P_a]) - [P_e] \\ &= \frac{n_{ir}^2 - 1}{n_{ir}^2 - 2} \cdot \frac{M}{d} - [P_e] \\ &= \frac{3M}{d}\frac{(n_{ir}^2 - n_e^2)}{(n_{ir}^2 + 2)(n_e^2 + 2)}\end{aligned} \tag{1.282}$$

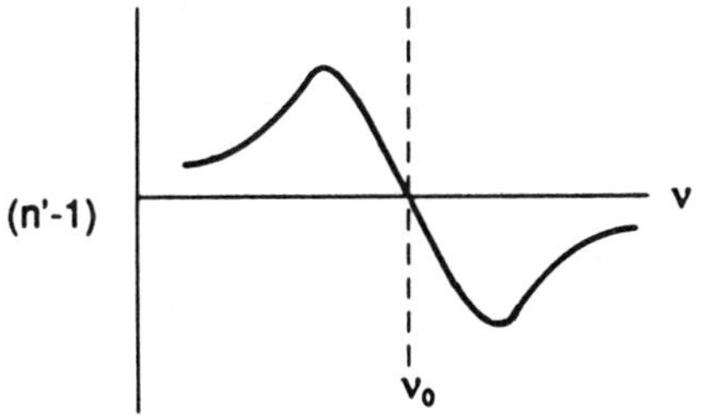

FIG. 1.66—*Idealized curves of* n" *and* (n' − *1*) *versus* ν *for the case of resonance absorption with damping.*

The unknown value of the damping factor, γ, in Eq 1.279, prevents the use of that equation for determining the value of n_{ir}. However, it is often observed in practice that the width of the absorption region is generally extremely narrow, and, as a consequence, one can consider on either side of ν_o the resonance phenomenon as if the damping were nonexistent, that is, $(n_{ir} - 1)$ essentially tends to $+\infty$ just before ν_o is reached and falls to $-\infty$ just on the other side of ν_o. With $\gamma = 0$, $(n_{ir}^2 - 1)$ may be expressed as [*198–201*]

$$(n_{ir}^2 - 1) = \sum \frac{N e^2}{\pi m_e} \cdot \frac{1}{(\nu_e^2 - \nu^2)} + \sum \frac{N e_a^2}{3\pi m_a} \frac{1}{(\nu_a^2 - \nu^2)} \tag{1.283}$$

where the first term on the right is the summation involving all the electron vibrational frequencies while the second term comprises the summation of all the atomic vibrational frequencies. Since n_{ir} normally is measured in the extreme or far infrared region at a frequency ν where the effects of dipole orientation contributions can be neglected to a first approximation, $\nu_e \gg \nu$ and $\nu_a \gg \nu$ so that Eq 1.283 reduces to the simpler form

$$(n_{ir}^2 - 1) = \sum \frac{N e^2}{\pi m_e \nu_e^2} + \sum \frac{N e_a^2}{3\pi \ m_a \nu_a^2} \tag{1.284}$$

which is often termed as the Sellmeier equation. The ratio of e_a^2/m_a will be generally quite different from the values taken for the individual atoms and must be regarded as an effective value [*197*]. The effective charge to atomic mass ratio may be estimated using the expression derived by Dennison [*202*]

$$\frac{e_a^2}{m_a} = \frac{3c}{\pi N} \int \frac{4\pi n' \kappa}{\lambda} d\nu \tag{1.285}$$

where c is the velocity of light, $n'\kappa$ is the absorption index, and λ the radiation wavelength. The integration must be carried out over the entire absorption band region. With the e_a^2/m_a ratio determined and the molecular weight of the liquid known, the value of n_{ir} may be obtained by means of Eq 1.284 and compared to its measured value.

The work of Cartwright and Errera [*197*] represents perhaps one of the more extensive attempts to determine information on the atomic and electronic polarization of a number of liquids in terms of the n_{ir} values measured. They determined the values of P_e and P_a in terms of Eqs 1.281 and 1.282, respectively. The index of refraction, n_e, due to electronic polarization alone was determined from the refraction of visible light by the liquid under test. In the visible light region, the frequencies are too high to induce any significant distortion of the nuclear positions of the molecules and too low to excite the electron eigen frequencies. The value of n_{ir} was obtained from measurement of the index of refraction in the infrared within the wavelength region extending from 52 μm (142 cm^{-1}) to 152 μm (66 cm^{-1}) where the overall index is contingent upon both the electronic and atomic polarization. The chemical technique of *Reststrahlen* was utilized by Cartwright and Errera [*loc. cit.*] to obtain the monochromatic radiation and care was exercised to carry out the measurements at frequencies sufficiently far removed from the resonance absorption frequencies to ensure that the measured index of refraction remained approximately equal to the square root of the static value of the dielectric constant, ϵ_s. Table 1.6 shows the results obtained by the authors on a number of nonpolar liquids. The tabulated value of 0.55 cm^{-3} for $[P_a]$ of benzene shows considerable disagreement with that of 1.5 cited by Petro and Smyth [*203*], which was determined in terms of the dielectric constant at low frequency with the usual extrapolation made to $1/T = 0$ to obtain $[P_e] + [P_a]$. Since the index of refraction can be found to an accuracy equal to at

TABLE 1.6—$[P_e]$ *and* $[P_a]$ *values for a number of nonpolar liquids (after Cartwright and Errera* [197]*).*[a]

Liquid	n_e	$[P_e]$, cm^3	n_{ir}	$[P_a]$, cm^3
Vaseline oil	1.459	0.321 M*	1.47	0.007 M*
Paraffin oil	1.470	0.317 M*	1.47	0.00 M*
Hexane	1.367	29.33	1.37	0.26
Carbon tetrachloride	1.446	25.78	1.47	1.2
Benzene	1.477	25.04	1.49	0.55
Carbon disulfide	1.577	19.83	1.66	2.3
Dioxane	1.412	21.25	1.77	2.6

[a]M* = molecular weight.

least $\pm 10^{-4}$, the optical method should provide a more accurate result than the appreciably less certain extrapolation method.

Table 1.7 provides data on a number of polar liquids, including water. In determining the n_{ir} value for polar liquids, the wavelength of measurement must be selected first such that, as in the case of the nonpolar liquids, the region be free of any spurious resonance absorption. In addition, the wavelength of measurement should be sufficiently large to exceed the wavelengths including all the vibrational frequencies so as to approximate the case of infinite wavelength, but it must be short enough so as to exclude the effect of permanent dipole orientation in order that the molar polarization due to dipole orientation, $[P_d]$ may be neglected. In the near infrared, resonance absorption both for polar and nonpolar molecules is usually due to vibrations within small groups of atoms and depends very little upon the structure of the entire molecule. For both nitrobenzene and pyradine, the molecular bending vibrations would tend to shift the eigenfrequencies into the far infrared regime. In general, the absorption bands for polar molecules ar stronger due possibly to the presence of significant bending vibrations in addition to the stretching vibrations within the small groups of atoms. Nevertheless, although the absolute magnitudes of $[P_a]$ for the polar molecules in Table 1.7 appear to be larger than those for the nonpolar molecules in Table 1.6, the ratio of $[P_s]/[P_e]$ for both types of molecules is decidedly very small. Evidently, irrespective of whether one deals with polar or nonpolar molecules the value of $[P_a]$ is always a fraction of $[P_e]$. However, in regards to water this deduction is incorrect because the value of 3.2 for $[P_a]$ as

TABLE 1.7—$[P_e]$ *and* $[P_a]$ *values for a number of polar liquids (after Cartwright and Errera* [197]*).*

Liquid	n_e	$[P_e]$, cm^3	n_{ir}	$[P_a]$, cm^3
Methyl alcohol	1.323	8.12	1.41	1.9
Ethyl alcohol	1.353	12.64	1.47	3.8
Amyl alcohol	1.397	29.20	1.47	4.6
Octyl alcohol	1.415	39.71	1.47	4.5
Glycerine	1.457	19.81	1.81	11.7
Diethyl ether	1.344	22.10	1.40	3.2
Dipropyl ether	1.370	30.80	1.44	5.0
Nitrobenzene	1.515	29.24	1.56	3.8
Pyridine	1.483	23.03	1.52	1.5
α-picoline	1.476	27.91	1.51	1.7
β-picoline	1.478	28.05	1.50	1.0
Toluene	1.473	29.69	1.49	0.9
Amyl acetate	1.391	35.50	1.43	3.0
Dimethyl sulfate	1.378	21.89	1.51	6.4
Acetone	1.349	15.80	1.37	0.9
Water	1.322	3.60	1.68	3.2

determined by Cartwright and Errera is in error. More accurate measurements carried out by Chamberlain et al. [*204,205*], using a Fourier interferometric type spectrometer that was capable of measurements down to 20 cm^{-1}, established the value of $[P_a]$ for water at 12.0 cm^3. Hence in the unique case for water, $[P_a]$ exceeds $[P_e]$ (equal to 3.60). This discrepancy in Cartwright's and Errera's results for water may be due to the absence of the longer measurement wavelengths in their *Reststrahlen* arrangements, which failed to take into account significant molecular vibration effects at the longer wavelengths.

1.8 Measurement Techniques

Dielectric materials may be viewed as lumped electrical circuit elements as long as the wavelength of the applied electromagnetic field is large compared with the physical dimensions of the dielectric specimen under test. When the wavelength of the electromagnetic field becomes smaller than the dimensions of the liquid specimen, then the specimen commences to act as a distributed parameter system and the spatial variation of the electromagnetic field over the specimen must be taken into account. The transition from lumped to distributed parameter behavior occurs generally in the region from 300 to 600 MHz. Hence from this frequency point onwards into the microwave and optical frequency regime, the dielectric measurement methods involve necessarily the relationship between the complex values of the propagation factor, γ^*, the permittivity, ϵ^*, and the index of refraction, n^*.

The difference in approach between lumped and distributed parameter measurement techniques can be perhaps best delineated by a comparison between resonant cavity and lumped resonant circuit test procedures. If the resonant frequency of a resonant cavity *in vacuo* is f_o, then upon filling the cavity with a dielectric liquid the resonant frequency will be found to shift to a value f with the real dielectric constant of the liquid given by

$$\epsilon' = \left(\frac{f_o}{f}\right)^2 \tag{1.286}$$

Because the quality factor, Q, of a cavity resonator containing a dielectric material is defined as [*206*]

$$Q = \omega\left[\frac{\text{energy stored in the resonator dielectric}}{\text{power loss in the resonator walls and dielectric}}\right] \tag{1.287}$$

then the dissipation factor, $\tan\delta$, of the liquid filled cavity is equal to $1/Q$. Subtracting from the latter the value $1/Q'$, where Q' designates the Q value of the resonant cavity filled with a loss free dielectric fluid of the same capacitance as the physically real dielectric fluid of finite loss so that the resonant frequency of the cavity remains unaltered, yields the $\tan\delta$ value of the dielectric fluid

$$\tan\delta = \frac{1}{Q} - \frac{1}{Q'} \tag{1.288}$$

The theoretical value of Q' of the hypothetical loss free fluid is given by [*206*]

$$Q' = \frac{al}{x(a + l)} \tag{1.289}$$

where l is the axial length of the resonator of radius a; x is the depth of current penetration in the walls at the resonant frequency and is given by $1/\sqrt{\pi\,\mu_m\,f\,\sigma_m}$ where μ_m is the permeability of metallic walls of conductivity σ_m (*S/m*) measured at the frequency f.

Now considering in analogy the case of the lumped-parameter dielectric fluid of capacitance C resonating in a parallel arrangement with a coil of inductance L, and series resistance R_L, we can say that the quality factor, Q, of the resonant circuit will be given by

$$Q_{lp} = \frac{\omega_o L}{(R_L + R_C)} = \frac{1}{\omega_o C(R_L + R_C)} \tag{1.290}$$

where the finite loss component in the dielectric field of capacitance C is taken into account by placing a small resistance R_C in series with C. If the dielectric fluid between the two measuring electrodes is removed and replaced by either vacuum or air, a perfect dielectric results and R_C is essentially reduced to zero. To reestablished resonance at the frequency ω_o, the two electrodes are moved closer until the capacitance between the electrodes becomes again equal to C at the resonant frequency. The value of Q is then given by

$$Q'_{lp} = \frac{1}{\omega_o R_L C} \tag{1.291}$$

Since for a lumped parameter system

$$Q = 2\pi \left[\frac{\text{energy stored in the dielectric per cycle}}{\text{energy dissipated in the dielectric per cycle}} \right] \tag{1.292}$$

and the energy stored in the capacitance C of the fluid is

$$W_s = \frac{1}{2} C V_m^2 \tag{1.293}$$

where V_m is the peak value of the alternating voltage; or alternatively, since $V_m = I_m/\omega C$, then it follows that

$$W_s = I_m^2/2\omega^2 C \tag{1.294}$$

where I_m is the peak value of the sinusoidal current. Because the power dissipated in a series RC circuit is I^2R or $(I_m/\sqrt{2})^2R$, this leads to an energy loss per cycle of

$$W_e = I_m^2 R/2f \tag{1.295}$$

Thus, it can be seen that for a series RC circuit

$$Q = 2\pi \left[\frac{W_s}{W_e} \right] = \frac{1}{\omega RC} \tag{1.296}$$

Since for an RC series circuit $\tan\delta = \omega RC$, then in terms of Eq 1.297, $\tan\delta = 1/Q$. Accordingly, using Eqs 1.291 and 1.292, the $\tan\delta$ value of the dielectric fluid in the lumped parameter case may be expressed as stated by Horner et al. [*206*]

$$\tan\delta = \frac{1}{Q_{lp}} - \frac{1}{Q'_{lp}} \tag{1.297}$$

Note that in the lumped parameter case, one need not invoke a quasi-fictitious value of Q'_{lp} as has been done with Q' in the distributed parameter case. In the lumped parameter case, the real value of the dielectric constant is determined simply from the ratio of the specimen capacitance to that of the parallel plate electrode geometry *in vacuo*. The specimen capacitance is obtained from the air capacitance value obtained as the electrodes are moved closer together until resonance is reestablished after removal of the liquid specimen.

From the foregoing considerations, it can be appreciated that notwithstanding some similarities between the low and high frequency testing approaches, there are substantial differences in both the apparatus employed as well as the experimental procedures. One of the major differences lies in the nature of the specimen holders employed. At high frequencies the specimen cell constitutes an integral part of say either the resonant cavity or the waveguide, whereas at low frequencies it is a containment vessel with suitable electrodes and connectors for insertion into the measuring circuit. In Vol. II-B of this monograph series, various test procedures already were described for application to solid dielectric specimens [*207*]. Since many of these are directly applicable to liquids, they shall not be repeated here; however, some aspects of these measurement techniques may be briefly recapitulated to emphasize certain general principles of measurement. The prime purpose of this section will be to describe fully the measurement cells used with liquids specimens and to present a complete description of test methods and instruments that have been either omitted inadvertently from Vol. II-B or have become available only recently.

1.8.1 Lumped-Parameter-Specimen Test Cells

The history of development of test cells for use with liquid specimens differs considerably from that concerning solid type dielectric specimens. Dielectric liquids have been used primarily in the insulation of high-voltage power apparatus, such as capacitors, transformers, and cables. Thus, their electrical properties had to be well defined at the power frequencies (50 or 60 Hz) within the operating temperature range, usually extending from 20 to 100°C inclusively. Accordingly, a great deal of effort has been expended to develop a suitable test cell within which the dielectric fluid could be evaluated at a single frequency up to a temperature of at least 100°C subjected to high voltage stress conditions. It can be appreciated that such a cell design had to meet a number of rather stringent requirements. Since the cell was to be employed for measurements at 50 or 60 Hz, an adequately shielded enclosure was necessary to screen out the extraneous 60 Hz interference from the surrounding power apparatus. This led to an all-enclosed concentric electrode metal enclosure design having an outside HV electrode. Furthermore, the relatively low value of the standard capacitor (*ca.* 100 pF) in the Schering bridge dictated a minimal specimen capacitance of approximately 40 to 50 pF in order to maintain an acceptable value of sensitivity for the capacitance and dissipation factor measurements. This requirement was met by using a concentric inner voltage electrode of a relatively large surface area fixed at an acceptably small distance from the concentric outer electrode. The small separation between the electrodes also served the purpose of reducing the volume of the specimen required for each measurement—an important consideration when a large number of measurements are to be performed in order to obtain a statistically correct average value of ϵ' and $\tan\delta$.

The development of the concentric electrode 50/60 Hz oil test cell began in the early 1930s and culminated in two rather similar designs, referred to in the early days as the American (Berberich) and the British (NPL) cells. The former was primarily the result of the work carried out by Berberich [*208*] and Balsbaugh et al. [*209–211*] and the latter was principally a design by Hartshorn and Rushton carried out at the National Physical Laboratory (NPL) and subsequently modified by Webb [*212*]. Both the Berberich and NPL cells were designed for easy mounting and disassembling so as to make all parts of the cell accessible for cleaning purposes; also both cells were designed to attain temperature equilibrium in a sufficiently short time as compared to the time that it would take for a given mineral oil to exhibit the commonly observed decrease of $\tan\delta$ as a function of time [*213*].

The antecedent designs of the Berberich and NPL cells have evolved into a cell construction that is now accepted as the norm by a number of standard organizations (see ASTM D 924, VDE 0303, and IEC 247). This cell construction is presently available from a number of commercial suppliers; Fig. 1.67 shows a three terminal guarded commercial cell conforming to the cited standards. The liquid cell, available from the Multi-Amp Company, has an overall length of 19.5 cm and a diameter of 6.0 cm. It has a standard interelectrode distance of 2.0 mm, with a nominal air capacitance of 60 pF. The required liquid test volume is 30 mL; the cell may be operated in the temperature range from 20 to 150°C and will withstand an alternating voltage of 1 kV while empty and approximately 2.5 kV when filled with oil. The electrode material consists of highly polished stainless steel and borosilicate glass is used for insulation support between the electrodes. The type of cell construction delineated in Fig. 1.67 is referred to commonly as the IEC type cell. It should

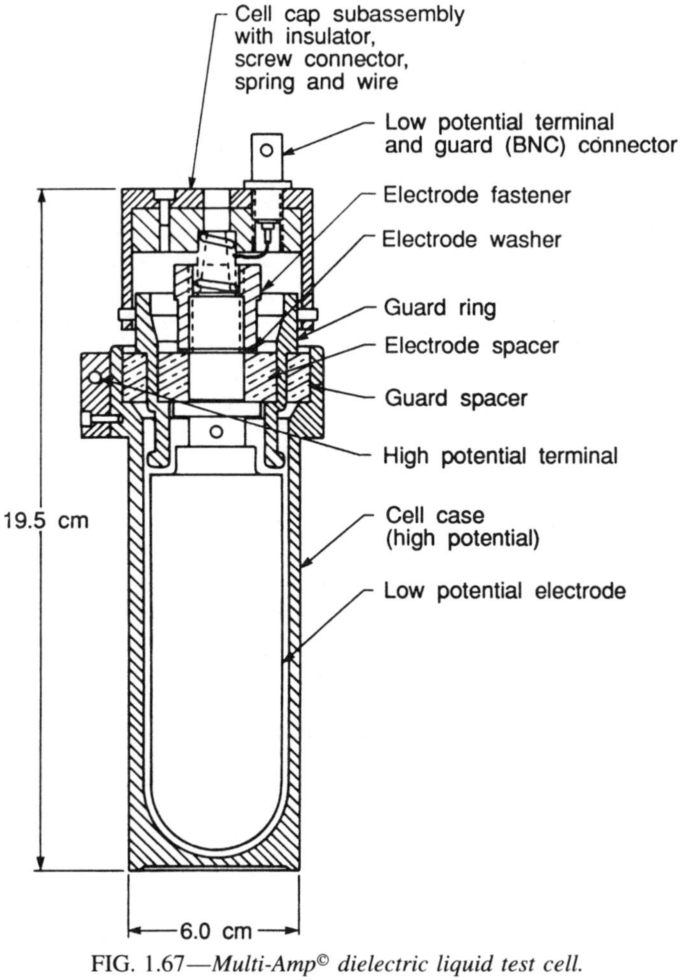

FIG. 1.67—*Multi-Amp© dielectric liquid test cell.*

be mentioned here that the concentric electrode cell designs may be altered to accommodate smaller test specimen volumes.[3]

In North America, a test cell which has been used extensively over the past few decades, and whose usage is still prevalent in some laboratories, is the Balsbaugh cell which is depicted in Fig. 1.68. The Balsbaugh type 3HV35 cell depicted in Fig. 1.68 is a metal mounted precision cell of air capacitance 35 pF suitable for measurements up to 5 kV with the liquid specimen *in situ.* To minimize the dielectric losses in its support element, it uses Teflon insulation bushings which, however, limit the temperature range to 150°C. Temperatures up to 500°C have been achieved using similar cell designs, but with ceramic (boron nitride) bushings replacing the Teflon bushings. In the Balsbaugh cell, the concentric electrodes consist of highly polished monel (70% nickel, 30% copper). Apart from the difference in the interelectrode spacing, the Balsbaugh cell differs from the IEC cell in that its inner electrode is hollow which accounts mainly for the larger specimen volume of 40 mL as compared to 30 mL for the IEC cell. Notwithstanding the differences in their construction, both cells when used with comparable measuring instruments yield essentially the same accuracy and precision in measurement. With both cells considerable care must be exercised when filling with the specimen fluid to ensure that no air bubbles become trapped between the closely spaced electrodes which have a considerable surface area. This is accomplished most effectively by filling the cells with the test fluid at some suitably higher temperature, such as for example at 50°C for fluids having a viscosity range of 10 to 500 cP at room temperature.

There have been a number of earlier two terminal oil test cells adopted for routine use that may be regarded as variations of the NPL and Balsbaugh's three terminal type cell designs. A modified two terminal test cell of the NPL construction is portrayed in Fig. 1.69 [*214*]. The two terminal cell primarily intended for measurements at 50 or 60 Hz, uses quartz for the isolation of the measuring electrodes that consist of rhodium plated invar. The electrode separation is fixed at 1 mm, and the total required liquid specimen volume is 16 mL. On the assumption that the quartz insulation may be represented by a constant resistance, R', Martin and Patterson [*214*] obtain a corrected value of the measured $\tan\delta$ value with the cell. If the empty cell is represented by an equivalent circuit of R' in parallel with C' and C_o, where C_o is the replaceable air capacitance of the cell and C' is the sum of the irreplaceable air capacitances of the cell (for example, quartz insulator capacitance, including the stray capacitances), then the dissipation factor of the empty cell may be expressed as

$$\tan\delta_e = \frac{1}{\omega(C' + C_o)R'} \tag{1.298}$$

Now, when the cell is filled with a liquid specimen having a real value of permittivity equal to ϵ' and an equivalent insulation resistance of R, then the equivalent circuit elements R, $\epsilon' C_o$, R', and C' appear all in parallel, leading to an overall dissipation factor value of

$$\tan\delta_f = \frac{1}{\omega(C' + \epsilon' C_o)}\left[\frac{1}{R'} + \frac{1}{R}\right] \tag{1.299}$$

Since the $\tan\delta$ value of the fluid specimen in a parallel equivalent circuit representation is by definition,

$$\tan\delta = \frac{1}{\omega\epsilon' C_o R} \tag{1.300}$$

[3]A description of a cell design, requiring a diminished liquid test volume of 24 mL, has been given by R. G. Heydon, *Transactions on Electrical Insulation,* Institute of Electrical and Electronic Engineers, Vol. 24, 1989, pp. 649–655.

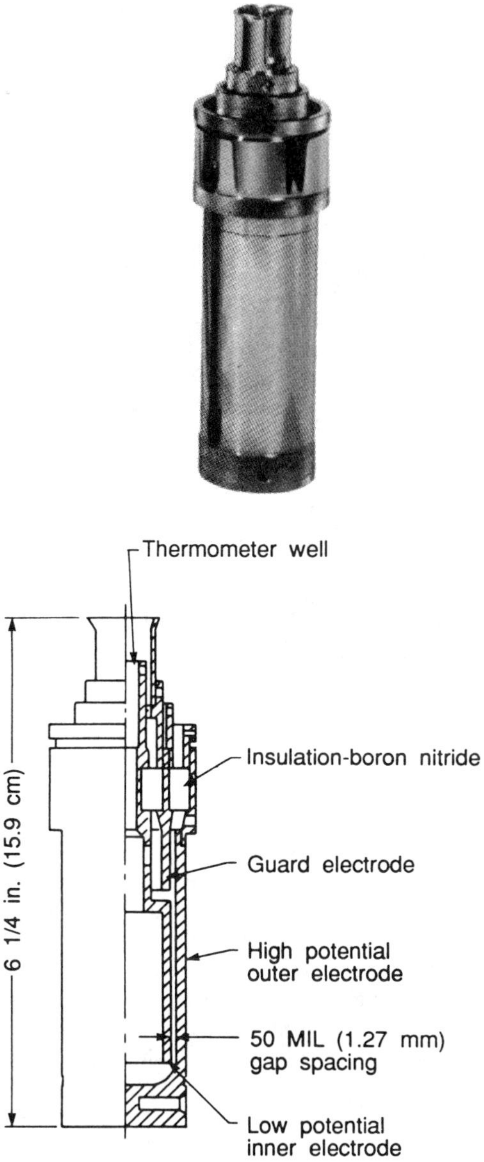

FIG. 1.68—*Balsbaugh type 3HV35 three terminal concentric electrode cell:* (a) *pictorial-view;* (b) *plan-view (after ASTM D 924).*

then in terms of the measured values of $\tan\delta_e$ (with empty cell) and $\tan\delta_f$ (with filled cell), the dissipation factor of the specimen liquid is given by [*214*]

$$\tan\delta = \tan\delta_f\left(1 + \frac{C'}{\epsilon' C_o}\right) - \tan\delta_e\left(\frac{C' + C_o}{\epsilon' C_o}\right) \qquad (1.301)$$

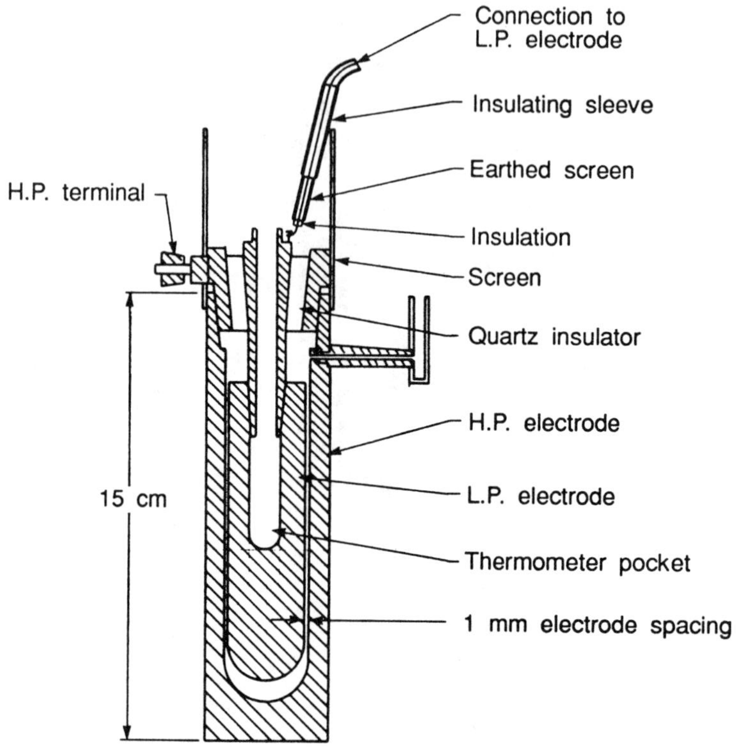

FIG. 1.69—*Modified two terminal N.P.L. cell (after Martin and Patterson* [214]*).*

Another two-terminal concentric-electrode cell that has been used extensively in the past is a Balsbaugh glass support type cell. One of the most common type cells normally used with mineral oil specimens, the Balsbaugh type 2TN50 measuring cell is depicted in Fig. 1.70. The unit consists of two parts; an outside member consisting of a pyrex-glass tube containing the liquid specimen and an inner member upon which is mounted the concentric electrode assembly. The inner member fits into the outer member via a glass-ground joint, thereby sealing the specimen from the outer atmosphere. The electrodes consist of concentric nickel cylinders, spot-welded to nickel-tungsten prongs, which are sealed in a solid pyrex-glass press to which are jointed two external terminals. The spacing between the concentric-cylinder electrodes is 50 mil (1.27 mm); an overall specimen volume of 40 mL is required. This type of cell has been used in the past for many research applications and can be operated to temperatures as low as −70°C as well as up to about 150°C. The temperature range is limited by the thermal expansion coefficients of the constituent materials and at the higher temperatures by the oxidation or corrosion conditions affecting the nickel electrodes. However, some Balsbaugh glass cell designs used platinum electrodes in order to circumvent the oxidation and corrosion induced effects at the electrodes. The measuring frequency of the glass cells extended from the power frequency to well into the audio frequency range, and, with a suitable substitution method, measurements could be extended even into the lower radio frequency range without serious loss in accuracy.

For research purposes where there is a considerable need for measurement over a very wide frequency range, the concentric electrode measuring assembly now rarely is used and preference is given to parallel-plane electrode systems. For precision measurements in the frequency range from 10^{-5} to 10^6 Hz, the three-terminal Balsbaugh model LD-3 cell has proven to be well suited. The cell, consisting of electrodes constructed of stainless steel contained within a rectangular housing of plated brass utilizing Teflon as an insulation support for the electrodes, is shown in Figs. 1.71*a* and *b*. The interelectrode spacing is continuously variable by means of a micrometer screw in the range from 0 to 300 mil ± 0.25 mil (0 to 0.75 mm ± 0.006 mm), thus allowing various liquid film thickness to be evaluated for their dielectric loss behavior. The maximum temperature of operation is limited to approximately 200°C by the type of Teflon insulation supports used.

Three terminal electrodes, that are employed normally to minimize the electrical field fringing effects, cannot be used for frequencies in excess of 1 MHz due to uncontrollable inductance effects. At the higher frequencies one is obliged to revert back to the two electrode system. The latter is essentially of the Hartshorn and Ward [*215*] design with the electrodes appropriately curved to form a containment vessel for the liquid specimen undergoing test. That is, the high potential electrode has a convex surface, while low potential electrode situated at the bottom of the test cell assembly is of a concave form for the containment of the liquid specimen. Evidently the surface contour of the convex underside of the electrode must match exactly the contour of the surface of the lower concave electrode. The liquid specimen is placed in the incurvation of the lower low potential electrode and the upper electrode is then lowered causing a radial displacement of the meniscus until the desired liquid thickness is obtained. The resolution of the main micrometer for specimen thickness adjustment is generally of the order of 10^{-3} to 10^{-4} mm. The flexible metallic bellows, as may be seen from Fig. 1.72, connect the upper electrode to the high potential terminal and are designed to provide a path of low impedance. Although the early design of Hartshorn and Ward used copper for the completely enclosed and shielded construction of their cell as well as the electrodes themselves, present designs tend to use brass and, in particular, gold-plated brass as has

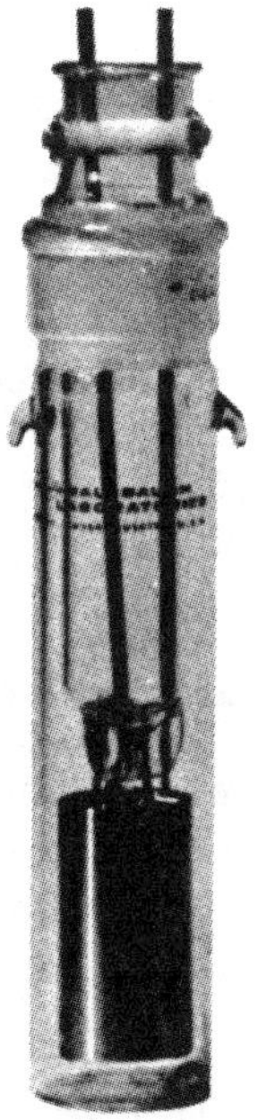

FIG. 1.70—*Balsbaugh type 2TN50 two terminal concentric electrode cell.*

FIG. 1.71*a*—*Pictorial view of the type LD-3 three terminal parallel-plane electrode Balsbaugh cell.*

been done by Reddish et al. [*216*]. The latter authors also improved upon the original usage of silica insulators by covering these with a coating of polytetrafluorethylene in order to reduce the surface conductivity of silica, which varies very appreciably with changes in the ambient humidity.

Since the Hartshorn and Ward cell is used in the susceptance variation method, it must necessarily possess the capability of a fine tuning adjustment. Fine tuning adjustment near the resonance point is made by means of an incremental capacitor, whose adjustment is controlled by means of the micrometer head, M_2 (Fig. 1.72). This micrometer is also used to check the square-law characteristics of the voltmeter. The incremental capacitor constitutes essentially a coaxial capacitor that is placed in parallel with the main electrode capacitance. Although in the original design the inner part rotated as it advanced into the cylindrical hole, this is not the case with the recent design by Buckingham and Billing [*214*] who employ a nonrotating plunger. They were able to ensure a high stability in its capacitance by maintaining a concentricity between the cylindrical bore and the advancing inner part axes to within 0.05 mm, providing a constant per unit length capacitance value of 0.3 pF/mm.

The unique features of the Hartshorn and Ward cell are that the inductances and resistances of the overall measuring assembly are maintained relatively constant, irrespective of whether the specimen is in or out of the circuit. The earlier designs of the cell required manipulation of the

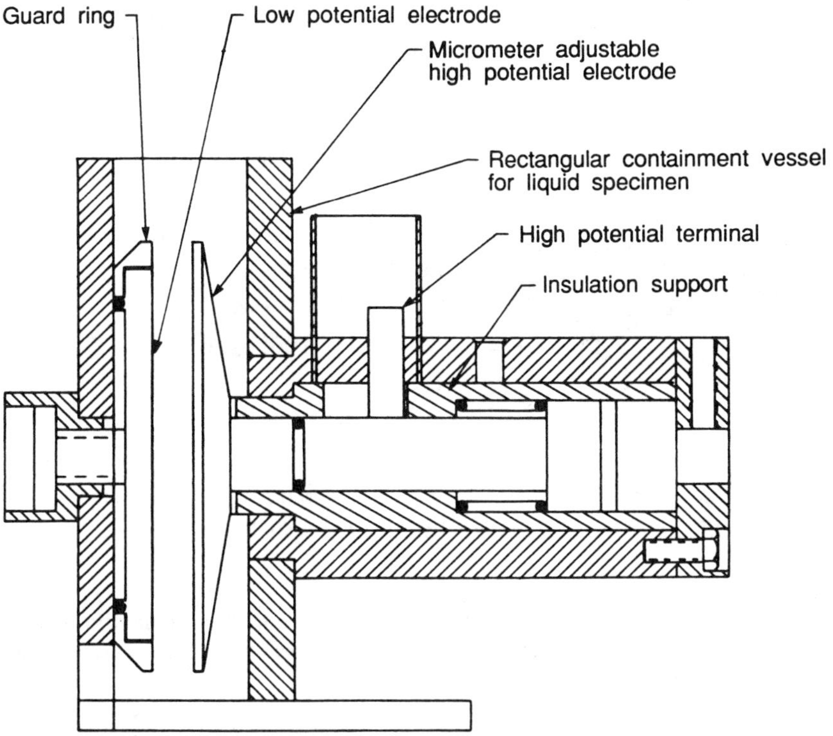

FIG. 1.71*b*—*Profile plan view of the type LD-3 three terminal parallel-plane electrode Balsbaugh cell.*

incremental capacitor micrometer via a long broomstick to minimize operator proximity effects, but the greater shielding effectiveness of the present cells no longer poses this requirement. Some modified Hartshorn and Ward cells eliminate the incremental capacitor micrometer entirely and achieve a fine tuning adjustment by means of the main specimen retaining capacitor as was first done by Works et al. [*218*] in their reentrant cavity technique, which is analogous to the susceptance variation technique of Hartshorn and Ward.

A recent design of a Hartshorn and Ward type specimen holder disposing of the use of the incremental capacitor micrometer is that reported by Kakimoto et al. [*219*], which makes use of an electrode shield and a high potential knife-edge electrode in order to reduce the edge capacitance of the electrodes.

1.8.2 Low Frequency Measurements

Very low frequency measurements within the range from 10^{-6} to 10^{-1} Hz are used extensively in research related studies to derive information on space charge behavior as well as provide a means to identify the nature of the conduction ions involved, as for example their type and mobility. At frequencies less than 0.1 Hz, steady state measurements by means of bridge type circuits become impractical because of the unduly long balancing times required. At this point measurements in the

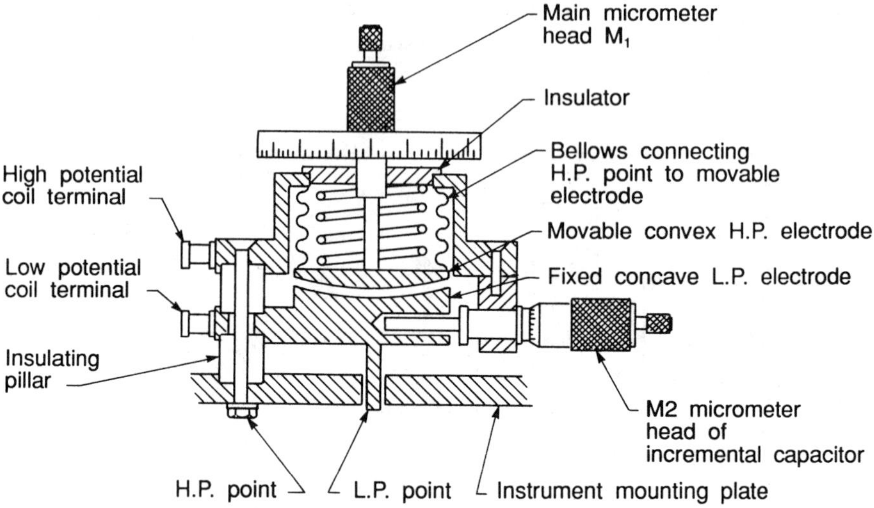

FIG. 1.72—*Hartshorn and Ward type cell for liquid specimens.*

frequency domain must be replaced by suitable measurements in the time domain from which the real and imaginary values of the permittivity, ϵ' and ϵ'', may be obtained as a function of frequency by applying Fourier transformation techniques.

In 1946, Whitehead [*220*] described a transient response method that could be used to derive information on the steady state response of dielectrics without the necessity of point by point measurement at the various frequencies of interest. This was soon followed by a contribution from Davidson et al. [*221*], who provided a detailed account of the apparatus and measurement procedure of the transient method. Their early apparatus is depicted in Fig. 1.73. The linear sweep generator provided a linearly rising ramp voltage of the form $V(t) = V_o(t/t')$, where t' represents the duration of the charging sweep voltage. The transient current response due to the ramp voltage is of the form [*221*]

$$i(t) = \frac{V_o}{t'}[\epsilon_\infty C_o + (\epsilon_s C_o - \epsilon_\infty C_o)\{1 - \exp(-t/\tau)\}] \tag{1.302}$$

where the time constant τ is given by

$$\tau = R(\epsilon_s C_o - \epsilon_\infty C_o) \tag{1.303}$$

Here R represents the equivalent resistance of the dielectric and C_o the capacitance of the dielectric *in vacuo.* The assumed equivalent circuit is delineated in Fig. 1.73 with the corresponding transient current response shown in Fig. 1.74. Note the analogy between the form of the transient current response and the polarization of an idealized dielectric given in Fig. 1.19; likewise, the similarity in form should be observed between Eqs 1.123 and 1.302. From Fig. 1.74, it is apparent that the values of the time constant or relaxation time, τ, as well as ϵ_s and ϵ_∞ may be obtained provided the capacitance *in vacuo,* C_o, is known. The real and imaginary permittivity values, ϵ' and ϵ'', may then be obtained in turn at different frequencies. In the foregoing method the duration time of the

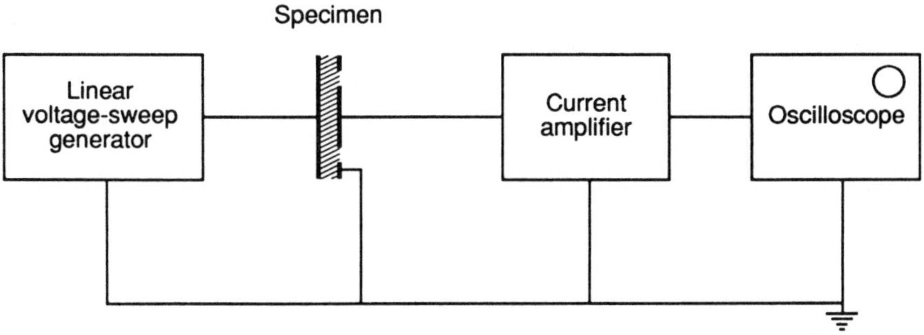

FIG. 1.73—*Schematic diagram of the early transient test circuit arrangement (after Davidson et al.* [221]*).*

charging sweep voltage must be much longer than the time constant $R\ C_o(\epsilon_s - \epsilon_\infty)$, if the whole of the trace of Fig. 1.74*a* is to be obtained. Similarly, the rate of the voltage rise V_o/t' must be sufficiently great to produce a current of adequate amplitude as concerns the sensitivity of the recording instruments.

Whereas the antecedent transient method was capable of descending down to 10^{-3} Hz, recent step and square pulse response methods go well beyond this frequency value down to 10^{-6} Hz [*222–224*]. The square pulse technique doubles the speed of measurement and is particularly useful for this reason when measurements are to encompass the 10^{-5} to 10^{-6} Hz range. Computer based numerical techniques are employed extensively to ease the tedium of calculation. For complete description of these methods the reader is referred to Vol. II-B of this series, since it is not intended to repeat their description here.

The advantage of numerical Fourier transformation procedures when applied to step or square pulse response measurements is that not only can they be carried out using computer-based calculation procedures, but, in addition, they do not invoke any *a priori* assumptions involving the form of the current response function. The only restriction on their validity, as is incidentally the case also for the current transient method itself, is that the superposition principle holds and the dielectric behaves linearly. The lowest practical frequency limit is imposed by the elapsed time of the experiment, that is, the longest time at which the smallest current value that the electrometer can

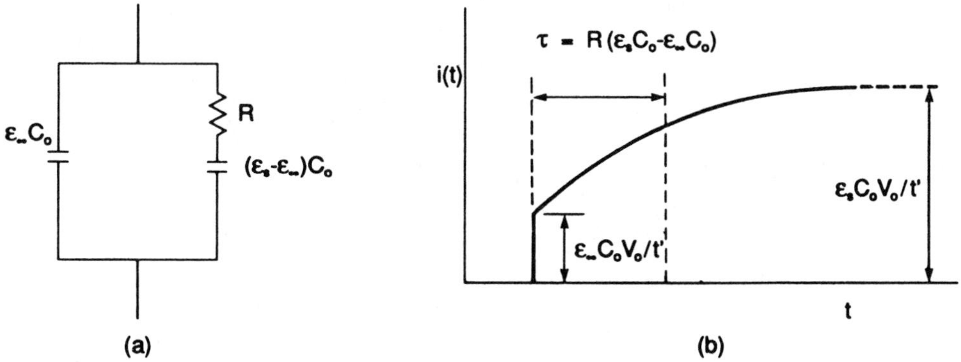

FIG. 1.74—*Equivalent circuit of dielectric* (a): *and the corresponding transient current response due to an applied ramp voltage* (b): *(after Davidson et al.* [221]*).*

record within reasonable accuracy under given ambient noise conditions (for example, 10^{-16} to 10^{-17} A). Whereas the highest frequency limit (the *Nyquist frequency*) is set by the time interval between two successive discharge or charge current values, which is a limit that essentially is determined by the rise time of the electrometer used (~1.0 s). It can be appreciated that the upper frequency limit is not a serious practical constraint, since there are a number of low frequency bridges that are capable of covering this frequency range (0.1 to 1.0 Hz) quite adequately.

The early bridge circuits designed to attempt low frequency measurements down to 0.1 Hz were of the capacitance-resistance type, using the Wagner earth connection [*226,227*]. The use of the Wagner earth connection proved to be particularly tedious for measurements at the lower frequency limits of the bridges. However, by utilizing the concept of a grounded detector in conjunction with a Schering bridge, Scheiber [*228*] was able to circumvent this tedium. This type of the Schering bridge arrangement, which was capable of measurements in the frequency range from 10^{-2} to 10^{2} Hz became known as the Scheiber bridge; its full description may be found in Vol. II-B of this book series.

Perhaps the most convenient low frequency bridge that may be used within the frequency regime encompassing 10^{-2} to 10^{4} Hz is the so called Thompson-Harris bridge [*229,230*]. Various arrangements of this bridge have been dealt with extensively in Vol. II-B and, consequently, will not be repeated here. Essentially the bridge consists of a highly regulated oscillator with quadrature voltage output in conjunction with an operational inverter amplifier and high precision variable air capacitors to balance the loss and capacitive current components of the dielectric specimen. Stability is achieved by making use of operational amplifiers with high feedback. The bridge is capable of an accuracy of 0.1%, with a resolution of better than 0.1%. The relatively high accuracy is attained because the balancing is carried using only precision air capacitance components.

Another bridge, which in principle of operation resembles somewhat the Thompson-Harris bridge in that a center tapped generator is used to provide in phase and 180° out of phase source voltages, has been proposed by Vince [*231*] in 1965. Since the bridge uses resistances as balancing elements, the accuracy of the bridge is of the order of ±1.0% when specimens having a capacitance of 100 *pF* and an equivalent resistance of $10^{12}\Omega$ are measured. Since the Vince low frequency bridge was omitted inadvertently from Vol. II-B, its description will be given here. The basic circuit

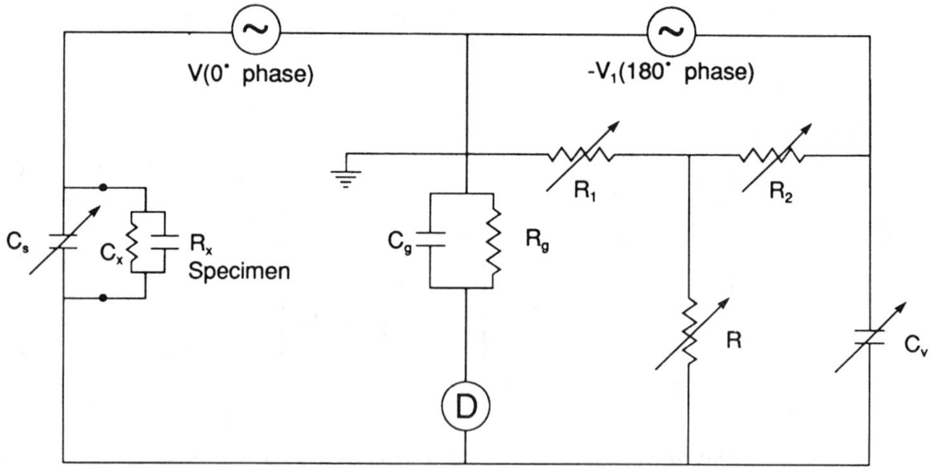

FIG. 1.75—*The Vince low frequency bridge* [231].

diagram of the bridge is delineated in Fig. 1.75. Here C_s and C_v represent variable capacitors, R_1, R_2, and R are variable resistors and R_g and C_g are the equivalent resistance and capacitance of the detector. The balance equations of the bridge are obtained by transforming the Y connections of R_1, R_2, and R into a delta arrangement as depicted in Fig. 1.76, where R_o, R_v, and R_d is the Δ equivalent of R_1, R_2, and R in the Y connection of Fig. 1.75. The balance condition is given by [*231*]

$$V[(1/R_x) + j\omega(C_x + C_s)] = V_1[(1/R_v) + j\omega C_v] \quad (1.304)$$

Equating the real terms yields the equivalent parallel resistance of the specimen

$$R_x = R_v (V/V_1) \quad (1.305)$$

but in terms of Fig. 1.75

$$R_v = R_2 + R\left(\frac{R_2}{R_1} + 1\right) \quad (1.306)$$

therefore

$$R_x = \left[R_2 + R\left(\frac{R_2}{R_1} + 1\right)\right]\frac{V}{V_1} \quad (1.307)$$

Equating the imaginary terms in Eq 1.304 yields

$$C_s + C_x = C_g(V_1/V) \quad (1.308)$$

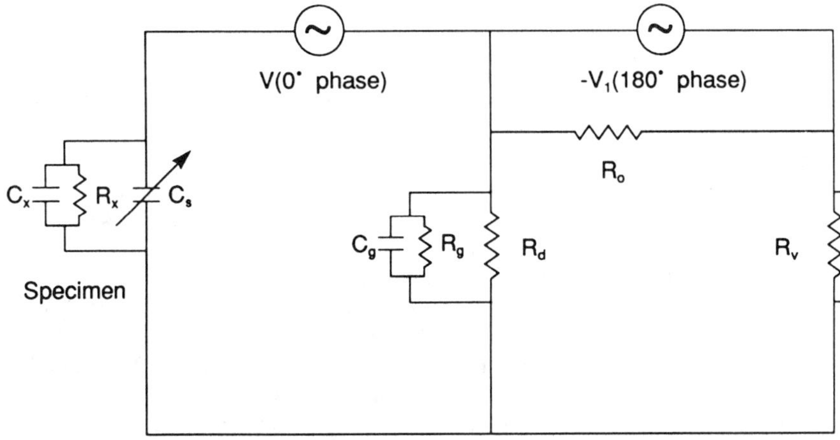

FIG. 1.76—∇-*transformed circuit of the Vince low frequency bridge* [231].

from which the capacitance of the specimen is given by

$$C_x = C_g(V_1/V) - C_s \tag{1.309}$$

and in terms of the parallel equivalent circuit the tanδ value of the specimen is found as

$$\tan\delta = \frac{1}{\omega R_x C_x} = \frac{1}{\omega\left[R_2 + R\left(\frac{R_2}{R_1} + 1\right)\right]\frac{V}{V_1}[C_g(V_1/V) - C_s]} \tag{1.310}$$

It is pointed out by Vince [*231*] that the accuracy may be improved by employing a substitution method of measurement. By setting the ratio of V/V_1 equal to unity, one can write that

$$C_x = C''_v - C'_v \tag{1.311}$$

and

$$\frac{1}{R_x} = \frac{1}{R''_v} - \frac{1}{R'_v} \tag{1.312}$$

where the double prime indicates the initial balance obtained with the specimen disconnected and the single prime applies to the balance situation with the specimen connected in the circuit. The frequency range of the Vince bridge may be extended into the audio frequency range by a number of suitable modifications, that is, reintroduction of the Wagner earth, use of a single sinusoidal oscillator, and the substitution of the out-of-phase oscillators by two equal resistances.

Recently, frequency response analyzer techniques have been increasing in popularity mainly because of the ease with which an extensively broad frequency range (10^{-4} to 10^4 Hz) may be covered. As reported by Pugh [*232*] in 1984, the measurement tedium is eliminated by the use of a computerized measurement procedure that involves correlation techniques. Pugh's approach is described in Vol. II-B and, therefore, will not be repeated here.

1.8.3 Measurements at Intermediate Frequencies

Here we shall consider the frequency régime within 50 Hz to 1 MHz, which constitutes the range most amenable to *RC*-type bridge measuring techniques. The field is dominated by the Schering and transformer ratio arm bridges that are readily capable of measurement to an accuracy of 0.1%. Voltage comparitors or admittance meters are also used over this frequency range, though their resolution and precision is considerably poorer. Once more the reader is referred to Vol. II-B for further details on both high and low voltage measurements over this frequency range.

A considerable amount of effort has been expended into automating dielectric measurements, particularly at the power frequency of 50 or 60 Hz by means of either the Schering or the transformer ratio arm bridge. Geyger [*233*] was the first to disclose in 1936 an automated capacitance and dissipation factor bridge, which was soon followed by Poleck's [*234*] description of partially automated Schering bridge in 1939. Finally in 1964, Petersons [*235*] reported on an automated transformer ratio arm bridge. The next logical step was to design a microcomputer-controlled automatic capacitance and dissipation factor bridge; this task was accomplished by Seitz and Osvath [*236*] and Osvath and Widmer [*237*] in 1984. The schematic circuit diagram of the computer controlled bridge is portrayed in Fig. 1.77. The coarse balance and fine balances of the bridge are effected by manipulating the number of turns in the current comparator windings, N_1, N_2 and N_3, N_4, respectively. The currents in the fine balancing coils N_3 and N_4 are proportional to the current of the standard capacitor, C_N, and are further controlled by the multiplying *D/A* converters α and β. Here N_1 constitutes the winding of the null detector; at balance the condition $I_i = 0$ must hold and at an applied voltage, V, the ampere turns equation may be expressed as [*237*]

$$V(G_x + j\omega C_x)N_1 = Vj\omega C_N(N_2 + \alpha RG_1N_4 - j\beta RG_2N_3) \tag{1.313}$$

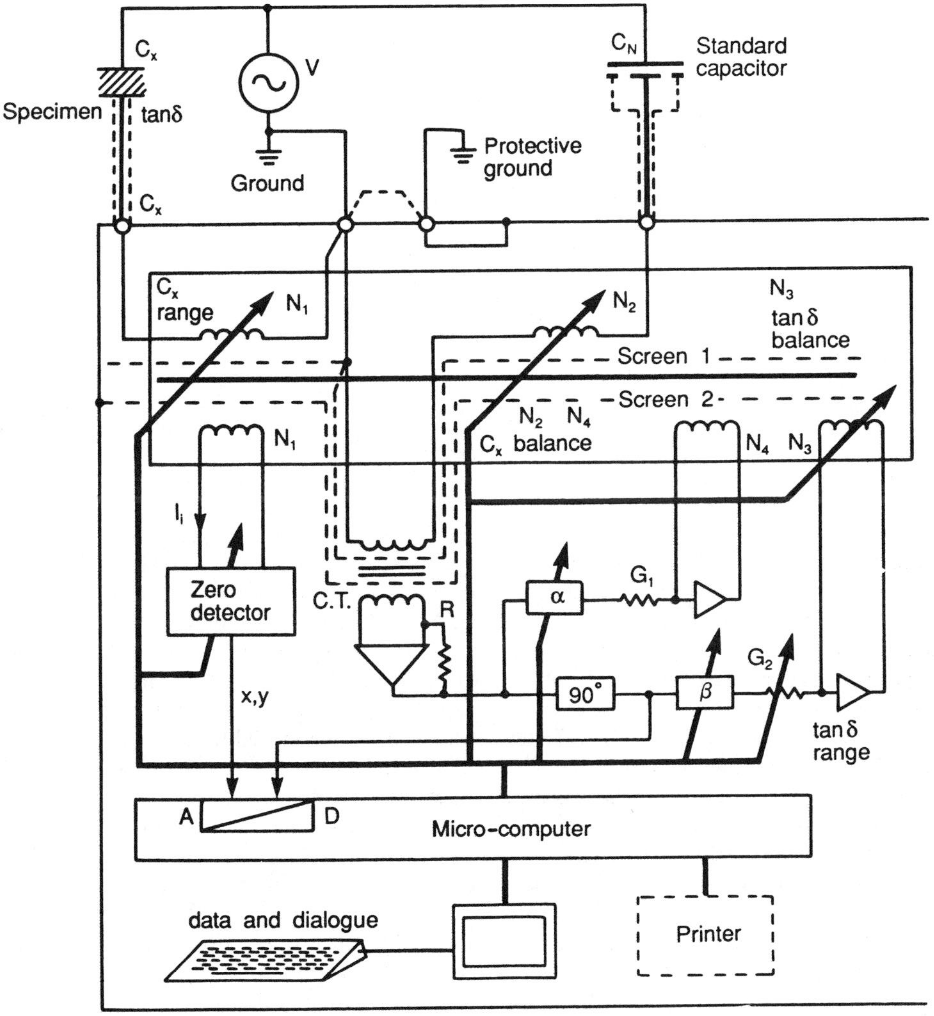

FIG. 1.77—*Microcomputer-controlled automated transformer ratio arm bridge (after Osvath and Widmer* [237]*).*

Equating the real terms yields the conductance of the specimen, G_x,

$$G_x = \frac{C_N(\beta RG_2N_3)}{N_1} \tag{1.314}$$

and equating the imaginary terms gives the capacitance of the specimen C_x

$$C_x = C_N\left\{\frac{N_2 + \alpha RG_1N_4}{N_1}\right\} \tag{1.315}$$

whence

$$\tan\delta = \frac{G_x}{\omega C_x}$$

$$= \frac{N_3}{(N_2 + \alpha RG_1 N_4)}[\beta RG_2] \tag{1.316}$$

Equations 1.315 and 1.316 are simplified by noting that in practice $RG_1N_4 \simeq 1$ and $N_3 = N_4$, such that

$$C_x = \frac{(N_2 + \alpha)C_N}{N_1} \tag{1.317}$$

and

$$\tan\delta = \frac{\beta RG_2}{1 + \dfrac{\alpha}{N_2}} \tag{1.318}$$

Although Osvath and Widmer [*237*] use a microcomputer to achieve balance of the ratio arm bridge, their arrangement bears considerable similarity to the automated bridge by Petersons [*235*]. The microcomputer carries out the balancing procedure and thence calculates the capacitance and tanδ values of the specimen in terms of the N_1, N_2, N_3, α, β, and G_2 variables. Since the bridge is intended for high voltage measurements, protection against breakdown surges of the computer equipment is achieved by the use of optical fiber cables. Automation of the bridge results in a somewhat reduced accuracy: 1×10^{-5} and 1×10^{-4}, respectively, for tanδ and the capacitance. With nonautomated ratio arm bridges, the tanδ value may be determined to within 10^{-7}.

Another microcomputer controlled automated bridge that is used at power frequencies and is designed specifically for oil specimens is the so-called Bauer complex compensation bridge [*238*]; in principle, this type of bridge is essentially a mutated version of the Thompson-Harris bridge in that in the nonautomated version potentiometers replace the variable capacitor elements, while in the automated version analog to digital (*A/D*) converters with corresponding resolution are employed as substitutes for the potentiometers. A schematic circuit diagram of the nonautomated Bauer complex compensation bridge is given in Fig. 1.78. If m and ℓ are the respective potentiometer readings at balance, then the capacitance of the specimen is given by [*238*]

$$C_x = \frac{C_N}{m} \tag{1.319}$$

and the dissipation factor by

$$\tan\delta = \frac{\ell I_N}{I_N} = \ell \tag{1.320}$$

where C_N represents the value of the standard capacitor. Figure 1.79 portrays the microcomputer controlled version of the Bauer type complex compensation bridge.

1.8.4 Measurements in the Upper and High Radio Frequency Range

The appearance of pronounced stray inductance effects renders bridge techniques unsuitable for measurements above 1 MHz. As has been pointed out in Vol. II-B of this series, the most commonly used measurement procedures in the frequency range encompassing 1 to 200 MHz are based on resonance methods, using either the Q-meter or susceptance variation techniques. The susceptance variation (Hartshorn and Ward) method as used in its present form comprises some improvements introduced by Reddish et al. [*216*] and Buckingham and Billing [*217*], following the earlier attempts made by Barrie [*239*] to improve the tanδ measurements by means of a photoelectric device for scale magnification. In the improved Hartshorn and Ward apparatus a high stability, high impedance, square law a-c to d-c converter is used to measure the resonant circuit voltage; the output from the a-c to d-c converter is applied to a potentiometer. The voltage difference between the latter and a stable fixed voltage is displayed on a recorder for direct reading and for monitoring the stability in the course of the measurement. Also a fixed resistance/capacitance unit is employed for periodic accuracy checks of the measurement. The improved Hartshorn and Ward setup is delineated in Fig. 1.80.

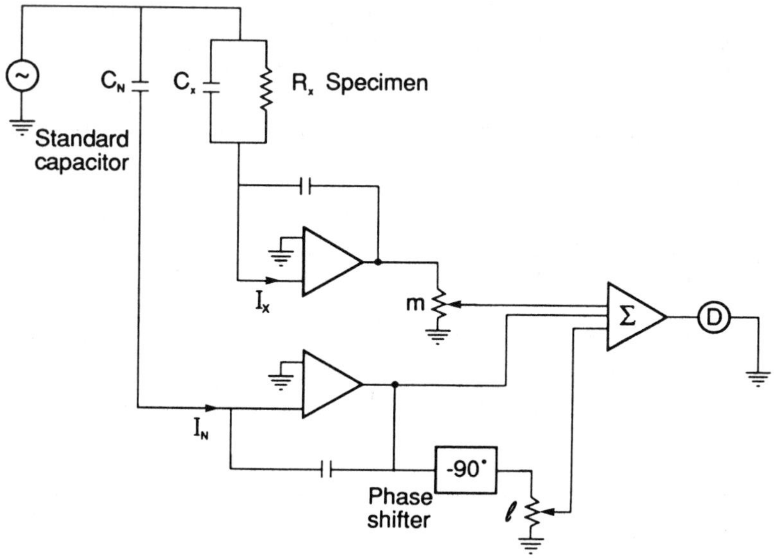

FIG. 1.78—*Bauer complex compensation bridge* [238].

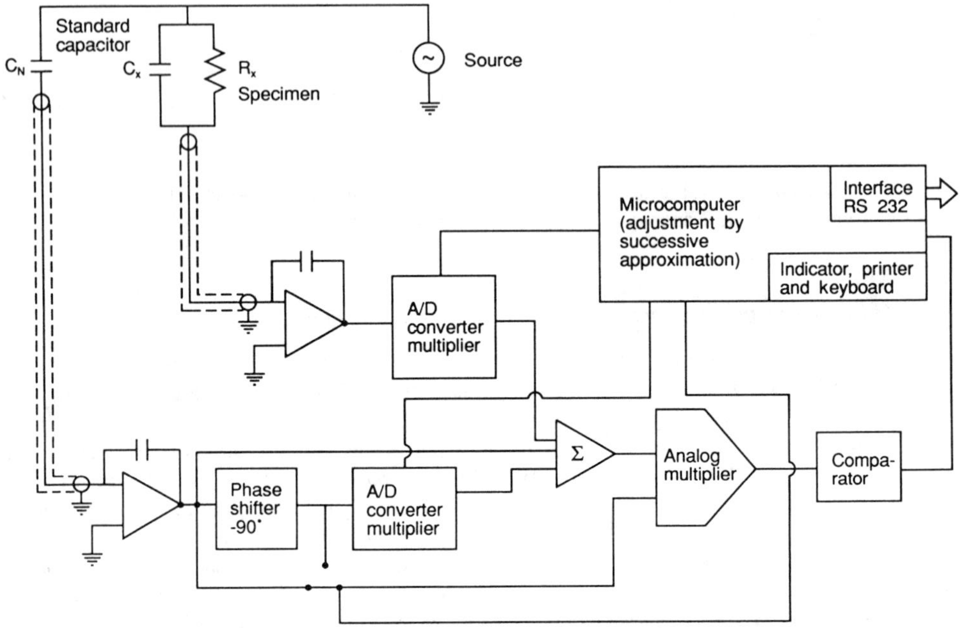

FIG. 1.79—*Automated Bauer complex compensation bridge with computer control* [238].

It will be recalled that in the Hartshorn and Ward circuit, the permittivity and the dissipation factor of a specimen are deduced from a change in the voltage across an *LC* circuit that is at resonance when the liquid dielectric is replaced by air in the curvilinear electrode system depicted in Fig. 1.72. First resonance is obtained with the specimen liquid between the electrodes, and the maximum voltage, e_1, of the a-c to d-c converter is recorded. The specimen liquid is then emptied from the specimen holder, and the electrodes cleaned with a suitable solvent and resonance is reestablished by reducing the air spacing between the electrodes. The voltage output, e_o, of this resonance maximum will be observed to be larger than that with the specimen liquid between the electrodes. From the calibrated main micrometer reading, the capacitance of this air gap (which is

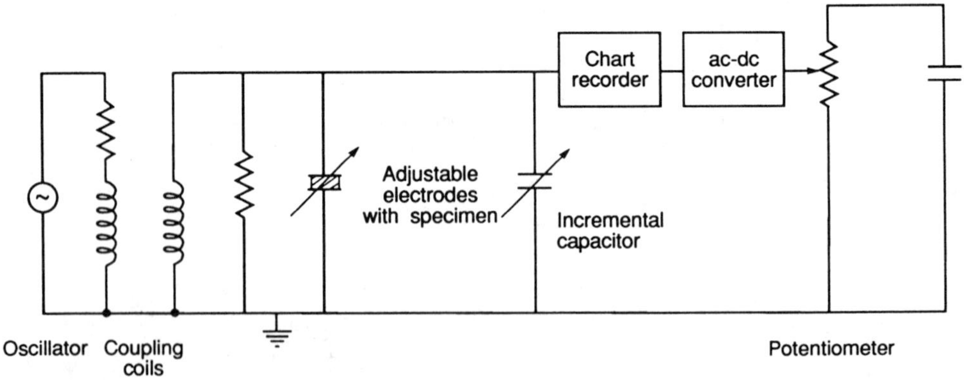

FIG. 1.80—*Modified Hartshorn and Ward susceptance variation circuit (after Reddish et al.* [216]*).*

numerically equal to the capacitance of the liquid specimen, C_x) is obtained. The subsequent step is to determine the value of ΔC_o, which represents the width of the resonance curve at the half power points, with the liquid specimen absent. This is accomplished by manipulating the micrometer head of the small incremental or vernier capacitor of the electrode assembly (see Fig. 1.72). The square law detection feature of the instrument ensures that the change in the calibrated incremental micrometer reading in moving from one half power point to the other is equal to the change in capacitance, ΔC_o. The dissipation factor of the specimen is thus given by [*216*]

$$\tan\delta = [(e_o/e_1)^{1/2} - 1]\frac{\Delta C_o}{2C_x} \tag{1.321}$$

where e_o is the output voltage of the a-c to d-c converter at resonance with the liquid specimen removed from the electrode holder, e_1 is the output voltage of the a-c to d-c converter at resonance with the liquid specimen present, ΔC_o is the capacitance width of the resonance curve at the half-power points with the specimen absent in *pF* and C_x is the air capacitance of the electrodes. Also since Δe is defined by

$$\Delta e = e_o - e_1 \tag{1.322}$$

then Eq 1.321 may be rearranged as

$$\tan\delta = \left[\left(1 + \frac{\Delta e}{e_1}\right)^{1/2} - 1\right]\frac{\Delta C_o}{2C_x} \tag{1.323}$$

furthermore, if Δe is negligibly small, then [*239*]

$$\tan\delta = \frac{1}{4}\left(\frac{\Delta e}{e_1}\right)\left(\frac{\Delta C_o}{C_x}\right) \tag{1.324}$$

Reddish et al. [*216*] have demonstrated that the ratios of $\Delta e/e_1$ and $\Delta C_o/C_x$ in Eq 1.324, that determine the resolution of the measurement when dielectrics with low losses are being evaluated, are related to different aspects of the instrument design. Obviously, amplification cannot be increased indefinitely to reduce the ratio of $\Delta e/e_1$ because of electronic noise problems. Improved electromagnetic shielding and electrical contacts have been shown to be effective. However, as regards to ratio $\Delta C_o/C_x$, it has been established that complete screening of the inductors affects adversely the *Q*-values of the circuit with the consequence of increasing ΔC_o. Increase in the air capacitance of the cell in order to increase C_x implies an increase in the area or diameter of the electrodes of the specimen holder. However, this affects adversely the upper frequency range of the instrument and tends to augment ΔC_o. For adequate resolution, the ratio of $\Delta C_o/C_x$ should fall within the limits of 5.0×10^{-3} to 2.5×10^{-2} with a corresponding minimum ratio of $\Delta e/e_1$ of 2.5×10^{-5}.

The real value of the permittivity or dielectric constant, ϵ', of the liquid specimen may be determined by dividing the value of C_x by the capacitance of an air film having the thickness of the liquid dielectric film used between the main electrodes. More precisely, it may be determined from [*216*]

$$\epsilon' = \frac{d_s}{d_s - \Delta d} \tag{1.325}$$

where d_s is the thickness of the liquid dielectric film used and Δd represents the decrease in separation between the two measuring electrodes in air that is required in order to restore the capacitance to the same value as that when the liquid specimen was present in the intervening electrode space. Equation 1.325 would be expected to be subject to some error due to fringing capacitance effects. The modified Hartshorn and Ward method when applied to liquid specimens should be capable of providing an accuracy of at least 1% on permittivity measurements, and it should be possible to measure $\tan\delta$ to approximately within $\pm 1.0 \times 10^{-6}$.

1.8.5 High Frequency Measurements

At frequencies above 300 MHz, the dielectric specimen begins to increasingly exhibit a distributed parameter behavior so that measurements intended for lumped circuit behavior become inapplicable. Depending upon the test frequency, the specimen under investigation may be inserted in an open or closed resonator, in a shorted coaxial line, a waveguide, or placed in a free space path. Each measurement technique requires that the specimen conform to certain geometrical shapes and dimensions to fit a particular specimen holder. Whereas this requirement may pose at times certain serious constraints with solid specimens, liquid specimens, in general, do not encounter difficulties as the liquid readily assumes the physical form of the cell container. High frequency measurements on dielectric liquids were performed first by Drude [*240,241*] using a parallel transmission line as early as 1895, long before any measurements were attempted at power and audio frequencies. In fact, Drude used the results of his experiments to demonstrate the relationship between the dielectric constant and the real value of the index of refraction, that is, $\epsilon' = (n')^2$.

The region within the frequency régime from 300 to 600 MHz constitutes a rather singular "grey" area where the specimen undergoes transition from a lumped parameter to a distributed parameter behavior. Within this region of frequencies, it is common practice to employ a reentrant cylindrical cavity measurement technique that can be viewed essentially as an extension of the Hartshorn and Ward method [*218,242–244*]. To illustrate this method of measurement, we shall consider the reentrant cavity resonator of the coaxial type employed by Parry [*244*], which is depicted in Fig. 1.81. The metallic parts of the cavity are constructed of machined solid brass, with the internal walls flashed with copper and silver plated. The upper electrode, having a diameter of 2.54 cm, is mounted on a steel spindle driven by a precision micrometer. The liquid dielectric specimen is placed within the polyethylene tray, which rests upon the lower electrode. The inside diameter of this tray is 2.7 cm; its wall thickness is 0.4 mm, and its thickness at the base is 1 mm. The tray is not constructed from metal because the latter would give rise to appreciable field distortion effects. As in the case of the Hartshorn and Ward specimen cell, the vernier capacitor is mounted perpendicular to the axis of the main electrode micrometer and the capacitance change per millimeter of movement is equivalent to a capacitance change of 0.005 *pF*. The calibration of the vernier capacitor is frequency dependent and, consequently, the reentrant cavity must be calibrated as a wavemeter, treating the main specimen capacitor as the prime frequency control device and the vernier capacitor as an incremental control device. In general, it is found that the resultant calibration plot of the change in resonant frequency versus the change in the vernier capacitor setting is nonlinear.

The calibration of the main micrometer, that is, the micrometer assembly containing the specimen, usually is made using a low frequency bridge. The capacitance between the two electrodes necessarily includes the edge capacitance contribution, which must be added to the normal geometrical capacitance. Since the upper electrode of the specimen holder must be immersed in the liquid specimen contained within the polyethylene cup, the effective "diameter" of the liquid specimen is essentially larger than the electrode diameter. Under such circumstances the edge correction, C_{es}, consists of two components: the edge correction with the upper electrode immersed in the liquid, C_{el}, and the edge correction with the electrodes in air, C_{ea}, such that [*244*]

$$C_{es} = C_{ea} + C_{el} \tag{1.326}$$

If the measured low frequency capacitance of the polyethylene of the bottom layer between two tin foil electrodes is designated as C_T, then the value of C_{el} may be estimated from

$$C_T = C_{pe} + C_{es} + (\epsilon' - 1)C_{el} \tag{1.327}$$

where C_{pe} is the calculated capacitance of the polyethylene disk, and C_{es} is the edge capacitance associated with the fringing field of the electrode system. Parry [*244*] estimates the value of C_{es} using Kirchhoff's edge correction formula [*245*]

$$C_{es} = \frac{r}{3.6\pi}\left[\ln\frac{16\pi r(b + d)}{d^2} + \frac{b}{d}\ln\frac{b + d}{b} - 3\right] \tag{1.328}$$

where the units of C_{es} are given in *pF*, *r* is the electrode radius in cm, *b* is the electrode thickness in cm, and *d* is the separation of the electrodes in cm. From Eq 1.327, the value of $C_{e\ell}$ is thus obtained in terms of

$$C_{el} = \frac{C_T - C_{pe} - C_{es}}{(\epsilon' - 1)} \tag{1.329}$$

Using the edge corrections the permittivity of the liquid dielectric specimen may be measured to an accuracy of approximately + 1%, which tends to decrease somewhat with specimen thickness or with frequency due to smaller capacitance of the specimen to be measured and increased inductance

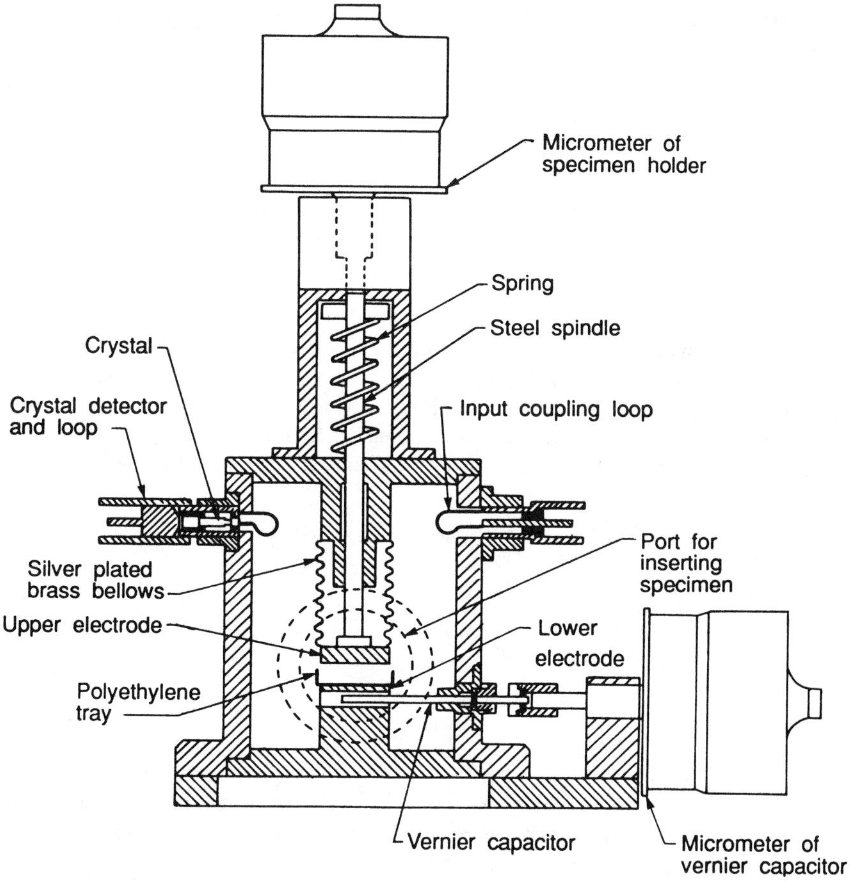

FIG. 1.81—*Re-entrant cavity measurement system (after Parry* [244]*).*

and edge corrections. It should be emphasized that the Kirchhoff edge correction formula is valid primarily for small specimen thicknesses, that is, when $d \ll r$.

In the measurement procedure, the tray is partially filled with the dielectric liquid and the main micrometer is screwed down into the liquid to obtain resonance at some suitable depth of the liquid between the electrodes. Under these circumstances the equivalent circuit consists of the capacitance of the liquid, C_s, of dissipation factor $\tan\delta_s$ in series with the capacitance of the polyethylene base of the cup C_{pe} of dissipation factor $\tan\delta_{pe}$. The equivalent capacitance of this circuit is thus

$$C_1 = \frac{C_s C_{pe}}{C_s + C_{pe}} \tag{1.330}$$

and the dissipation factor is

$$\tan\delta_1 = \frac{C_1}{C_s}\tan\delta_s + \frac{C_1}{C_{pe}}\tan\delta_{pe} \tag{1.331}$$

The second step consists in removing the liquid specimen from the cup and placing the cleaned empty cup on the lower electrode and then restoring resonance by the variation of the main electrode. Now the liquid specimen is replaced by an equivalent air gap of capacitance, C_{ag}, and negligibly small dissipation factor, $\tan\delta_{ag}$. The capacitance, C_2, and dissipation factor, $\tan\delta_2$, of this equivalent circuit are given respectively by

$$C_2 = \frac{C_{ag} C_{pe}}{C_{ag} + C_{pe}} \tag{1.332}$$

and

$$\tan\delta_2 \simeq \frac{C_2}{C_{pe}}\tan\delta_{pe} \tag{1.333}$$

For both of the foregoing cases, the variation of the vernier capacitor at resonance yield the respective widths of the resonance curve, Δf_1 and Δf_2 for any values of the ordinate m, which is defined as the ratio e_r^2/e^2, where e_r is the voltage at resonance indicated by the crystal detector (see Fig. 1.81) and e is the voltage at the half power points where the width Δf (in frequency units) is measured. In terms of the equivalent circuits at the resonant frequency, f_r, [244]

$$\begin{aligned} \frac{\Delta f_1 - \Delta f_2}{f_r(m-1)^{1/2}} &= \tan\delta_1 - \tan\delta_2 \\ &= \frac{C_1}{C_s}\tan\delta_s + \frac{C_1 - C_2}{C_{pe}}\tan\delta_{pe} \\ &\simeq \frac{C_1}{C_s}\tan\delta_s \end{aligned} \tag{1.334}$$

With the empty tray or cup removed from the reentrant cavity, resonance is once more reestablished by varying the main micrometer and the returned capacitance, C_m, is obtained in terms of the calibrated micrometer setting at low frequency such that [244]

$$C_1 = \left[\frac{\tan\left(\frac{\omega_r l_2}{c}\right)}{\tan\left(\frac{\omega_r l_1}{c}\right)}\right] C_m - [C_{es} + (\epsilon' - 1)\,C_{el}]$$
$$= \frac{C_s C_{pe}}{C_s + C_{pe}} \qquad (1.335)$$

in which the first bracketed expression represents an inductance correction, where l_1 is the overall length of the two rods of the main micrometer to which the electrodes are attached when the liquid specimen is inserted and l_2 is the equivalent length with the liquid specimen removed at the resonance frequency $\omega_r = 2\pi f_r$ and c is the velocity of light *in vacuo.* In Eq 1.335, the capacitance C_{pe} is determined from the dimensions of the base of the polyethylene cup and the edge corrections C_{es} and C_{el} in terms of Eqs 1.328 and 1.329, respectively. The capacitance contribution of the walls of the cup is included in the edge correction component $\epsilon' C_{el}$, where ϵ' is the dielectric constant of the liquid specimen. Parry [*244*] justifies the latter by the fact that the thickness of the polyethylene wall is normally only one half of that of the liquid situated within the fringing electric field region. The dissipation factor and permittivity of the liquid specimen are determined by calculating the values of $\tan\delta_s$, C_1, and C_s in terms of Eqs 1.330 to 1.335. Since in the measurement procedure for obtaining the dissipation factor, $\tan\delta_s$, of the liquid specimen a tin foil counter electrode was not applied onto the base of the polyethylene cup, a correction must be applied to the $\tan\delta_s$ value; that is, it must be multiplied by the ratio of the true specimen capacitance to the apparent specimen capacitance (obtained without the counter electrodes). The accuracy of the $\tan\delta$ measurements would be expected to be somewhat less than $\pm 2 \times 10^{-5}$. Figure 1.82 depicts a $\tan\delta$ versus frequency characteristic obtained by Parry [*244*] on a transformer oil of an unspecified viscosity; the experimental points within the frequency range of 10^2 to 10^3 MHz were obtained by means of the reentrant cavity method and are seen to complete a continuous absorption curve in conjunction with the experimental points obtained with other test methods.

For measurements within the frequency range of 300 MHz to 5 GHz, it is common practice to employ transmission line methods, whose origins are associated with the earlier classical work of

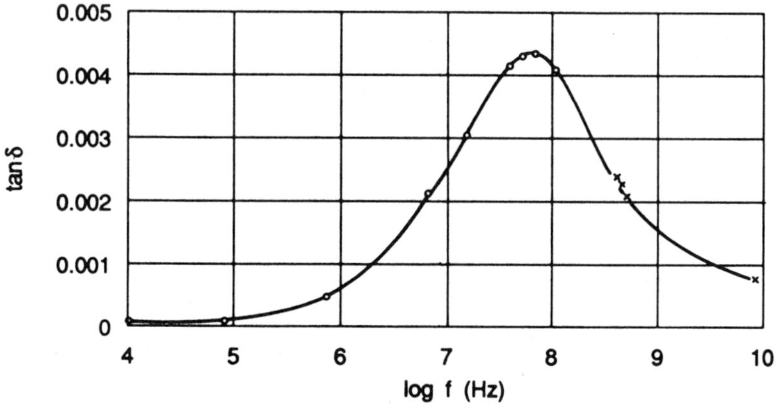

Legend: □—Schering bridge; ○—Hartshorn and Ward method; ×—re-entrant cavity method; △—H_{01} resonator.

FIG. 1.82—*Absorption characteristic of a transformer oil at room temperature (after Parry* [244]*).*

Drude [*241,242*] with the exception that in lieu of the parallel transmission line approach applied earlier by Linhart [*246*] the coaxial line method variation by Roberts and von Hippel [*247*] is employed. The advantage of the coaxial line-waveguide method is that the electrical field is confined within the hollow waveguide geometry (circular or rectangular), thus facilitating precision measurement in the absence of all boundary and stray inductance and capacitance effects. The Roberts and von Hippel technique has been adopted for dielectric liquids by Williams [*248*], who utilized a coaxial transmission line arrangement, such that the thickness of the liquid could be varied within the coaxial line by altering the vertical position of a glass beaker containing the liquid specimen as depicted in Fig. 1.83.

The Roberts-von Hippel method involves the measurement of the standing wave pattern within the waveguide or coaxial line that results from a reflection of the incident wave at the short circuit termination or plate adjacent to the liquid specimen (see Fig. 1.83). If V_{min}/V_{max} denotes the inverse voltage standing wave ratio, then [*247*]

$$\frac{\tanh \gamma_2 d_s}{\gamma_2 d_s} = \frac{\dfrac{V_{min}}{V_{max}} - j\tan\beta_o\chi_o}{j\beta_o d\left(1 - j\left(\dfrac{V_{min}}{V_{max}}\right)\tan\beta_o\chi_o\right)} \tag{1.336}$$

where γ_2 represents the propagation constant of the dielectric liquid of depth d_s, χ_o is the distance of the first mode of the standing wave patterns in air from the surface of the liquid specimen, and the phase constant β_o for the incident wave is given by

$$\beta_o = 2\pi/\lambda_o \tag{1.337}$$

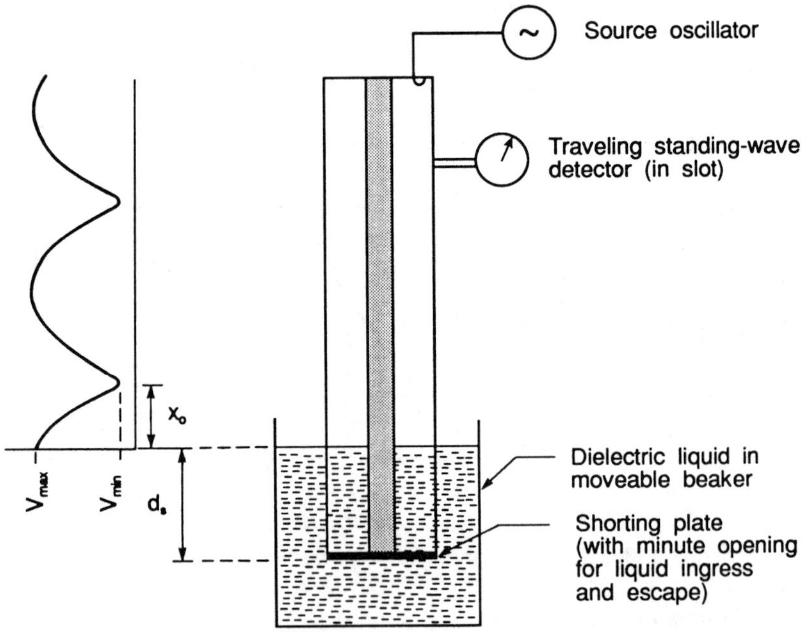

FIG. 1.83—*Roberts-von Hippel apparatus as applied to liquids (after Williams* [248]*).*

where λ_o is the wavelength of the electromagnetic radiation in free space. If the denominator of Eq 1.336 is rationalized, then Eq 1.336 may be written in the form

$$\frac{\tanh \gamma_2 d_s}{\gamma_2 d_s} = A_o \exp[j\zeta] \tag{1.338}$$

when A_o represents the magnitude of the phasor and ζ its angle. The value of the latter is given by

$$\zeta = \tan^{-1}\left[\frac{\dfrac{V_{\min}}{V_{\max}}(1 + \tan^2 \beta_o \chi_o)}{\left\{\left(\dfrac{V_{\min}}{V_{\max}}\right)^2 - 1\right\}\tan\beta_o\chi_o}\right] \tag{1.339}$$

The function A_o exp $[j\zeta]$ is found by measurements of d_s, λ_o, $V_{\min}/V_{\max}$, and χ_o with appropriate substitution into Eq 1.336. Equation 1.338 can be further rewritten as [*247*]

$$\frac{\tanh \gamma_2 d_s}{T[\exp\ j\tau]} = A_o \exp[j\zeta] \tag{1.340}$$

where by definition $T[\exp j\tau] = \gamma_2 d_s$. In order to determine the value of $T[\exp j\tau]$ and thence the value of $\gamma_2 d_s$, Roberts and von Hippel developed charts based on a series approximation of the function defined by the function given in Eq 1.340 in which the argument τ is plotted as the ordinate and the absolute value of T as the abscissa while A_o and ζ are parameters of the intersecting curves. Since the hyperbolic functions are multivalued, measurements carried out with a single specimen thickness, d_s, will result in an uncertainty as concerns the value of γ_2. However, this can be resolved by using two different specimen thicknesses whereby only one set of T/d_s and τ values will be found to satisfy by experimental results. Since the propagation constant in the liquid specimen may be expressed as

$$\gamma_2 = \alpha_2 + j\beta_2 \tag{1.341}$$

where α_2 is the attenuation constant in the liquid, then [*248*]

$$\alpha_2 = \frac{T\cos\tau}{d_s} \tag{1.342}$$

and

$$\beta_2 = \frac{T\sin\tau}{d_s} \tag{1.343}$$

In terms of which the real and imaginary values of the permittivity may be expressed respectively as [*207*]

$$\epsilon' = (n')^2[1 - \alpha_2^2/\beta_2^2] \tag{1.344}$$

and

$$\epsilon'' = 2(n')^2[\alpha_2\beta_2] \tag{1.345}$$

where the real value of the index of refraction for the liquid specimen is given by [*248*]

$$n' = \lambda_o\beta_2/2\pi \tag{1.346}$$

Corfield et al. [*249*] have described a variation of the Roberts and von Hippel apparatus that uses a short circuit section of variable length whereby the length of the specimen may be adjusted to an integral number of half wavelengths. Their approach reduces the length of the measurement time very appreciably. The method is applicable primarily to low loss specimens and is capable of an accuracy of $\pm 2\%$ for permittivity measurements and a $\pm 1 \times 10^{-3}$ or $\pm 5\%$ accuracy (whichever is greater) on $\tan\delta$ measurements.

According to Dakin and Works [*250*] for low loss medium dielectrics, Eq 1.336 may be modified to

$$\frac{\tan\beta_2 d_s}{\beta_2 d_s} = -\left(\frac{\lambda_o}{2\pi d_s}\right)\tan(\beta_o \chi_o) \tag{1.347}$$

Separation of real and imaginary terms, then yields [*249*]

$$\tan\delta = \left(\frac{\Delta\chi}{d_s}\right)\left[\frac{\beta_2 d_s (1 + \tan^2 \beta_o \chi_o)}{\beta_2 d_s (1 + \tan^2 \beta_2 d_s) - \tan\beta_2 d_s}\right] \tag{1.348}$$

where $\Delta\chi$ is the contribution of the dielectric fluid to the width of the voltage minimum (at twice the minimum power) of the standing wave in air. The expressions for $\tan\delta$ and ϵ' simplify very appreciably, however, if the length of the specimen, d_s, is adjusted to correspond to an integral number of half wavelengths [*249,251*]

$$\tan\delta = \frac{\Delta\chi}{d_s} \tag{1.349}$$

and

$$\epsilon' = \left(\frac{m\lambda_o}{2d_s}\right)^2 \tag{1.350}$$

where m denotes the number of half wavelengths in the dielectric liquid of depth d_s. When the number of wavelengths in the specimen is odd, then the foregoing equations are modified to

$$\tan\delta = \frac{\Delta\chi}{\epsilon' d_s} \tag{1.351}$$

and

$$\epsilon' = \left(\frac{m'\lambda_o}{4d_s}\right)^2 \tag{1.352}$$

where m' represents now the odd number of wavelengths in the specimen.

The schematic diagram of the arrangement employed by Corfield et al. [*249*] is depicted in Fig. 1.84. The oscillator applies a sinusoidal signal across a horizontally situated slotted line having a characteristic impedance of 50 Ω. A right angle joint is used for the attachment of the coaxial line containing the liquid, placed in a vertical position as shown. The standing wave pattern is obtained by means of a moveable probe in the slotted line in accordance with the classical arrangement employed by Roberts and von Hippel [*247*]. A polyethylene diaphragm with a thickness of 0.3 mm is utilized to separate the liquid from the air portion of the coaxial line and is held *in situ* by means

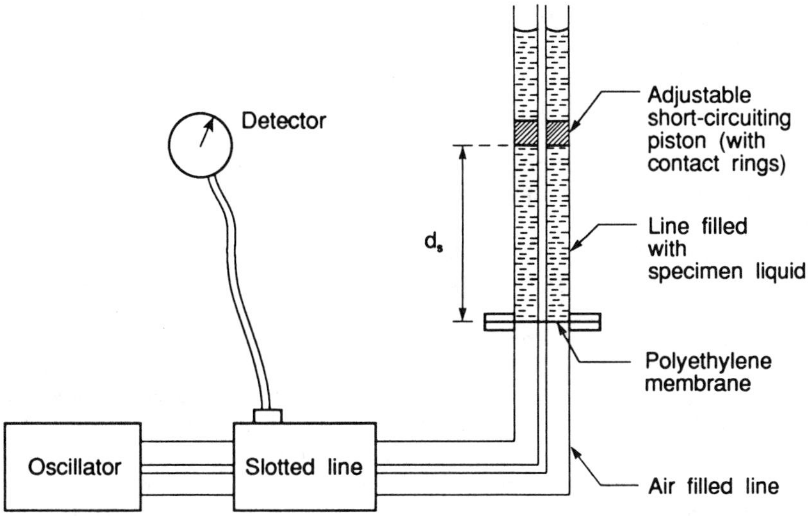

FIG. 1.84—*Adjustable coaxial line apparatus for the measurement of permittivity and loss of dielectric liquids* [249,252].

of flanges of the outside conductor. A coupling stud, traversing a small aperture in the diaphragm, connects the inner conductors in the air and liquid portions of the coaxial line. The short-circuit piston, made of phosphor bronze within the coaxial line, makes electrical contact with the inner and outer conductors by means of two phosphor-bronze ring-springs. Its movement is controlled by coarse and fine adjustment mechanisms and is observed by means of a cathetometer focussed on a reference mark. Displacement of the dielectric liquid, while the short circuiting piston is in motion, occurs via a small vent drilled in the piston.

When half wavelength measurements are carried out, after the specimen liquid is placed in the container, the short circuit piston is moved first downward until it is in contact with the diaphragm. At this point, the position of the voltage minimum of the standing wave in the slotted line and its width $\Delta\chi_o$ at twice the minimum power are determined by means of the detector probe. The value of $\Delta\chi_o$ represents the attenuation between the probe and the bottom of the liquid column in the measuring system. Subsequently, the shorting piston is retracted upwards through the dielectric fluid until the probe detector indicates another voltage minimum; this vertical distance of the liquid column, d_s, represents a half wavelength of the radiation in the liquid and is the value that must be substituted in Eq 1.350 to obtain the real value of the permittivity, ϵ', of the dielectric fluid. Once again the width of the voltage minimum at twice minimum power, $\Delta\chi_1$, is measured at this point. The value of $\Delta\chi$ in Eq 1.349 is then given by

$$\Delta\chi = \Delta\chi_1 - \Delta\chi_o \tag{1.353}$$

Substitution of $\Delta\chi$ in Eq 1.349 thus yields the $\tan\delta$ value of the specimen liquid. The accuracy of the $\tan\delta$ value is improved, if correction is made for the wall losses.

When measurements are performed at quarter wavelengths, the probe indicator is placed at an integer number of quarter wavelengths from the diaphragm with the cell filled with the specimen fluid. The position of the shorting piston is adjusted to yield again a voltage minimum. An identical procedure is then followed as in the half wavelength measurement procedure to obtain the vertical distance, d_s, between the diaphragm and the shorting piston; its value is then inserted in Eq 1.352 to obtain the real value of the permittivity, ϵ', of the liquid. Identical steps are followed to obtain the value of $\Delta\chi_1$ which when substituted in Eq 1.351 yields the value of tanδ.

The foregoing measurement technique developed by Corfield et al. [*249*] works very well with low loss fluids, but for lossy fluids it must be modified as shown by Bennett and Calderwood [*250*]. Their method, however, is inapplicable for liquids with tanδ values beyond 0.8791 nor does it work well for liquids with large dielectric constants. Like the Roberts-von Hippel method (which is readily applicable to highly lossy liquids with high dielectric constants), it involves considerable computation.

For measurements beyond 10 GHz, resonant cavities must be utilized; their useful frequency range extends up to approximately 60 GHz. A cavity resonator is equivalent to a transmission line that is short-circuited at both ends, the latter being separated by an arbitrary multiple of one half the working wavelength. Insertion of a dielectric material within the cavity will thus alter the wavelength with the result that the required frequency or dimensional shift necessary to reestablish resonance and the original value of the Q-factor will be related to the dielectric properties of the specimen material. Since resonators may have Q-values as high as 10 000, it is apparent that they constitute an effective means for measuring low dielectric loss materials. Resonators are designed generally with a particular geometry suitable for a specific liquid specimen thickness with anticipated dielectric properties over the frequency range of interest. Essentially, resonators tend to be narrow frequency band devices whose frequency range is determined by the resonator geometry, the properties of the liquid specimen, and its means of and position of placement within the resonant cavity. The upper frequency limit is imposed by the physical dimensions of the resonant cavity, which become smaller with increasing frequency: each octave increase requires a successively smaller size of cavity. Equations 1.286 and 1.289 indicate the relationship between the working frequency of the resonant cavity and Q-value and the corresponding values of the real permittivity and dissipation factor of the liquid specimen.

A cavity may be resonated either by varying the frequency of the source oscillator or by varying the physical dimensions of the cavity at a fixed frequency. In most instances, the practical dimension to be varied is evidently the axial length of the cavity. However, a cavity system operating in the E_{010} mode would be resonated by varying the frequency, since variation of cavity radius would not be feasible [*206*]. Whereas, as pointed out by Horner et al. [*206*], with coaxial line and H_{01N} mode system resonators, the most efficient means of resonating is by varying the axial length for which purpose one end of the resonator assumes the form of a sliding short-circuiting piston. It will be recalled that as concerns the subscript designations of the electric and magnetic field modes, the first subscript refers to the number of half-waves across the shorted cavity, the second subscript designates the number of half-waves from top to bottom of the cavity, and the last subscript represents the odd number of half waves along the cavity.

In the case of an air-filled cavity having a radius r, the mode H_{01N} would occur for excitation wavelengths less than the critical or cutoff value defined by $\lambda_c = 1.64r$ [*206*]. The cavity would undergo resonance when its axial length becomes equal to an integer value of $\lambda_g/2$, where λ_g is the wavelength of the cavity which is related to the energizing or free space wavelength, λ_o, given by the expression [*206*]

$$1/\lambda_g^2 = 1/\lambda_o^2 - 1/\lambda_c^2 \tag{1.354}$$

A resonant cylindrical cavity operating in the H_{01} mode, that was originally designed by Lamb [*253*] for operation of the frequency range from 10 to 30 GHz, is shown in Fig. 1.85. The cavity is of brass with an inside diameter of 1.96 cm, having a lower variable noncontact piston mounted on the shaft of a micrometer and a removable upper tight-fitting piston. The two coupling holes drilled antipodally on the sides of the cavity permit connection to the oscillator and detector via rectangular sections of an H_{01} copper waveguide such that the transverse magnetic field direction in the guide coincides with that of the longitudinal magnetic field in the resonator. A brass cup, placed on

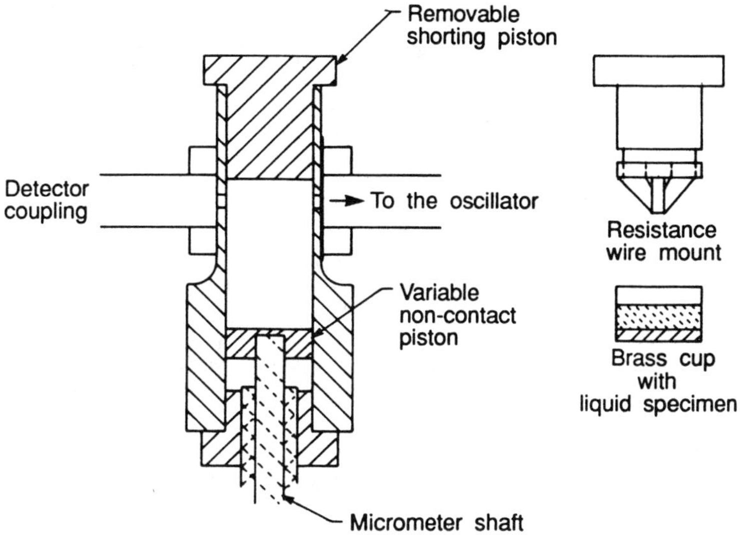

FIG. 1.85—H_{01} *resonant cavity assembly (after Lamb* [253]*).*

variable piston, is used as a containment vessel for the liquid specimen. A pyramid of suitably mounted resistance wires mounted on the removable upper piston may be used to surpress extraneous modes possessing either radial or longitudinal components of the electric field. The presence of the brass cup, which is fitted smoothly into the resonator so as to allow its unobstructed movement along the vertical axis, has no measurable effect upon the wavelength or *Q*-factor of the air-filled cavity.

The resonance condition of Lamb's cavity is given by [*206*]

$$\frac{\tan\beta_1 d_s}{\beta_1} + \frac{\tan\beta_o(l_r - d_s)}{\beta_o} = 0 \tag{1.355}$$

where d_s is the thickness of the liquid specimen and l_r is the resonant length of the cavity, so that by definition $(l_r - d_s)$ is equivalent to the distance required for the piston to move in order to reestablish resonance following the removal of the dielectric fluid. Since $\beta_o = 2\pi/\lambda_{go}$ is the phase constant in air, its value can be deduced from the wavelength of the air portion of the cavity, λ_{go}. Then substitution of the appropriate values in Eq 1.355, will yield the quantity of the phase constant in the dielectric fluid, β_1. The real permittivity or dielectric constant of the dielectric fluid follows from [*206*]

$$\epsilon' = \frac{\beta_1^2 + (3.832/r)}{\beta_o^2 + (3.832/r)} \tag{1.356}$$

where *r* is the radius of the resonant cavity.

Further, it is demonstrated by Horner et al. [*206*] that the *Q*-factor of the resonant cavity containing the specimen fluid may be expressed as

$$Q = \frac{\{p(2d_s - s) + \epsilon'[2(l_r - d_s) - q]\}}{\frac{r'}{r}\left[\frac{1}{\beta_1^2 + 3.832/r}\right]\left\{\left(\frac{3.832}{r}\right)^2 p[2d_s - s] + 2(l_r - d_s) - q + 2r(p\beta_1^2 + \beta_o^2)p(2d_s - s)\tan\delta_1\right.} \tag{1.357}$$

where r' represents the depth of current penetration in the walls of the waveguide and

$$p = \frac{\sin^2 \beta_o (l_r - d_s)}{\sin^2 \beta_1 d_s} \tag{1.358}$$

$$q = \frac{\sin 2\beta_o (l_r - d_s)}{\beta_o} \tag{1.359}$$

and

$$s = \frac{\sin 2\beta_1 d_s}{\beta_1} \tag{1.360}$$

with the specimen fluid removed, $\tan\delta_1 = 0$, and for this condition Eq 1.358 defines the Q-factor of the empty cavity which is designated as Q'. From these two conditions, the dissipation factor of the fluid assumes the value

$$\tan\delta_1 = \frac{\{p(2d_s - s) + \epsilon'[2(l_r - d_s) - q]\}}{p(2d_s - s)} \left\{\frac{1}{Q} - \frac{1}{Q'}\right\} \tag{1.361}$$

It was shown by Lamb [*253*] that the accuracy for the dielectric constant and dissipation factor for liquid specimens was somewhat less than $\pm 1.5\%$ and $\pm 4.0\%$ (for $0.0008 > \tan\delta < 0.006$), respectively.

When low loss fluids are investigated, the cavity, in lieu of being only partially filled with the liquid, may be filled completely as has been done by Bleaney et al. [*254*]. Then the dielectric constants of the filled and unfilled cavity, ϵ' and ϵ'_a, respectively, are given by

$$\epsilon' = \lambda_o^2 \left\{\frac{1}{\lambda_{g1}^2} + \frac{1}{\lambda_c^2}\right\} \tag{1.362}$$

and

$$\epsilon'_a = \lambda_o^2 \left\{\frac{1}{\lambda_{go}^2} + \frac{1}{\lambda_c^2}\right\} \tag{1.363}$$

where λ_o is the wavelength in free space, λ_{go} is the wavelength in the cavity containing only air, λ_{g1}, is the wavelength in the cavity completely filled with the liquid specimen, and λ_c is the cutoff or critical wavelength of the cavity. It will be recalled that in a resonant cavity, resonance will take place as the variable piston controlled length of the cavity becomes an integral number of the wavelengths of the waveguide; this permits the determination of λ_{g1} and λ_{go}.

As Barlow and Cullen [*255*] have pointed out, with the completely liquid filled cavity, the $\tan\delta$ value may be determined using once more the relation $\tan\delta = (1/Q) - (1/Q')$ though the value Q' for the cavity filled with the specimen fluid must be corrected for changes in the wall current distribution due to the alteration of the piston position to restore resonance after the fluid is admitted into the cavity. With high loss fluids the low resulting Q-values after the filling of the cavity would lead to asymmetrical resonance curves, thereby disqualifying the measurement procedure on theoretical grounds [*255*].

Time domain spectrometry (TDS) techniques of the type described in Vol. II-B of this series may also be used on liquids to overlap the frequency range ordinarily covered by the transmission line and resonant cavity methods, though usually at an appreciable sacrifice of both precision and accuracy. The convenience in such a replacement lies primarily in providing a quick frequency scan of the dielectric properties. With the development of the two cell technique by Cole and co-workers [*256–258*], a considerable improvement in the accuracy of the TDS technique may be achieved. One cell is used for the specimen, while the other contains a reference liquid, for example, the specimen cell may contain a dilute solution of polar molecules, while the reference cell may contain simply the solvent. The sum and difference of the reflected signals from the two cells are then transformed from the time to the frequency domain by means of Fourier analysis to obtain the real and imaginary permittivities; this procedure is readily computerized.

With increasing frequency beyond 60 GHz, the unduly small size of the required resonant cavities renders them impractical for most applications. Over this régime of frequencies free wave methods, which are analogous the optical methods, employing the classical Fresnel equations, must be used. The optical method was first applied successfully to microwave measurements by Bäz [*259*] using spherical and plane wave fronts. It may be well to emphasize here that microwave methods, utilizing optical techniques, are referred to as quasi-optical methods or often simply as optical methods (see Vol. II-B of this series). In the free wave methods the incident power supplied by the source to the dielectric surface is compared with that transmitted through or reflected by the dielectric specimen.

Garg et al. [*260*] have made use of a microwave analogue of the Michelson optical type interferometer to carry out precise measurements of the permittivity and dielectric loss in liquids to an accuracy of 0.2 and 2% or ± 0.0002 whichever is higher respectively at a frequency of ca. 150 GHz. The permittivity and loss are determined in terms of a power standing wave pattern resulting from the superposition of two waves, one being reflected from the reference arm of the interferometer and the other being a reflection from a variable-length cell containing the liquid specimen. The standing wave pattern is measured as a function of the liquid specimen thickness. The schematic diagram of the apparatus is shown in Fig. 1.86.

Power from the microwave generator enters the hybrid ring where it is divided into two equal parts: one half proceeds via the transmitting horn out into free space directed towards a dielectric lens whose function is to produce a collimated beam. The latter then traverses the liquid specimen and is in turn reflected from the movable piston and on its return is focussed by the lens towards the horn. Quarter wave grooves are machined into the lens in order to match to air. The other half of the power from the hybrid ring enters the reference arm of the interferometer, passing via a waveguide through a variable attenuator and then, upon reaching an adjustable shorting piston, is reflected. Thus, power is returned to the hybrid junction from both of the interferometer arms; the power from each arm undergoes division, with one half directed to the detector and the other half to the power source. Interference then occurs within the waveguide leading to the detector between the two beams reflected from the reference arm and from the variable length cell containing the liquid specimen. The detector compares the amplitude and phase of the plane wave reflected from the variable length specimen cell with that reflected from the reference arm. By means of moving the adjustable piston, the phase of the reflected wave can be varied in relation to the phase of the reference wave. The detected power can be varied thereby through maxima and minima, so that the wavelength of the signal in the liquid is then determined in terms of the micrometer readings between the piston positions corresponding to the extrema points. A similar procedure is employed to determined the signal wavelength in air, with the specimen cell empty.

Figure 1.87 depicts the variable length specimen cell used in conjunction with the Michelson interferometer in which the reflecting piston [2.54 cm (1.00 in.) diameter] is free to move within a conducting carbon impregnated Teflon cylinder contained within a brass cylinder having an inside and outside diameter of 4.445 cm and 7.64 cm (1.75 in. and 3.00 in.), respectively. The purpose of the carbon impregnated Teflon cylinder is to present an absorbing medium for the suppression of the higher-order mode propagation. A constant temperature bath, containing circulated water, is utilized for controlling the temperature of the specimen fluid. The thickness of the Teflon window at

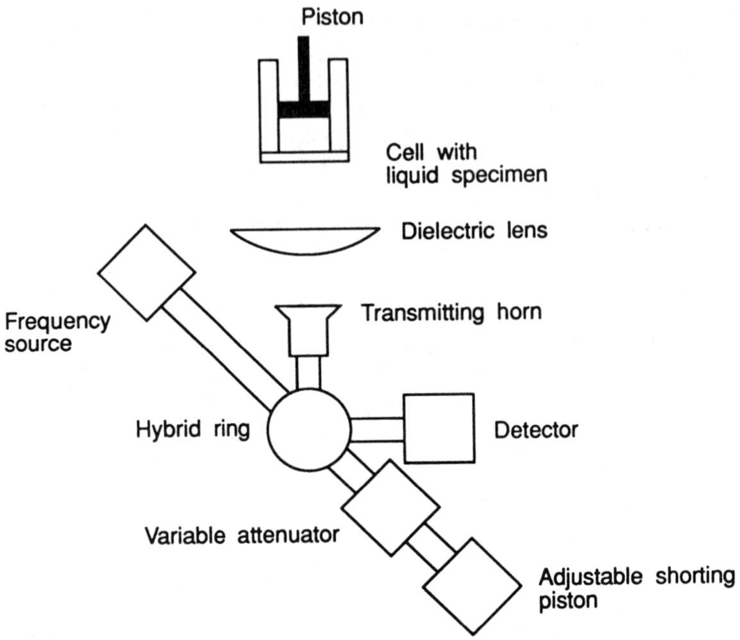

FIG. 1.86—*Michelson optical type interferometer arrangement for measurements at 150 GHz (after Garg et al.* [260]*).*

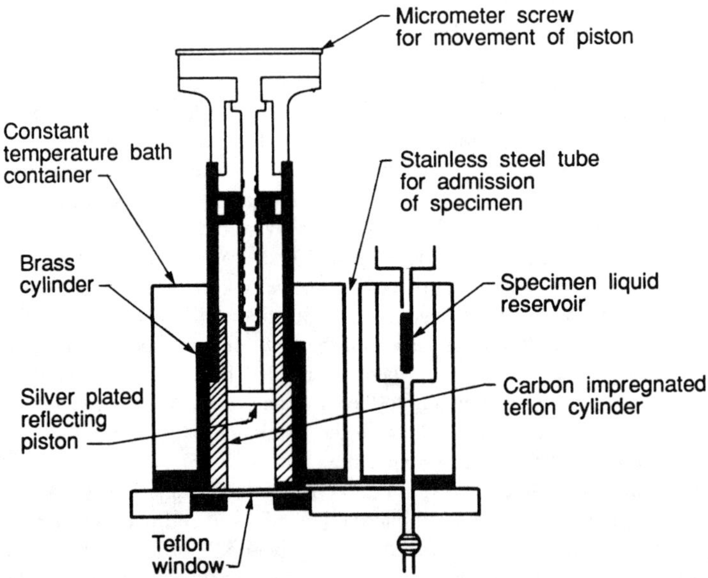

FIG. 1.87—*Variable length liquid cell for use in conjunction with a Michelson optical type interferometer (after Garg et al.* [260]*).*

the bottom of the cell, whereby the radiation beam enters and is subsequently reflected by the optically silver-plated movable piston, must be carefully determined. Teflon exhibits no dispersion at microwave frequencies and, consequently, may be made completely transparent at a particular wavelength for a specific thickness. For a parallel-plane film of a dielectric with zero dispersion loss, the condition for a minimum in the reflection coefficient may be expressed as [*260*]

$$\beta_2 d = m\pi \tag{1.364}$$

where m is an integer, d is the thickness of the dielectric film, and β_2 is the propagation constant of the dielectric film; for a dielectric with a real permittivity of ϵ', β_2 is given by $(2\pi\lambda_o)(\epsilon')^{1/2}$, where λ_o is radiation wavelength in free space. Accordingly, the condition for zero reflection may be expressed as [*260*]

$$d = m\lambda_o/2(\epsilon')^{1/2} \tag{1.365}$$

Garg et al. [*260*] selected a value of $m = 3$ for a dielectric film thickness of 2.175 ± 0.001 mm at the operational wavelength of $\lambda_o = 2.1$ mm. They were obliged to make a minor adjustment in their frequency of operation in order to reduce the reflections from the Teflon window to an insignificantly low value.

The distance between the Teflon window and the radiating horn must be adjusted carefully to prevent shifts in the diffraction pattern with respect to the horn as the position of the reflecting piston is altered. Also the orientation of the cell with respect to the beam of the incident radiation requires careful adjustment in order to attain sharp fringes for different positions of the reflecting piston. The value of the real and imaginary permittivity, ϵ' and ϵ'', respectively, are determined in terms of the real refractive index, n', and the absorption index κ, using the relations

$$\epsilon' = (n')^2 - \kappa^2 \tag{1.366}$$

and

$$\epsilon'' = 2n'\kappa \tag{1.367}$$

The value of the absorption index, κ, is obtained using a method of successive approximations. First Garg et al. [*260*] make the approximation that

$$\kappa = \frac{1}{2\beta_2(d_{s2} - d_{s1})} \ln\left[\frac{I_{\text{max}1} - I_{\text{min}1}}{I_{\text{max}2} - I_{\text{max}2}}\right] \tag{1.368}$$

where I_{max} and I_{min} are the maximum and minimum radiation intensity values at the detector corresponding to two separate specimen thickness, d_s settings of the cell micrometer for positions 1 and 2 (used as subscripts for the quantities in Eq 1.368). A plot of ln $[I_{\text{max}} - I_{\text{min}}]$ versus d_s yields a straight line with a slope equal to $-2\pi\kappa/n'$. The value of κ determined from the slope is employed as an estimate in the relation [*260*]

$$[I_{\text{max}} - I_{\text{min}}] = p_o E_o^2 \left\{ \frac{8n'[(1+n') - (1-n')\exp(4\beta_2\kappa d_s)]\exp(-2\beta_2\kappa d_s)}{[(1+n')^2 - (1-n')^2\exp(-4\beta_2\kappa d_s)]} \right\} \tag{1.369}$$

to obtain ln $[I_{\text{max}} - I_{\text{min}}]$ for different values of d_s, κ is then varied to receive a best fit with the experimental data. The calculations are carried out by means of a computer. In Eq 1.369, E_o represents the electric field at a reference plane near the power source and $p_o = p_1^2 = p_2^2$, where p_1 designates the amplitude and phase of the field E_o which reaches the detector directly and p_2 refers to the magnitude and phase of the field after the radiation passes the specimen cell. The condition $p_o = p_1^2 = p_2^2$ is obtained by moving the adjustable piston of the specimen cell to the Teflon window ($d_s = 0$) at which the value of the electric field intensity at the detector is minimum.

Optical methods for the determination of the real and imaginary indices of refraction, n' and n'', over the visible and infrared spectrum have been described in Vol. II-B of this series, and, consequently, this description will not be repeated here. It is necessary to add, however, that with the improvements in dispersive Fourier transform spectrometry (DFTS), it is now possible to characterize materials over a continuous frequency spectrum extending from 150 GHz to 11 THz. This approach has been most effectively utilized by Afsar et al. [*261*] on both dielectric liquids and solids by means of a polarization interferometer with a parallel-wire grid arrangement in lieu of the conventional Michelson interferometer, whose transmissivity would tend to be strongly frequency dependent due to multiple beam interference within the beam divider.

In the DFTS technique, a specimen is introduced in one arm of a double beam interferometer and the dispersive voltage interferogram, $V(x)$ is recorded as a function of the path difference, x. It may be expressed as [*261*],

$$V(x) = V_r(x) + V_t(x) + V_m(x) \tag{1.370}$$

where $V_r(x)$ is the interference signature reflected from the surface of the dielectric specimen, $V_t(x)$ represents the portion of the interference signature transmitted through the specimen and $V_m(x)$ is the radiation multiply reflection from the specimen. The value of $V(x)$ is thus compared to the nondispersive, single signature voltage interferogram obtained without the specimen, $V_o(x)$, so that the ratio of the full Fourier transforms of $V(x)$ and $V_o(x)$ is be definition the complex positional insertion loss of the specimen. Transparent specimens are evaluated in the transmission mode while with opaque specimens an interferometer mirror is replaced by the specimen and the evaluation is carried in terms of the reflection mode. In principle the liquid specimen holder employed by Afsar et al. [*261*] is similar to that originally utilized by Garg et al. [*260*], with the exception that for transmission measurements the preferred material for the window consists of quartz having a low refractive index while for reflection measurements silicon with a high refractive index is utilized. The described technique may be extended below 100 GHz to permit a comparison of the measurements made over this region using microwave techniques.

References

[*1*] Daniel, V. V., *Dielectric Relaxation,* Academic Press, London, 1967.

[*2*] Kirkwood, J. G., *Journal of Chemical Physics,* Vol. 7, 1939, pp. 919–925. Also see Green, H. S., in *Handbuch der Physik,* Vol. X, Structure of Liquids, S. Flügge, Ed., Springer Verlag, Berlin, 1960.

[*3*] Zernike, F. and Prins, J. A., *Zeitschrift für Physik,* Vol. 41, 1927, pp. 184–194.

[*4*] Debye, P. and Mencke, H., *Physikalische Zeitschrift,* Vol. 31, 1930, pp. 797–798.

[*5*] Mencke, H., *Physikalische Zeitschrift,* Vol. 33, 1932, pp. 593–604.

[*6*] Debye, P., *Physikalische Zeitschrift,* Vol. 13, 1912, pp. 97–100.

[*7*] Debye, P., *Physikalische Zeitschrift,* Vol. 21, 1920, pp. 178–187.

[*8*] Debye, P., *Physikalische Zeitschrift,* Vol. 22, 1921, pp. 302–308.

[*9*] Debye, P., *Berichte der deutschen physikalischen Geselschaft,* Vol. 15, 1913, pp. 777–793.

[*10*] Debye, P., *Polare Molekeln,* Hirzel, Leipsig, 1928.

[*11*] *The Dipole Moment and Chemical Structure,* P. Debye, Ed., Blackie and Son Ltd., London, 1931.

[*12*] *Engineering Dielectrics, Vol. IIA, Electrical Properties of Solid Insulating Materials: Molecular Structure and Electrical Behavior, STP 783,* R. Bartnikas and R. M. Eichhorn, Eds., American Society for Testing and Materials, Philadelphia, 1983.

[*13*] Dekker, A., *Solid State Physics,* Prentice-Hall, Englewood Cliffs, NJ, 1963.

[*14*] Langevin, P., *Journal de Physique,* Vol. 4, 1905, pp. 678–693.

[*15*] *Dielectric and Waves,* A. von Hippel, Ed., Wiley, New York, 1954.

[*16*] *Dielectric Materials and Applications,* A. von Hippel, Ed., MIT Press, Boston, 1954.

[*17*] Böttcher, C. J. F., *Theory of Electric Polarization,* Vol. I, Elsevier, Amsterdam, 1973.

[*18*] Böttcher, C. J. F. and Bordewijk, P., *Theory of Electric Polarization,* Vol. II, Elsevier, Amsterdam, 1978.

[*19*] Mosotti, O. F., *Memoria di Mathematica e di Fisica in Modena,* Vol. 24, 1850, p. 49.

[*20*] Clausius, R., *Die Mechanische Wärmetheorie,* Vieweg Verlag, Braunschweig, 1879.

[21] Lange, L., *Zeitschrift für Physik,* Vol. 33, 1925, pp. 169–182.
[22] Pauli, W., *Zeitschrift für Physik,* Vol. 6, 1921, pp. 319–327.
[23] Mensing, L. and Pauli, W., *Physikalische Zeitschrift,* Vol. 27, 1926, pp. 509–512.
[24] Manneback, C., *Physikalische Zeitschrift,* Vol. 27, 1926, pp. 563–569.
[25] Van Vleck, J. H., *Physical Review,* Vol. 29, 1927, pp. 727–744.
[26] Van Vleck, J. H., *Physical Review,* Vol. 30, 1927, pp. 31–54.
[27] Van Vleck, J. H., *The Theory of Electric and Magnetic Susceptibilities,* Oxford, London, 1932.
[28] Müller, F. H., *Physikalische Zeitschrift,* Vol. 34, 1933, pp. 689–710.
[29] Onsager, L., *Journal of the American Chemical Society,* Vol. LVIII, 1936, pp. 1486–1493.
[30] Lorenz, L., *Annalen der Physik,* Vol. 11, 1880, pp. 70–103.
[31] Lorentz, H. A., *Annalen der Physik,* Vol. 9, 1880, pp. 641–665.
[32] Maxwell, J. C., *Treatice on Electricity,* Vols. I and II, Oxford, London, 1881.
[33] Bell, R. P., *Transactions of the Faraday Society,* Vol. 27, 1931, pp. 797–802.
[34] Oster, G., *Journal of the American Chemical Society,* Vol. 68, 1946, pp. 2036–2041.
[35] Van Vleck, J. H., *Journal of Chemical Physics,* Vol. 5, 1937, pp. 556–568.
[36] Kirkwood, J. G., *Journal of Chemical Physics,* Vol. 7, 1939, pp. 911–919.
[37] Kirkwood, J. G., *Transactions of the Faraday Society,* Vol. 42A, 1946, p. 7.
[38] Oster, G. and Kirkwood, J. G., *Journal of Chemical Physics,* Vol. 11, 1943, pp. 175–178.
[39] Fröhlich, H., *Transactions of the Faraday Society,* Vol. 44, 1948, pp. 238–243.
[40] Fröhlich, H. and Sack, R., *Proceedings of the Royal Society,* A, Vol. 182, 1944, pp. 388–403.
[41] Fröhlich, H., *Theory of Dielectrics,* Oxford University Press, Oxford, 1958.
[42] Smyth, C. P., *Dielectric Behavior and Structure,* McGraw-Hill, New York, 1955.
[43] Brown, W. F. in *Handbuch der Physick,* Vol. XVII, Dielectrics, S. Flügge, Ed., Springer Verlag, Berlin, 1956.
[44] Cole, R. H., *Journal of Chemical Physics,* Vol. 27, 1957, pp. 33–35.
[45] Buckingham, A. D. in *Physical Chemistry—An Advanced Treatice,* H. Eyring, D. Henderson, and W. Jost, Eds., Academic Press, New York, 1970.
[46] Wilson, J. N., *Chemical Reviews,* Vol. 25, 1939, pp. 377–406.
[47] Scholte, T. G., *Physica's Grav,* Vol. 15, 1949, pp. 437–450.
[48] Ross, I. G. and Sack, R. A., *Proceedings of the Physical Society of London, B,* Vol. 63, 1950, pp. 893–906.
[49] Abott, J. A. and Bolton, H. C., *Transactions of the Faraday Society,* Vol. 48, 1952, pp. 422–430.
[50] Omini, M., *Physica,* Vol. 83A, 1976, pp. 431–453.
[51] Omini, M., *Physica,* Vol. 84A, 1976, pp. 129–142.
[52] Omini, M., *Physica,* Vol. 84A, 1976, pp. 492–506.
[53] Omini, M., *Physica,* Vol. 88A, 1977, pp. 478–496.
[54] Omini, M., *Transactions on Electrical Insulation,* Institute of Electrical and Electronics Engineers, Vol. EI-20, 1985, pp. 965–973.
[55] Kielich, S. in *Dielectrics and Related Molecular Processes,* Vol. I, M. Davies, Ed., The Chemical Society, Burlington House, London, 1972.
[56] Adams, D. J., *Proceedings of the Royal Society of London,* Vol. A394, 1984, pp. 137–160.
[57] Berne, B. J. and Pechukas, P., *Journal of Chemistry and Physics,* Vol. 56, 1972, pp. 4213–4216.
[58] Van Vleck, J. H., *Molecular Physics,* Vol. 24, 1972, pp. 341–348.
[59] Dekker, A. J., *Electrical Engineering Materials,* Prentice-Hall Inc., Englewood Cliff, NJ, 1959.
[60] McAlpine, K. B. and Smyth, C. P., *Journal of Chemical Physics,* Vol. 3, 1935, pp. 55–57.
[61] McAlpine, K. B. and Smyth, C. P., *Journal of the American Chemical Society,* Vol. 55, 1933, pp. 453–462.
[62] Smyth, C. P., *Journal of Chemical Physics,* Vol. 1, 1933, pp. 247–250.
[63] Mathes, K. N., *Transactions on Electrical Insulation,* Institute of Electrical and Electronics Engineers, Vol. EI-2, 1967, pp. 24–32.
[64] Jeffries, M. J. and Mathes, K. N., *Transactions on Electrical Insulation,* Institute of Electrical and Electronics Engineers, Vol. EI-5, 1970, pp. 83–91.
[65] Errera, M. J., *Journal de Physique et le Radium,* Vol. 6, 1925, pp. 390–396.
[66] Wesson, K. G., *Tables of Electric Dipole Moments,* MIT Technology Press, Cambridge, MA, 1948.
[67] Allen, P. W. and Sutton, L. E., *Acta Crystallographica,* Vol. 3, 1950, pp. 46–72.
[68] Stuart, H. A., *Molekülstructur,* Springer, Berlin, 1934.
[69] Smyth, C. P. and Walls, W. S., *Journal of Chemical Physics,* Vol. 3, 1935, pp. 557–559.
[70] Groves, L. G. and Sugden, S., *Journal of the Chemical Society,* Part II, 1934, pp. 1094–1098.
[71] Bartnikas, R., *Transactions on Electrical Insulation,* Institute of Electrical and Electronics Engineers, Vol. EI-2, 1967, pp. 33–54.

[72] Mulliken, R. S., Rieke, C. A., and Brown, W. G., *Journal of the American Chemical Society,* Vol. 63, 1941, pp. 41–56.
[73] Bartnikas, R. in *Power Cable Engineering,* R. Bartnikas and K. D. Srivastava, Eds., Sandford Educational Press, Waterloo, Ontario, 1987.
[74] Smyth, C. P. and Stoops, W. N., *Journal of the American Chemical Society,* Vol. 51, 1929, pp. 3312–3341.
[75] Kurtz, S. S. and Martin, C. S., *India Rubber World,* Vol. 126, 1952, pp. 495–499.
[76] Lipkin, M. R., Sankin, A., and Martin, C. C., *Analytical Chemistry,* Vol. 20, 1948, pp. 598–606.
[77] Martin, C. C. and Sankin, A., *Analytical Chemistry,* Vol. 25, 1953, pp. 206–214.
[78] van Nes, K. and van Westen, H. A., *Aspects of the Constitution of Mineral Oils,* Elsevier, New York, 1951.
[79] Kurtz, S. S., King, R. W., Stout, W. J., and Peterkin, M. E., *Analytical Chemistry,* Vol. 30, 1958, pp. 1224–1235.
[80] King, R. W., Kust, M. A., and Kurtz, S. S., *Analytical Chemistry,* Vol. 32, 1960, pp. 733–745.
[81] Kurtz, S. S., King, R. W., Stout, W. J., Partikian, and Strabek, E. A., *Analytical Chemistry,* Vol. 28, 1956, pp. 1928–1936.
[82] Hückel, W., *Theoretischen Grundlagen der Organischen Chemie,* Vol. I, Elsevier, Amsterdam, 1955.
[83] Hückel, W., *Theoretischen Grundlagen der Organischen Chemie,* Vol. II, Elsevier, Amsterdam, 1958.
[84] Barry, A. J., *Journal of Applied Physics,* Vol. 17, 1946, pp. 1020–1024.
[85] Sauer, R. O. and Mead, D. J., *Journal of the American Chemical Society,* Vol. 68, 1946, pp. 1794–1797.
[86] Baker, E. B., Barry, A. J., and Hunder, M. J., *Industrial Engineering Chemistry,* Vol. 38, 1946, p. 1117.
[87] Dasgupta, S. and Smyth, C. P., *Journal of Chemical Physics,* Vol. 47, 1967, pp. 2911–2916.
[88] Mark, J. E. and Flory, P. J., *Journal of the American Chemical Society,* Vol. 86, 1964, pp. 138–141.
[89] Crescenzi, V. and Flory, P. J., *Journal of the American Chemical Society,* Vol. 86, 1964, pp. 141–146.
[90] Flory, P. J., Crescenzi, V., and Mark, J. E., *Journal of the American Chemical Society,* Vol. 86, 1964, pp. 141–146.
[91] *Tables of Interatomic Distances and Configurations in Molecules and Ions,* L. E. Sutton, Ed., The Chemical Society, London, 1958.
[92] Eyring, H., *Physical Review,* Vol. 39, 1932, pp. 746–748.
[93] Mark, J. E., *Journal of Chemical Physics,* Vol. 49, 1968, pp. 1398–1402.
[94] Sutton, C. and Mark, J. E., *Journal of Chemical Physics,* Vol. 54, 1971, pp. 5011–5014.
[95] Clark, F. M., *Insulating Materials for Design and Engineering Practice,* Wiley, New York, 1962.
[96] Siemens, W., *Annalen der Physik und Chemie,* Vol. 125, 1865, pp. 137–139.
[97] Hess, A., *Journal de Physique,* Vol. 2, 1893, pp. 145–160.
[98] Rowland, H. A., *Philosophical Magazine,* Vol. XLV, 1898, pp. 66–85.
[99] Hopkinson, J., *Philosophical Magazine,* Vol. II, 1876, pp. 314–315.
[100] Tank, F., *Annalen der Physik,* Vol. 48, 1915, pp. 307–359.
[101] Wagner, K. W., *Annalen der Physik,* Vol. 40, 1913, pp. 817–855.
[102] Pellat, H., *Annales de Chimie et de Physique,* Vol. 18, 1899, pp. 150–181.
[103] Pellat, H., *Journal de Physique,* Vol. 9, 1900, pp. 313–323.
[104] Whitehead, J. P., *Lectures on Dielectric Theory and Insulation,* McGraw-Hill, New York, 1927.
[105] Cole, K. S. and Cole, R. H., *Journal of Chemical Physics,* Vol. 9, 1941, pp. 341–353.
[106] Fischer, E., *Physikalische Zeitschrift,* Vol. 40, 1939, pp. 645–663.
[107] Hill, N. E., *Proceedings of the Physical Society of London,* Vol. 67B, 1954, pp. 149–158.
[108] Gierer, A., Spernol, A., and Wirtz, K., *Zeitschrift für Naturforschung,* Vol. 8a, 1953, pp. 522–532.
[109] Glasstone, S., Laidler, K. J., and Eyring, H., *The Theory of Rate Processes,* McGraw-Hill, New York, 1941.
[110] Bayer, E., *Cahiers de Physique,* Vol. 20, 1944, p. 1.
[111] Bartnikas, R., *Transactions on Electrical Insulation,* Institute of Electrical and Electronics Engineers, Vol. EI-5, 1970, pp 36–41.
[112] Jackson, W., *Proceedings of the Royal Society of London,* Vol. 153A, 1935–36, pp. 158–165.
[113] Frank, F. C., *Transactions of the Faraday Society,* Vol. 32, 1936, pp. 1634–1647.
[114] Baker, W. O. and Smyth, C. P., *Journal of Chemical Physics,* Vol. 5, 1937, pp. 113–124.
[115] Kono, R., McDuffie, G. E., and Litovitz, T. A., *Journal of Chemical Physics,* Vol. 44, 1966, pp. 965–970.
[116] Hakim, R., *Transactions on Electrical Insulation,* Institute of Electrical and Electronics Engineers, Vol. EI-6, 1971, pp. 158–164.
[117] Hill, N. E., Vaughan, W. E., Price, A. H., and Davies, M., *Dielectric Properties and Molecular Behavior,* Van Nostrand Reinhold Company, London, 1969.
[118] Glarum, S. H., *Journal of Chemical Physics,* Vol. 33, 1960, pp. 639–643.
[119] Anderson, J. E. and Ullman, R., *Journal of Chemical Physics,* Vol. 47, 1967, pp. 2178–2184.

[*120*] Johari, G. P. and Dannhauser, W., *Journal of Chemical Physics,* Vol. 50, 1969, pp. 1862–1876.
[*121*] Whitehead, J. P., *Impregnated Paper Insulation,* Wiley and Sons, New York, 1935.
[*122*] Bönning, P., *Elektrische Isolierstoffe: Ihr Verhalten anf Grund der Ionenadsorption an Inneren Grenzflächen,* Vieweg and Sohn, Braunschweigh, 1938.
[*123*] Garton, C. G., *Journal,* Institution of Electrical Engineers, Vol. 88, Part II, 1941, pp. 103–120.
[*124*] Mead, D. J. and Fuoss, R. M., *Journal of the American Chemical Society,* Vol. 62, 1940, pp. 1720–1723.
[*125*] Piper, J. D. in *Dielectric Materials and Applications,* A. von Hippel, Ed., Wiley and Sons, New York, 1954, pp. 156–167.
[*126*] Creighton, H. J., *Principles and Applications of Electrochemisty,* Vol. I, Chapman and Hall Ltd., London, 1943.
[*127*] Dunkley, J. and Sillars, R. W., *Proceedings,* Institution of Electrical Engineers, Vol. 100, Part II-A, 1953, pp. 89–93.
[*128*] Clarke, S. A. and Reynolds, E. H., *Dielectrics,* Vol. 1, 1963, pp. 26–44.
[*129*] Nederbragt, G. W., *Journal,* Institution of Electrical Engineers, Vol. 79, 1936, pp. 282–290.
[*130*] Ruhle, F., *Archiv für Electrotechnik,* Vol. 35, 1941, pp. 490–501.
[*131*] Angerer, L., *Dielectrics,* Vol. 2, 1964, pp. 234–235.
[*132*] Dimeler, G. R., Mills, I. W., and Melchiore, J. J., *Transactions on Electrical Insulation,* Institute of Electrical and Electronics Engineers, Vol. EI-4, 1969, pp. 7–12.
[*133*] Fessler, W. A., Nichols, F. S., and Rouse, T. O., *1978 IEEE International Symposium on Electrical Insulation,* Institute of Electrical and Electronics Engineers, Conference Record No. 78CH 1287-2-EI, Philadelphia, 12–14 June 1978, pp. 265–274.
[*134*] Melchiore, J. J. and Mills, I. W., *Journal of Electrochemical Society,* Vol. 112, 1965, pp. 390–395.
[*135*] Lamarre, C. and Crine, J.-P., *Transactions on Electrical Insulation,* Institute of Electrical and Electronic Engineers, Vol. EI-20, 1985, pp. 639–641.
[*136*] Krishnamoorthy, P. R., Vijayakumari, S., and Sankralingam, S., *Transactions on Electrical Insulation,* Institute of Electrical and Electronics Engineers, Vol. 27, 1992, pp. 271–277.
[*137*] Forster, E. O., *1972 Annual Report, Conference on Electrical Insulation and Dielectric Phenomena,* National Academy of Sciences, Washington, 1973, pp. 30–36.
[*138*] Sillars, R. W., *Electrical Insulating Materials and their Application,* Peter Peregrinus Ltd., (IEE), Stevenage, Herts, 1973.
[*139*] Lidiard, A. B. in *Handbuch der Physik,* Vol. 20, *Electrische Leitungsphänomene II,* S. Flügge, Ed., Springer Verlag, Berlin, 1957.
[*140*] Stevels, J. M. in *Handbuch der Physik,* Vol. 20, *Elektrische Leitungsphänomene II,* S. Flügge, Ed., Springer Verlag, Berlin, 1957.
[*141*] Taylor, H. E., *Journal of the Society of Glass Technology,* Vol. 44, 1957, pp. 350T–382T.
[*142*] Hakim, R. M., *Transactions on Electrical Insulation,* Institute of Electrical and Electronics Engineers, Vol. EI-7, 1972, pp. 185–195.
[*143*] Tasköprülli, N., Barlow, A., and Lamb, J., *Journal of the Acoustical Society of America,* Vol. 33, 1961, pp. 278–285.
[*144*] Coelho, R., *Second International Conference on Conduction in Low Mobility Materials,* Taylor and Francis Press, London, 1971, pp. 439–451.
[*145*] Hakim, R., Oliver, R. G., and St-Onge, H., *Transactions on Electrical Insulation,* Institute of Electrical and Electronics Engineers, Vol. 12, 1977, pp. 360–370.
[*146*] Vincent, G., *1974 Annual Report, Conference on Electrical Insulation and Dielectric Phenomena,* National Academy of Sciences, Washington, DC, 1975, pp. 559–566.
[*147*] Crine, J.-P., Grob, R., Casanovas, J., and Barbay, H., *Transactions on Electrical Insulation,* Institute of Electrical and Electronics Engineers, Vol. EI-21, 1986, pp. 233–237.
[*148*] Wien, M., *Physikalische Zeitschrift,* Vol. 29, 1928, pp. 751–781.
[*149*] Wien, M., *Physikalische Zeitschrift,* Vol. 32, 1931, pp. 545–547.
[*150*] Nikuradse, A., *Das flüssige Dielectrikum,* Julius Springer, Berlin, 1933.
[*151*] Plumley, H. J., *Physical Review,* Vol. 59, 1940, pp. 200–207.
[*152*] Onsager, L., *Journal of Chemical Physics,* Vol. 2, 1934, pp. 599–615.
[*153*] von Smoluchowski, M., *Physikalische Zeitschrift,* Vol. 15, 1916, pp. 585–599.
[*154*] Bjerrum, N., *Kgl, Danske Vid, Selskab, Math.-fys.medd.,* Vol. 7, 1926, p. 9.
[*155*] Brière, G. B., *British Journal of Applied Physics,* Vol. 15, 1964, pp. 413–417.
[*156*] Mead, D. J. and Fuoss, R. M., *Journal of the American Chemical Society,* Vol. 61, 1939, pp. 2047–2053.
[*157*] Eicke, H. F., *Berichtungen der Bunsengeselschaft, Physikalische Chemie.,* Vol. 71, 1967, p. 384.
[*158*] Nackaerts, R., DeMaeyer, M., and Hellemans, L., *Journal of Electrostatics,* Vol. 7, 1979, pp. 169–186.
[*159*] Denat, A., Gosse, B., and Gosse, J. P., *Journal of Electrostatics,* Vol. 7, 1979, pp. 205–255.
[*160*] Denat, A., Gosse, B., and Gosse, J. P., *Journal of Electrostatics,* Vol. 11, 1982, pp. 179–194.
[*161*] Denat, A., Gosse, B., and Gosse, J. P., *Journal of Electrostatics,* Vol. 11, 1982, pp. 197–205.

[162] Aij, A., Denat, A., Gosse, J. P., and Gosse, B., *Transactions on Electrical Insulation,* Institute of Electrical and Electronics Engineers, Vol. EI-20, 1985, pp. 221–231.
[163] Sharbaugh, A. H. and Watson, P. K. in *Progress in Dielectrics,* Vol. 4, J. B. Birks and J. Hart, Eds., Heywood and Company Ltd., London, 1962.
[164] Yasufuku, S., Umemura, T., and Tanii, T., *Transactions on Electrical Insulation,* Institute of Electrical and Electronics Engineers, Vol. EI-14, Feb. 1979, pp. 28–35.
[165] Shimokawa, H., Ohashi, A., and Ueda, M., *Journal of Physics D,* Vol. 13, 1980, pp. 107–116.
[166] Tsuchida, N. and Ueda, M., *Journal of Physics D,* Vol. 13, 1980, pp. 1681–1687.
[167] Casanovas, J., Grob, R., Chemin, A., Guelfucci, J. P., and Crine, J.-P., *Transactions on Electrical Insulation,* Institute of Electrical and Electronics Engineers, Vol. EI-20, 1985, pp. 143–146.
[168] Casanovas, J., Chemin, A., Guelfucci, J. P., Grob, R., and Crine, J.-P., *Revue Physique Appliqué,* Vol. 22, 1987, pp. 1037–1042.
[169] Liebscher, F., *Elektrotechnische Zeitschrift (ETZ),* Vol. 64, 1943, pp. 425–427.
[170] Liebscher, F., *Elektrotechnische Zeitschrift (ETZ),* Vol. 64, 1943, pp. 450–453.
[171] Schumann, W. O., *Zeitschrift für Physik,* Vol. 79, 1932, pp. 532–549.
[172] Böning, P., *Elektrische Isolierstoffe: Ihr Verhalten auf Grund der Ionenadsorption an inneren Grenzflächen,* Vieweg und Sohn, Braunschweigh, 1938.
[173] Garton, C. G., *Journal,* Institution of Electrical Engineers, Vol. 88, Part III, 1941, pp. 23–40.
[174] Jaffé, G., *Annalen der Physik,* Vol. 16, 1933, pp. 249–284.
[175] Jaffé, G., *Physical Review,* Vol. 85, 1952, pp. 354–363.
[176] MacDonald, J. R., *Physical Review,* Vol. 92, 1953, pp. 4–17.
[177] Friauf, R. J., *Journal of Chemical Physics,* Vol. 22, 1954, pp. 1329–1338.
[178] Trukhan, E. M., *Soviet Physics—Solid State,* Vol. 4, 1962, pp. 2560–2570.
[179] Stern, F. and Weaver, C., *Journal of Physics C,* Vol. 3, 1970, pp. 1736–1746.
[180] Held, W. and Wenzel, K., *Elektrotechnische Zeitschrift (ETZ),* Vol. 81, 1960, pp. 121–127.
[181] Schmidt, W. F. and Allen, A. O., *Journal of Chemical Physics,* Vol. 52, 1970, pp. 4788–4794.
[182] Schmidt, W. F., Bakale, G., and Sowada, V., *Journal of Chemical Physics,* Vol. 61, 1974, pp. 5275–5278.
[183] Bagley, B. G., *Solid State Communications,* Vol. 8, 1970, pp. 345–348.
[183a] Kok, J. A., *Electrical Breakdown of Insulating Liquids,* Eindhoven, Philips Technical Library, 1961.
[184] von Schweidler, E., *Annalen der Physik,* Vol. 24, 1907, pp. 711–770.
[185] Ulhmann, D. R. and Hakim, R. M., *Journal of Physics and Chemistry of Solids,* Vol. 32, 1971, pp. 2652–2655.
[186] Davidson, D. W. and Cole, R. H., *Journal of Chemical Physics,* Vol. 18, 1950, pp. 1417–1422.
[187] Nee, T. W. and Zwanzig, R., *Journal of Chemical Physics,* Vol. 52, 1970, pp. 6353–6363.
[188] Tanke, J., Litovitz, T. A., and Macedo, P. B., *Journal of the American Ceramic Society,* Vol. 51, 1968, pp. 158–163.
[189] Reed, C. W. in *Dielectric Properties of Polymers,* F. E. Karasz, Ed., Plenum Pres, New York, 1972.
[190] Rieche, H., *Zeitschrift für Physik,* Vol. 95, 1935, pp. 158–178.
[191] Ramu, T. S., *Transactions on Electrical Insulation,* Institute of Electrical and Electronics Engineers, Vol. EI-14, 1979, pp. 136–141.
[192] Higasi, K., Bergmann, K., and Smyth, C. P., *Journal of Physical Chemistry,* Vol. 64, 1960, pp. 880–883.
[193] Lui, Q., Wan, R., Wang, Y., and Chang, X., *Ninth International Conference on Conduction and Breakdown in Dielectric Liquids,* Institute of Electrical and Electronics Engineers, Conference Record, Salford, U.K., 27–31 July 1987, pp. 198–202.
[194] Mizutani, T., Ieda, M., Ochai, S., and Ito, M., *Journal of Electrostatics,* Vol. 12, 1982, pp. 427–433.
[195] Lee, C. L., Johannson, O. K., Flaningham, O. L., and Hahn, P., *Polymer Reprints,* Vol. 10, pp. 1311–1326.
[196] Sugden, S., *Transactions of the Faraday Society,* Vol. 30, 1934, pp. 734–739.
[197] Cartwright, C. H. and Errera, J., *Proceedings of the Royal Society,* Vol. 154A, 1936, pp. 138–157.
[198] Martens, F. F., *Annalen der Physik,* Vol. 6, 1901, pp. 603–640.
[199] Paschen, F., *Annalen der Physik,* Vol. 26, 1908, pp. 120–138.
[200] Born, M., *Annalen der Physik,* Vol. 55, 1918, pp. 177–240.
[201] Herzfeld, K. F. and Wolf, K. L., *Annalen der Physik,* Vol. 28, 1925, pp. 35–56.
[202] Dennison, D. M., *Philosophical Magazine,* Vol. 1, 1926, pp. 195–218.
[203] Petro, A. J. and Smyth, C. P., *Journal of the American Chemical Society,* Vol. 80, 1958, pp. 73–76.
[204] Chamberlain, J. E., Findlay, F. D., and Gebbie, H. A., *Nature,* Vol. 206, 1965, pp. 886–887.
[205] Chamberlain, J. E., Chantry, G. W., Gebbie, H. A., Stone, N. W. B., Taylor, T. B., and Wyllie, G., *Nature,* Vol. 210, 1966, pp. 790–791.
[206] Horner, F., Taylor, T. A., Dunsmuir, R., Lamb, J., and Jackson, W., *Journal,* Institution of Electrical Engineers, Vol. 93, Part III, 1946, pp. 53–68.

[207] Bartnikas, R. in *Engineering Dielectrics, Vol. II-B, Electrical Properties of Solid Insulating Materials: Measurement Techniques, ASTM STP 926,* American Society for Testing and Materials, Philadelphia, 1987.
[208] Barberich, L. J., *Electrical Engineering,* Vol. 55, 1936, pp. 264–268.
[209] Balsbaugh, J. C., Kenney, N. D., and Herzenberg, A., *Electrical Engineering,* Vol. 54, 1935, pp. 272–279.
[210] Balsbaugh, J. C., Larsen, R. G., and Oncley, J. L., *Industrial and Engineering Chemistry,* Vol. 30, 1938, pp. 287–293.
[211] Balsbaugh, J. C. and Assaf, A. G., *Transactions,* American Institute of Electrical and Electronics Engineers, Vol. 62, 1943, pp. 311–322.
[212] Webb, J. K., *Journal,* Institution of Electrical Engineers, Vol. 87, 1940, pp. 689–691.
[213] Stark, K. H., *Proceedings,* Institution of Electrical Engineers, Vol. 100, Part II-A, 1953, pp. 89–93.
[214] Martin, R. G. and Patterson, E. A., *Proceedings,* Institution of Electrical Engineers, Vol. 100, Part II-A, 1953, pp. 68–72.
[215] Hartshorn, L. and Ward, W. H., *Journal,* Institution of Electrical Engineers, Vol. 79, 1936, pp. 597–609.
[216] Reddish, W., Bishop, A., Buckingham, K. A., and Hyde, P. J., *Proceedings,* Institution of Electrical Engineers, Vol. 118, 1971, pp. 597–609.
[217] Buckingham, K. A. and Billing, J. W., *Third International Conference on Dielectric Materials, Measurements and Applications,* Birmingham, 10–13 Sept. 1979, pp. 392–395.
[218] Works, C. N., Dakin, T. W., and Boggs, F. W., *Proceedings,* Institute of Radio Engineers, Vol. 33, 1945, pp. 245–254.
[219] Kakimoto, A., Ogawa, I., and Matsushita, T., *Review of Scientific Instruments,* Vol. 48, 1977, pp. 1570–1575.
[220] Whitehead, S., *Transactions of the Faraday Society,* Vol. 42A, 1946, pp. 66–78.
[221] Davidson, D. W., Anty, R. P., and Cole, R. H., *Review of Scientific Instruments,* Vol. 22, 1951, pp. 678–682.
[222] St-Onge, H., *Transactions on Electrical Insulation,* Institute of Electrical and Electronics Engineers, Vol. EI-11, 1976, pp. 20–27.
[223] Mopsik, F. I., *Review of Scientific Instruments,* Vol. 55, 1984, pp. 79–87.
[224] Takeishi, S., Nozaki, R., Yagihara, S., and Mashimo, S., *Review of Scientific Instruments,* Vol. 54, 1983, pp. 639–640.
[225] *Journal of the Chemical Society, Faraday Transactions II,* Vol. 68, 1972, pp. 1890–1896.
[226] Volger, J., Stevels, J. M., and van Amerongen, C., *Philips Research Reports,* Vol. 8, 1953, pp. 452–470.
[227] Nakajima, T. and Saito, S., *Journal of Polymer Science,* Vol. 31, 1958, pp. 423–436.
[228] Scheiber, D. J., *Journal of Research of the National Bureau of Standards,* Vol. 65C, 1961, pp. 23–42.
[229] Thompson, A. M., *Proceedings,* Institution of Electrical Engineers, Vol. 103, Parts B and C, 1955, pp. 704–709.
[230] Harris, W. P., "Operators Procedure Manual for the Harris Ultra Low Frequency Impedance Bridge," National Bureau of Standards Report No. 9627, Washington, DC, Dec. 1968.
[231] Vince, P. M., *Proceedings of the Institute of Electrical Engineers,* Vol. 112, 1965, pp. 432–440.
[232] Pugh, J. in *Fourth International Conference on Dielectric Materials, Measurements and Applications,* Institution of Electrical Engineers' Conference Publication No. 239, 10–13 Sept. 1984, Lancaster, England, pp. 247–249.
[233] Geyger, W., ATM., No. J924-1, 1936.
[234] Poleck, H., *Wissenschaftliche Veröffnungen des Siemens Konzerns,* Vol. 18, 1939, pp. 129–147.
[235] Petersons, O., *Transactions on Instrumentation and Measurement,* Institute of Electrical and Electronics Engineers, Vol. IM-13, 1964, pp. 216–224.
[236] Seitz, P. and Osvath, P., *Third International Symposium on High Voltage Engineering,* Milan, 1979.
[237] Osvath, P. and Widmer, S., *Transactions on Instrument and Measurement,* Institute of Electrical and Electronics Engineers, Vol. IM-35, 1986, pp. 19–23.
[238] Krüger, M., "Testing of Dielectric Properties of Insulating Liquids," Baur Technical Report No. 8602e, Sulz, Austria, 1986.
[239] Barrie, I. T., *Proceedings,* Institution of Electrical Engineers, Vol. 112, pp. 408–415.
[240] Drude, P., *Annalen der Physik,* Vol. 55, 1895, pp. 633–655.
[241] Drude, P. P., *Zeitschrift der Physikalische Chemie,* Vol. 23, 1897, pp. 267–325.
[242] Borguis, F., *Naturwissenschaften,* Vol. 29, 1941, p. 516.
[243] Works, C. N., *Journal of Applied Physics,* Vol. 18, 1947, pp. 605–612.
[244] Parry, T. V. L., *Journal,* Institution of Electrical Engineers, Vol. 98, 1951, Part III, pp. 303–311.
[245] Kirchhoff, G., *Monatsberichte der Königlich Preussischen Akademie der Wibenschaften zu Berlin,* 1877, pp. 144–162.
[246] Linhart, H., *Zeitschrift der Physikalische Chemi,* Vol. B38, 1937, pp. 23–45.

[*247*] Roberts, S. and von Hippel, A., *Journal of Applied Physics,* Vol. 17, 1946, pp. 610–616.
[*248*] Williams, G., *Journal of Physical Chemistry,* Vol. 63, 1959, pp. 534–537.
[*249*] Corfield, M. G., Horzelski, J., and Price, A. H., *British Journal of Applied Physics,* Vol. 12, 1961, pp. 680–682.
[*250*] Dakin, T. W. and Works, C. N., *Journal of Applied Physics,* Vol. 18, 1947, pp. 789–796.
[*251*] von Hippel, A. in *Dielectric Materials and Applications,* A. von Hippel, Ed., John Wiley, New York, 1954.
[*252*] Bennett, R. G. and Calderwood, J. H., *Proceedings,* Institution of Electrical Engineers, Vol. 112, 1965, pp. 416–420.
[*253*] Lamb, J., *Journal,* Institution of Electrical Engineers, Vol. 93, Part III-A, 1946, pp. 1447–1451.
[*254*] Bleaney, B., Loubser, J. H. N., and Penrose, R. P., *Proceedings of the Physical Society,* Vol. 59, 1947, pp. 185–199.
[*255*] Barlow, H. M. and Cullen, A. L., *Microwave Measurements,* Constable and Company Ltd., London, 1966.
[*256*] Cole, R. H., *Journal of Physical Chemistry,* Vol. 79, 1975, pp. 1469–1974.
[*257*] Cole, R. H., Mashimo, S., and Winsor, P., *Journal of Physical Chemistry,* Vol. 84, 1980, pp. 786–793.
[*258*] Nakamura, W., Mashimo, S., and Wada, A., *Japanese Journal of Applied Physics,* Vol. 21, 1982, pp. 467–474.
[*259*] Bäz, G., *Physikalische Zeitschrift,* Vol. 40, 1939, pp. 394–404.
[*260*] Garg, S. K., Kilp, H., and Smyth, C. P., *Journal of Chemical Physics,* Vol. 43, 1963, pp. 2341–2346.
[*261*] Afsar, M. N., Birch, J. R., and Chamberlain, J., Conference of Dielectric Materials, Measurements and Applications, Institution of Electrical Engineers, 21–25 July 1975, Cambridge, U.K.

W. F. Schmidt[1]

Chapter 2

Conduction Mechanisms in Liquids

2.1 Introduction

Dielectric liquids have served as insulators or dielectrics in electrical equipment for over one hundred years already. Equally long in duration is the history of research on the electrical conduction of these liquids. The results are not too impressive if we compare the state of knowledge with that which has been achieved for gases or solids. This situation is due to the fact that in most applications of insulating liquids in electrotechnology, mixtures of liquids are being used which in addition are in contact with a variety of materials. Transformer oil contains well over one hundred chemical compounds. In a transformer it is in contact with organic and inorganic materials. Traces of these materials will leak into the liquid, and it is often that the presence of these traces determines the overall properties of the liquid insulation in the particular apparatus. This complex situation impeded the progress in the understanding of the basic conduction mechanisms in dielectric liquids because in fundamental studies it was attempted to incorporate as many conditions as possible under which actual liquid insulations were employed in industry. The result is an enormous amount of empirical information which is difficult to understand or to put into a proper perspective. Many books or reviews published during the last fifty years reflect this situation [*1–7*]. This lack of understanding of the fundamental properties of liquid insulations may be one of the reasons for their gradual replacement in electrotechnical apparatus. Large power transformers, railway transformers, power capacitors, and paper/oil insulated high voltage cables are nowadays the major electrotechnical devices in which mineral oil and other liquids are used as insulants and coolants. Judging from the developments in gaseous and solid state electronics, it can be predicted that a better understanding of the basic conduction processes in liquids will also lead to new applications.

At present, water filled discharge lines for pulsed power are under intensive investigation and development. In the area of high voltage equipment, fluorocarbons, although relatively expensive, and their use recently discouraged because of their effects on the ozone layer, are (or were) on their way to replace mineral oil in applications where weight and safety requirements impose stringent restrictions. In the fields of chemical engineering or surface coating processes, atomization of liquids by electric charging could lead to the improvement of existing processes or to the development of new systems. Machining of metals by electrical sparks in a dielectric liquid (known as electroerosion or spark machining) are fields where progress has been made in recent years. Laser-induced breakdown at a liquid/solid interface may have also a potential in materials processing. The study of electronic conduction processes in ultrapure liquids [*8*] has led to the development of liquid ionization chambers, a new class of detectors for high energy radiation. The ultrapurification of nonpolar liquids may also lead to new applications of these liquids in other fields of science and engineering.

[1]Hahn-Meitner-Institut Berlin, Abteilung Strahlenchemie, D14109 Berlin 39, Germany.

In the present chapter we shall attempt to give a brief description of the basic physical principles of electrical conduction in nonpolar and polar liquids. The basic situation is depicted in Fig. 2.1. Two metal electrodes are immersed in a dielectric liquid. This arrangement represents a capacitance C. Voltage is applied and a current flows in the external circuit. Several mechanisms may govern the magnitude and the temporal evolution of this current: bulk conductivity of the liquid, injection at one or both electrodes, or blockage of the current at one or both electrodes. If an alternating voltage, V, is applied to C a current I is flowing which is given by

$$I = C\,dV/dt + V/R \tag{2.1}$$

R is the ohmic resistance of the dielectric liquid. This current consists of two contributions, a conduction current, V/R and a dielectric displacement current, CdV/dt. In vector notation the current density, $\bar{J}$, in SI units is given as

$$\bar{J} = \sigma\bar{E} + \epsilon_r\epsilon_o\frac{\partial\bar{E}}{\partial t} \tag{2.2}$$

where $\sigma\bar{E}$ is the conduction current density component and $\epsilon_r\epsilon_o\partial\bar{E}/\partial t$ is the displacement current density component. $\bar{E}$ represents the electric field vector, ϵ_r is the relative permittivity, and ϵ_o is the permittivity *in vacuo*. With a d-c voltage applied the conduction current is observed, only, while

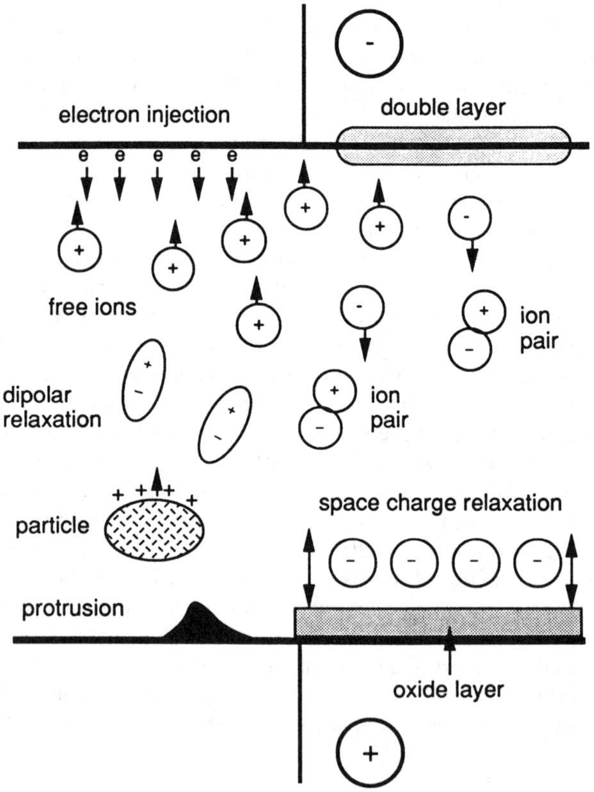

FIG. 2.1—*Conduction mechanisms in dielectric liquids.*

with an a-c voltage both currents contribute to the current observed in the external measurement circuit. In this chapter both mechanisms will be discussed although the main emphasis is on the conduction current. More detailed discussions on the displacement current in liquid insulation can be found in Chapter 1 of this volume. In the present chapter, emphasis is put on the elementary processes in chemically pure liquids with examples extended to practical insulating liquids. It is hoped that a deeper understanding of the conduction phenomena in these liquids will also lead to a better understanding of the processes occurring in actual insulation systems with dielectric liquids. The various conductivity mechanisms treated here are depicted schematically in Fig. 2.2.

The chapter is organized as follows: Section 2.2 deals with the chemical structure, the dielectric and electronic properties of the liquids and with their thermophysical properties. Section 2.3 describes the major mechanisms leading to charge carrier formation or injection in dielectric liquids. In Section 2.4, a summary of the transport properties of charge carriers in dielectric liquids is given. Section 2.5 summarizes briefly the major areas of engineering liquid dielectrics. In Section 2.6 the conductivity of nonpolar liquids is discussed. Included is also a short subsection on the chemical transformations induced by the passage of electrical current. Section 2.7 treats the conduction of water and other polar liquids. Breakdown and prebreakdown mechanisms are discussed in Section 2.8. Here, we could be brief, since these phenomena are treated extensively in Chapter 3. In Section 2.9 other applications of and developments with dielectric liquids in electrotechnology and related areas are presented.

The literature references selected for the different topics were chosen with the intention in mind of providing the reader with good introductory papers for a given subject and to give an overview of the various scientific fields involved in dielectric research. With the help of modern data banks, retrieval of more information on a particular topic should represent no problem. The series of Conference Records of the International Conference on Conduction and Breakdown of Dielectric Liquids (ICDL) provides another source of up to date information [*9–13*]. The tables quoted are to serve for orientation, only. For the most accurate values of certain quantities collections of reference data should be consulted. All equations in this chapter are written in the International System of Units (SI-units), unless indicated otherwise.

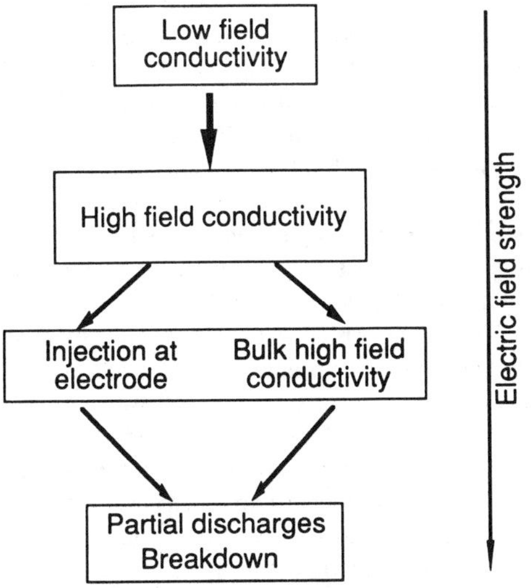

FIG. 2.2—*Basic conduction mechanisms as a function of electric field strength.*

2.2 Properties of Liquids

2.2.1 General Remarks

Most of the liquids which are employed in dielectric engineering are, from a chemical point of view hydrocarbons and their derivatives, siloxanes, simple atomic and molecular cryogenic liquids, and water and aqueous solutions. There exist many more dielectric liquids. A better understanding of the basic principles of electrical conduction and of the influence of the molecular structure on their electrical properties may lead to new and more unconventional applications.

With respect to the application of insulating liquids in electrotechnical equipment the following electrical, chemical and thermophysical properties are important: **electrical properties**—electrical conductivity, dielectric constant and dielectric losses, electrical breakdown strength, electrical arc extinguishing properties; **chemical properties**—chemical stability of the liquid under elevated temperature and electrical stress, gas generating and gas absorbing properties, carbon formation, fire resistance, environmental compatibility; **thermo-physical properties**—thermal conductivity, specific heat, viscosity, density.

All of these properties can never be optimized with one particular liquid. Recent developments in liquid insulation attempt to incorporate as many as possible of the good properties in one liquid for a specific application. Especially, environmental requirements recently have placed the focus on the nonelectrical properties. The treatment of these problems is beyond the scope of this chapter. The emphasis in this chapter is more on conductivity mechanisms such as conductivity and breakdown strength as Chapter I deals specifically with dielectric constant and loss. In the following sections we give an overview of the atomic and molecular properties which determine the electrical properties of insulating liquids.

2.2.2 Chemical Bonds

The molecular liquids we are concerned with here are composed of molecules in which the atoms are bound by covalent chemical bonds. The simplest example is the hydrogen molecule in which two protons are bound together by two electrons. In order to achieve a better understanding of this chemical bond we will recapitulate some facts from the quantum mechanics of atoms and molecules.

The movement of electrons around nuclei is described by quantum mechanics. Each electron occupies a certain orbital which is characterized by three quantum numbers, n, 1, and m_l. The main quantum number n determines essentially the energy of the bound electron with respect to the ground state, and it determines the spatial extension of the orbital. The quantum number 1 is related to the total angular momentum, and it can take values only given by $1 < (n-1)$. The orbital with $1 = 0$ is called *s*-orbital, with $1 = 1$ *p*-orbital, with $1 = 2$ *d*-orbital, and so on. m_l is the magnetic quantum number which becomes important when the atom is placed in a magnetic field. m_l can take values from -1 to 0 to $+1$. In the simplest atom, the isolated hydrogen atom, the electron is spread out in a spherically symmetric distribution of charge around the proton. This state is called the 1s atomic orbital of the hydrogen atom. It is characterized by the following quantum numbers, $n = 1$, $1 = 0$, $m_l = 0$. The same orbital can be occupied by two electrons with their spins antiparallel. An example is the helium atom where two *s*-electrons occupy the lowest orbital with $n = 1$. These two electrons form a closed electronic shell which is also called K-shell. Additional electrons can only be accommodated on orbitals with the main quantum number $n = 2$. For $1 = 0$, the *s*-orbital can accommodate two electrons. For $1 = 1$, three *p*-orbitals are possible with $m_l = -1, 0, +1$. Each *p*-orbital can accommodate two electrons. A new electron shell, the L-shell is formed, which contains 2 *s*-electrons and 6 *p*-electrons. The charge distributions of the three *p*-orbitals extend into the *x*-, *y*-, and *z*-direction (Fig. 2.3).

When two hydrogen atoms combine, a molecular orbital is formed which is occupied by the two *s*-electrons. Their spins are antiparallel. The electronic charge distribution is of rotational symmetry

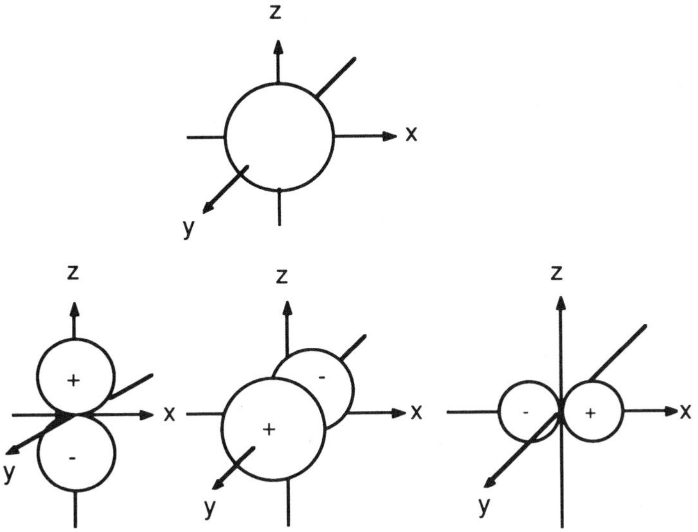

FIG. 2.3—*Atomic orbitals for s-electrons and p-electrons.*

around the axis which connects the two atoms. This molecular orbital is called σ-orbital. The chemical bond produced by this orbital is called, σ-bond. Since two electrons are involved in the formation of the bond one can also say that the bond is formed by an electron pair (Fig. 2.4). Atoms with more than one electron in their outer shell can form several bonds.

In the case of carbon, the atom has in addition to the two *s*-electrons of the inner K-shell, four outer electrons on the L-shell, two *s*-electrons and two *p*-electrons. Combination of atomic *p*-orbitals leads to molecular orbitals which are called π-orbitals. The corresponding electronic charge distribution has nodes in a plane connecting the atoms. The binding strength of a molecular π-orbital is smaller than that of a σ-orbital. In the carbon atom a hybridization of the *s*- and *p*-orbitals takes place, so that with respect to hydrogen all four electrons are equivalent. In methane (CH_4), for instance, all hydrogen atoms are bound with the same strength to the carbon atom. The same holds for single bonds between carbon atoms. If double or triple bonds are formed between carbons atoms then the σ- and π-character of the bonds becomes apparent again. A double bonds consists of a σ- and a π-bond, while in a triple bond two bonds have π-character and one bonds has σ-character. The atomic separation of two carbon atoms decreases with the number of bonds which are acting between them (Table 2.1). In the case of the hydrogen molecule the electrons are distributed equally between the two atoms. If two different atoms form a chemical bond the electron distribution may be shifted with its center towards one of the atoms. In the limit of large separation of positive and negative charge an ionic bond is formed, as for instance in the case of sodium chloride. The covalent bond exhibits partially an ionic character. The strength of this effect is measured as the electronegativity of the particular atoms involved in the bond. Halogens, oxygen, and nitrogen have large values on the electronegativity scale, which means that they attract electrons, hydrogen and semimetals have medium values while metals have small values. The partial ionic character of a covalent bond is approximately given by the difference of the electronegativity of the atoms involved [*14*]. Partially ionic covalent bonds are in general stronger than pure covalent bonds due to the additional electrostatic attraction. Examples of bond energies are compiled in Tables 2.2 and 2.3. These data give only a general indication of the strengths of the various bonds. In real polyatomic molecules, the other atoms bound to the two atoms which form the bond influence the energy necessary to break the bond between these two atoms.

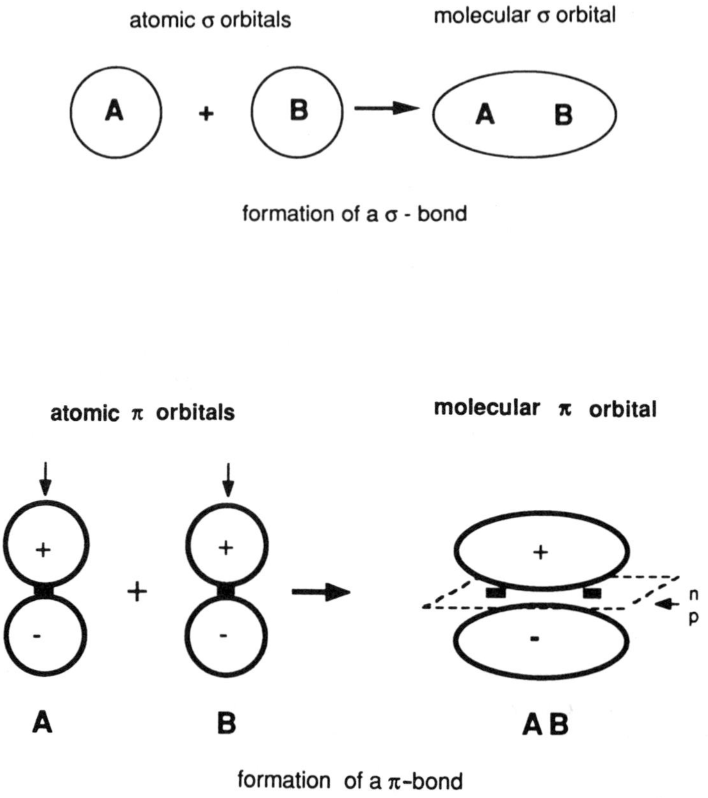

FIG. 2.4—*Molecular orbitals in a σ-bond and a π-bond.*

TABLE 2.1—*Bond dissociation energy and bond length for carbon* [14,15].

Bond	Energy eV	Energy kJ/mol	Bond Length, nm
C − C	3.6	347	0.154
C = C	6.3	610	0.134
C ≡ C	8.7	836	0.120

2.2.3 Dielectric Constant

One of the important physical quantities of the liquids used in electrotechnology is their dielectric constant. The dielectric constant is intimately related to the electronic structure of the atoms or molecules comprising the liquid [*16*]. Its magnitude is determined by the fact whether the molecules possess an electric dipole moment or not. The liquefied rare gases and molecular liquids with molecules which do not have a dipole moment are called nonpolar. In the atoms or molecules of these liquids the centers of the positive and negative charge distributions coincide. In large molecules, certain groups of atoms may exhibit dipole moments (microdipoles), the sum of which cancels. For instance, in a n-hexane molecule the C—H group is polarized with the hydrogen being more negative than the carbon atom; still, the total dipole moment of the molecule is zero. These

TABLE 2.2—*Bond energies of single bonds at 298 K* [14].

	Energy			Energy	
Bond	eV	kJ/mol	Bond	eV	kJ/mol
H—H	4.5	436	C—H	4.3	413
C—C	3.6	347	C—O	3.6	351
Si—Si	1.8	176	C—Si	3.0	290
O—O	1.4	139	C—F	4.6	441
N—N	1.7	160	C—Cl	3.4	328
F—F	1.6	155	C—N	3.0	291
Cl—Cl	2.5	242	Si—O	3.8	369
			Si—H	3.1	294
			Si—F	5.6	540
			Si—Cl	3.7	358
			N—H	4.1	390

TABLE 2.3—*Average bond energies* [15].

	Single Bond		Double Bond		Triple Bond	
Bond	eV	kJ/mol	eV	kJ/mol	eV	kJ/mol
C—C	3.6	347	6.3	610	8.7	836
N—N	1.7	167	3.9	376	9.8	940
O—O	1.4	138	5.1	489		
C—O	3.6	351	7.5	719		
C—N	3.3	318	6.3	606	9.3	890

nonpolar liquids are characterized by a relative dielectric constant ϵ_r of around 2. A few illustrative examples are summarized in Fig. 2.5. If such atoms or molecules are placed in an electric field, a separation of the centers of the positive and negative charge distributions results, and the atoms or molecules become polarized (electronic polarization or induced polarization). The corresponding dipole moment is proportional to the electric field strength. The proportionality constant is called polarizability, α of the atom or molecule and is defined by Eq 1.8 in Chapter 1. A molecule in which in the absence of an external electric field the centers of the positive and negative charge distributions do not coincide has in addition a permanent dipole moment, μ. Permanent dipoles are connected with asymmetries of the molecular structure or with the presence of electronegative atoms in the molecule. Some examples which are relevant in the context of this chapter are shown in Fig. 2.6. Thermal motion leads to a random orientation of the molecular dipoles so that the liquid does not exhibit a spontaneous polarization. On a microscopic scale association of two or more dipoles may occur, which gives rise to particular physico-chemical properties of the liquid. In an electric field, orientation of the dipoles in field direction is counterbalanced by the thermal motion. The resulting specific polarization is given by Eq 1.12*b* (in Chapter 1). Since in addition the external electric field induces also a displacement of the electron shell with respect to the positive nuclei (electronic polarization, as in the case of the nonpolar molecules), the total specific polarization is then given by Eq 1.13 in Chapter 1. The relation between polarization and relative permittivity or dielectric constant, ϵ_r is given by the Clausius-Mossotti equation

$$\frac{\epsilon_r - 1}{\epsilon_r + 1} = \frac{n}{3\epsilon_o}\left(\alpha + \frac{\mu^2}{3kT}\right) \tag{2.3}$$

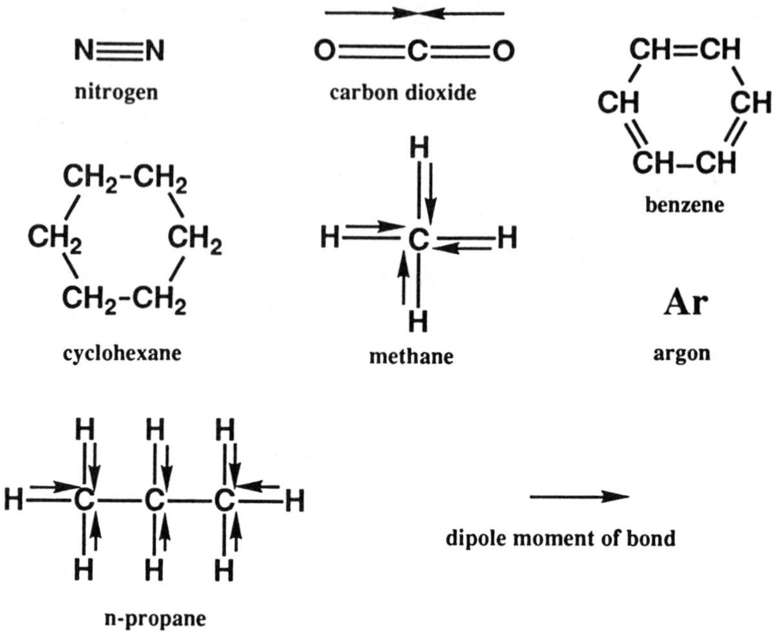

FIG. 2.5—*Examples of nonpolar molecules.*

k denotes the Boltzmann constant, T is the absolute temperature, and n is the number density. Polar liquids exhibit a wide range of dielectric constants depending on the magnitude of the dipole moment. Examples of dielectric constants, polarizabilities, and dipole moments are compiled in Table 2.4. The relative dielectric constant ϵ_r is a function of temperature. Of special importance are

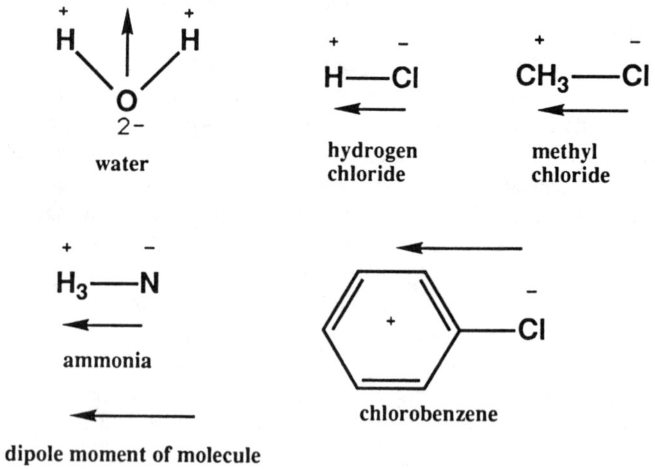

FIG. 2.6—*Examples of polar molecules (the dipole moment is directed from the negative charge to the positive charge).*

TABLE 2.4—*Dielectric constant, ϵ_r, electronic polarization, α, and dipole moment, μ* [18,19].

Liquid	t°C	$T(K)$	ϵ_r	$\alpha\ 10^{24}$ (cm^3)	$\mu\ 10^{30}$ (cm)
Transformer oil	20	293	2.2		0
Methane	−173	100	1.70	2.6	0
n-hexane	20	293	1.89	11.78	0
n-decane	20	293	1.99	19.1	0
Cyclohexane	20	293	2.9	10.8	0
Benzene	20	293	2.28	10.3	0
Carbon tetrachloride	20	293	2.24	10.5	0
Methanol	25	298	32.63	3.9	1.7
Ethanol	25	298	24.30	5.4	1.7
Glycerol	25	298	42.5	9.9	2.7
Ammonia	−33.4	239.6	22.4	2.4	1.44
Helium	−268.8	4.2	1.048	0.215	0
Argon	−191	82	1.53	1.64	0
Xenon	−161	112		4.0	0
Hydrogen	−252	20.4	1.23	0.79	0
Nitrogen	−196	77		1.76	0
Carbon dioxide	20	293	1.60	2.9	0
Sulfur	118	391	3.52		0
Nitrobenzene	25	298	34.82	14.7	4.23
Propylene carbonate	25	298	63		5.0
Water	25	298	78.5	1.5	1.84

the dielectric properties of water. The temperature dependence of the static dielectric constant of water, $\epsilon_r(w)$ between 0 and 100 degrees was found to follow the equation [*17*],

$$\epsilon_r(w) = 87.74 - 0.40008\theta + 9.398\ 10^{-4}\theta^2 - 1.41\ 10^{-6}\theta^3 \tag{2.4}$$

with θ in degrees Celsius. In electrolytic aqueous solutions a lowering of the dielectric constant occurs. Up to concentrations of 2 mol/L, the dielectric constant of the aqueous solution decreases linearly with solute concentration according to [*19*]

$$\epsilon_r = \epsilon_r(w) - (\delta_+ + \delta_-)c \tag{2.5}$$

where c is the concentration in moles per liter and δ_+, δ_- are the specific values given in Table 2.5.

Measurement of the dielectric constant is usually performed with an a-c Wheatstone bridge. Since the electric field is varying in time Eq 2.2 describes the current in the measurement cell. The measurement cell is treated as a lossy capacitor of capacitance C and conductance G (Fig. 2.7). The complex admittance Y is then given as

$$Y = G + j\omega C \tag{2.6}$$

where $\omega = 2\pi f$, with f as the frequency of the a-c voltage, and

$$Y = 1/Z \tag{2.7}$$

where Z is the complex impedance. The capacitance C of the test cell filled with the dielectric liquid is given as

$$C = \epsilon' C_o \tag{2.8}$$

TABLE 2.5—*Dielectric constant of aqueous electrolytic solutions; the lowering of the dielectric constant of water per mole of ions per liter of solution* [19].

Cations	δ_+	Anions	δ_-
H^+	17	F^-	5
Li^+	11	Cl^-	3
Na^+	8	I^-	7
K^+	8	OH^-	13
Rb^+	7	$SO_4^=$	7
Mg^{++}	24		
Ba^{++}	22		
La^{+++}	35		

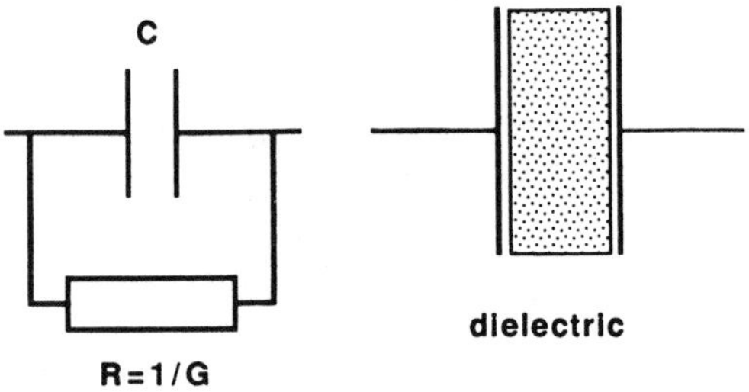

FIG. 2.7—*The equivalent circuit of a test cell of capacitance* C, *filled with a liquid dielectric of conductance 1/*R.

where C_o represents the capacitance of the empty test cell and ϵ' is the real part of the complex dielectric constant, thus ϵ^* defined as

$$\epsilon^* = \epsilon' - j\epsilon'' \tag{2.9}$$

where ϵ'' describes the energy losses in the liquid dielectric due to electronic, atomic and dipolar orientational polarization. In addition, Joule heat is dissipated in the liquid due to the presence and movement of ionic charge carriers. If an ionic conductivity, σ_{ion}, exists, then the complex dielectric constant can be written as

$$\epsilon^* = \epsilon' - j(\epsilon'' + \sigma_{ion}/\epsilon_o\omega) \tag{2.10}$$

The complex capacitance of the test cell can then be written as

$$C^* = C_o[\epsilon' - j(\epsilon'' + \sigma_{ion}/\epsilon_o\omega)] \tag{2.11}$$

For the admittance Y we write

$$Y = j\omega C^* \tag{2.12}$$

Comparison of Eqs 2.6, 2.11, and 2.12 yields for the real part of the capacitance

$$C = \epsilon' C_o \tag{2.13}$$

and for the conductance G,

$$G = \omega C_o(\epsilon'' + \sigma_{ion}/\epsilon_o\omega) \tag{2.14}$$

which consist of two contributions, the relaxation losses and the losses due to ionic conductivity. The admittance Y is best measured with a Wheatstone bridge (Fig. 2.8). Each of the four arms contains an impedance Z for which at balance holds

$$Z_1/Z_2 = Z_3/Z_4 \tag{2.15}$$

The unknown admittance $Y = 1/Z_1$ is then given as

$$Y = Z_3Z_4/Z_2 \tag{2.16}$$

At low frequencies where $1/f < \tau$ (τ-relaxation time), the contribution of the relaxation processes to G can be neglected, and G is proportional to the ionic conductivity σ_{ion}. At these frequencies the a-c and d-c conductivities have the same value.

Above a few hundred megahertz, measurement systems with coaxial lines or wave guides have to be employed. The test cell consists of a coaxial line (Fig. 2.9) which is closed at one end by a flat metal plate. This represents a short circuit where the electromagnetic wave is reflected. A standing wave is generated in the wave guide. The wavelength and amplitude of this wave depends on the complex dielectric constant of the liquid. If an electromagnetic wave with an electric field strength E_o travels through the coaxial line filled with a dielectric liquid, then attenuation occurs, and, after a distance d, the field strength is reduced to a value $E(d)$ given as

$$E(d) = E_o \exp(-j\omega t) \exp(-\gamma d) \tag{2.17}$$

Here γ denotes the complex propagation constant

$$\gamma = \alpha + j\beta \tag{2.18}$$

where α is the attenuation coefficient, and β denotes the phase constant. α is measured as the ratio of the two signals V which are proportional to the electric field strengths at 0 and d

$$\alpha = (1/d) \ln (V(d)/V(0)) \tag{2.19}$$

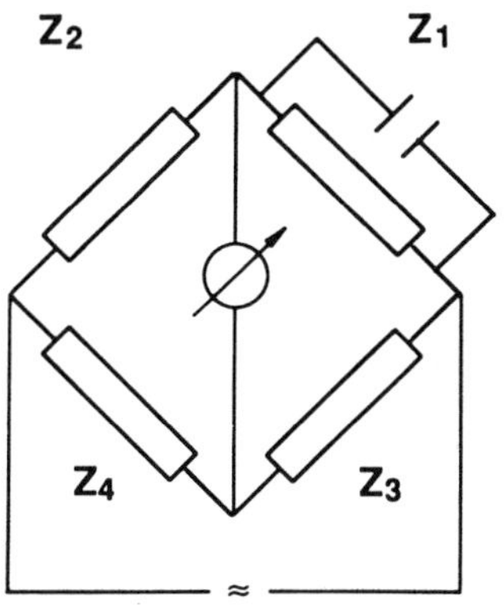

FIG. 2.8—*Schematic diagram of an a-c Wheatstone bridge.*

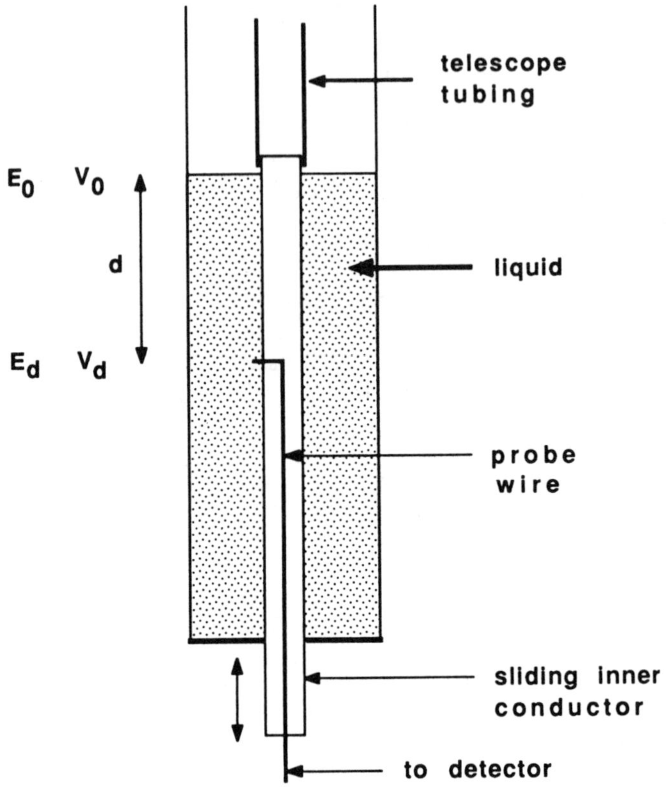

FIG. 2.9—*Coaxial test cell for the measurement of the complex dielectric constant of liquids by the standing wave method.*

where α is expressed in neper cm^{-1}; alternatively. α in decibel cm^{-1} leads to the equation

$$\alpha = (1/d)20 \ln (V(d)/V(0)) \tag{2.20}$$

Conversion is achieved by

$$\alpha[\text{neper cm}^{-1}] = 0.1151\alpha[db\ \text{cm}^{-1}] \tag{2.21}$$

The wavelength with a liquid specimen within the wave guide, λ_{liq}, is related to the phase constant β by

$$\lambda_{liq} = (2\pi/\beta) \tag{2.22}$$

The propagation constant is approximately given as

$$\gamma = \frac{2\pi}{\lambda_o}\sqrt{-\epsilon^*} \tag{2.23}$$

where λ_o denotes the wave length in the wave guide with vacuum. Expansion of this complex equation into the real and the imaginary parts yields

$$\epsilon' = \frac{\lambda_o^2}{4\pi^2}(\beta^2 - \alpha^2) \tag{2.24}$$

and

$$\epsilon'' = \frac{\alpha\beta\lambda_o^2}{2\pi^2} \tag{2.25}$$

For a loss-less liquid ($\alpha = 0$), λ_{liq} becomes

$$\lambda_{\text{liq}} = \lambda_o\sqrt{\epsilon'} \tag{2.26}$$

The wave length of the standing wave is measured by means of a movable probe protruding from the inner conductor (see Fig. 2.9).

The quantities α and β are related to the index of refraction, n, and to the absorption index, κ, by the following relations

$$\kappa = \alpha/\beta \tag{2.27}$$

and

$$n = \lambda_o/\lambda_{\text{liq}} \tag{2.28}$$

ϵ^*, n, and κ are connected by Eqs 1.366 and 1.367 in Chapter 1.

2.2.4 Structure

The liquid state is characterized on the molecular level by an absence of long range forces and long range order and by the presence of short range forces and short range order. Generally, the following elementary attractive interactions between two molecules (or atoms) can be distinguished [*20*]: (*a*) dipole-dipole interactions, (*b*) dispersion interaction, and (*c*) hydrogen bonds. At very short distance between two atoms or molecules a repulsive force sets in (*d*).

2.2.4.1 Dipole-Dipole Interaction—Two polar molecules of dipole moments μ_1, μ_2 at a distance r are bound by an average energy, E_{dip} given as (absolute cgs-system)

$$\overline{E}_{\text{dip}} = -\frac{2}{3}\frac{\mu_1^2\mu_2^2}{r^6}\frac{1}{RT} \tag{2.29}$$

Dipole-dipole interactions decrease rapidly with distance (r^{-6}) and increasing temperature T. R denotes the universal gas constant. The interaction leads to the formation of stabile molecular clusters in the liquid. Examples for this process are water and alcohols, which are also called associative liquids.

2.2.4.2 Dispersion Interaction (van der Waals Interaction)—This force between two nonpolar molecules or atoms is caused by the fast motion of inner shell electrons. At any moment in time the molecule (or atom) exhibits a fast rotating dipole which interacts with other molecules (or atoms). The strength of the interaction can be calculated quantum mechanically. It is proportional to some power of the electronic polarizability of the molecule (or atom). The range varies with r^{-6}.

2.2.4.3 Hydrogen Bonds—Another force which is important in our context is the hydrogen bridge. Interaction between the OH- or NH-groups of one molecule with the electronegative atoms

O, N, F, or Cl of another molecule leads to formation of a temporary bond. This type of interaction is connected with the high mobility of protons in a molecule.

2.2.4.4 Attraction and Repulsion—Between nonpolar and weakly polar molecules the dispersion is the major force. Between small molecules with a large dipole moment, as water and methanol, the dipole-dipole interaction is predominant. At very short distances between molecules or atoms repulsive forces set in. Attraction and repulsion can be described by a potential proposed by Lennard and Jones. The interaction energy is given as

$$\overline{E}_{dip} = -\underbrace{\frac{M}{r^6}}_{\text{attraction}} + \underbrace{\frac{N}{r^q}}_{\text{repulsion}} \quad (2.30)$$

where $q = 12$ is a good approximation for small molecules. E_{dip} as a function of r is shown in Fig. 2.10. At an equilibrium distance, r_{min}, the energy E_{dip} is a minimum.

2.2.4.5 Short Range Order—The forces discussed here lead to the formation of a short range order which extends over dimensions of a few molecular diameters. The short range order depends on the molecular shape. Benzene molecules, for instance can form stacks consisting of several molecules. Short paraffin molecules can arrange themselves as parallel rods. Polar molecules may form clusters of defined size, especially if hydrogen bridges are participating (Fig. 2.11). Due to the lack of long range order, vacancies exist between molecules which change continuously in shape and volume as a result of the thermal motion of the molecules. Energy fluctuations lead then to a jump-like migration of individual molecules. With increasing temperature more and more molecules obtain higher kinetic energies which allows them to overcome their mutual attraction. Low density regions and finally a vapor phase are formed [*21*].

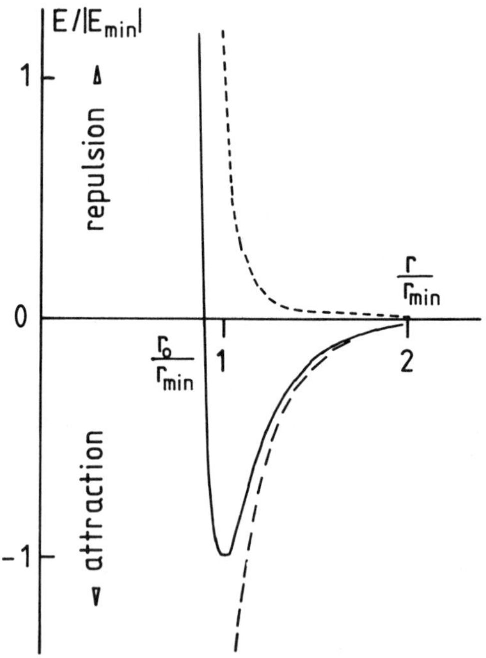

FIG. 2.10—*Lennard-Jones potential of interaction between two atoms or nonpolar molecules.*

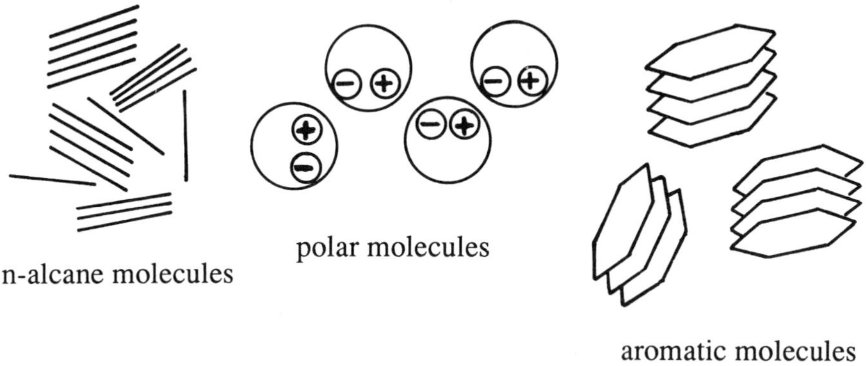

FIG. 2.11—*Arrangement of molecules in the liquid phase.*

2.2.5 Electronic Properties

In the pure state, nonpolar liquids are excellent insulators. An idea on the generation of electron/ion or electron/hole pairs can be obtained by applying a model developed for intrinsic semiconductors. Although liquids do not exhibit a lattice arrangement of the atoms or molecules (lack of long-range order), it has been shown that near the melting point a microscopic short-range order exits. In semiconductors the electrical conductivity decreases by a factor of two, only, if the material changes from solid to liquid [*22*]. This indicates that the basic physical properties governing the electrical conductivity are affected only moderately. We may assume the existence of a valence band and a conduction band separated by a band gap, I_{liq}. A simplified energy diagram is shown in Fig. 2.12. The ionization energy of atoms or molecules in the liquid phase, I_{liq}, has been obtained from measurements of the photo conductivity in the vacuum ultraviolet (uv) spectral range. A collection of values is given in Table 2.6. The values range from 6.8 eV to over 10 eV which is an indication that these liquids should behave as ideal insulators, or in other words that thermal generation of electron/hole pairs at room temperature should be negligible. The ionization energy I_{liq} for most liquids is reduced as compared to the gas phase ionization energy, I_{gas} on account of three contributions: V_o, the electron affinity of the liquid (also known as the energy of the conduction band with respect to vacuum), P_+, the polarization energy of the positive charge, and E_v, a broadening of the levels of the valence electrons due to intermolecular forces. Values of $V_o <$ 0 indicate that energy is released if an electron is traversing from the vacuum into the liquid, $V_o > 0$ means that energy has to be provided to the electron in order for it to enter from the vacuum into the liquid. The energy P_+ associated with the polarization exerted on the liquid by the positive charge carrier can be estimated by means of Born's formula [*23*]

$$P_+ = -\frac{e_o^2}{8\pi\epsilon_o R}\left[1 - \frac{1}{\epsilon_r}\right] \tag{2.31}$$

Here R denotes the ionic radius, e_o is the electronic charge. P_+ is always negative. The broadening of the valence levels for nonpolar liquids is small, of the order of -0.1 eV. The ionization energy in the liquid phase is then given as

$$I_{\text{liq}} = I_{\text{gas}} + V_o + P_+ + E_v \tag{2.32}$$

The quantity V_o can be measured by a photoelectric method [*24*]. Values obtained for pure liquids are also included in Table 2.6. The polarization energy P_+ can be estimated with the help of Eq

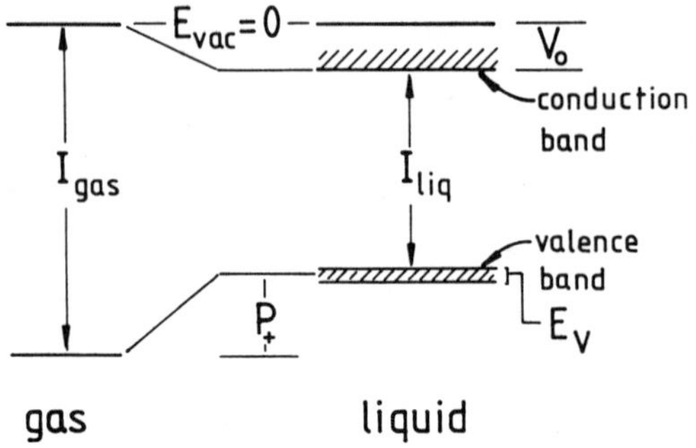

FIG. 2.12—*Simplified diagram of electronic energy levels in a dielectric liquid.*

2.31, or it can be obtained from measured values of I_{liq}, I_{gas}, and V_o from Eq 2.32. The application of Eq 2.31 requires knowledge of the ionic radius, for which in first approximation the hard core radius as determined from viscosity or diffusion measurements may be taken. This approach fails if large molecules are involved. For most liquids $I_{liq} < I_{gas}$ has been measured; however, in liquid helium and in liquid neon on account of their low dielectric constants and of the high positive values of V_o, the liquid ionization energy is greater than the gas phase ionization energy [*25*].

TABLE 2.6*a*—*Ionization energies,* I_{liq} *in nonpolar liquids* [30–35].

Liquid	$T(K)$	I_{liq} (eV)
Xenon	161	9.2
Krypton	121	11.56
Tetramethylsilane	294	8.1 ± 0.05
		8.05 ± 0.05
Tetramethylgermanium	294	7.6 ± 0.05
Tetramethyltin	294	6.9 ± 0.1
2,2-dimethylpropane	294	8.85 ± 0.05
(Neopentane)		8.55 ± 0.05
Methylbutane	294	9.15 ± 0.05
(Isopentane)		
n-pentane	294	9.15 ± 0.1
		8.85 ± 0.05
2,2-dimethylpropane	294	8.73 ± 0.05
(Neohexane)		8.50 ± 0.05
n-hexane	294	8.6 ± 0.05
3-methylpentane	294	8.85 ± 0.1
2,2,4-trimethylpentane	294	8.3 ± 0.05
(Isooctane)		
2,2,4,4-tetramethylpentane	295	8.2
n-tridecane	294	9.25 ± 0.05
Tetrakis(dimethylamino)ethylene	295	3.77 ± 0.02
n-pentene-1	294	8.33 ± 0.05
Tetramethylethylene	294	6.80 ± 0.05
Cyclopentane	294	8.80 ± 0.05
Cyclohexane	294	8.75 ± 0.1
		8.43 ± 0.05
Hexamethyldisilane	294	6.75 ± 0.1
Triethylsilane	294	8.25 ± 0.1
Bistrimethyl silylethane	295	6.85 ± 0.05
Polydimethy siloxane (50 cSt)	296	7.29 ± 0.05

TABLE 2.6*b*—*Electronic energy level* V_o *in nonpolar liquids* [24–29].

Liquid	$T(K)$	V_o (eV)
Helium	4.2	+1.05
Neon	25	+0.67
Argon	84	−0.20
Krypton	116	−0.40
Xenon	161	−0.67
Methane	100	−0.25
Ethane	100	+0.20
n-pentane	296	0.0
n-hexane	296	+0.04
Neopentane	296	−0.43
Neohexane	296	−0.15
Isooctane	296	−0.18
Tetramethylsilane	296	−0.62
Tetramethyltin	296	−0.7
Polydimethyl-siloxane[a]	296	−1.0

[a]Recent measurement by Koizumi et al., *Journal of Physics D*, 1992.

TABLE 2.6*c*—*Energy levels in polar liquids* [25].

Liquid	$T(K)$	I_{liq} (eV)	I_{val} (eV)	V_o (eV)
Water[a]	300	6.5 ± 0.5		
	294	6.7	6.3	−1.3
Ethanol	294		7.4	−0.65
Methanol	294			−1.0

I_{val} = energy of the valence level with respect to the vacuum.
[a]Sander et al., *Berichte der Bungesellschaft für Physikalische Chemie*, Vol. 97, No. 953, 1993, have received recently the energy levels of liquid waters.

2.2.6 *Dielectric Losses*

The subject of dielectric losses constitutes the main topic of Chapter 1 of this volume. Here we are concerned with aspects relating to electrical conductivity. In Section 2.2.3 we discussed the static dielectric constant which is measured when the liquid dielectric is in equilibrium with the externally applied electric field. Such is the case with d-c or slowly varying a-c fields. With electric fields of higher frequency the response of the polarization of the dielectric liquid may lag behind the electrical forces it is subjected to by the electric field. This lag in time usually is called relaxation. It is due to the frictional resistance of the liquid to the change in molecular orientation. The time scale where this relaxation is active in a dielectric liquid depends on the type of polarization and on the structure of the liquid. Relaxation of dipoles in a liquid, which is connected with a rotation of the dipole, occurs in a lower frequency range than electronic polarization which involves the displacement of the electronic shell relative to the positive nuclei. Another relaxation is produced by the atomic polarization, the displacement of the atoms in a molecule relative to one another. Table 2.7 gives an overview of the ranges of relaxation times due to these different polarization mechanisms. At the interface between two different materials an additional type of polarization occurs which is connected with the accumulation of charges at the interface. In test cells for liquid studies these interfaces may appear at the metal electrodes which are covered usually by an oxide layer.

The relaxation time is defined as the time in which the polarization is reduced to $1/e$ of its initial value when the externally applied electric field is suddenly removed. Each of these polarization

TABLE 2.7—*Relaxation times and polarization.*

Polarization	Relaxation time, s	Liquid
Electronic	10^{-15}	all liquids
Atomic	10^{-14} to 10^{-12}	all molecular liquids
Orientation	10^{-12} to 10^{-11}	small molecules, low viscosity
	10^{-6}	large molecules, high viscosity
Space charge	$>10^{-1}$	nonpolar liquids

processes leads to dissipation of energy in the dielectric liquids. They manifest themselves by a conductance component in phase with the applied electric field.

A slab of liquid insulation with two parallel electrodes can be represented as an ideal capacitor, *C*, and a parallel resistor, *R* (Fig. 2.13). If an a-c voltage, *V*, is applied, a current flows, which consists of the dielectric displacement current density J_c and the conduction or leakage current density J_1. While J_c precedes the voltage by a phase difference of $\pi/2$, the conduction current density is in phase with *V*. For a very good liquid dielectric J_c is very small and the total phase difference is close to $\pi/2$. The complementary phase angle *d* is then given as

$$\tan\delta = J_1/J_c = (1/R)/(\omega C) \tag{2.33}$$

Dielectric losses are produced by the movement of charges in the alternating electric field. Several contributions can be distinguished: (*a*) movement of ions or charged particles in the liquid, (*b*) rotation of permanent dipoles, (*c*) rotation of induced dipoles, (*d*) interfacial relaxation, and (*e*) space charge relaxation.

2.2.6.1 Movement of Charges—For parallel plates of area *A* separated by *d*, the capacitance *C* is given as

$$C = \epsilon_r \epsilon_o A/d \tag{2.34}$$

The conductance $1/R$ produced by the presence of free ions is given as

$$1/R = \sigma A/d \tag{2.35}$$

where σ denotes the specific conductivity. Then, for $\tan\delta$ we obtain

$$\tan\delta = \sigma/(\epsilon_r \epsilon_o \omega) \tag{2.36}$$

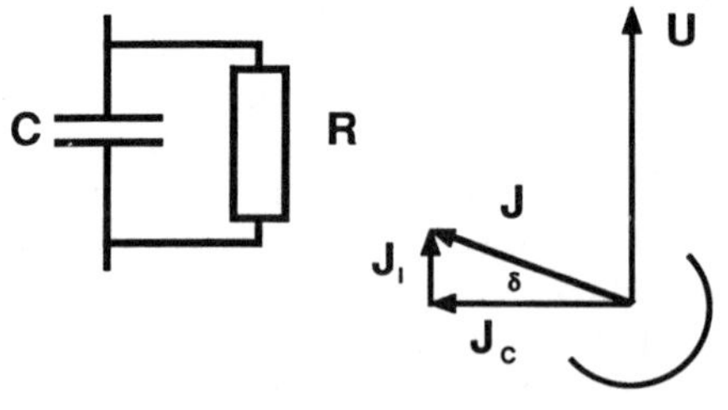

FIG. 2.13—*Conduction current density* J_L *and displacement current density* J_C *for a test cell filled with a conducting dielectric liquid.*

In liquids, where the specific conductivity σ is constant the loss tangent is inversely proportional to the frequency ω of the voltage. An example is shown in Fig. 2.14 [*36*]. The losses are probably due to impurity ions dissolved in the liquid.

2.2.6.2 Dipolar Relaxation—The presence of impurities with permanent dipoles in a nonpolar liquid or the response of polar liquids to an alternating electric field manifests itself in a maximum of the loss tangent as a function of frequency. A molecule with a permanent dipole moment undergoes orientational changes under the influence of the alternating electrical field. A phase shift may occur between the polarization and the electric field due to the viscous coupling of the dipolar movement to the liquid. As already demonstrated in Chapter 1, this will lead to a loss tangent given by

$$\tan\delta = \epsilon''/\epsilon' \tag{2.37}$$

where the real and imaginary permittivities are given by

$$\epsilon' = \epsilon_\infty + \frac{(\epsilon_r - \epsilon_\infty)}{(1 + \omega^2\tau^2)} \tag{2.38}$$

and

$$\epsilon'' = \frac{(\epsilon_r - \epsilon_\infty)\omega\tau}{(1 + \omega^2\tau^2)} \tag{2.39}$$

where τ is the relaxation time. For small and large values of $\omega\tau$, ϵ'' approaches zero. It reaches a maximum at $\omega\tau = 1$. Debye calculated the relaxation time for a dipolar molecule of radius a embedded in a viscous liquid of viscosity η to be

$$\tau = (4\pi\eta a^3)/(kT) \tag{2.40}$$

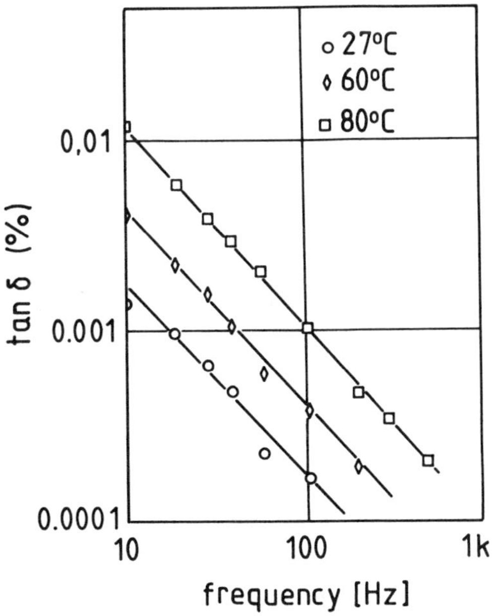

FIG. 2.14—*Tanδ as a function of frequency and temperature for diarylalkanes (after Ref* 36*).*

or if we introduce the hypothetical volume V_D of N_A (Avogadro number) dipole molecules

$$V_D = (4\pi a^3 N_A)/3 = (\tau RT)/(3\eta) \tag{2.41}$$

where R is the universal gas constant. With typical values for η between 1 and 10 cP and molecular radii between 2 and 10 Å, values for τ between 10^{-11} and 10^{-8} follow. Generally, a distribution of relaxation times has to be considered. In such a case, the measured data deviate from the semicircular Cole-Cole plot (see Chapter 1).

The relaxation time τ of water decreases with increasing temperature approximately exponentially. At 0°C a value of τ = 18 ps is measured which falls to τ = 4 ps at 60°C. The temperature dependence of τ as an Arrhenius plot is shown in Fig. 2.15. The data do not follow a straight line but they can be approximated by a dependence

$$\tau = \tau_o T^{-n} \exp(\Delta H/kT) \tag{2.42}$$

with n between 0 and 1 and ΔH between 15.3 and 18.3 kJ/mol.

Increase of dielectric losses of mineral oil may occur due to oxidation. Oxidation occurs in hydrocarbons due to the presence of air. The result is the formation of water, organic dipolar molecules, and organic acids. All of these substances lead to an increase of the dielectric losses. Antioxidants (AH) convert radicals of high reactivity, R·, ROO·, into radicals of low reactivity, A·

$$R\cdot + AH \longrightarrow RH + A\cdot \tag{2.43}$$

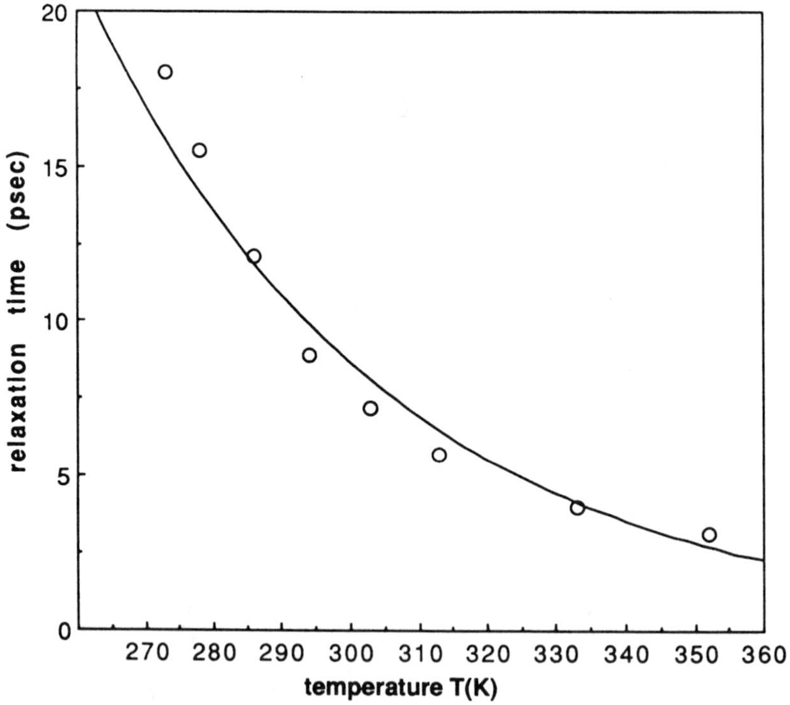

FIG. 2.15—*Dielectric relaxation time of water as a function of temperature.*

or

$$ROO\cdot + AH \longrightarrow ROOH + A\cdot \tag{2.44}$$

Two groups of antioxidants can be distinguished. The first group inhibits free radicals, R·, and it comprises phenols, amines, and aminophenols. The second group decomposes peroxides, and it comprises compounds containing sulfur. Most of them exhibit dipolar molecules and thus lead to an increase of the loss tangent. Some may even dissociate into ions and increase the conductivity of the transformer oil.

At high frequencies, only the electronic shell of the molecules can follow the electric field. The dielectric constant and the loss tangent reach a limiting value.

2.2.6.3 Induced Dipoles—Losses due to induced dipoles appear at very high frequencies of the a-c voltage or at high voltages. In nonpolar liquids, small dielectric losses are observed in the microwave range and in the far infrared which are probably caused by temporary dipole moments. These dipole moments are of the order of 0.1 *D* and they arise from the mutual interaction of colliding molecules. During the collision a distortion of the molecular framework takes place. Measurements of the complex dielectric constant, ϵ^* yields information on the losses [*37*]

$$\epsilon^* = \epsilon'(1 - j\tan\delta) \tag{2.45}$$

The dielectric loss factor as a function of the microwave frequency is shown in Fig. 2.16. From the maxima of $\tan\delta_{max}$ an effective dipole moment, μ_{eff} has been calculated by using the expression

$$\epsilon''_{max} = \epsilon_\infty \tan\delta_{max} \tag{2.46}$$

and

$$\epsilon''_{max} = \frac{2\pi N}{3k_BT}\mu^2_{eff}\left[\frac{n^2+2}{3}\right]^2 \tag{2.47}$$

n is the refractive index, N the number of molecules per cubic centimeter. The results are compiled in Table 2.8. The existence of a mean value of $\mu_{eff} = 0.063$ *D* for all alkanes seems to indicate that the temporary dipole moment is localized at both ends of the molecule.

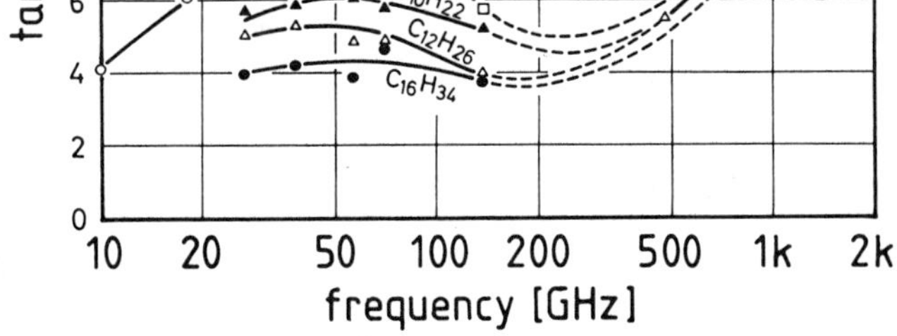

FIG. 2.16—*Tanδ as a function of frequency for n-alkanes (after Ref* 37).

TABLE 2.8—*Effective dipole moments in Debye units and dielectric relaxation times of n-alkanes at room temperature* [37].

Liquid	μ_{eff} (D)	t_{relax} (ps)
n-hexane	0.061	2.3
n-heptane	0.061	2.9
n-nonane	0.066	3.5
n-decane	0.061	3.8
n-dodecane	0.063	4.5
n-hexadecane	0.063	3.7

2.2.6.4 Interfacial Relaxation (Maxwell-Wagner)—This type of dielectric loss takes place when the dielectric in the test capacitor consists of two different materials with dielectric constants ϵ_1 and ϵ_2 and electrical conductivities σ_1 and σ_2 (Fig. 2.17). If a voltage step is applied, the initial distribution of the potential is determined by the dielectric constants, while the steady-state distribution is determined by the conductivities. Since these two distributions are not the same, application of an a-c voltage leads to a relaxation process and dielectric losses. Examples for this type of situation are solid/liquid insulation systems in transformers, oxide layers on electrodes immersed in liquids, and blocking electrodes.

2.2.6.5 Space Charge Polarization and Relaxation—Space charge formation takes place near electrodes which are unable to discharge the ions arriving at them. Reversal of the electric field leads to a redistribution of the space charge and to the evolution of a new equilibrium. Under the influence of a constant applied electric field a layer of material is macroscopically polarized. The charge density in the vicinity of the electrodes depends on the x coordinate only. A dipole moment is produced given as

$$\mu = \int \rho(x)\, dx \tag{2.48}$$

where $\rho(x)$ is the charge density. It is obvious that the relaxation time for the reversal of the polarization in an alternating electric field is related to the ion drift velocity. An approximate treatment of this problem is given by Coelho [38].

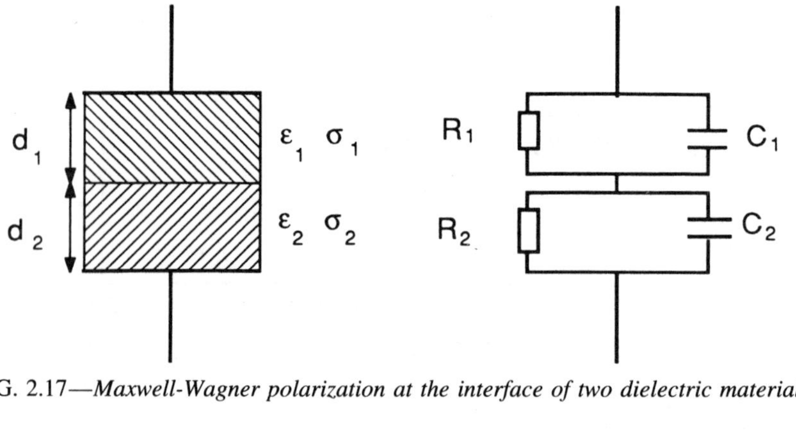

FIG. 2.17—*Maxwell-Wagner polarization at the interface of two dielectric materials.*

2.2.7 Thermo-Physical Properties

The thermo-physical properties of a liquid depend strongly upon the interaction of the molecules in the liquid state and on the presence of electrolytes. Nonpolar or weakly polar liquids do not support large concentrations of electrolytes and do not exhibit associative effects of the molecules. The thermo-physical properties of liquids comprised of polar molecules are influenced by strong dipole-dipole interactions. In this section we have compiled those thermo-physical data which relate to the conduction mechanisms. The data have been collected from several sources [*39–41*].

The thermo-physical state of a liquid can be described by an equation of state

$$f(V_\upsilon, p, T) = 0 \tag{2.49}$$

where V_υ denote the volume, p the pressure, and T the absolute temperature, respectively. Experiments show that the volume decreases with increasing pressure (an exception is water) and that the volume increases with temperature. The following coefficients can be defined:

the thermal expansion (p = const)

$$\alpha_{\text{cub}} = \frac{1}{V_\upsilon}\left(\frac{\partial V_\upsilon}{\partial T}\right) \tag{2.50}$$

the isothermal compressibility (T = const)

$$\beta_{\text{ic}} = -\frac{1}{V_\upsilon}\left(\frac{\partial V_\upsilon}{\partial p}\right) \tag{2.51}$$

and the mechanical tension (V_υ = const)

$$\chi = \frac{1}{p}\left(\frac{\partial p}{\partial T}\right) \tag{2.52}$$

Between these three quantities the following relation exists

$$\beta_{\text{ic}} = \frac{\alpha_{\text{cub}}}{p\chi} \tag{2.53}$$

In Table 2.9 values for the coefficients of thermal expanion and isothermal compressibility are compiled. Generally, both coefficients grow with increasing temperature. Nonassociative liquids can be described by the van der Waals equation although quantitative agreement is not very good. Of special importance are the critical quantities, $V_{\upsilon,\text{crit}}$, p_{crit}, and T_{crit}, respectively. Above the critical temperature, the liquid phase does not exist any longer. In Table 2.10 some characteristic critical data are compiled. They apply to one mole of the substance.

Another important quantity is the heat of evaporation Q_{bp} which defines the amount of energy necessary for conversion of a mole of liquid into a mole of gas. It consists of two portions, the inner heat which is used to overcome the attraction between the molecules or atoms in the liquid state, and an external portion which is due to the increase in volume. Usually, the external portion is roughly 10% of the inner portion. A large value of Q_{bp} is indicative of a strong interaction between the molecules or atoms of the liquid. The heat of evaporation decreases with increasing temperature, and it becomes zero at the critical temperature. For organic liquids it was found that the ratio of heat of evaporation, Q_{bp} and boiling point temperature, T_{bp}, is roughly constant (rule of Pictet

TABLE 2.9—*Thermal expansion and isothermal compressibility at 18°C.*

Liquid	α_{cub} 10^4 (degree^{-1})	β_{is} 10^{11} (Pa^{-1})
n-pentan	16.0	242
n-hexan	13.5	150
n-heptan	12.44	120
Carbon tetrachloride	12.2	110.5
benzene	10.6	95.4
methanol	11.9	120
ethanol	11.0	114
acetone	14.3	125.6
glycerol	5.0	21.7
water	1.8	45.9

TABLE 2.10—*Critical data of some liquids.*

Liquid	T_{crit} (*K*)	p_{crit} (atm)	V_{crit} (cm^3)
Helium	5.3	2.26	58
Hydrogen	33.3	12.8	65.02
Nitrogen	126.1	33.5	90.08
Acetone	508.2	47	216.72
Methanol	513.2	99	89.5
Ethanol	516.3	63.1	164.54
Water	647.4	217.5	54.76

and Trouton [*16*]). Some data are compiled in Table 2.11. In a certain temperature range, liquid and vapor phase coexist. The vapor pressure of a liquid is a strong function of temperature. The dependence can be written in first approximation as

$$\log p = -a/T + b \tag{2.54}$$

Values of a and b for organic liquids can be found in Weissberger [*39*]. Intimately connected with the interaction of the molecules or atoms in the liquid state are the surface energy σ (surface tension) and the viscosity η of a liquid. The surface energy gives the energy which is necessary to bring a molecule or atom from the bulk of the liquid to its surface. It decreases with increasing

TABLE 2.11—*Boiling point and heat of vaporization.*

Liquid	T_{bp} (*T*)	Q_{bp} kcal/mol	Q_{bp} kJ/mol
Helium	4.22	0.024	0.10
Hydrogen	20.4	0.225	0.94
Nitrogen	77.3	1.333	5.57
Carbon tetrachloride	349.8	6.99	29.22
Benzene	353.4	7.38	30.85
Ethanol	351.5	9.55	39.92
Water	373.1	9.71	40.59

temperature, and it becomes zero at the critical temperature. Typical data are given in Table 2.12. The viscosity is a measure of the resistance which has to be overcome if two layers of liquids are to move relative to each other. The viscosity decreases with increasing temperature approximately as

$$\log \eta = A/T + B \tag{2.55}$$

Typical values are compiled in Table 2.13. Application of an electric field to pure liquids or to solutions or suspensions can lead to an increase of their viscosity (electroviscous effect [*42,43*]).

The dissipation of energy in a defined volume leads to an increase in temperature. Specific heat and thermal conductivity are the important quantities. The specific heat gives the amount of energy necessary to increase the temperature of 1 kg of liquid by 1°. Data are given in Table 2.14.

The thermal conductivity of most liquids has a value of around 0.3 to 0.45 10^{-3} cal $s^{-1}cm^{-1}K^{-1}$. Only water and other associative liquid exhibit higher values (Table 2.15). A description of thermal conduction on the basis of molecular theory is still missing. A close

TABLE 2.12—*Surface energy of liquids.*

Liquid	*T(K)*	$\sigma\ 10^3$ [J m^{-2}]
Benzene	293	29
Acetone	293	23.7
Ethanol	293	22.8
Water	293	72.8
Glycerol	293	59.4

TABLE 2.13—*Viscosity of liquids.*

Liquid	*T(K)*	η (Pa s)
n-pentane	291	2.4 10^{-4}
Benzene	293	6.5 10^{-4}
Water	293	1.0 10^{-3}
Ethanol	293	1.2 10^{-3}
Glycerol	293	1.5
Castor oil	293	0.96
Transformer oil	293	0.22
Polydimethyl siloxane		
M_w 6 10^3	293	0.10
M_w 6 10^4	293	100

TABLE 2.14—*Specific heat of liquids.*

Liquid	*T(K)*	C_p (kJ $kg^{-1}K^{-1}$)
Nitrogen	77	1.04
Argon	85	1.13
Helium	4.2	5.02
Methane	110	3.47
Acetone	293	2.18
Glycerol	293	2.4
Ethanol	293	2.43
Water	288	4.18

correlation with the velocity of sound was found. There is a slight temperature dependence of the thermal conductivity in such a way that it decreases with increasing temperature by approximately 0.1% per degree.

The condensation of liquids is accompanied by the heat of condensation which is equal to the heat of vaporization. The transfer of this heat to a metallic plate which is part of a heat exchanger can be augmented by the application of an electric field. Electrohydrodynamic instabilities determine the magnitude of the heat transfer coefficient [*44*]. In spite of the interaction between the molecules and atoms in a liquid, molecules or atoms change their place by jumps. This is due to the existence of holes (also called voids or defects) in the liquid structure [*21*]. Approximately 10^8 changes in position take place for each molecule in liquid at room temperature. At higher temperature, the frequency increases. This motion is random. On a macroscopic scale it is described by the diffusion coefficient D. The order of magnitude of D in liquids at room temperature is 10^{-5} cm^2/s. The temperature dependence of the diffusion coefficient can be described as

$$\log D = -A/T + B \qquad (2.56)$$

Increase of static pressure on a liquid eventually leads to solidification. The pressure can be applied hydrostatically or also by electrostriction in high electric fields. The solidification pressure can be determined by various methods. For polar liquids or nonpolar liquids containing small amounts of polar molecules, the measurement of the dielectric loss (see Section 2.2.6) can be employed as indicator of the solidification.

2.2.8 Physico-Chemical Classification of Liquids

The liquids which are of interest in the context of this chapter can be classified with respect to the mechanism of intrinsic charge carrier generation. Table 2.16 gives a possible classification. *Aprotic*

TABLE 2.15—*Thermal conductivity of liquids at 20°C.*

Liquid	Λ 10^3 (cal/s cm K)
n-pentane	0.32
n-hexane	0.33
Benzene	0.37
Acetone	0.43
Methanol	0.51
Glycerol	0.68
Water	1.4

TABLE 2.16—*Classification of dielectric liquids.*

Type of Liquid	Examples
Aprotic	hydrocarbons, chlorocarbons, fluorocarbons, nitrobenzene, propylene carbonate
Amphiprotic	methanol, ethanol, other alcohols
Protogenic	acetic acid
Protophilic	amines
Aqueous	water and aqueous solutions
Inert	liquefied rare gases, liquefied simple diatomic gases

liquids contain no ionizable proton in the molecule. Charge carrier generation proceeds via ionization. Typical examples are paraffins, chlorinated or fluorinated hydrocarbons, nitrobenzene, and propylene carbonate. With proper purification, samples of very low selfconductivity can be prepared. *Amphiprotic* liquids dissociate into cations and anions. Alcohols are typical examples, where charge carrier formation is thought to occur in the following way

$$R{-}OH + R{-}OH \longrightarrow R{-}OH_2^+ + RO^- \tag{2.57}$$

Protogenic liquids are the carbonic acids which form protons and anions. As an example we give acetic acid which dissociates according to

$$CH_3COOH \longrightarrow CH_3COO^- + H^+ \tag{2.58}$$

Protophilic liquids form ion pairs by proton transfer from one molecule to another. Examples are the amines, for instance trimethylamine

$$(CH_3)_3N + (CH_3)_3N \longrightarrow (CH_3)_3NH^+ + (CH_3)_2N\,CH_2^- \tag{2.59}$$

Water and *aqueous solutions* are characterized by the presence of hydronium ions (hydrated proton) and hydroxyl ions

$$2\,H_2O \longrightarrow H_3O^+ + OH^- \tag{2.60}$$

In addition, in solutions of electrolytes cations and anions are present. Charge carrier generation in inert liquids can only proceed via thermal excitation, irradiation, dissolution of an electrolyte, or injection at electrodes (see Section 2.3).

2.2.9 Liquids Used in Engineering Dielectrics

The main groups of liquids used in dielectric engineering are: *mineral oils, synthetic hydrocarbons, silicone oils, perfluorocarbons, esters, liquid helium and nitrogen, water and aqueous solutions, propylene carbonate, other polar organic liquids.* For any given dielectric application the following physical properties are of importance: *dielectric strength, volume resistivity, loss tangent, dielectric constant, thermal conductivity, specific heat, boiling point and heat of vaporization, density, viscosity and pour point.*

2.2.9.1 Mineral Oils—Mineral oils are derived from crude petroleum and they contain various alkanes, cycloalkanes, and aromatic compounds. Structure formulae for the three groups are shown in Fig. 2.18. Besides these main components they contain small amounts of organic compounds containing sulfur and nitrogen. These impurities influence in a decisive way the long-term behavior of the liquid insulation. In order to improve the performance of a given oil, additives are added in order to inhibit oxidation processes, improve gas absorbing, and pour point characteristics. Many impurities dissolved in the oil and sometimes, in connection with water, lead to an increase of the volume conductivity and to a high loss tangent. Drying and purification from solid and liquid impurities are important steps in the preparation of mineral oil for use in electrical equipment. Details are discussed in the pertinent literature [*45*].

2.2.9.2 Perfluorocarbons—Perfluorocarbons are completely fluorinated organic compounds. They are produced by electrochemical fluorination of hydrocarbons in liquid hydrogen fluoride. This process proceeds with the expenditure of a large amount of electrical energy (Fig. 2.19). The perfluorocarbons are nonpolar, and they have very little solvent power for impurities. The volume resistivity of commercially available liquids is already of the order of $10^{-15}\ \Omega^{-1}cm^{-1}$. They are

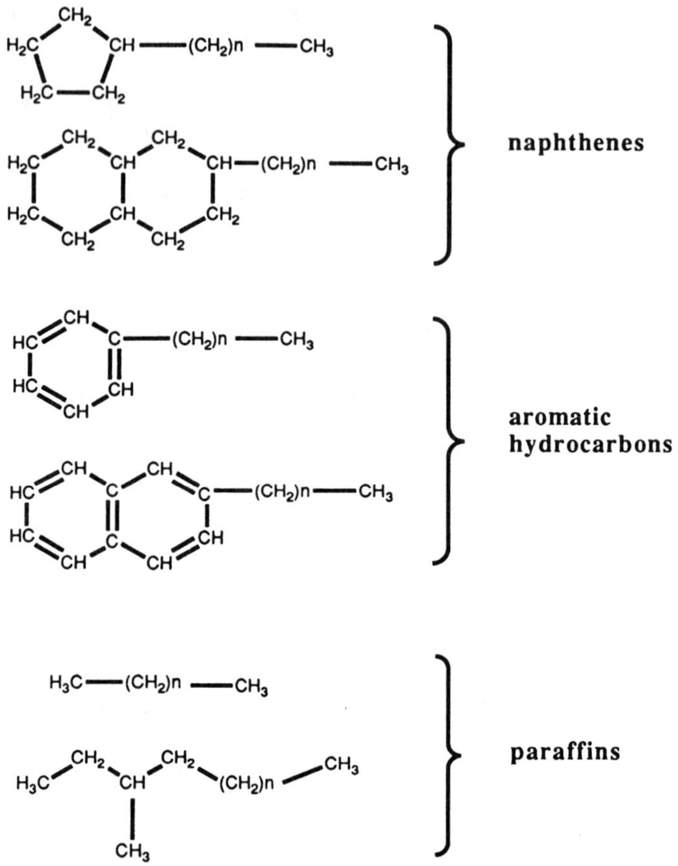

FIG. 2.18—*Structure formulae of the main constituents of mineral oil.*

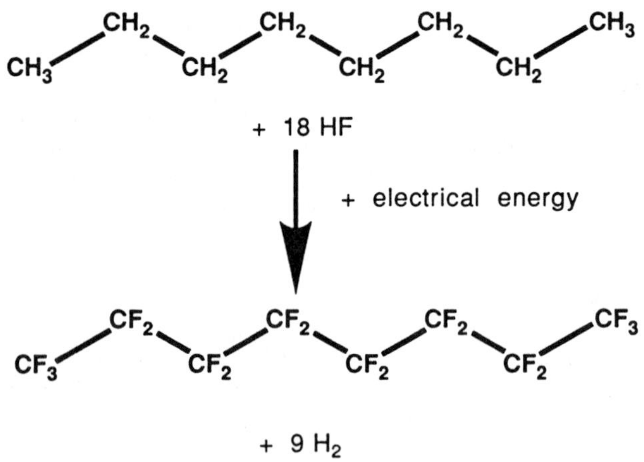

FIG. 2.19—*Synthesis of perfluorocarbons.*

colorless, odorless, low in toxicity, and nonflammable. They have excellent thermophysical properties. A certain disadvantage is their higher density of 1.7 to 2 g/cm^3 as compared to hydrocarbons which have densities between 0.7 and 1 g/cm^3.

2.2.9.3 Silicone Oils—Silicone oils are produced by chemical synthesis. Depending on the conditions of manufacture, their molecular weights vary from 162 to over 160 000, or in other words their chain length can be varied from a few units to molecules with more than 1000 units (Fig. 2.20). With increase of chain length the viscosity of the oil increases. For electrical insulation in transformers, silicone oils with approximately 30 units per molecule and a viscosity of 40 cSt at 40°C are used. The silicone oils have higher flash point, fire point, and auto ignition temperatures than mineral oils which makes them especially suitable in application were a high degree of environmental protection is required.

2.2.9.4 Synthetic Hydrocarbons—Synthetic hydrocarbons are produced by chemical synthesis. Although they are derived from petroleum they are more homogeneous in composition as compared to mineral oil. Essentially, two classes of compounds are used as insulating liquids, olefins, and alkyl aromatics. Olefins are unsaturated hydrocarbons containing one double bond. Long chain molecules containing a double bond at the end can be produced. The most prominent compounds are the polybutenes (Fig. 2.21). Alkyl aromatics are composed of benzene rings with a paraffin chain attached to them (see Fig. 2.21). Dodecylbenzene or diaryl alkanes have found applications as impregnating fluids for capacitors.

2.2.9.5 Organic Esters—Organic esters are synthesized by the reaction of alcohols and acids. The general reaction can be described as follows

$$\underset{\text{alcohol}}{\text{R—OH}} + \underset{\text{acid}}{\text{H—X}} \longrightarrow \underset{\text{ester}}{\text{RX}} + \underset{\text{water}}{\text{H}_2\text{O}} \tag{2.61}$$

The final quality of the product depends on the degree of removal of water, acid, and alcohol.

2.2.9.6 Liquid Helium and Liquid Nitrogen—Cryogenic liquids require large amounts of electrical energy in the refrigeration process. Helium is obtained from wells of natural gas, and it is compressed and liquefied by the Joule Thompson process. Nitrogen is obtained from the atmosphere and is then condensed together with the other atmospheric gases. Separation occurs by

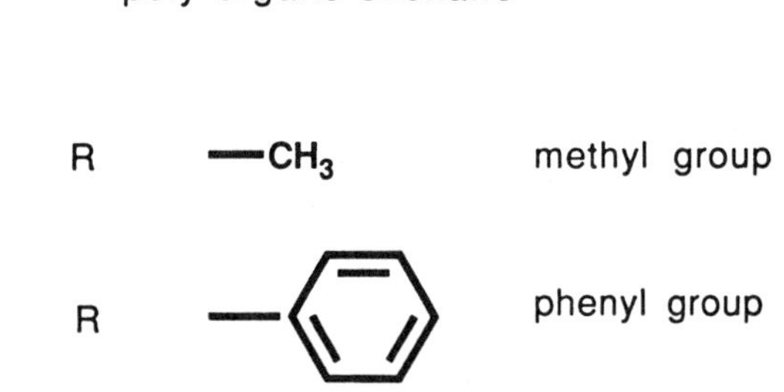

FIG. 2.20—*Structure formulae of silicone oils (polysiloxanes).*

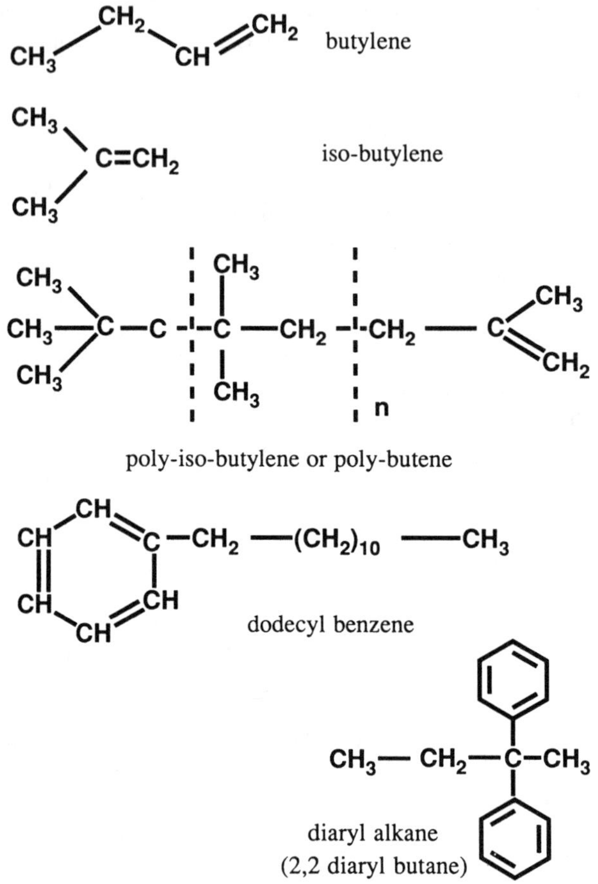

FIG. 2.21—*Structure formulae of synthetic hydrocarbons used as insulating oils.*

fractionated distillation. Traces of oxygen remain in the final product, the level of impurities being dependant upon the effort expended in the fractionation process. Air pollution in the vicinity of the liquefaction plant may introduce impurities usually not present in the air; in addition, lubricants of compressors and pumps represent another source of contaminants.

2.2.9.7 Water and Aqueous Solutions—Water and various aqueous solutions are used in experimental physics in pulsed power discharge lines. In the water purification process, tap water is deionized usually by reverse osmosis and ion exchange resins then UV irradiation is used to kill bacteria and filtration removes particulate matter (see also Section 2.7). For work at temperatures below 0°C, ethylene glycol (or glycerol) is added; ethylene glycol is produced either by reaction with bromine and replacement of the bromine by OH groups or by direct oxidation of ethylene (in an aqueous solution of permanganate)

$$CH_2{=}CH_2 \longrightarrow HO{-}CH_2{-}CH_2{-}OH \tag{2.62}$$

This compound is also used as antifreeze for car engines. In its pure state, it is colorless.

2.2.9.8 Propylene Carbonate—Propylene carbonate has found wider applications in lithium batteries. Its properties make it a candidate for other applications in electrotechnology. Propylene carbonate is produced by the addition of CO_2 to propylene oxide (Fig. 2.22). Removal of traces of water and other electrolytes is the main object of the purification procedures employed [*46*].

2.3 Elementary Processes of Charge Carrier Generation

2.3.1 General Remarks

Charge carriers are produced in dielectric liquids by a variety of external agents. Temperature is the most prominent, but light and high energy radiation are convenient means to generate defined charge carrier pairs and to study their physical and chemical properties. Application of a high electric field strength also enhances the conductivity of a liquid. Besides its influence on charge carrier generation processes, it influences also the transport properties of the charge carriers, and, sometimes, it is difficult to separate these processes. In polar liquids ionizable solutes lead to an increase of the conductivity. In this section a review of the various methods will be given which have been applied to the study of the properties of charge carriers in dielectric liquids.

2.3.2 Thermal Excitation

In Section 2.2.5 we described the electronic energy levels in a liquid. Thermal excitation of charge carriers is related to the energy gap between valence and conduction band, I_{liq}. For pure, crystalline insulators the concentration of electron/hole pairs, n (m^{-3}) due to thermal ionization has been derived as [*47*]

$$n = 2\left[\frac{2\pi kT}{h^2}\right]^{3/2}(m_e^* m_p^*)^{3/4}\exp\left(-\frac{I_{liq}}{2kT}\right) \tag{2.63}$$

Here m_e^* and m_p^* are the effective masses of the electron and hole, respectively; h is Planck's constant, and k denotes the Boltzmann constant. For an estimate of the intrinsic charge carrier concentration n in a liquid insulator, we may assume the effective masses of electron and hole to be equal to the free electron mass ($0.91\ 10^{-30}$ kg). Equation 2.63 then becomes

$$n[m^{-3}] = 4.83\ 10^{21}\ T^{3/2}\exp\left(-\frac{I_{liq}}{2kT}\right) \tag{2.64}$$

At room temperature (T = 295 K) we obtain with Eq 2.64 the following charge carrier densities with I_{liq} = 6 eV, n = $1.1\ 10^{-26}\ m^{-3}$ and with I_{liq} = 9 eV, n = $2.3\ 10^{-52}\ m^{-3}$. Even if the assumed values of the effective masses are too small by one order of magnitude, the intrinsic concentration will remain unmeasurably small. In other words, under all practical conditions

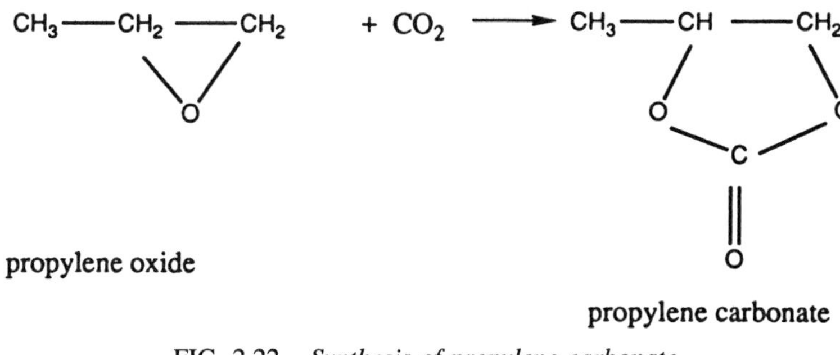

FIG. 2.22—*Synthesis of propylene carbonate.*

electron/hole generation by thermal ionization of the liquid itself can be neglected. Smaller energies are involved if ionic dissociation takes place. This process plays an important role in most of the polar liquids if the cations are protons. There much higher intrinsic charge carrier concentrations are produced by thermal excitation.

2.3.3 Photoionization and Photoeffect

Photoionization of liquids can be effected by single photon or by multiphoton absorption. Since the energy necessary to generate one electron ion pair is above 6 eV, vacuum ultraviolet light is necessary in the case of single photon ionization. Gas discharge lamps were the only sources available for a long time. The advance of electron storage rings as sources of synchrotron light has facilitated studies of the photoionization process in the liquid phase. Double photon ionization can be achieved with pulse lasers. The total energy of the two photons has to exceed the ionization energy of the liquid. Dye lasers provide tunable sources for this type of measurements. In recent years photoconductivity studies with nonpolar liquids have been carried out, and information on the ionization energy in the liquid state, I_{liq} could be obtained [*30–35*]. A typical setup is shown in Fig. 2.23. Monochromatic light falls through a window of lithium fluoride or magnesium fluoride. The window is covered with a thin (5 nm) layer of gold on the side in contact with the liquid. Ionization occurs within a very thin layer of the liquid since the absorption coefficient is of the order of 10^5 cm^{-1}. A typical photo current spectrum is shown in Fig. 2.24. From the onset of the photocurrent, the ionization energy can be obtained. Data have been presented already in Table 2.6. An estimate of I_{liq} can be obtained for liquids where this quantity is not available by making use of Eq 2.39.

Emission of electrons from metal electrodes in liquids can be achieved by the photoelectric effect [*24,27*]. The work function of a metal electrode Φ is reduced or increased in the liquid by V_o (the energy of the electronic conduction level in the liquid, see Fig. 2.12)

$$\Phi\,(\text{liq}) = \Phi\,(\text{vac}) + V_o \tag{2.65}$$

If Φ (liq) $>$ Φ (vac), then V_o is the energy necessary to introduce electrons from the vacuum into the electronic conduction level of the liquid. Φ (liq) $<$ Φ (vac) means that the energy of the electronic conduction level in the liquid is lower by V_o as compared to the vacuum. Measurements

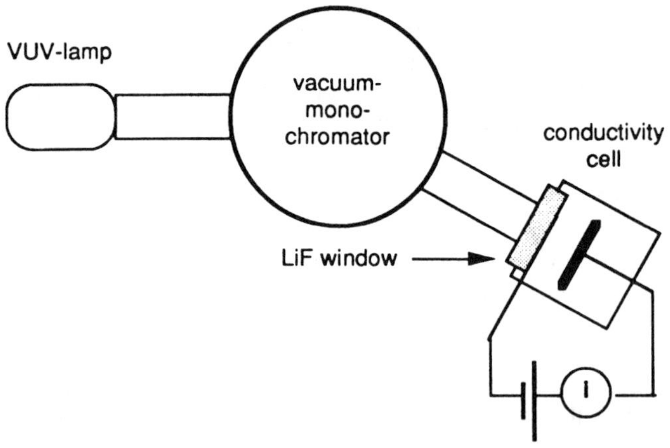

FIG. 2.23—*Experimental set-up for the study of the photoconductivity of nonpolar liquids induced by vacuum ultraviolet light.*

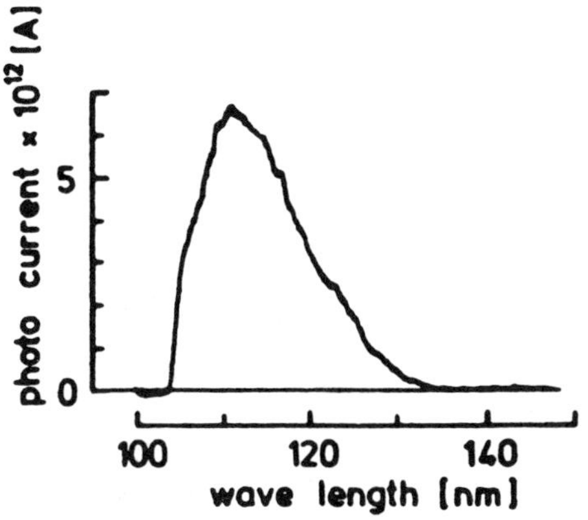

FIG. 2.24—*Photoconductivity spectrum of n-tridecane at 295 K (after Ref* 35*).*

of V_o have been carried out for many liquids, and the data are presented in Table 2.6. From the quantum yield of the photoelectric current, information on the range of the injected photo electrons as a function of their kinetic energy can be obtained. These data may have some bearing on the early processes of the breakdown process [*25*].

2.3.4 High Energy Radiation

One of the fundamental interactions of high energy radiation with matter is the process of ionization. In collisions of high energy elementary particles or photons with the electron shell of the atoms or molecules, electrons are ejected and positively charged atoms or molecules (ions or holes) remain behind. Charge carrier separation results. In gases and semiconductor solids this forms the basis of measurement instruments for ionizing radiation or for the triggering of gas discharge lamps by cosmic radiation. Irradiation of liquids with high energy radiation produces ionizations as well. The detailed action depends, however, on the type of radiation and on the liquid. From a fundamental point of view, γ-radiation and X-rays are most important since spatially uniform distributions of the ionization events in a given volume can be produced. From a practical point of view γ-radiation, X-rays, and fast neutrons are important when oil filled transformers are placed near X-ray sources or nuclear reactors.

High energy X-rays and γ-rays are radiations with low ionizing power (low LET, linear energy transfer). The microscopic deposition of energy and the resulting spatial distribution of ionization events is stochastic. A schematic picture of the possible distributions of ionization events along the track of a high energy electron is shown in Fig. 2.25 [*48*]. Most of the energy is deposited in spurs where approximately 100 eV of energy are absorbed by the microscopic region of the liquid. The result are three to four ionizations. The spurs are widely spaced apart. From time to time an energetic electron is kicked out of a molecule (if the energy of the electron is of the order of 10 keV, it is called a delta ray. It has a much greater ionizing power, and it produces a little side track where the ionization events are grouped much denser. At the end of such a side track the ionization becomes very intense indeed, and a blob of ionizations is formed. The relative frequency of these three different kinds of spatial distributions of ionizations are shown in Fig. 2.26. The mean separation distance between a positive and a negative charge carrier in a liquid is only of the order of several nanometer. Most of the originally produced pairs recombine (geminate recombination). Only a very small percentage is separated widely enough to escape their mutual attraction. They escape into the bulk of the liquid. A spatially homogeneous distribution of charge carriers on a

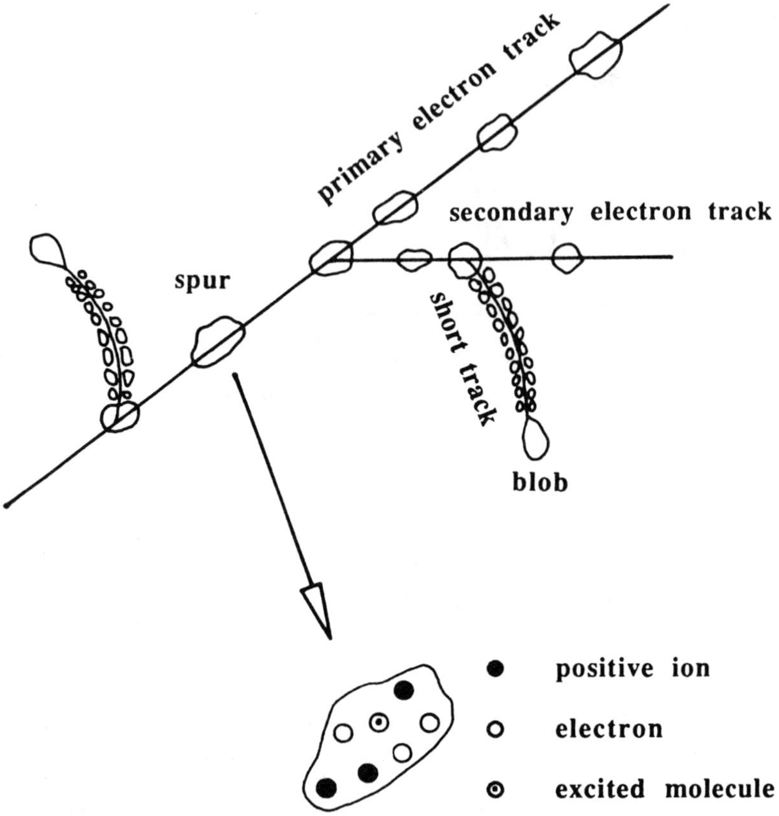

FIG. 2.25—*Schematic representation of the ionization and excitation events produced by the passage of a high energy electron through a dielectric liquid.*

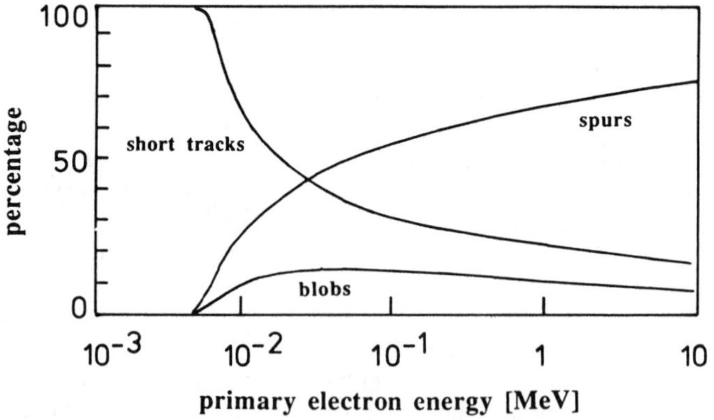

FIG. 2.26—*Frequency of spurs, blobs, and short tracks as a function of the primary electron energy (after Ref* 48*).*

macroscopic scale is produced. Application of an electric field leads to drift movement of the charge carriers, and an electric current is detected in the external circuit, the ionization current. If no electric field is applied, these charge carriers will disappear eventually by volume recombination. Depending on the experimental conditions a variety of processes can be studied. They are summarized in Fig. 2.27. This forms the basis for the determination of many important fundamental quantities.

Densely ionizing particles as protons, alpha particles, or other heavy ions produce tracks of ionization where the individual ionization events overlap. The track consists of a column of positive ions surrounded by a cylinder of electrons or negative ions. Alpha particles emitted by radioactive isotopes have energies of a few MeV. The tracks produced by them in a liquid hydrocarbons have a length of 15 to 20 μm. An alpha source deposited on an electrode of a parallel plate cell forms a thin sheath of ionization from which electrons or ions can be extracted into the bulk of the liquid. Details on the ionization of liquids by radiation can be found in the specialized literature [*49,50*].

2.3.5 Solution of Ionizing Compounds

The conductivity of polar liquids can be varied over a wide range by the solution of ionizable compounds (electrolytes). These may be acids, bases, or salts. The specific conductivity is proportional to the amount of substance dissolved in the polar liquid. Substances which dissociate completely upon dissolution are called strong electrolytes, while compounds of which a fraction dissociates into ions only are called weak electrolytes. The conductivity σ is given as

$$\sigma = \sum_{1=1}^{n} (Z_i e_o c_i \lambda_i) \tag{2.66}$$

where $Z_i e_o$ denotes the charge of the ions of type i in C, c_i is the concentration in moles per liter, and λ_i is the ionic mobility in Ω^{-1} cm^2 val^{-1}.

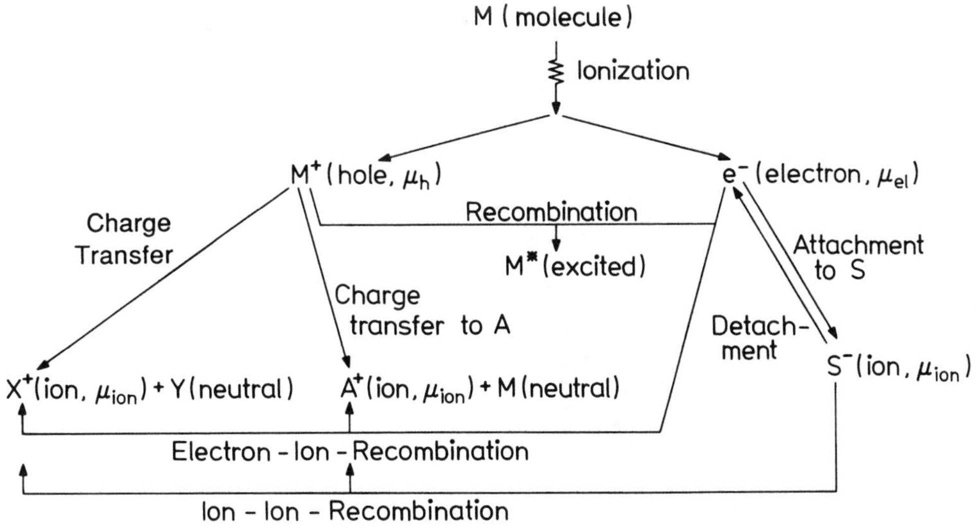

FIG. 2.27—*Reactions occurring in a dielectric liquid after ionization of a molecule by a high energy electron or photon.*

In nonpolar liquids, the solubility of many ionizable compounds is limited, and, furthermore, the degree of dissociation is usually very small. In fundamental studies on the effect of an electric field on the dissociation various organic compounds (ammonium salts of organic acids, aerosols or surfactants, for instance di-2-ethylhexylsulfosuccinate) have been used as weak electrolytes in liquid hydrocarbons and weakly polar liquids [*51–55*]. The presence of water in nonpolar liquids has been connected always with an increase of the electrical conductivity. It is not entirely clear, however, if this is due to dissociation of water molecules in the solution or due to ion injection at the electrodes.

The solution of an electrolyte in a nonpolar or weakly polar liquid leads to the formation of free ions which contribute to the conductivity. This is a two step process involving as intermediates the formation of ion pairs

$$\underset{\text{undissociated molecule}}{AB} \longleftrightarrow \underset{\text{ion pair}}{(A^+B^-)} \longleftrightarrow \underset{\text{free ions}}{A^+ + B^-} \qquad (2.67)$$

Application of an electric field or increase of the temperature shifts the equilibrium to the right-hand side. More free ions are produced and the conductivity increases (see also Section 2.6).

2.3.6 Injection at Electrodes

Injection of charge carriers at electrodes is one of the major mechanisms for electrical conductivity in nonpolar and also in some polar liquids, especially when ionizable solutes are present. At sufficiently high electric field strengths, electron transfer at the electrodes becomes the dominating mechanism. The presence of a weak electrolyte may enhance the injection process due to the formation of a double layer.

2.3.6.1 Electric Double Layer—The first double layer model proposed is due to Helmholtz who depicted the double layer as a condensor with opposite charges fixed at a distance *d* (Fig. 2.28). Gouy and Chapman extended this model by taking into account the diffusive motion of the counter ions in the liquid phase (Fig. 2.29). While they are attracted by the coulomb force towards the fixed

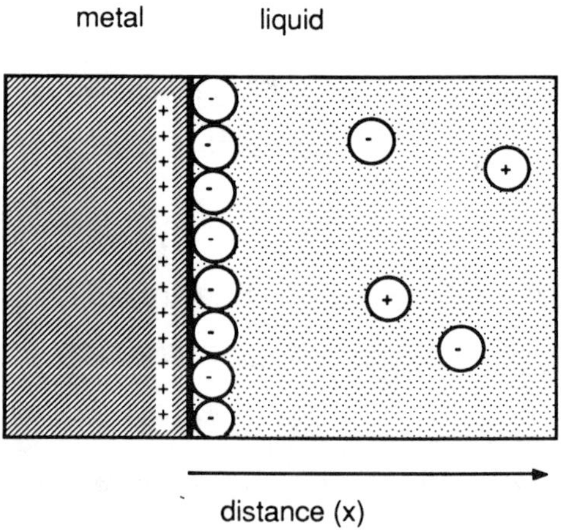

FIG. 2.28—*Schematic representation of an electric double layer (Helmholtz layer).*

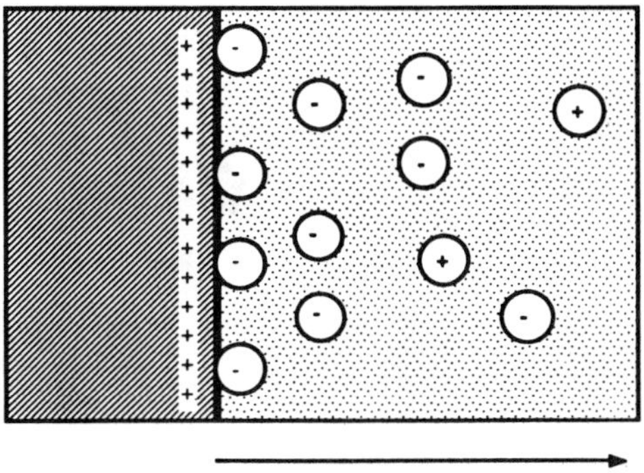

FIG. 2.29—*Schematic representation of an electric double layer (Guoy-Chapman).*

counter ions at the electrode, thermal motion makes them move into the bulk of the liquid. The electric potential Ψ in the diffusive layer can be calculated by taking into account the Poisson equation and the Boltzmann equation

$$\Delta\psi = -\frac{\rho}{\epsilon\epsilon_o} \tag{2.68}$$

and

$$n_+ = n_{+o}\exp\left(\frac{-z_+e\psi}{kT}\right) \tag{2.69}$$

$$n_- = n_{-o}\exp\left(\frac{z_-e\psi}{kT}\right) \tag{2.70}$$

Here n_+ and n_- denote the number density of ions in the diffuse layer, n_{+o} and n_{-o} are their number densities in the bulk of the liquid, respectively. For a symmetrical electrolyte ($n_{+o} = n_{-o} = n$, and $z_+ = z_- = z$), the Poisson equation becomes

$$\rho = ze(n_+ - n_-) \tag{2.71}$$

or

$$\rho = 2nze\sinh\frac{ze\psi}{kT} \tag{2.72}$$

Since the problem can be treated as one dimensional, Eq 2.68 becomes

$$\frac{d^2\psi}{dx^2} = \frac{2nze}{\epsilon\epsilon_o}\sinh\frac{ze\psi}{kT} \tag{2.73}$$

If $ze\psi \ll kT$, Eq 2.73 simplifies to

$$\frac{d^2\psi}{dx^2} = \kappa^2\psi \tag{2.74}$$

with

$$\kappa^2 = \frac{2nz^2e^2}{\epsilon\epsilon_o kT} \tag{2.75}$$

κ^{-1} is the socalled Debye screening length. Integration of Eq 2.74 yields

$$\psi = 2\frac{kT}{ze}\ln\frac{1 + \gamma\exp(-\kappa x)}{1 - \gamma\exp(-\kappa x)} \tag{2.76}$$

with

$$\gamma = \frac{\exp(Z/2) - 1}{\exp(Z/2) + 1} \tag{2.77}$$

and

$$Z = \frac{ze\psi_o}{kT} \tag{2.78}$$

Ψ_o is the surface potential and x is the distance from the surface. A refinement of this model is due to Stern (see [*16*]) who took into account that the ions are of finite size and that in a layer of thickness δ from the metal electrode no ions can exist. The image force is so large that electron transfer neutralizes the ion. The thickness of this layer is of the order of several Angstrom units.

In low conductivity liquids (liquids with a small concentration of ion pairs) we have to assume the existence of two layers at an electrode. From 0 to δ the Stern layer exists where no free charges are present, and, for distances greater δ, we have the diffusive layer of the Gouy-Chapman theory. In the Stern layer the potential varies linearly with distance as

$$\Psi(x = 0) = \Psi_o \text{ and } \Psi(x = \delta) = \Psi_\delta \tag{2.79}$$

The situation at the cathode is depicted in Fig. 2.30. Fully solvated positive ions form an immobile layer near the cathode from which they are separated by a layer of neutral liquid molecules. Further away from this layer a concentration gradient of the positive ions exists (diffusive layer) and in the bulk of the liquid electro-neutrality is maintained. A strong potential gradient exists between the cathode and the fixed ion layer which is also called Helmholtz layer. This is followed by a more gradual change of the potential due to the diffusive layer.

2.3.6.2 Injection—In very pure nonpolar liquids, that is, in liquids with a very low self-conductivity of less than $10^{-14}\Omega^{-1}\text{cm}^{-1}$, not enough ions may be present to set up this layer near the electrodes. Increase of the applied electric field strength, however, may lead to a situation where electron injection or positive ion formation at the electrodes becomes the major source of the electric current through the liquid. Injection at the electrodes due to the formation of double layers is more important in liquids of higher dielectric constant which also exhibit higher self conductivities.

In propylene carbonate (PC), injection of electrons at the cathode was found to contribute predominantly to the conductivity according to the following mechanisms [*56*],

$$\text{PC} + 2e \longrightarrow CO_3^{2-} + CH_3\text{—}CH\text{=}CH_2 \tag{2.80}$$

The formation of the radical anion presents another possibility

$$\text{PC} + e \longrightarrow \text{PC}^- \tag{2.81}$$

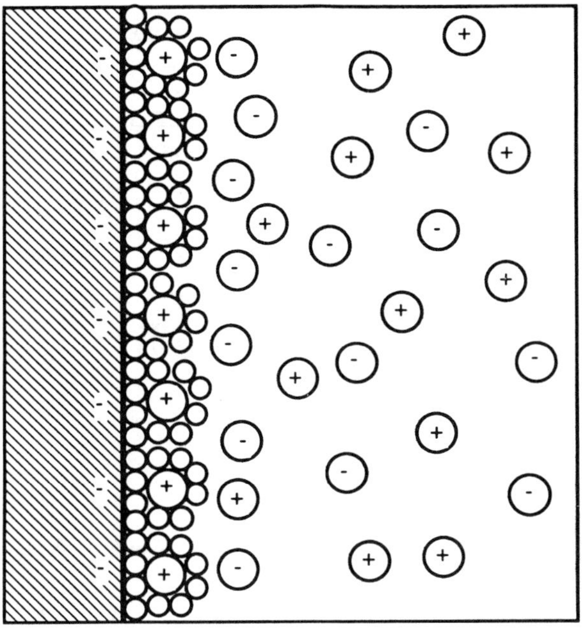

FIG. 2.30—*Schematic representation of an electric double layer consisting of a fixed layer (Helmholtz layer) and a diffuse layer (Guoy-Chapman).*

A similar mechanism was assumed for the formation of negative ions in nitrobenzene [*57*]. The conductivity of moist benzene seems to be governed entirely by this injection at the electrodes [*58–60*]. The nature of the charge carriers remained unclear. The onset of the injection process is characterized by a strong increase of the electric current as a function of applied voltage

$$i = AU^x \tag{2.82}$$

For benzene $x = 1.4$ was found. An exponent of $x = 1.5$ can be explained by space charge limitation of the injection process [*58,59*]. The current voltage characteristic observed in liquid sulfur seems also to indicate injection. Although electron multiplication was inferred [*61*], the electron mobility measured in liquid sulfur is rather small. It seems unlikely that the electrons can gain enough energy from the field to produce collisional ionization.

2.3.6.3 Electro-Optical Measurements—Besides electrical measurements for the detection of injection processes, the Kerr-effect has been used extensively by Zahn et al. [*62*], to study charge injection and transport in water and aqueous solutions. Linearly polarized light of wavelength λ_o which incidents on the medium under 45° with respect to the direction of the electric field is split into two beams (Fig. 2.31*a*). One beam is polarized parallel to the direction of the electric field, the other perpendicular. After travelling through a layer of thickness l, the two light beams have acquired a difference in optical path given by

$$\Delta\lambda = l(n_1 - n_2)/\lambda_o = BlE^2 \tag{2.83}$$

where n_1 denotes the index of refraction for the polarized light parallel to the electric field and n_2 for light perpendicular to the electric field lines, respectively. B is the Kerr constant which may be

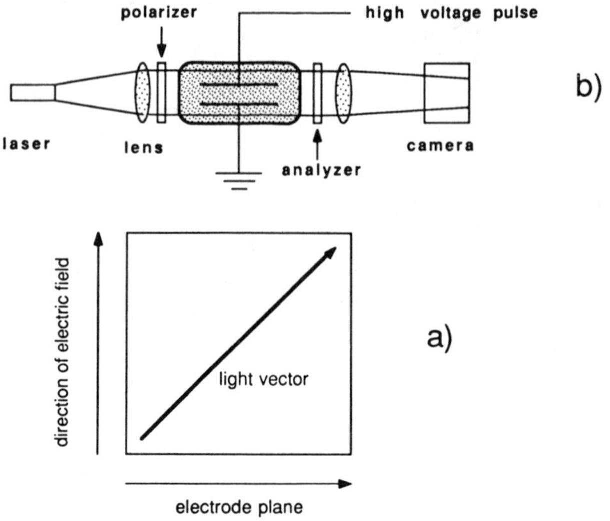

FIG. 2.31—(a) *Orientation of the light vector in a Kerr effect experiment;* (b) *Schematic of a setup for Kerr effect measurements.*

either positive or negative. The medium is said to exhibit birefringence. This optical anisotropy appears in liquids of molecules with different polarizabilities in at least two different perpendicular molecular axes. The electric field induces a dipole. The water molecule, in addition, has a permanent dipole, and the resulting dipole tries to align with the electric field. Thermal motion is opposing this effect.

The phase shift $\Delta\Phi$ connected with the difference in optical path

$$\Delta\Phi = 2\pi lBE^2 \tag{2.84}$$

can be detected with a pair of crossed polarizers (Fig. 2.31*b*). With no phase shift, no light passes through the crossed polarizers. With increasing $\Delta\Phi$ the intensity of the light behind the second polarizer increases. Injection at the electrodes leads to a spatial modification of the electric field by space charge. A change of the birefringence results. Different optical setups can be employed to convert the light signal into patterns depicting the actual field distribution [*63–65*].

2.3.7 Field Emission and Field Ionization

The process of charge carrier generation at the metal/liquid interface under the influence of a high electric field strength is best studied with tip-plane geometry. The generation process is then defined to a small volume around the tip. The situation with an externally applied electric field is shown in Fig. 2.32. With negative polarity at the tip, a barrier is set up between the Fermi level of the electrons in the metal and the bottom of the conduction band in the liquid. The electrons can leave the metal by tunnelling through the barrier of height ΔE_b and width Δx. The probability is an exponential function of the product. In the case of the metal positive, the barrier is set up between the valence electrons of the atoms or molecules of the liquid and the Fermi level in the metal. Electrons from the atoms or molecules of the liquid can tunnel into the metal once the barrier given by ΔE_b and Δx becomes small enough. In nonpolar liquids, the field strengths required for this processes to take place are of the order of 15 to 20 MV/cm. For many nonpolar liquids and electrode materials, the Fermi level is roughly in the middle between the conduction band and the valence band of the liquid. Exceptions are LHe and LNe, where the energy difference ΔE_b is much

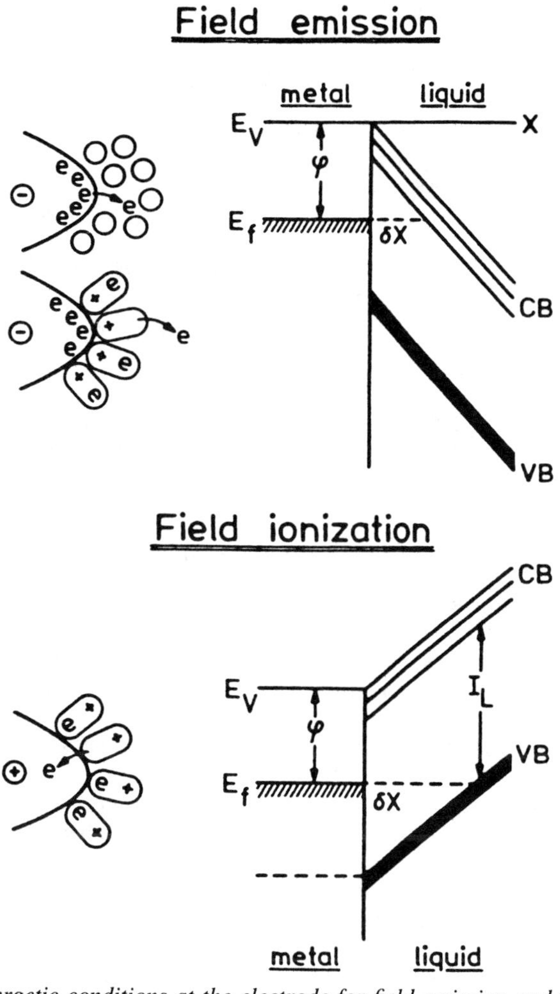

FIG. 2.32—*Energetic conditions at the electrode for field emission and field ionization.*

smaller than $\Delta\Phi$, $\Delta W \ll \Delta\Phi$, due to the high value of the ionization potential of the atoms. This property leads to specific effects in the breakdown of LHe (see Section 2.8.2).

2.3.8 Particles

All commercial liquids contain solid matter in the form of microscopic and submicroscopic particles. Filtering the liquid through sintered glass filters or through membrane filters removes a certain fraction of these, but smaller size particles are thought to be present even below 10 nm of diameter. If these particles are of electrically conducting material (metals or semiconductors) then they will get charged once they come into contact with an electrode. The total charge q they can transport depends on their radius and on the external electric field E. It is given after Maxwell *[66]* as

$$q = (2/3)4\pi^4\epsilon_o\epsilon_r R^2 E \qquad (2.85)$$

ϵ_r is the dielectric constant of the liquid. The current induced by the movement of this particle to the electrode of opposite polarity is given by

$$i = q/t_d \tag{2.86}$$

where t_d denotes the drift time, which is inversely proportional to the external field E. Then, the current increases with E^2 (see Section 2.6.4).

2.3.9 Collisional Ionization

Electrons injected into a gas volume which is under the influence of a high electric field strength can ionize the molecules or atoms of the gas by collisions. This process takes place when the energy received by the electron from the field E between two subsequent collisions $\Delta\epsilon$

$$\Delta\epsilon = eE\Lambda \tag{2.87}$$

is comparable to the ionization energy of the molecules or atoms I_{gas}. Λ is the mean free path between collisions. If n_o electrons are injected at the cathode by photoeffect collisional ionization leads to a multiplication of the number of electrons (Fig. 2.33). After traversing the distance x their number $n(x)$ is given as

$$n(x) = n_o \exp(-\alpha_T x) \tag{2.88}$$

α_T is the first Townsend coefficient. Such a charge multiplication process has also been assumed to be operative in liquids near the breakdown threshold. The only liquid where this process has been observed unambiguously is liquid xenon. Derenzo et al. [*67*] investigated the electron impulses produced by the passage of ionizing particles in a Geiger counter which was filled with LXe. With increasing applied voltage the electron pulse amplitude increased. From the analysis of the pulse height as a function of the coaxial field strength they were able to deduce average values for α as a function of field strength (Table 2.17).

Estimates of α for liquid alkanes have been attempted by analyzing time lag to breakdown measurements in n-hexane [*68*] and by interpreting prebreakdown phenomena in point-plane test cells filled with cyclohexane or propane [*69*].

2.4 Transport Properties of Charge Carriers

2.4.1 General Remarks

The transport properties of charge carriers in dielectric liquids are of interest from a fundamental point of view, since the conductivity of a liquid sample is due to the motion of numerous charge carriers each of which interacts in a specific way with the neutral atoms or molecules of the liquid. From a practical point of view, in any application which involves electrical fields varying in time, the knowledge of the charge carrier transport properties is a prerequisite for the design of the specific application. In this section, we are concerned with electrons, positive holes, negative and positive ions, with charged particles, and molecular aggregates. The following transport properties are important: drift mobility, electrohydrodynamic mobility, recombination, electron attachment and detachment, and charge transfer.

2.4.2 Measurement Methods

Various measurement methods have been developed which either make use of an external agent for the generation of the charge carriers or an electric impulse field is applied to the liquid via electrodes. From the relaxation of the current in time, information on the desired quantities can be extracted. As external agents, light or high energy radiation may be employed. Methods based on

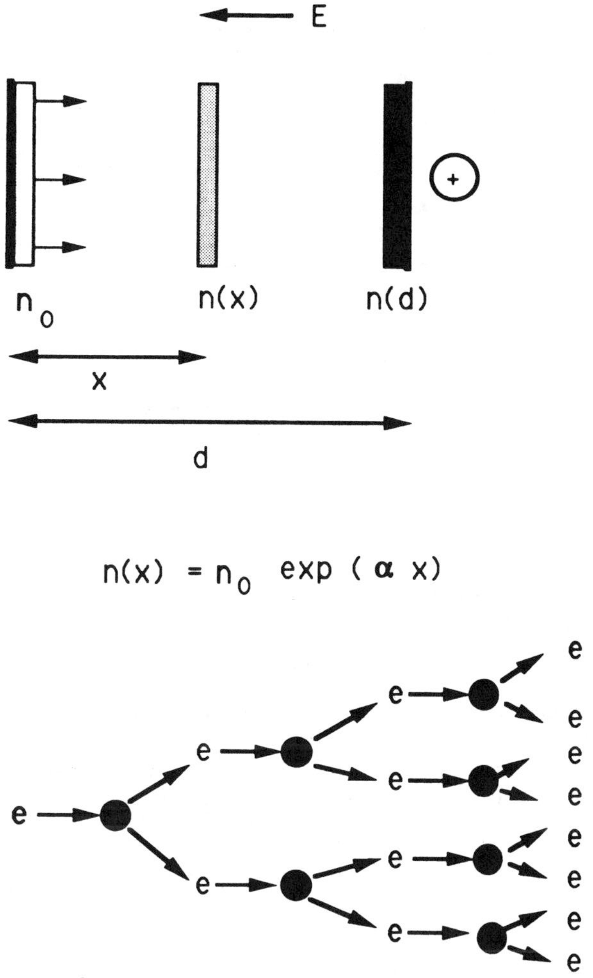

FIG. 2.33—*Formation of an electron avalanche according to Townsend.*

these forms of excitation have been reviewed repeatedly [*49,70*]. Basically, either pulsed radiation sources (accelerators, X-ray tubes, radioactive isotopes) or pulse lasers and flash lamps are employed. The detection methods for the charge carriers are mainly those previously developed for gases and solids [*71,72*].

As an example we shall describe briefly a method using a pulsed accelerator. Charge carrier pairs are produced homogeneously in the liquid volume by a nanosecond pulse of high energy X-rays. The setup is schematically shown in Fig. 2.34. The processes following the ionization of a molecule are schematically depicted in Fig. 2.27. Depending on the experimental conditions, electron and ion mobilities, electron attachment and detachment, and charge carrier recombination can be measured. An example of the oscilloscopic traces of the ionization current produced by a short pulse of high energy X-rays is shown in Fig. 2.35*a–c*. Figure 2.35*a* refers to the measurement of electron and ion mobilities. The duration of the X-ray flash was short compared to the drift time t_d of the charge carriers. On a time scale of 100 ns, a linear decay of the ionization current is observed which

TABLE 2.17—*First Townsend coefficient* α *(cm*$^{-1}$*) in LXe at 161 K as a function of the electric field strength* E *(kV/cm)* [67].

E (kV/cm)	α (cm^{-1})
400	470 $^{+600}_{-470}$
600	8 050 + 3500
800	2 550 $^{+7000}_{-2550}$
1000	8 300 $^{+3400}_{-4700}$
1200	15 000 ± 6000
1400	27 500 ± 6000
1600	33 500 ± 4500
1800	38 400 ± 4500
2000	44 700 ± 2600

corresponds to the motion of electrons, while on one the time scale of milliseconds a second linear decay is observed which corresponds to the motion of the positive ions. The mobility is obtained as

$$\mu = (d^2)/(t_d V) \tag{2.89}$$

where *d* is the plate separation of the measurement cell and *V* is the applied voltage.

If an intense X-ray pulse is applied to the liquid at a low applied voltage, then volume recombination of charge carrier pairs is the predominant process, and, from the decay of the current in time *i*(*t*), the recombination rate constant k_r can be obtained

$$\frac{1}{i(t)} = \frac{1}{i_o} \frac{k_r}{(\mu_+ + \mu_-)} \frac{d}{qV} t \tag{2.90}$$

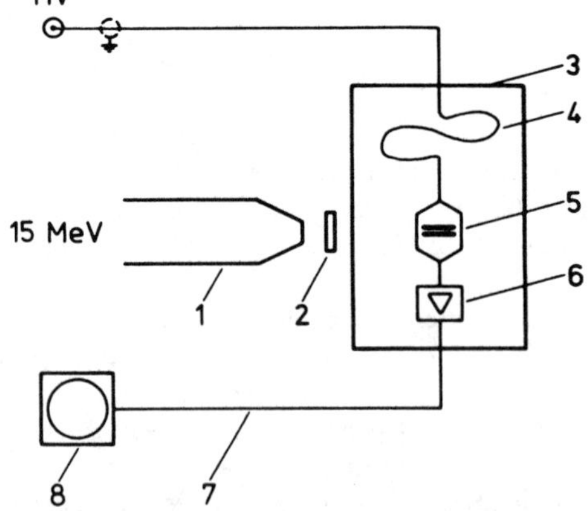

FIG. 2.34—*Schematic representation of an experimental setup for measurements of transport properties of charge carriers in nonpolar liquids; 1–linear electron accelerator, 2–target, 3–shielded cage, 4–coaxial cable, 5–measurement cell, 6–preamplifier, 7–50 Ω signal cable, 8–oscilloscope.*

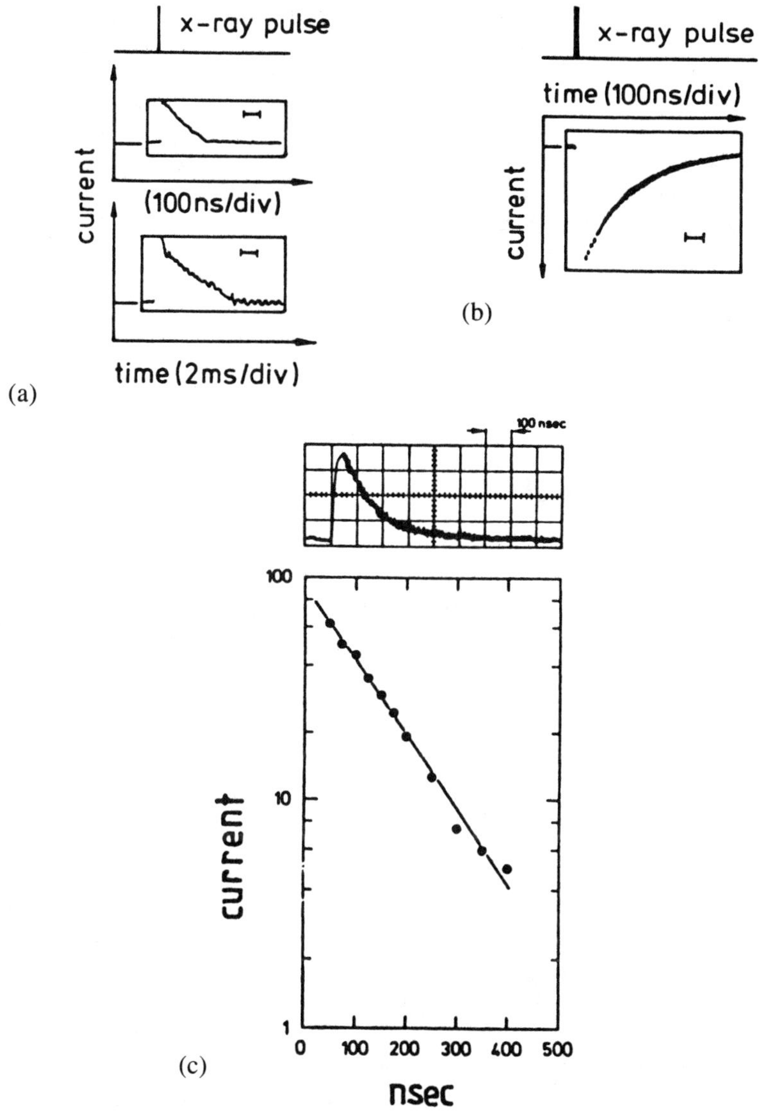

FIG. 2.35—(a) *Measurement of mobility: upper trace electron signal, lower trace ion signal in liquid methane at 110 K;* (b) *measurement of electron/ion recombination in neopentane at 295 K;* (c) *measurement of electron attachment of oxygen in liquid methane at 110 K (after Ref* 25*).*

where q is the cross section of the electrode. This hyperbolic decay in time of the ionization current is shown in Fig. 2.35*b*.

If the liquid contains an electron attaching compound, then an exponential decay in time of the current is observed from which the attachment rate constant k_s can be obtained (Fig. 2.35*c*). The current as a function of time is given as

$$i(t) = i_o(t) \exp(-tk_s n_s) \tag{2.91}$$

where n_s denotes the concentration of the electron attaching impurity.

In another group of methods, a voltage pulse is applied to the test cell and the evolution of the current as a function of time is observed [*53,58,73*]. Figure 2.36*a* shows the schematic of the setup. Electrochemical generation of charge carriers at one or both electrodes leads to injection currents similar to those in solids. An example of the temporal variation of the conduction current in cyclohexane solution is shown in Fig. 2.36*b*. After application of the voltage, the current rises linearly until the ions which are injected at one electrode have reached the counter electrode. The transit time decreases with increasing electric field strength.

Besides these major groups, a variety of other methods have been employed [*72*]. Flowing liquids were used to transport electrons or ions [*74,75*].

2.4.3 Electron Mobility

Electron and hole mobility measurements require extremely pure liquids in order for the electron or hole to survive attachment or charge transfer, respectively. Naturally, the degree of purity

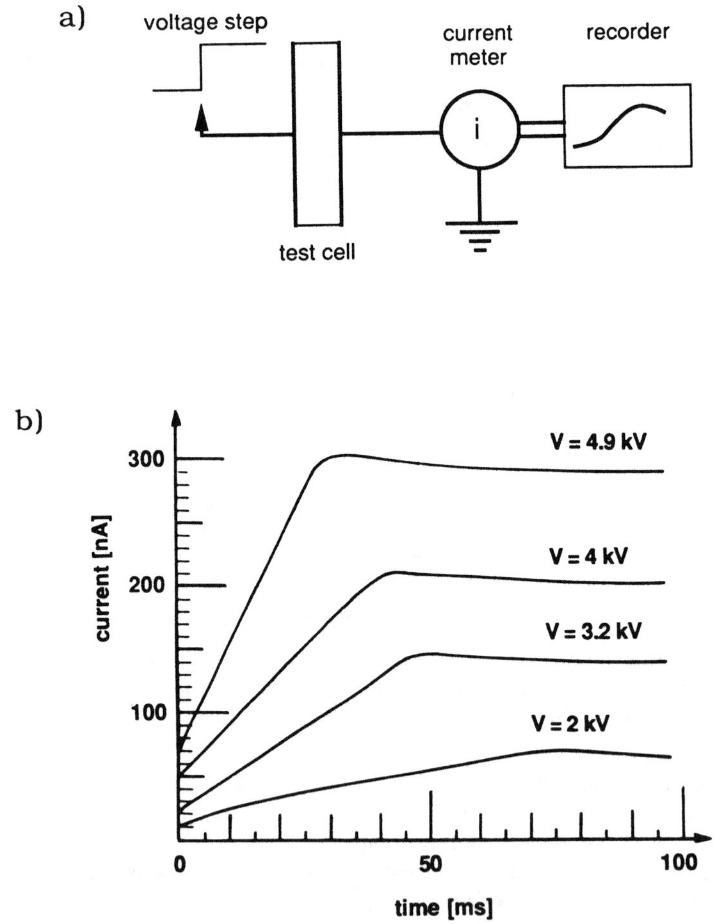

FIG. 2.36—(a) *Electric circuit for ion mobility measurement;* (b) *temporal variation of the electric current in a cyclohexane/surfactant solution under the condition of ion injection at one electrode (after Ref* 52).

required depends on the dimensions of the experiment. If electron drift is to be observed over large distances (say several centimeters) the purity has to be much greater than if the characteristic drift distance is only a few micrometers. In other words, although the results which will be discussed here have been obtained in test cells of a few millimeter drift distance, the data may be also of relevance in problems where, in not so pure liquids on a microscopic scale, electronic processes may be precursors of other electrical phenomena. Transport properties of electrons have been investigated for many nonpolar and some polar liquids.

Electron mobilities in nonpolar liquids were studied as a function of temperature, of pressure, of the applied electric field strength, and in the presence of a magnetic field [*8,50,76–84*]. There exists much less information on hole mobilities. This may be due mainly to the fact that the requirements for purity are much more difficult to fulfill in this case.

2.4.3.1 Effect of Structure—One of the interesting features of electron mobility in nonpolar liquids is its large variation with the molecular structure. In Table 2.18 some examples are given for liquid hydrocarbons and liquefied rare gases. Although the comparison of the mobility data at a particular temperature is somewhat ambiguous, some general features can be extracted by the inspection of Table 2.18. In normal alcanes the electron mobility decreases with increasing chain length at room temperature. For $n > 7$ (n-heptane), the values seem to level off. This could serve as an indication that the mode of electron transport is similar in the paraffins. In Fig. 2.37 we plotted the temperature necessary to give a mobility of 0.1 cm^2/Vs as a function of the chain length n (number of carbon atoms). Branching of the hydrocarbon chain leads to higher values of the mobility. Molecules that contain a quarternary C-atom (this is a C-atom which is connected to four other C-atoms) exhibit the highest values (neopentane, neohexane, isooctane). Also, if the quarternary C-atom is replaced by silicon, germanium, or tin, higher values of the electron mobility are observed as compared to the normal alcanes. The liquefied rare gases can be divided into two groups, in LHe and LNe low electron mobilities are observed, more like that of ions, while in LAr, LKr, and LXe the electron mobility approaches values of semiconductors.

2.4.3.2 Effect of Temperature—The electron mobility in liquid hydrocarbons usually exhibit positive temperature coefficients (Fig. 2.38). Exceptions are the highly mobility liquids, as for instance methane, neopentane, tetramethyl silane, but also LAr, LKr, and LXe where a strong variation of the electron mobility with temperature has been found in the vicinity of the critical temperature. Most normal alcanes show over a limited range an Arrhenius type dependence with typical activation energies between 0.1 and 0.2 eV.

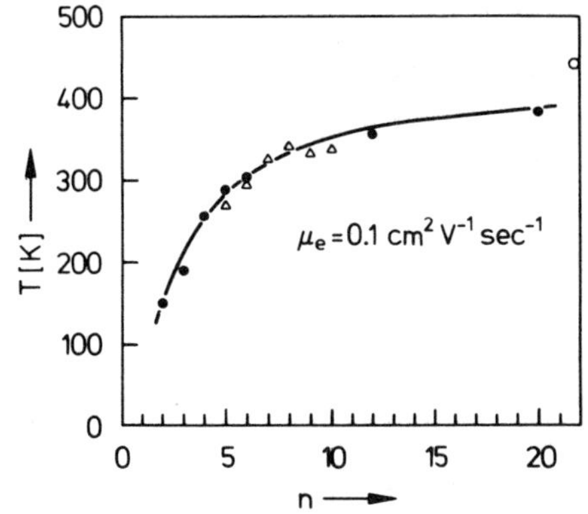

FIG. 2.37—*Electron mobility in n-alkanes: the temperature necessary to obtain an electron mobility of 0.1 cm^{2}V^{-1}s^{-1} as a function of the number of carbon atoms n, in the molecule (after Ref 85).*

TABLE 2.18—*Electron mobilities in nonpolar liquids* [76,77,83].

Liquid	Chemical formula	$T(K)$	μ_{el} ($cm^2V^{-1}s^{-1}$)
Helium	^4He	4.2	$2\ 10^{-2}$
Neon	Ne	25	$1.6\ 10^{-3}$
Argon	Ar	85	475
Krypton	Kr	117	1800
Xenon	Xe	163	2200
Methane	CH_4	111	400
Ethane	C_2H_6	111	$1.3\ 10^{-3}$
n-butane	C_4H_{10}	296	0.4
n-pentane	C_5H_{12}	296	0.16
n-hexane	C_6H_{14}	296	0.1
n-octane	C_8H_{18}	296	0.04
n-decane	$C_{10}H_{22}$	296	0.036
Isooctane	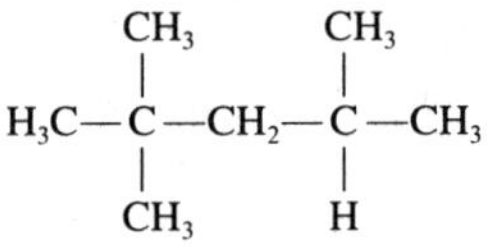	296	7
Neohexane	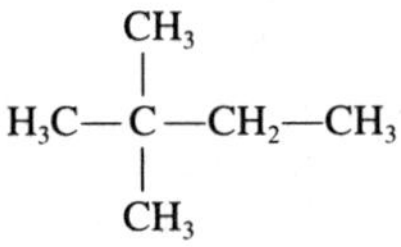	296	12
Neopentane	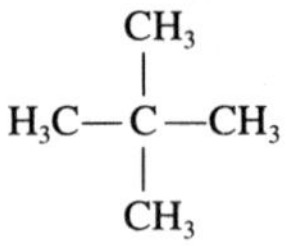	296	65
Tetramethyl silane	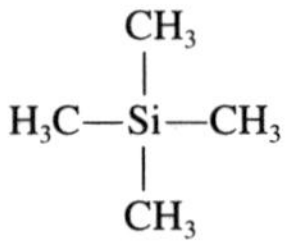	296	95
Tetramethyl germane	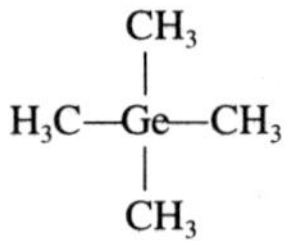	296	90
Tetramethyl tin	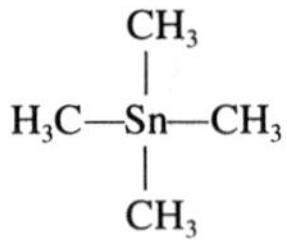	296	70
Cyclohexane	C_6H_{12}	296	0.4
Benzene	C_6H_6	296	0.13
			0.6

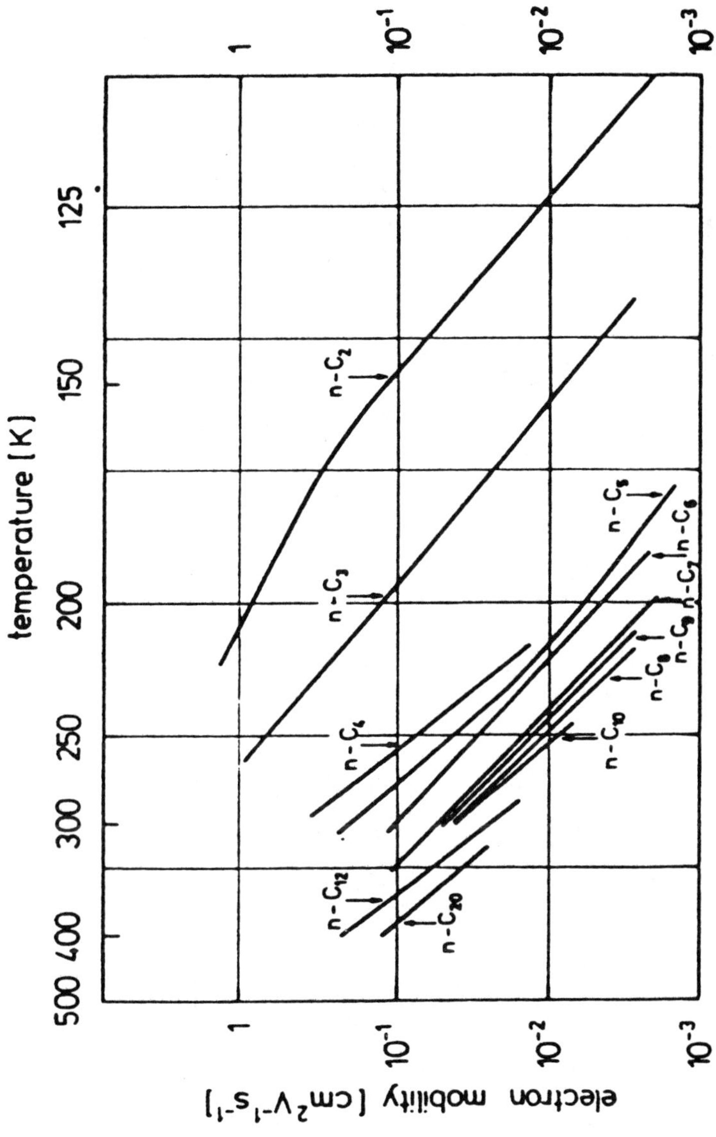

FIG. 2.38—*Arrhenius plot of the electron mobility in n-alkanes (after Ref 76).*

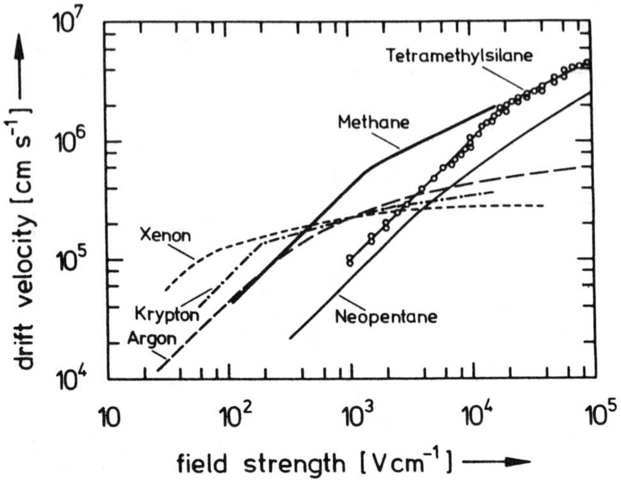

FIG. 2.39—*Electron drift velocity as a function of the electric field strength in nonpolar liquids (after Ref* 86*).*

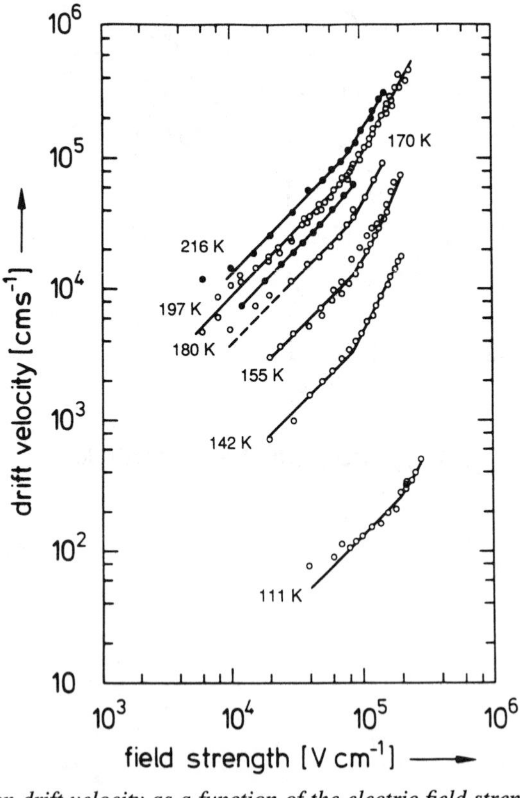

FIG. 2.40—*Electron drift velocity as a function of the electric field strength and of the temperature in liquid ethane (after Ref* 87*).*

2.4.3.3 Electric Field Effects—For the liquid hydrocarbons, at low electric field strength strict proportionality between drift velocity and field strength is observed. The mobility obtained is referred to as low field mobility. At higher values of the electric field strength deviations from this proportionality are observed. At temperatures where the mobility is >10 cm^2/Vs a sublinear dependence of υ_d on E is observed (Fig. 2.39). If the mobility is <1 cm^2/Vs a superlinear dependence is observed (Fig. 2.40). In the mobility range 1 to 10 cm^2/Vs a superlinear dependence is followed by a sublinear dependence [*81*].

The liquefied rare gases and the cryogenic molecular liquids, as LN_2 or LH_2 exhibit a field dependence of the electron drift velocity which again depends on the magnitude of the low field electron mobility. LAr, LKr, and LXe show a sublinear dependence of υ_d on E which at higher values of E shows saturation with respect to E (see Fig. 2.40). The low mobility liquids show a superlinear dependence of υ_d on E the same way as is exhibited by ethane in Fig. 2.40.

2.4.3.4 High Pressure—Very little work has been carried out so far on the influence of high pressure on the electron mobility. Some recent data are shown in Fig. 2.41. The explanation for these variations is still under discussion in the literature [*78*].

2.4.3.5 Magnetic Field—Mobility measurements with magnetic field applied have been carried out for electrons. The Hall mobilities measured for high mobility liquids agreed with the drift mobilities which indicates that the influence of traps is negligible. No Hall mobility measurements exists for low mobility liquids [*80,82*].

The measurement of electron mobilities in polar liquids is difficult due to the inevitable presence of ions and due to reaction of the electrons with ions or neutral molecules of the liquid. The unambiguous assignment of a measured effect as being caused by electrons, exclusively, is often not possible. Electron mobilities in some polar liquids are compiled in Table 2.19.

2.4.4 Ion Mobility and Particle Motion

Although a large body of information on ionic mobilities in nonpolar liquids has been accumulated [*3*], few studies have been reported where ions of known molecular structure could be

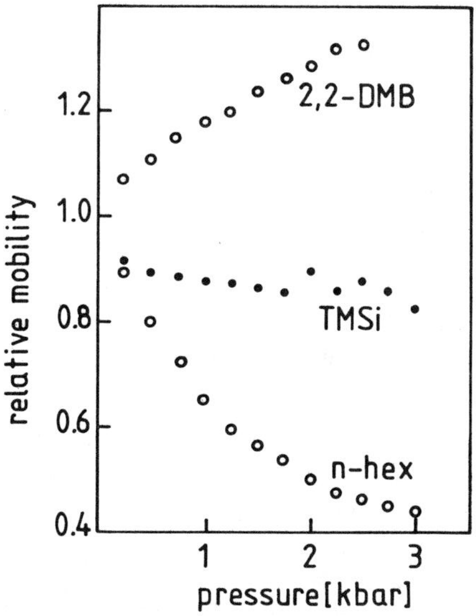

FIG. 2.41—*Relative change of the electron mobility as a function of pressure:* ○ *2-2-dimethyl butane;* ● *tetramethyl silane;* ○ *n-hexane (after Ref* 78*).*

TABLE 2.19—*Electron mobility in polar liquids* [83,84].

Liquid	$T(K)$	ϵ_r	μ_{el} ($\times 10^{-4} cm^2V^{-1}s^{-1}$)
Water	293	80.4	18
Methanol	293	33.6	5.9
Ethanol	293	24	2.5
Isopropanol	293	19	5.1
n-butanol	293	18	7.4
n-hexanol	293	12	4.3
Ethylene glycol	293	39	2.8
Tributylamine	293	5	20
Methylethyl ketone	293	21	15
Ammonia	213	22	100

identified. Quite a different situation exists for polar liquids, especially water, where the nature of the ions observed is well defined.

2.4.4.1 Ions in Nonpolar Liquids—Although the investigation of ion mobilities in hydrocarbons and liquefied gases has received a lot of attention [*3,7*], the experimental values are characterized by a large degree of scattering. On the theoretical side, virtually only one model for ionic transport has been developed. It assumes that the motion of the ions is that of a sphere of radius R through a viscous medium of viscosity η. The electric force exerted by the externally applied field E is balanced by a viscous force (Stokes law). The mobility μ_{ion} is obtained from (cgs-units)

$$e_oE = 6\pi\eta R v_d \tag{2.92}$$

as

$$\mu_{\text{ion}} = e_o/(6\pi\eta R) \tag{2.93}$$

This equation is also known as Walden's rule

$$\mu_{\text{ion}}\eta = \text{const} \tag{2.94}$$

While this rule is approximately fulfilled for negative ions in hydrocarbons, for positive ions a relationship

$$\mu_{\text{ion}}\eta^{3/2} = \text{const} \tag{2.95}$$

was found [*88*]. Ion mobility and ionic diffusion coefficient, D are related by the Nernst-Einstein equation

$$\frac{D}{\mu_{\text{ion}}} = \frac{k_BT}{e_o} \tag{2.96}$$

The temperature dependence of ionic mobilities follows an Arrhenius type dependence

$$\mu_{\text{ion}} = \mu_o \exp(-W/kT) \tag{2.97}$$

The activation energies W are in the range of 0.1 to 0.2 eV. The ion mobility in n-hexane was found to be independent of the applied electric field strength up to 500 kV/cm [*89*]. A collection of typical data for ionic mobilities in nonpolar liquids is given in Table 2.20. Data on ionic mobilities in nonpolar liquids suffer from the fact that in most cases the identity of the ions was not known. The presence of impurities may lead to the formation of impurity ions or to formation of cluster ions.

2.4.4.2 Ions in Polar Liquids—The ion mobility depends both on the viscosity of the solvent and on the size of the ion [*91*]. The ionic radius R_s can be estimated from the Stokes' equation given as

TABLE 2.20—*Ionic mobilities in nonpolar liquids* [3,70,76,90].

Liquid	T(K)	Cation	Anion
Nitrogen	63		6.5×10^{-4}
	73		1.3×10^{-3}
	77	1×10^{-2}	2.5×10^{-3}
		2×10^{-3}	
		9×10^{-4}	
Oxygen	62		4.5×10^{-4}
	77	8×10^{-3}	
		1×10^{-3}	
Hydrogen	21	8.3×10^{-3}	8.6×10^{-3}
		4.5×10^{-2}	
Methane	111	2.0×10^{-3}	
Carbon monoxide	69		8.7×10^{-4}
Sulfur	393		1.0×10^{-4}
Polydimethyl siloxane, 50 cSt	r.t.	7.3×10^{-5}	7.2×10^{-5}
n-hexane	297	6.8×10^{-4}	1.3×10^{-3}
1,4 dioxane	297	8.0×10^{-4}	4.8×10^{-4}
Carbon tetrachloride	297	4.0×10^{-4}	3.3×10^{-4}

$$R_s = Z_i F^2/(6\pi\eta\Lambda_i N_A) \tag{2.98}$$

where Z_i is the ionic charge, expressed as number of elementary charges, F denotes the Faraday constant, η is the viscosity of the solvent, and Λ_i is the mobility (equivalent conductance) of the ion, and N_A stands for Avogardo's constant. With the relevant constants from Eq 2.98, R_s can be obtained in Angstroms as

$$R_s(\text{Å}) = \frac{0.82Z_i}{\Lambda_i(\text{cm}^2\Omega^{-1}\ \text{val}^{-1})\eta(\text{Poise})} \tag{2.99}$$

The ion solvent interaction is reflected in the solvation number N_s given as

$$N_s = (4\pi/3)(R_s^3 - R_c^3)/\upsilon_s \tag{2.100}$$

where υ_s is the molecular volume and R_c is the crystallographic ionic radius. Ionic mobilities in water or aqueous solutions are determined usually by measuring the equivalent conductivity Λ as a function of solute concentration. The equivalent conductivity is the ratio of the specific conductivity and the solute concentration in moles per cubic centimeters. It is assumed that all solute molecules are dissociated into singly charged positive ions (cations) and negative ions (anions). In the case of multiply charged ions, the concentration in moles divided by the number of elementary charges $Z_i e_o$ per ion is taken. The measured equivalent conductivity is then given as the sum of the equivalent conductivities of the anions, $\Lambda_i = \Lambda_\infty(-)$ and of the cations, $\Lambda_i = \Lambda_\infty(+)$

$$\Lambda_\infty = \Lambda_\infty(-) + \Lambda_\infty(+) \tag{2.101}$$

The unit of Λ is $\text{cm}^2\text{val}^{-1}\Omega^{-1}$. One val is equal to one mole divided by the number of elementary charges $Z_i e_o$ of the ion. In other words, the equivalent conductivity is the conductivity produced by one mole of elementary charges. The equivalent conductivity Λ_∞ of an ion is also called its mobility, μ_{ion}, since the mobility is given as the ratio of the specific conductivity σ and the product of ion concentration n_{ion} and number of elementary charges per ion $Z_i e_o$

$$\mu_{ion} = \sigma/(n_{ion} Z_i e_o) \tag{2.102}$$

In physics, the mobility is related to one ion and it is measured in $cm^2V^{-1}s^{-1}$. Since one mole of singly charged ions is equivalent to 96 500 C, the two mobilities are converted via

$$\mu_{ion}[cm^2V^{-1}s^{-1}]\,96\,500 = \Lambda[\Omega^{-1}cm^2val^{-1}] \tag{2.103}$$

A compilation of ion mobilities in some polar liquids is given in Tables 2.21*a–c*.

2.4.4.3 Particles—The transport of particles in an insulating liquid under the influence of an electric field is called electrophoresis. It can be treated like the motion of ions [*92*]. One has to take into account that the charges residing on the solid particle attract a cloud of opposite charges from the liquid. A double layer is set up (see Fig. 2.29). This double layer takes part in the movement of the particle. The first theoretical treatment of such double layer was due to Helmholtz who considered as model two planes with uniform surface charge density, Σ, separated by a distance *d*.

TABLE 2.21—*Ionic mobilities in polar liquids.*

(a) ION MOBILITIES IN *WATER* AT 298 K [*16*]

Cation	$\Omega^{-1}cm^2val^{-1}$	$cm^2V^{-1}s^{-1}$	Anion	$\Omega^{-1}cm^2val^{-1}$	$cm^2V^{-1}s^{-1}$
H^+	349.8	$3.62\ 10^{-3}$	OH^-	198.0	$2.05\ 10^{-3}$
Na^+	50.1	$5.2\ 10^{-4}$	Cl^-	76.3	$7.9\ 10^{-4}$
NH_4^+	73.4	$7.6\ 10^{-4}$	NO_3^-	71.5	$7.4\ 10^{-4}$
Ag^+	61.9	$6.4\ 10^{-4}$	SO_4^-	80.0	$8.3\ 10^{-4}$
Cu^{2+}	53.6	$5.5\ 10^{-4}$	CO_3^-	69.3	$7.2\ 10^{-4}$

(b) ION MOBILITIES, STOKES' RADII AND SOLVATION NUMBER IN *PROPYLENE CARBONATE* AT 30°C [*93*]

Ion	$\Omega^{-1}cm^2val^{-1}$	*R* (nm)	N_s
Li^+	8.3	0.48	3.3
Na^+	9.1	0.46	2.8
K^+	15.2	0.36	1.3
NH_4^+	20.1	0.32	0.9
$(C_2H_5)N^+$	13.7	0.40	0
ClO_4^-	21.2	0.31	0.6
Cl^-	12.5	0.41	2.2
Br^-	20.2	0.38	0.7
BF_4^-	25.7	0.32	

(c) IONIC MOBILITIES IN *METHANOL* AT 25°C [*94*]

Cation	$\Omega^{-1}cm^2val^{-1}$	Anion	$\Omega^{-1}cm^2val^{-1}$
Li^+	39.82	Cl^-	52.38
Na^+	45.22	Br^-	65.55
K^+	52.40	I^-	62.75

If the permittivity of the medium is ϵ_s then the capacity per unit area is given as

$$C = \epsilon_r\epsilon_o/d \tag{2.104}$$

The potential difference across the layer is

$$\zeta = \Sigma d/\epsilon_r\epsilon_o \tag{2.105}$$

This is the electrokinetic potential or zeta-potential. Equation 2.105 requires modification for the fact that the counterions in the liquid form a diffusive layer (see Fig. 2.31). The migration of charged particles in an electric field can be treated theoretically in an elementary way by assuming that the force exerted by the external electric field on the surface charge is balanced by the viscous drag of the particle with drift velocity v_d through the liquid

$$\Sigma E = \eta v_d/d \tag{2.106}$$

Combination of Eqs 2.105 and 2.106 yields for the drift velocity

$$v_d = \Sigma Ed/\eta = \epsilon_r\epsilon_o E\zeta/\eta \tag{2.107}$$

Taking into account the spherical shape of the particle yields

$$v_d = \epsilon_r\epsilon_o E\zeta/(6\pi\eta) \tag{2.108}$$

or for the mobility of the particle

$$\mu_p = \frac{\epsilon_r\epsilon_o\zeta}{6\pi\eta} \tag{2.109}$$

These derivations assume that the shape of the electric double layer is not disturbed during the motion of the particle in the electric field. At high electric field strengths, meaning high particle velocities, this may not be the case any longer.

In dielectric liquids metallic, semiconducting or insulating particles may be present. Metallic particles are generated from the electrodes when breakdown processes occur. Semiconducting particles are the result of the chemical decomposition of molecules containing carbon during electrical breakdown. Insulating particles may be introduced during the production of the liquid or they also can be formed in prebreakdown and breakdown processes. Another group of particle like matter are micelles [*4*], microscopic droplets of water which are suspended in nonpolar liquids by means of soap molecules (Fig. 2.42).

In experiments for the measurement of the ion mobility where liquid motion is allowed to develop (see Section 2.6.7) an electrohydrodynamic mobility μ_{EHD} is measured rather than the true mobility of the ions in the liquid. Application of the principle of energy conservation yields [*95*],

$$\mu_{\mathrm{EHD}} \approx \sqrt{\frac{\epsilon_o\epsilon_r}{\rho}} \tag{2.110}$$

where ρ denotes the density of the liquid.

2.4.5 Recombination, Attachment, Charge Transfer

Charge carrier recombination, electron attachment to impurity molecules, and charge transfer from positive ions to impurity molecules are processes which have to be taken into account in the modelling of prebreakdown and breakdown processes. Furthermore, they are of great importance in the design and operation of liquid ionization chambers (see Section 2.9).

2.4.5.1 Recombination—In nonpolar liquids recombination takes place between positive and

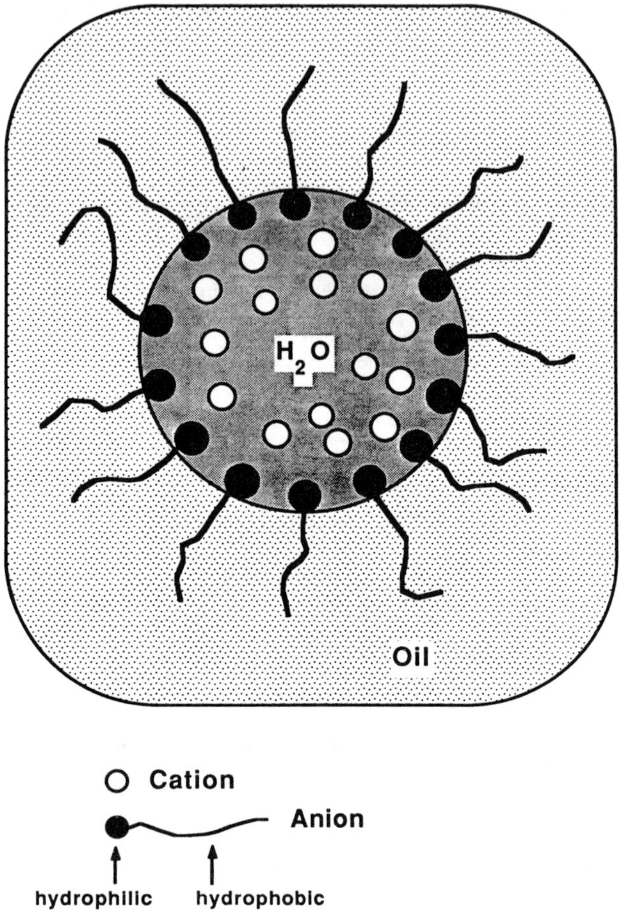

FIG. 2.42—*Schematic diagram of a micelle.*

negative charge carriers

$$A^+ + B^- \longrightarrow AB^* \longrightarrow AB \tag{2.111}$$

The speed of recombination, dn/dt is determined by their concentration n (ip/cm^3) and the recombination rate constant k_r. It is given as

$$dn/dt = -k_r n^2 \tag{2.112}$$

A half-life, $\tau_{1/2}$ can be defined after which a given concentration of ion pairs has been reduced to one half of its initial value by recombination. This half-life is given as

$$\tau_{1/2} = (k_r n)^{-1} \tag{2.113}$$

The half-life increases with decreasing ion pair concentration. Recombination rate constants (also known as recombination coefficient) have been measured, and for ionic charge carriers the Debye formula was found to hold. It relates the recombination rate constant with the mobilities of the positive and negative ions, respectively, as

$$k_r/(\mu_+ + \mu_-) = e/(\epsilon_o \epsilon_r) \tag{2.114}$$

where e is the electronic charge, ϵ_o is the permittivity of the vacuum, and ϵ_r is the relative dielectric constant of the liquid. The Debye formula was found to hold also for electron/positive ion recombination as long as the electron mobility did not exceed 10 $cm^2/V\ s$ [*96*]. At greater values of the electron mobility deviations occur due to the fact that not each encounter between an electron and an ion leads to recombination [*97*].

2.4.5.2 Attachment—Electron attachment takes place when an electron encounters an electronegative atom or molecule S

$$e + S \longrightarrow S^- \tag{2.115}$$

The speed of attachment is characterized by an attachment rate constant, k_s which depends on the type of electronegative molecule and on the liquid. Many values have been measured and a collection of data is given in Table 2.22. The life time τ of electrons with respect to attachment is given as

$$\tau = (k_s n_s)^{-1} \tag{2.116}$$

Here, n_s denotes the concentration of the attaching molecules. Under certain conditions electron detachment can occur also, either by increase of temperature or due to absorption of light [*98,99*].

2.4.5.3 Charge Transfer—An important process which affects the nature of the positive charge carrier is the process of charge transfer. If a positive ion A^+ encounters a neutral molecule M which has a lower ionization potential, then electron transfer from M to A^+ takes place

$$A^+ + M \xrightarrow[k_{ct}]{} A + M^+ \tag{2.117}$$

The original ion A^+ is converted into an impurity ion M^+. This process terminates any hole transport in a given liquid since M^+ cannot transfer back its positive charge to molecules A. Transfer rate constants, k_{ct} were measured, and it was found that in alkanes maximum values of $k_{ct} = 10^{10}$ [$M^{-1}s^{-1}$] were observed when the difference in gas phase ionization potential between the molecules of the liquid and of the impurity was 2 eV [*100*]. The lifetime with respect to charge transfer depends on the concentration of molecules M, n_M, and on the rate constant k_{ct} (see Eq 2.116).

2.4.6 Hole Mobility

Much less is known about hole transport in liquid hydrocarbons or liquefied rare gases. The experimental detection of holes is rendered difficult by the very high purity requirements. Thus far,

TABLE 2.22—*Rate constants for electron attachment* [84].[a]

Liquid	Solute	$T(K)$	k_s ($M^{-1}s^{-1}$)
Methane	SF_6	110	$4.0\ 10^{14}$
	O_2	110	$8.4\ 10^{11}$
n-hexane	SF_6	295	$1.9\ 10^{12}$
	CO_2	293	$1.8\ 10^{12}$
	O_2	293	$1.5\ 10^{11}$
Cyclohexane	SF_6	294	$4.0\ 10^{12}$
	CCl_4	294	$2.7\ 10^{12}$
	O_2	293	$1.7\ 10^{11}$
Isooctane	SF_6	293	$5.8\ 10^{13}$

[a]M = moles per liter.

positive hole transport has been observed only in cycloalkanes with six membered rings (cyclohexane and derivatives). Results obtained by de Haas et al. [*101*] are compiled in Table 2.23. In the liquefied rare gases only in LXe positive hole migration seems possible if calculations done on solid xenon are extrapolated to the liquid phase [*102*].

2.4.7 Models for Electron Transport

In theories on electron mobility in crystals, two limiting cases are distinguished: (i) electron motion in a delocalized state. The electron is treated quantum mechanically as a plane wave which is scattered by thermal oscillation of the atoms or molecules comprising the lattice. Impurities and lattice defects provide additional scattering centers. The observed mobility is determined by all these scattering processes. This case is characterized by a large electron mobility ($\gg$1 cm^2/V s). Electrons in LAr, LKr, LXe, and liquid methane fall into this category and (ii) the second case is that of low electron mobilities, $\ll$1 cm^2/V s. Here, one has to assume that the motion of the electron is slowed down by residence in localized states (traps). The electron migrates by jumping from one trap to the other. The residence time in a trap is much longer than the time it takes to jump. Increase of temperature and of the electric field strength increase the jump frequency and in this way increase the electron drift velocity.

These models (schematically represented in Fig. 2.43), originally developed for solids, also have been applied for the treatment of the electron mobility in liquids [*103–107*]. In the delocalized state the electron mobility is given by

$$\mu = \frac{e}{m^*}\tau \qquad (2.118)$$

where m^* denotes the effective mass and e is the elementary charge. τ is the relaxation time for the collisions with the atoms or molecules comprising the liquid. τ incorporates the liquid structure. At low applied electric field strength, the electrons are in thermal equilibrium with the atoms or molecules of the liquid. They exhibit a random velocity v_{th}. The relaxation time τ then can be expressed by a mean free path between two collisions Λ_1 as

$$\tau = \frac{\Lambda_1}{v_{th}} \qquad (2.119)$$

Introduction of a Maxwellian velocity distribution for v_{th} yields for the mobility

$$\mu = \frac{2}{3}\left(\frac{2}{\pi mkT}\right)^{0.5} e\Lambda_1 \qquad (2.120)$$

A decrease of the mobility with increasing temperature is predicted. This is observed experimentally over a limited range in the high mobility liquids. At higher values of the electric field strength the condition of thermal equilibrium cannot be fulfilled anymore. The electrons are picking up more

TABLE 2.23—*Hole mobility as a function of temperature; $\mu_h \times 10^3$ cm^{2}V^{-1}s^{-1}* [25,101].

	Temperature, °C									
Liquid	−26	−15	2	19	22	38	56	78	82	108
Cyclohexane				10.5		8.4	9.0	9.7		
Methylcyclohexane					5.8					
Transdecalin	14.4	8.2	10.2			8.8	7.2		7.9	7.7

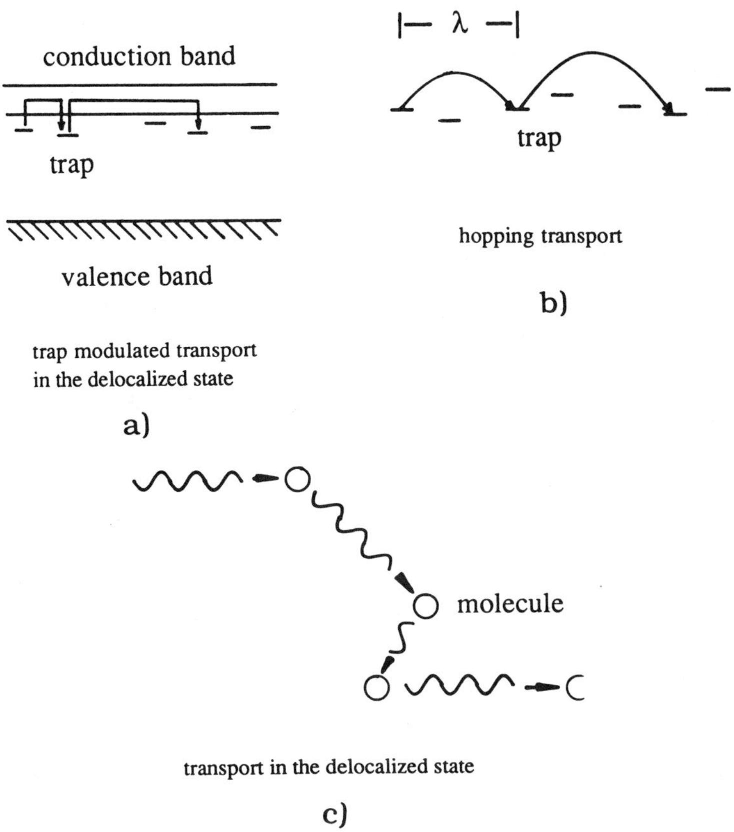

FIG. 2.43—*Schematic representation of various modes of electron transport in nonpolar liquids:* (a) *trap modulated transport,* (b) *hopping transport, and* (c) *transport in a delocalized state.*

energy from the electric field between a collision than they can loose in the subsequent collision. The electron mean energy increases. In Eq 2.119, v_{th} must be replaced by v_{el}, the electron mean velocity. It is obvious then, that μ decreases with increasing electric field strength (see Fig. 2.39).

Transport via localized states is characterized by low electron mobilities, a superlinear dependence on the electric field strength, and a positive temperature coefficient of the electron mobility. Motion occurs either by tunneling from one trap to a neighboring one with the participation of thermal energy (hopping transport), or by thermal activation into a delocalized state and recapture by another trap (trap modulated band motion) (see Fig. 2.43). The mobility is an exponential function of the temperature given as

$$\mu_{el} = A \exp\left(-\frac{E_a}{kT}\right) \tag{2.121}$$

If the electron is promoted into a delocalized state, A is identified usually as the electron mobility in this state. If the electron is hopping between localized states then A is proportional to the product of jump frequency and the square of the mean jump distance λ. Increase of field strength leads to a field dependent mobility [*106*] given by

$$\frac{\mu(E,T)}{\mu(0,T)} = \frac{2kT}{\lambda eE}\sinh\left(\frac{\lambda eE}{2kT}\right) \tag{2.122}$$

Comparison of Eq 2.122 with experimental data gave jump distances of 1 to 4 nm [*105*]. If the data are analyzed with respect to Eq 2.121, a mobility in the delocalized state of several hundred cm^2/V s results [*107*].

2.5 Engineering Liquid Dielectrics in Electrotechnology

Several areas of electrotechnology and adjacent fields can be identified in which dielectric liquids play an important role. The relationships between the main property of the liquid employed and a particular application are depicted in the Fig. 2.44.

The traditional area of application for dielectric liquids is in the field of electrical power, especially in high voltage equipment. Liquids are used in transformers and cables as insulating and cooling media. In switches, the liquid is to extinguish the electric arc which forms when the switch is opened, and the liquid is to maintain the isolating state of the opened switch.

In experimental physics two main areas have developed which make use of insulating liquids. The first field is the generation of high voltage pulses of short rise times. Here liquid insulated coaxial discharge lines are employed. Mainly water or aqueous solutions are used. The liquid serves to increase the capacity of the line by its dielectric constant. A low conductivity is required to maintain the charged state of the line for certain periods of time. Electrostatic high voltage generators also rely on liquids with a low conductivity and a high dielectric constant. Here, propylene carbonate fulfills these requirements. The second field of application is the use of ultrapure nonpolar liquids as detection media in ionization chambers for high energy radiation. Different types of applications are being developed. In high energy physics, calorimeters containing many thousand ionization chambers intercalated with plates of high atomic number materials are used. In medical physics and materials testing, liquid ionization devices may supplement the existing imaging techniques. The methods developed and the experience gained in this research and development effort will be of use for the application of these liquids in other fields of electrotechnology, as for instance semiconductor processing.

Besides these main lines of applications various areas exist where dielectric liquids play an important role, as, for instance, in the electrostatic separation of mineral particles, electrostatic spraying, and atomization of liquids. Machining of metal for tool making is done by means of electroerosion. Spark discharges in water provide powerful sources of ultrasound for medical applications.

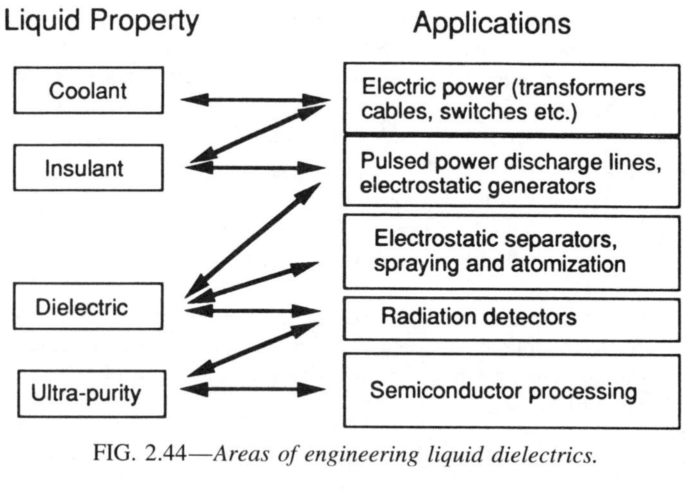

FIG. 2.44—*Areas of engineering liquid dielectrics.*

2.6 Electrical Conduction in Nonpolar Liquids

2.6.1 General Remarks

For more than sixty years it has been realized that nonpolar liquids and some polar liquids are characterized by a type of universal conduction characteristic [*1*]. An example obtained with transformer oil and stainless steel electrodes is shown in Fig. 2.45*a* [*108*]. Generally, with a parallel plate measurement cell three regions, as indicated in Fig. 2.45*b*, can be distinguished: first, a low field region where the current density J_c is proportional to the electric field strength E (ohmic region, I); second, a region where the current density with respect to the electric field strength seems to saturate (saturation region, II); and third, a region where the current density increases strongly with applied electric field strength (high field conductivity region, III). Further increase of the electric field strength leads to breakdown (breakdown region, IV). While regions (I) and (III) always are observed, region (II) is sometimes not apparent.

The ohmic region is characterized by a constant conductivity σ which in most cases is due to impurity ions present in the liquid. σ is given as

$$\sigma = \sum_i \sigma_i \tag{2.123}$$

with

$$\sigma_i = n_i z_i e_o \mu_i \tag{2.124}$$

Here, n_i denotes the concentration of the ith type of ion, $z_i e_o$ its electric charge, and μ_i its mobility. In this general formulation we allow ions of different types to contribute to the conductivity. Charge conservation requires that the total number of negative charges equals the total number of positive charges. If only single charged ions are present ($z_+ = 1, z_- = -1$), then the conductivity is given as

$$\sigma = ne_o(\mu_+ + \mu_-) \tag{2.125}$$

where μ_+ and μ_- are the positive and negative ion mobility, respectively. The extrinsic conductivity of a chemically pure liquid, for instance n-hexane, can vary over several orders of magnitude, reflecting the variation in concentration of ionizable impurities. Careful purification with chemical methods, especially treatment with powerful drying agents (phorphor pentoxide, sodium metal), and multistep distillation, yields liquid samples with conductivities of 10^{-19} to 10^{-20} $\Omega^{-1}\text{cm}^{-1}$. Since ionic mobilities are of the order of 10^{-3} to 10^{-5} $\text{cm}^2\text{V}^{-1}\text{s}^{-1}$ (see Table 2.20), the lowest conductivities obtained correspond to a concentration of several hundreds free charge carrier pairs per cubic centimeter. Part of such a low conductivity is caused by cosmic radiation. Nikuradse [*1*] found that shielding the sample with lead plates reduced the conductivity even further. Sharbaugh and Barker [*109*] concluded from an exhaustive review of the self-conductivities of different nonpolar and polar liquids as a function of the dielectric constant that the most likely impurity giving rise to free ions in these liquids was water. Naturally, any other impurity which is able to dissociate into ions can lead to an increase of the self-conductivity. The energy of dissociation is usually smaller than the ionization energy. In a liquid of low dielectric constant the dissociable impurity behaves as a weak electrolyte. The concentration of dissociated species is much smaller than the concentration of undissociated impurity molecules. Application of an electric field leads to an extraction of ions from the bulk of the liquid and the current density is determined by the rate of dissociation. This is the condition of region (II). Prolonged application of an electric field may lead to a permanent reduction of the electric conductivity due to the electrolysis of the impurity ions. This effect is utilized in the preparation of liquid samples of extremely low intrinsic conductivity [*49*]. If the concentration of undissociated impurity molecules is sufficiently high then increase of the electric field strength will lead to an enhancement of the dissociation process. More free ion

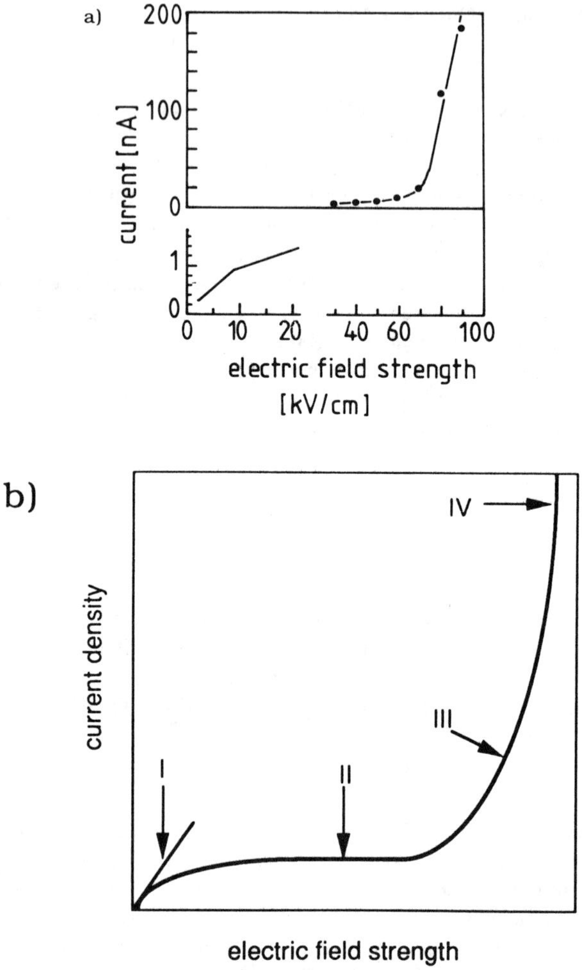

FIG. 2.45—*Current as a function of electric field strength:* (a) *in transformer oil, stainless steel electrodes,* d = *6 mm (after Ref* 108*), and* (b) *idealized schematic representation of the different regions.*

pairs are produced and the conductivity increases. This may be one of the possible reasons for region (III).

There exists another mechanism by which ions are injected into the liquid. Reduction or oxidation of impurity molecules at the electrodes may yield negative or positive ions. This effect requires a high electric field strength, but in the presence of a double layer the electric field strength at the electrode may be high enough already under the conditions of region (II). At further increase of the electric field strength, ion injection from the electrodes may become the dominating process. These ions are either produced from impurities or at very high field strength the processes of electron field emission and field ionization set in (see Section 2.3.7). Very little quantitative work on impurity conduction in liquids of low dielectric constants is available.

2.6.2 Intrinsic Conductivity of Pure Nonpolar Liquids

In Section 2.3.2 we calculated the stationary charge carrier concentration produced in a nonpolar liquid by thermal excitation and found that it would be unmeasurably small for all practical purposes. Nonpolar liquids can be considered as ideal insulators for which thermal excitation of charge carriers in the bulk can be neglected. Injection of charge carriers at the electrodes may contribute to the intrinsic conductivity of nonpolar liquids. At the liquid/metal interface, surface states of the liquid may accept electrons from the metal (Fig. 2.46). Although it has been demonstrated experimentally by Rose-Innes et al. that solidified rare gases do not exhibit surface states [*110*], molecules with an anisotropy of the electrical polarizability may give rise to surface states of the metal liquid interface. Most of the alkanes, naphthenes, and aromatic liquids meet this condition. Under the influence of the applied electric field, surface states occupied by electrons would convert into bulk states occupied by electrons. These would then form a virtual cathode bound to the metal cathode by the image force. Thermal excitation and electric field would help these electrons to escape into the bulk of the liquid. The theoretical one-dimensional problem of electron escape from a cathode under the influence of image potential, externally applied electric field and temperature has been treated by several authors [*111–113*].

Baker and Boltz [*114*] postulated that thermionic emission was responsible for the transfer of electrons into the liquid. The current would be governed by the Richardson equation [*47*]. The barrier for injection, Φ_{liq} is given as the work function of the metal in the vacuum, Φ_{vac} and the energy of the electronic conduction level, V_o, which can be positive or negative, such that

$$\Phi_{liq} = \Phi_{vac} + V_o \qquad (2.126)$$

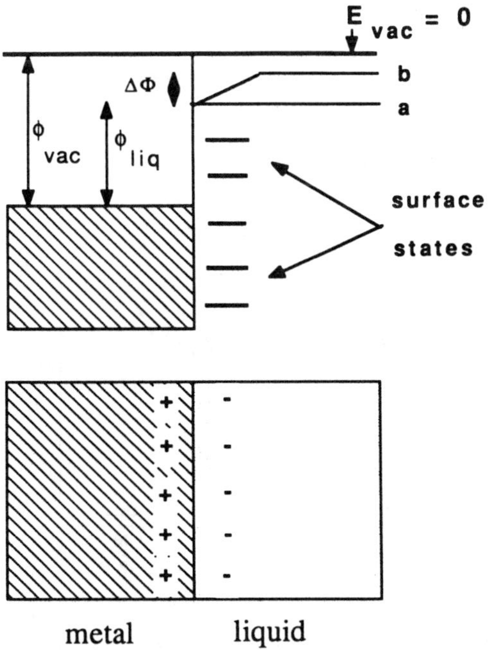

FIG. 2.46—*Electron traps at the liquid/metal interface.*

In addition a lowering of the barrier by the Schottky effect occurs (Fig. 2.47). The reduction, $\Delta\Phi_S$ is given as

$$\Delta\phi_S = \epsilon_o \left[\frac{e_o E}{4\pi\epsilon_o\epsilon_r} \right]^{1/2} \tag{2.127}$$

The effective barrier, Φ_{eff} for electron transfer into the liquid is then given as

$$\Phi_{eff} = \Phi_{liq} - \Delta\Phi_S \tag{2.128}$$

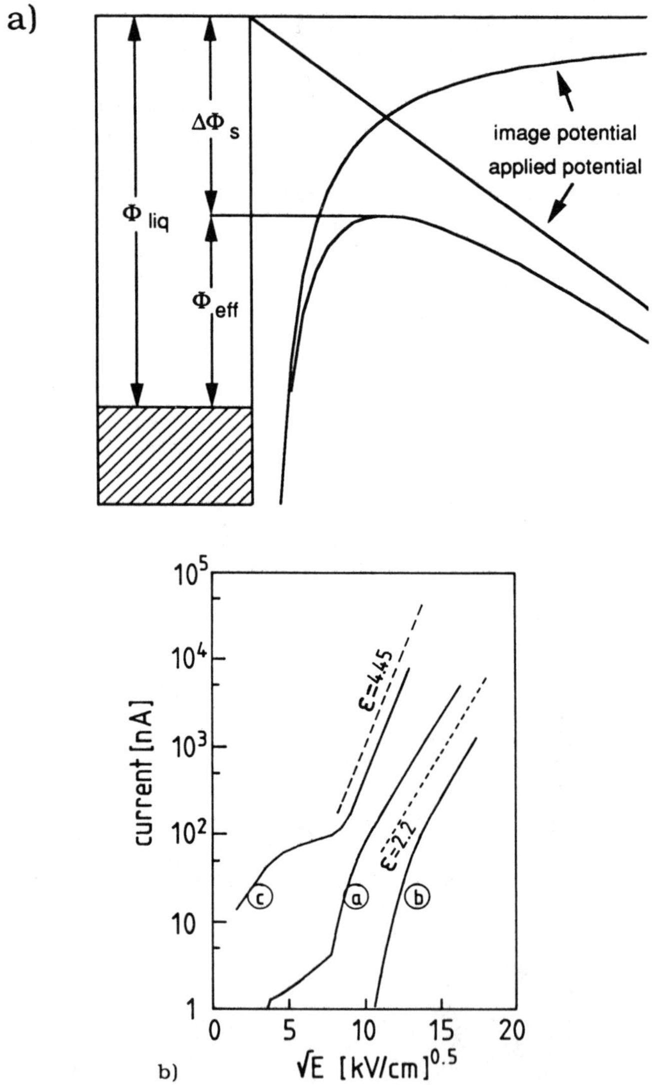

FIG. 2.47—*Schottky effect:* (a) *potential near the electrode and* (b) *current as a function of electric field strength for liquids with different dielectric constant (after Ref* 108*); (i) transformer oil, (ii) alkylbenzene, (iii) nonflammable oil (a mixture of 60% pentachlorodiphenyl and 40% trichlorobenzene).*

Introducing Eq 2.128 into the Richardson equation yields for the current $I(E)$

$$i(E) = AT^2 \exp\left\{ -\left[\frac{\phi_{\text{liq}}}{kT} - \frac{\Delta\phi_s}{kT}\right]\right\} \tag{2.129}$$

or

$$i(E) = I_o \exp\left\{\frac{e_o\left[\dfrac{e_oE}{4\pi\epsilon_o\epsilon_r}\right]^{1/2}}{kT}\right\} \tag{2.130}$$

I_o is the thermionic injection current given by the Richardson equation. Barker and Boltz found the current of a conductivity cell with carefully cleaned platinum-iridium electrodes and carefully purified toluene ($\epsilon_r = 2.38$) to follow the dependence on the electric field strength E given by Eq 2.130. Measurements on liquids of different dielectric constant confirmed the validity of this model, too [*108*]. A quantitative agreement between measurement and this model requires knowledge of the work function, Φ_{liq}. It has been shown that on a metal surface regions may exist which *in vacuo* have much lower work functions than usually are considered for thermionic emission [*115*].

In nonpolar, inert liquids a high electric field strength can promote electron emission at the cathode (field emission) or injection of positive charge carriers at the anode (field ionization), respectively. The classical work on this topic has been carried out by Halpern and Gomer [*116–118*] who investigated experimentally and theoretically the injection of electrons and the generation of positive ions in cryogenic liquids. Their contribution triggered several other investigations in which their approach was extended to liquid hydrocarbons and other liquids [*119–129*]. The energetic situation for field emission and field ionization was discussed in Section 2.3.7. It is shown in Fig. 2.33. Experimentally, a steep rise of the current is observed when a negative voltage is applied to the tip once the voltage exceeds a critical value (Fig. 2.48*a*). This is followed by a more gradual increase at higher values of the current. The steep rise of the current with voltage is due to field emission of electrons from the metal tip electrode. At higher values of the emission current, the limited mobility of the electrons leads to formation of a space charge around the tip and the current becomes space charge limited. Field emission of electrons in a vacuum has been treated theoretically by Fowler and Nordheim [*118*] who derived the dependence of the emission current, i as a function of the applied voltage, V as

$$i = 6.2\,10^{-6}\,S\frac{\sqrt{E_F}}{(\Phi + E_F)\sqrt{\Phi}}E_t^2 \exp\left\{-6.8\,10^7\frac{\sqrt{\Phi}}{E_t}\right\} \tag{2.131}$$

where Φ represents the work function of the tip metal and E_F is the Fermi energy. $E_t = \beta V$ is the electric field strength at the surface of the emitter in volts per centimeter while S is the emitter area in square centimeters. β is a factor relating the field strength at the tip to the applied voltage. This equation was found to describe also the injection of electrons nonpolar liquids with a modified work function. A test consists in a plot of $\ln(i/V^2)$ versus $1/V$ which should give a straight line. Experimental data were found to follow this relationship (see Fig. 2.48*b*).

Recently, Denat et al. [*122,123*] found that in cyclohexane, only for tip radii smaller 0.5 μm, Eq 2.131 was obeyed, while for larger radii the field emission consisted of current pulses. The experimental investigation of the field emission charge carrier generation regime is complicated by very stringent requirements on the purity of the liquid in order to observe the low injection current levels. The self-conductivity of the liquid leads to a current between the stem of the tip electrode and the counter electrode which is superimposed on the injection current. Only in carefully purified samples of very low self-conductivity the field emission regime can be observed over a sufficient

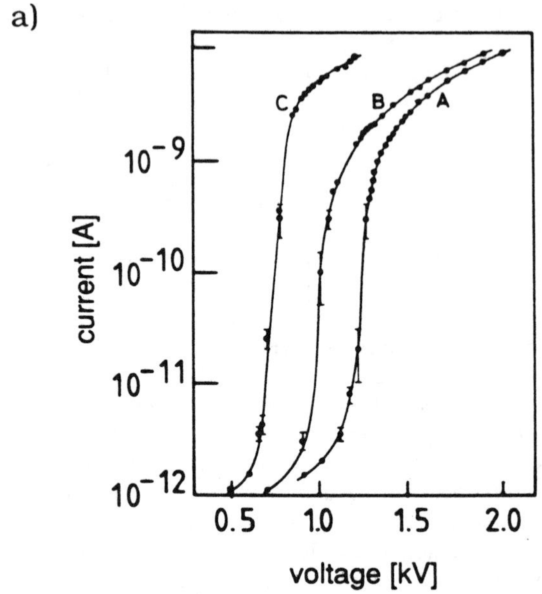

FIG. 2.48—(a) *Field emission of electrons in liquids; A n-hexane, B isooctane, C tetramethyl silane;* (b) *Fowler Nordheim Plot of the data of Fig.* (a), *after Ref* 120; *and* (c) *field ionization in liquid hydrogen, 20.4 K (after Ref* 117*).*

voltage interval. For the space charge limitation regime, Halpern and Gomer [*116*] derived an equation which was applied for the determination of charge carrier mobilities

$$V = V_{\text{vac}} + 98\left(\frac{R}{\epsilon_r \mu_- \alpha}\right)^{1/2} i^{1/2} \tag{2.132}$$

The electrode separation is R in centimeter, the mobility μ_- is given in $cm^2V^{-1}s^{-1}$, the current in *nA* and the voltage is in *V*. V_{vac} is the voltage which in a vacuum would produce the current i. $\alpha\pi$ is the solid angle into which emission occurs. Using Eq 2.132 for the determination of electron mobilities requires some caution since the presence of electronegative impurities in the liquid may lead to the formation of negative ions by electron attachment. Only an apparent mobility then can be extracted [*121*]. Another complication arises from the fact that charge injection leads to momentum transfer to the neutral liquid. A liquid flow away from the tip is set up. Under equilibrium conditions an electrohydrodynamic mobility, μ_{EHD} of the charge carriers results which is given by Eq 2.110.

In the case of the tip electrode positive, charge carrier formation is due to field ionization. From a molecule or atom near the tip an electron tunnels into the metal and a positive ion is formed. This regime also can be observed at low injection current levels, only. At higher levels of the injection current space charge limitation sets in, and again Eq 2.132 applies with the positive ion mobility. Halpern and Gomer reported on the field ionization regime in liquid hydrogen [*117*]. Denat et al. [*122,123*] observed field ionization currents in cyclohexane with very fine tips (radius of curvature smaller 0.5 μm). Schmidt and Schnabel [*128*] found field ionization in aromatic liquids, Rabe et al. [*129*] investigated silicone oils. The influence of aromatic side groups in these oils on the field ionization is especially apparent as is shown in Fig. 2.49. While the AK oil (polydimethyl siloxane) does not contain phenyl groups, AP and AS (methylphenyl siloxane) contain different amounts of

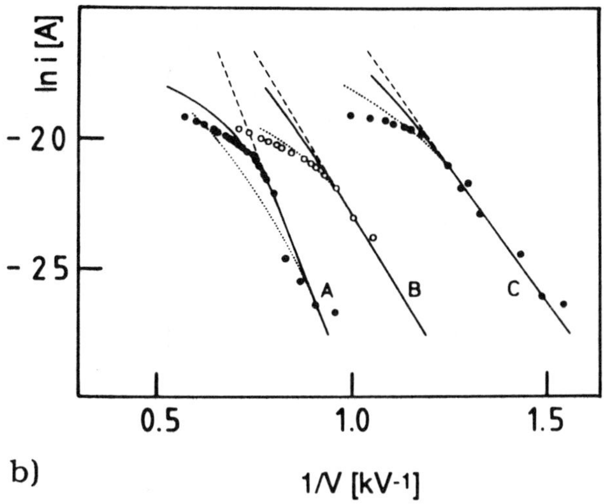

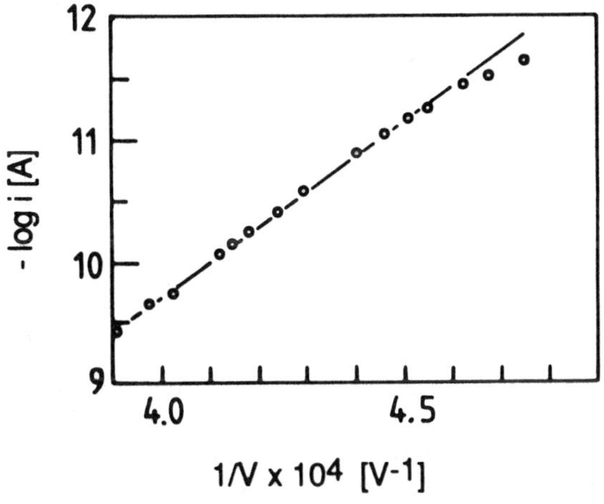

FIG. 2.48—*Continued.*

phenyl groups which have a lower ionization energy than the methyl groups of the AK oil. More positive ions are produced at a given applied voltage and a higher injection current results.

The spatial distribution of the emission current from a tungsten tip in liquid helium was measured by McClintock et al. in a test cell consisting of a spherical cavity with the tip in the center [*126*]. The cavity walls formed the anode which was fragmented into 16 ring electrodes insulated from each other by mylar sheets (Fig. 2.50*a* and *b*). The angular distribution of the emission current for positive and negative tips are shown in Fig. 2.50*c* and *d*. From the emission pattern it was concluded that the current was injected from a microdischarge enveloping a significant length of the emitter's shank (Fig. 2.50*d*). Such effect may be limited to low boiling simple inert liquids.

Liquids comprised of electronegative molecules, as perfluorocarbons or halogenated hydrocarbons provide molecular electron traps. Electrons from the metal can tunnel directly into these traps

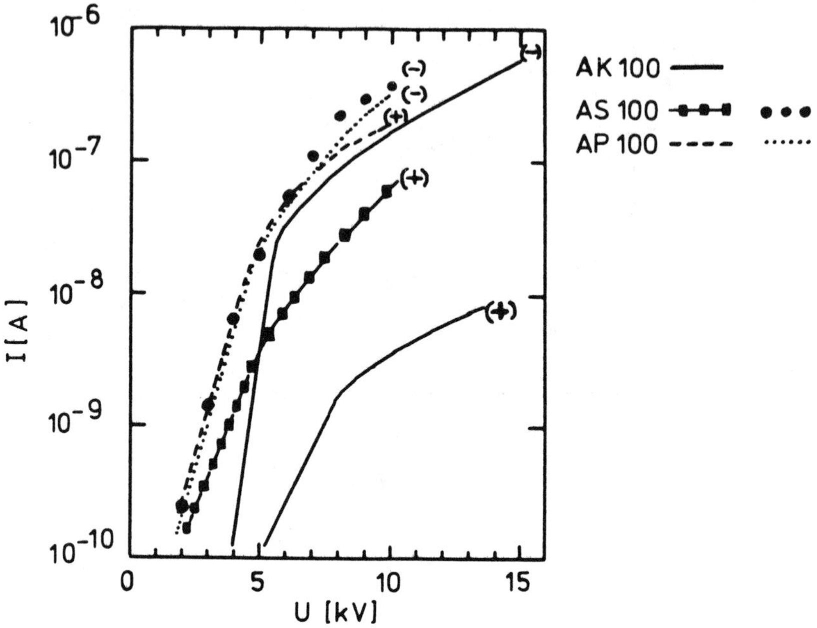

FIG. 2.49—*Injection currents in silicone oils (after Ref* 129)*; the polarity of the tip electrode is indicated.*

which are present the bulk of the liquid. This concept is also corroborated by the work on contact charging of solidified rare gases where addition of chlorine to solid Xe facilitates electron transfer from the metal into the insulator [*110*].

2.6.3 Extrinsic Conductivity of Nonpolar Liquids

Most commercially obtained liquids exhibit a conductivity caused by molecular impurities or particulate matter. In the following subsections some properties of this extrinsic conductivity will be discussed.

2.6.3.1 Electrolytic Conduction—Most chemically pure inert liquids contain enough impurities so that their conductivity is primarily due to the dissociation of these impurities. Two processes can be distinguished. The thermal dissociation of solutes which have an ionization energy I_{sol} much smaller than that of the liquid yields electron/ion pairs

$$S \longrightarrow S^{+} + e^{-} \tag{2.133}$$

As in the case of pure liquids, I_{sol} is smaller in solution as compared to the gas phase. Equation 2.32 is valid, accordingly

$$I_{sol}(\text{liq}) = I_{sol}(\text{gas}) + P_{+} + V_{o} \tag{2.134}$$

where P_{+}, V_{o} are determined by the properties of the liquid. The other process is dissociation of solute molecules into oppositely charged ions via the formation of ion pairs

$$AB \Longleftrightarrow (A^{+}B^{-}) \underset{k_b}{\overset{k_f}{\Longleftrightarrow}} A^{+} + B^{-} \tag{2.135}$$

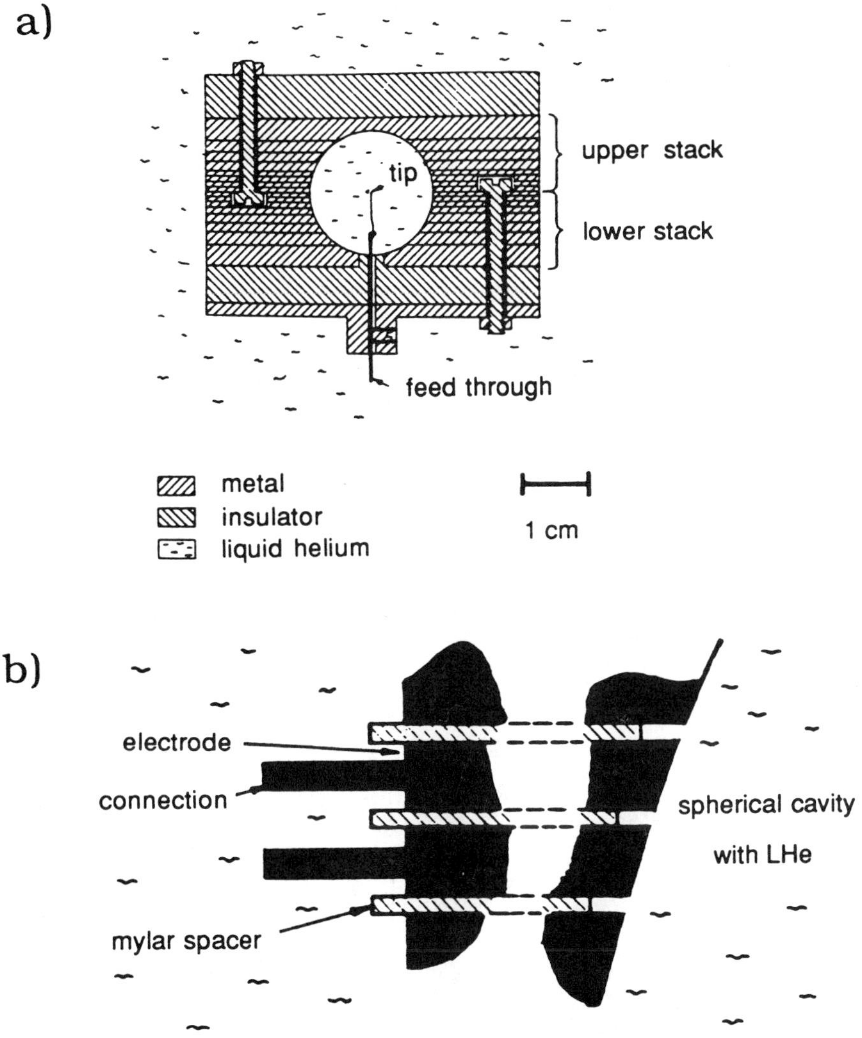

FIG. 2.50—*Angular distribution of the emission current (after Ref* 126*):* (a) *spherical test cell for the measurement of the angular distribution of the current emitted from a fine tungsten tip in liquid helium;* (b) *detail of the fragmented anode;* (c) *angular distribution of the positive current emitted into liquid helium at 1.69 K. The current density is presented as a symmetrical polar histogram with the current density in units of* 10^{-8} *A* cm^{-2}*;* (d) *angular distribution of the negative ion current emitted into liquid helium at 1.52 K. The current density is presented as a symmetrical polar histogram with the current density in units of* 10^{-9} *A* cm^{-2}*; and* (e) *formation of a microdischarge around the tip.*

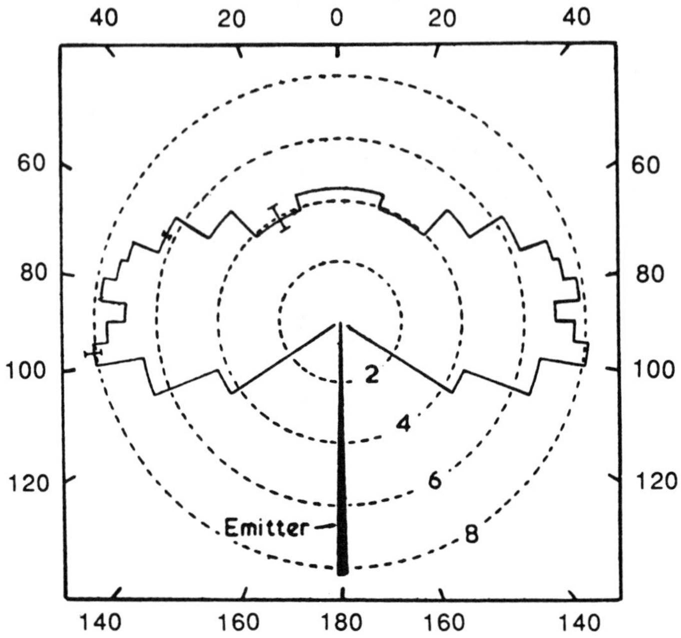

c)

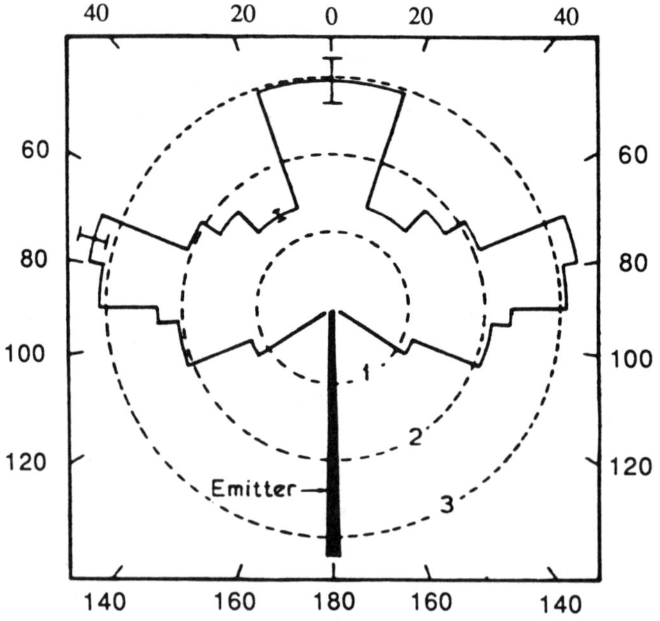

d)

FIG. 2.50—*Continued.*

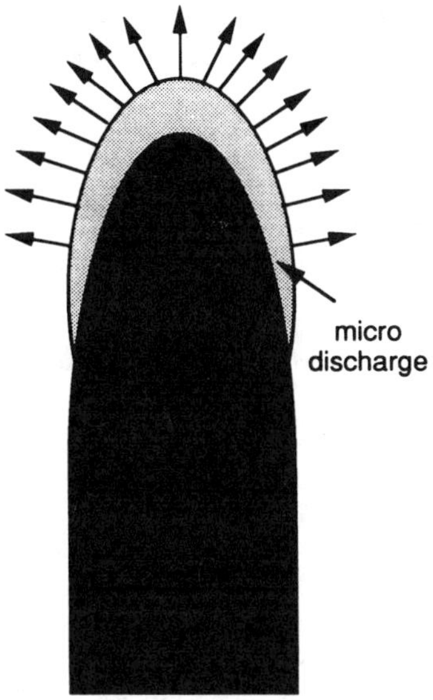

e)

FIG. 2.50—*Continued.*

Ion pairs are defined by the condition that the distance between the oppositely charged ions is smaller than the *Bjerrum distance,* r_B which is defined by the condition that the energy of electrostatic attraction is balanced by twice the value of the thermal energy kT. r_B is then given as

$$r_B = \frac{z_A e_o z_B e_o}{8\pi\epsilon_o\epsilon_r kT} \tag{2.136}$$

where z_A and z_B are the ionic valencies. Application of an external electric field shifts the equilibrium between ion pairs and free ions in the direction of the free ions. An increase of the electrical conductivity results. Onsager [*130*] calculated the effect of the electric field strength on the equilibrium constant K

$$K = k_f/k_b \tag{2.137}$$

as

$$K(E)/K(0) = 1 + 2\beta r_B + \tfrac{1}{3}(2\beta r_B)^2 + \tfrac{1}{18}(2\beta r_B)^3 + \cdots \tag{2.138}$$

The quantity 2β has the dimension of a reciprocal length and it is given by

$$2\beta = \frac{|e_o(z_A\mu_A - z_B\mu_B)E|}{kT(\mu_A + \mu_B)} \tag{2.139}$$

where μ_A and μ_B are the mobilities of the positive and negative ions, respectively. E is the electric field strength. For a moderate electric field strength, Eq 2.138 can be approximated by

$$K(E)/K(0) = 1 + 2\beta r_B \tag{2.140}$$

The increase of the ion concentration, c_{ion} due to this field dissociation effect is given by the law of mass action

$$\frac{c\alpha^2}{(1-\alpha)} = K \tag{2.141}$$

where α denotes the degree of dissociation, which is defined as

$$\alpha = c_{\text{ion}}/c \tag{2.142}$$

and c denotes the total concentration of the solute. The conductivity is proportional to the degree of dissociation. An increase in the degree of dissociation by the influence of the electric field on K is reflected in an increase $\Delta\sigma$ of the conductivity σ given as

$$\frac{\Delta\sigma}{\sigma} = \frac{\Delta\alpha}{\alpha} = \frac{(1-\alpha)}{(2-\alpha)}\frac{\Delta K}{K} = \frac{(1-\alpha)}{(2-\alpha)}(2\beta r_B) \tag{2.143}$$

The relative change in conductivity due to the electric field is largest when the dissociation in free ions is very small (that is, for $\alpha = 0$). This condition is found in most highly insulating liquids. The relative change of the conductivity is proportional to βr_B which is inversely proportional to the dielectric constant and the square of the absolute temperature.

One of the most prominent impurity in nonpolar liquids is water. If its concentration exceeds the solubility limit, microscopic water droplets will form. Below this limit, it may be assumed that the water molecules are dissolved individually. Ionic dissociation leads to an increased conductivity of the liquid above its intrinsic value (see Section 2.3.2). It could be shown, that the self-conductivity of many nonpolar and polar liquids could be explained by the presence of water [*109*].

A systematic investigation of the influence of water on the conductivity of cyclohexane is due to Staudhammer and Seyer [*131*]. At low electric field strength, three effects contribute to the magnitude of the observed current. First, free ion pairs are generated by dissociation,

$$k_{\text{ion}} = \frac{dc_{\text{ion}}}{dt} \tag{2.144}$$

where k_{ion} denotes the rate of dissociation in ion-pairs per cubic centimeters and seconds. Second, ions recombine and the change in concentration c_{ion} is given by

$$\frac{dc_{\text{ion}}}{dt} = -k_r c_{\text{ion}}^2 \tag{2.145}$$

where k_r denotes the recombination coefficient. Third, neutralization at the electrodes removes some ions from the liquid volume

$$\frac{dc_{\text{ion}}}{dt} = -\mu_{\text{ion}}\frac{Ec_{\text{ion}}}{d^2} \tag{2.146}$$

where $E = V/d$ denote the applied electric field strength and μ_{ion} the ionic mobility, respectively. Combining Eqs 2.144, 2.145, and 2.146 yields

$$-\frac{dc_{\text{ion}}}{dt} = k_r c_{\text{ion}}^2 + \mu_{\text{ion}}\frac{Ec_{\text{ion}}}{d^2} - k_{\text{ion}} \tag{2.147}$$

In the steady state, $dc_{\text{ion}}/dt = 0$ and Eq 2.147 becomes

$$\alpha c_{\text{ion}}^2 + \mu_{\text{ion}} \frac{Ec_{\text{ion}}}{d^2} - k_{\text{ion}} = 0 \tag{2.148}$$

Generally, the current density of an electrolyte is given as

$$J = \mu_{\text{ion}} c_{\text{ion}} E \tag{2.149}$$

Combining Eqs 2.148 and 2.149 yields for J

$$J = \frac{\mu_{\text{ion}}^2 E^2}{2d^2 k_r} + E \sqrt{\left(\frac{\mu_{\text{ion}}^4 E^2}{4d^6 k_r^2} + \frac{k_{\text{ion}} \mu_{\text{ion}}^2}{k_r d^2} \right)} \tag{2.150}$$

At low electric field strength, when the generation term is greater than the neutralization term, J can be written as

$$J = E \frac{\mu_{\text{ion}}}{d} \sqrt{\frac{k_{\text{ion}}}{k_r}} \tag{2.151}$$

The conductivity obeys Ohm's law, $J \sim$ E. This is the region (I) of Fig. 2.45*a*. The same equation describes the case of the radiation induced conductivity (see Section 2.6.6). When E becomes large, neutralization at the electrodes is the main cause for ion removal, and J can be approximated by a series

$$J = k_{\text{ion}} d - \frac{1}{E^2} \frac{d^5 k_{\text{ion}}^2 k_r}{\mu_{\text{ion}}^2} + \frac{1}{E^4} \frac{2d^9 k_{\text{ion}}^2 k_r^2}{\mu_{\text{ion}}^4} - \cdots \tag{2.152}$$

For $E \rightarrow \infty$

$$J = k_{\text{ion}} d \tag{2.153}$$

When the current density is independent of the electric field strength, this relation describes region (II) of Fig. 2.45*b*. Measurements of the cyclohexane/water system showed that the low field conductivity increased proportionally to the water concentration (Fig. 2.51). Taking into account the variation of the liquid viscosity with temperature, from the temperature dependence of the conductivity an activation energy of 19 kcal/mol (that is, 0.83 eV) was estimated for the process of ionic dissociation [*131*].

2.6.3.2 Time Dependence of the Current—If a voltage is applied to electrodes immersed in a nonpolar liquid, a current decaying in time is observed [*132–134*]. Generally, the temporal variation of the current can be approximated by

$$i(t) = i_o t^{-n} \tag{2.154}$$

where the exponent n decreases with time. Over a period of minutes to hours a stationary value, i_s, is reached. When this value is obtained, removal of the applied voltage leads to a discharge current flowing in the opposite direction. This transit phenomenon may be explained on the basis of the electrode polarization due to impurity ions or charged particles present in the liquid. It requires that the rate with which these ions or particles are discharged at the electrodes is smaller than the rate of

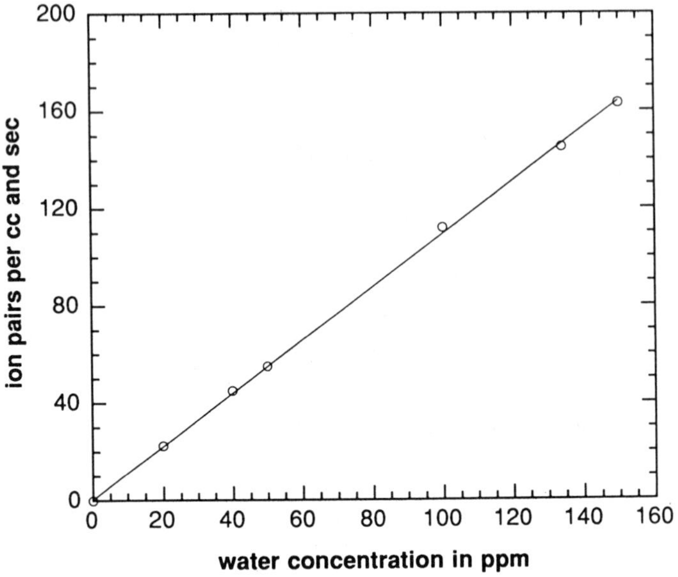

FIG. 2.51—*Ionic dissociation of water in cyclohexane (after Ref* 131*)*.

their arrival so that a space charge layer is set up in front of the electrodes. In polydimethyl siloxane measurements with different electrode separations and different electrode materials showed that the decay in time was influenced by the bulk of the liquid, while the final stationary state was influenced by the electrode material [*133*]. It was found, however, by Kahlstatt and Wollers [*135*] that at sufficiently high electric field strengths in liquid benzene, the injection at the electrodes takes over, and then no further decrease of the current with time is observed. The current did sometimes increase with time and show statistical fluctuations.

2.6.3.3 Space Charge Relaxation—Another situation arises when a space charge density ρ_o is created in the liquid, for instance at a metal/liquid interface. The conductivity σ of the liquid leads to a dissipation (relaxation) of the space charge. Two limiting cases can be distinguished: (*a*) the conductivity σ of the liquid is not augmented by the introduction of the space charge (Ohmic theory) and (*b*) the conductivity of the liquid is very small and the introduction of the space charge increases the conductivity of the liquid (Hyperbolic law) [*136*]. For the case (*a*) the relaxation of the space charge $\rho(t)$ is described by an exponential dependence

$$\rho(t) = \rho_o \exp\left(-t\Big/\left(\frac{\epsilon_o\epsilon_r}{\sigma}\right)\right) \tag{2.155}$$

with a relaxation time τ given as

$$\tau = \frac{\epsilon_o\epsilon_r}{\sigma} \tag{2.156}$$

In case (*b*), it may be assumed that the conductivity is given by

$$\sigma = \mu\rho \tag{2.157}$$

where μ denotes the mobility of the positive or negative space charge carriers, respectively. The relaxation is then described by

$$\rho(t) = \frac{\rho_o}{1 + \frac{\mu\rho_o}{\epsilon_r\epsilon_o}t} \tag{2.158}$$

and the time $t_{1/2}$, at which $\rho(t) = (\frac{1}{2}), \rho_o$ is given as

$$t_{1/2} = \frac{\epsilon_r\epsilon_o}{\mu\rho_o} \tag{2.159}$$

The space charge relaxation times given by Eqs 2.156 and 2.159 are important in the design of installations for the safe handling of flammable liquids [*173*].

2.6.3.4 Charge Injection and Fluid Motion—Injection of charge carriers from tips or edges is connected with the induction of liquid motion. The liquid flow which is set up in the test cell has a decisive influence on the current-voltage characteristics, which are observed. A detailed experimental and theoretical investigation of the injection of negative charges from a needle electrode and a blade electrode in transformer oil was carried out by Takashima et al. [*138*]. Experimentally, for the needle-plane electrode arrangement they found space charge limited currents which exhibited the voltage dependence given by Eq 2.132. The liquid motion induced by the injected negative ions (transformer oil contains enough impurities so that the initially injected electron is quickly converted into a negative ion) was taken into account by solving the Navier-Stokes equations of liquid flow numerically. For the blade-plane electrode arrangement, they observed and calculated

$$V = kI^{1/3} + V_o \tag{2.160}$$

2.6.3.5 Particle Induced Conduction—Since all liquids contain some solid matter in the form of microscopic particles, it has been suggested that the high field conduction is essentially due to the movement of these particles. They are charged at one electrode and discharged and recharged at the other. Krasucki [*139*] assumed metallic particles to be present in n-hexane, which would promote such charge transfer. A conducting spherical particle of radius r in contact with a plane electrode acquires a charge Q given as (in cgs units)

$$Q = \frac{\pi^2}{6}r^2\epsilon_r E \tag{2.161}$$

where E is the electric field applied across the parallel plane electrodes. The current produced by the movement of this particle is

$$i = \frac{Q}{\tau} \tag{2.162}$$

where τ is the transit time of the particle. In an experiment where the electric field lines are parallel to the direction of the gravitational force, the transit time τ is given as the arithmetic mean of the transit time τ_1 when the motion is opposed by gravitation and τ_2 when the motion is assisted by gravitation. The equation of motion of the particle is obtained from the following considerations. The driving force is the electric force F_E given as

$$F_E = QE \tag{2.163}$$

This is opposed by the inertia force of the particle given as

$$F_I = \frac{2}{3}\pi r^3(2\rho_s + \rho_1)\frac{dv}{dt} \tag{2.164}$$

and by the viscosity force given as

$$F_\eta = 6\pi\eta r v \tag{2.165}$$

and it is assisted or opposed by the gravitational force given as

$$F_g = \frac{4}{3}\pi r^3(\rho_s - \rho_1)g \tag{2.166}$$

where ρ_s and ρ_1 are the densities of the particle and the liquid, respectively. g is the gravitational acceleration at the surface, while v denotes the velocity of the particle.

At any given time, the equation of motion holds

$$F_E = F_I + F_\eta \pm F_g \tag{2.167}$$

Integration over time yields expressions for τ_1 and τ_2 or τ. The current due to the motion of a particle of radius r is then obtained as a function of E, η, electrode separation d and solid and liquid density ρ_s and ρ_1, respectively. In a model experiment, with one steel sphere, the validity of the theoretical approach could be demonstrated. Comparison with measured data of n-hexane could be rationalized with the present model by the assumption of a certain number of conductive particles. Later, the idea of impurity particle conduction was developed further by Molinari and Viviani who calculated the expected current for a variety of conditions [*140–143*].

Generalization of such a model led to the notion that all high field conduction in nonpolar liquids might be due to particles. It may be certainly assumed that particulate matter dispersed in the liquid will participate in the current transport. Ultracentrifugation of cyclohexane and n-hexane led to a reduction of the self-conductivity of these liquids. In addition the temporal dependence of the current after application of the voltage (see Eq 2.154) was changed. Untreated cyclohexane showed a strong decay of the current with time while after centrifugation the current changed little with time [*144*]. Since these particles may be also of semiconducting or insulating material, prolonged action of an electric field will lead to a removal of these particles by adhesion to the electrodes and the electric current will then decrease in time.

2.6.3.6 Thin Liquid Films—In order to exclude electrohydrodynamic effects on the conduction process, thin films and voltage step impulses were used. Shimokawa et al. [*133,134*] studied the time dependence of the electrical conduction in polymethyl siloxanes films of a few micrometers thickness. The scheme of their experimental setup is shown in Fig. 2.52. Under such conditions electron injection from the cathode is the dominating process. The conductivity as a function of electric field strength and temperature is shown in Fig. 2.53. The electron motion across the electrode gap is by hopping transport as was inferred from the electric field dependence (100 kV/cm to 1 MV/cm) and from the temperature dependence (room temperature to 150°C) (see also Section 2.4.3). Since the liquid layer is so thin, the influence of electron attaching impurities can be considered as being negligible. The results obtained might refer to the electronic properties of pure silicone oil.

2.6.4 High Pressure Conductivity

Application of high hydrostatic pressures on a liquid gives rise to two effects. First, a change of the electronic energy levels results, and, second, the charge carrier transport properties, as mobility, attachment, recombination, etc. are modified. Application of high pressure on the atoms or molecules of the liquid leads to an increased overlap of the electron orbitals and to a decrease of the energy gap, I_{liq} between valence and conduction levels. For solid xenon it has been demonstrated experimentally that at sufficiently high pressure (>300 kbar), valence and conduction band start to

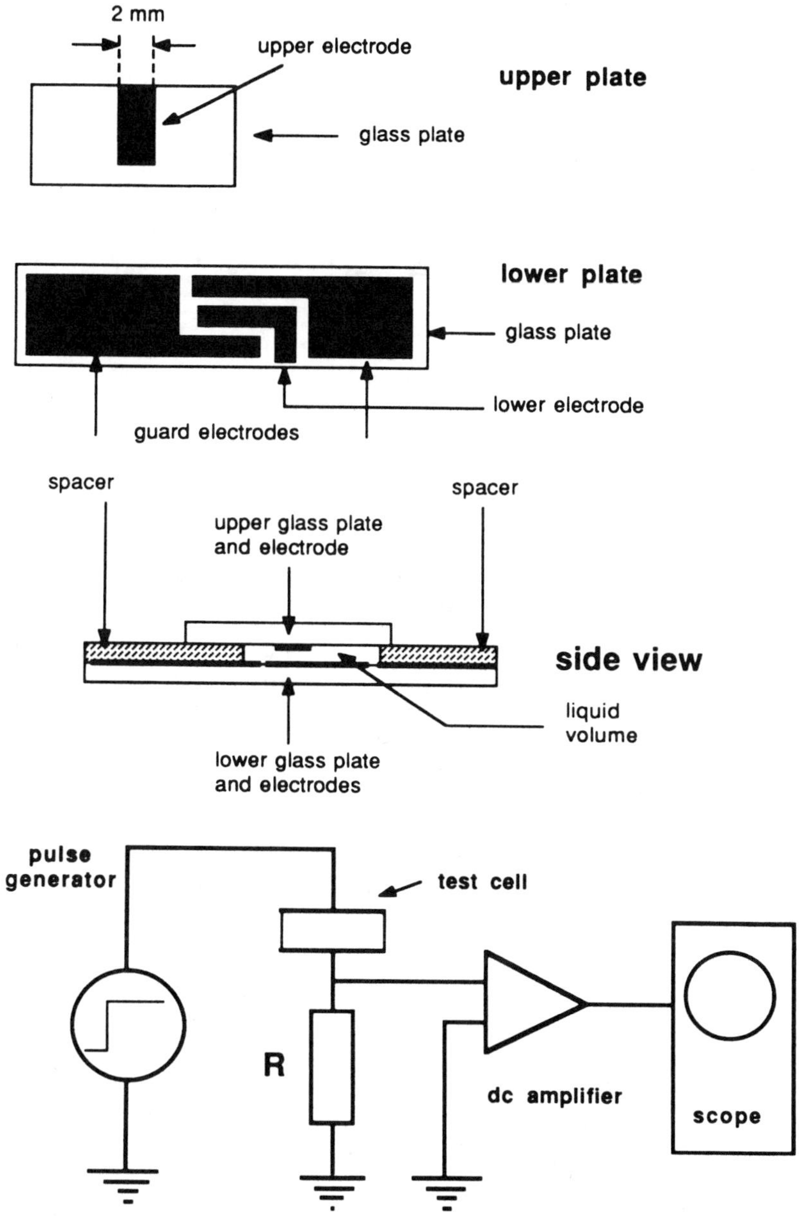

FIG. 2.52—*Thin liquid film measurement cell and circuit (after Ref* 133*).*

overlap, and the xenon sample exhibits metallic conductivity [*145*]. The accessible static pressure range in liquids is limited to lower pressures, since they solidify once a threshold pressure is exceeded. Higher pressures can be applied by means of shock waves which, however, in addition produce a rise in temperature that in turn also influences the conductivity. The electrical conductivity of liquid nitrogen subjected to shock waves has been reported [*146*]. Shock wave pressures of 20 to 30 GPa (200 to 300 kbar) were generated with a two stage light gas gun. In the interpretation of

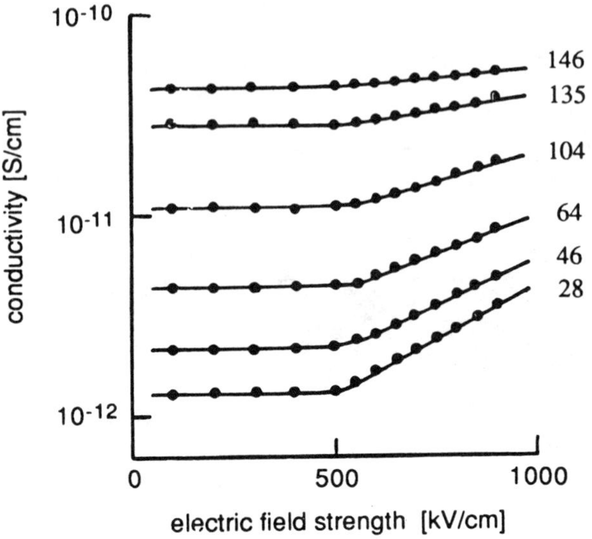

FIG. 2.53—*Initial conductivity of 10 cSt polysiloxane after application of a voltage step as a function of temperature and electric field strength; parameter-temperature in °C (after Ref* 134).

the conductivity results, a semiconductor model was employed (see Section 2.3.2, Eq 2.63). With the help of calculated shock temperatures, the authors estimated an effective electronic energy gap of 5 eV at 20 GPa with an electronic carrier concentration of the order of two 10^{18} cm^{-3} corresponding to about one carrier per 10^4 nitrogen molecules. The ionization energy at 10^5 Pa (= 1 bar) can be estimated from Eq 2.32 (see Section 2.2.5) as $I_{liq} \geq 12$ to 13 eV.

The work reported on hydrocarbons and similar compounds was done at a much lower pressure range. The self-conductivity of n-hexane, diethyl ether and castor oil under pressure up to 5 kbar was studied at a frequency of 1 kHz [*147*]. While the conductivity increased in diethyl ether almost 100 times compared to that at atmospheric pressure, it decreased in n-hexane by roughly one order of magnitude. In castor oil the conductivity decreased up to 2.5 kbar and then increased again.

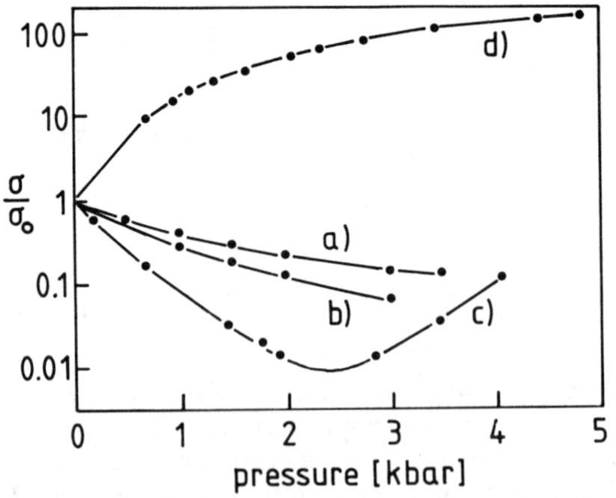

FIG. 2.54—*Relative variation of the self conductivity of n-hexane* (a,b), *diethyl ether* (c), *and castor oil* (d) *as a function of hydrostatic pressure (after Ref* 147).

These results are summarized in Fig. 2.54. A decrease of the conductivity can be explained by the increase of viscosity under pressure and its influence on the ionic motion (see Section 2.4.4). The observed increase may be due to an influence of the pressure on electrolytic dissociation of impurities present in the liquid (see p. 643 of Ref *94*). The d-c conductivity of benzene up to 5 kbar was found to increase with pressure [*148*], the relative increase being greatest at 220°C (Fig. 2.55). This finding seems to support the contention that at high pressure a reduction of the energy necessary for the creation of charge carriers takes place. From the previous discussion on LN_2, it seems unlikely that the pressures employed were sufficient to reduce the value of I_{liq} of benzene substantially. The remnant impurity content may have been responsible for the conduction. Somewhat contradictory results on benzene were obtained by the measurement of the a-c conductivity up to 3.4 kbar [*149*]. A decrease of the conductivity with increasing pressure was observed. The radiation-induced conductivity of several hydrocarbons under pressure up to 4 kbar showed a decrease of the conductivity with increasing pressure [*150*].

2.6.5 Radiation Induced Conductivity

Irradiation of gases with penetrating radiation produces charge carrier pairs which can be collected with a sufficiently high electric field yielding the so-called saturation current i_s. This current does not increase with further increase of the applied field (collisional ionization excluded) as all the charge carrier pairs produced per second are collected. The saturation current is, therefore, proportional to the radiation intensity (dose rate). In contrast, in nonpolar liquids quite a different behavior is observed where the current measured at a given electric field strength and at a given dose rate is much smaller than in gases and it increases with applied field strength. But even at the highest values of field thus far applied in most nonpolar liquids (exceptions being LAr, LKr, and LXe) no saturation current has been observed. The reason rests in the close spatial correlation of the charge carrier pairs produced by the high energy radiation, which favors recombination (geminate recombination, see Fig. 2.56).

When a nonpolar liquid is irradiated with high energy X-rays or γ-rays, an increase of the electrical conductivity results. In a parallel plate measurement cell of electrode area A and electrode separation d the radiation-induced conductivity σ_{irr} is given as

$$\sigma_{irr} = \frac{id}{VA} \tag{2.168}$$

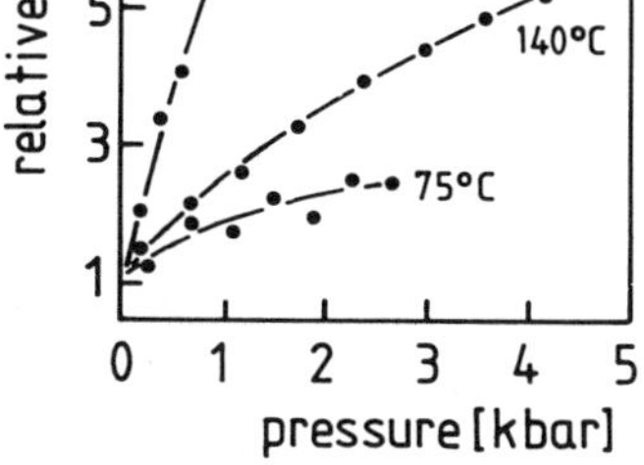

FIG. 2.55—*Relative variation of the self conductivity of benzene as a function of pressure and temperature (after Ref* 148*).*

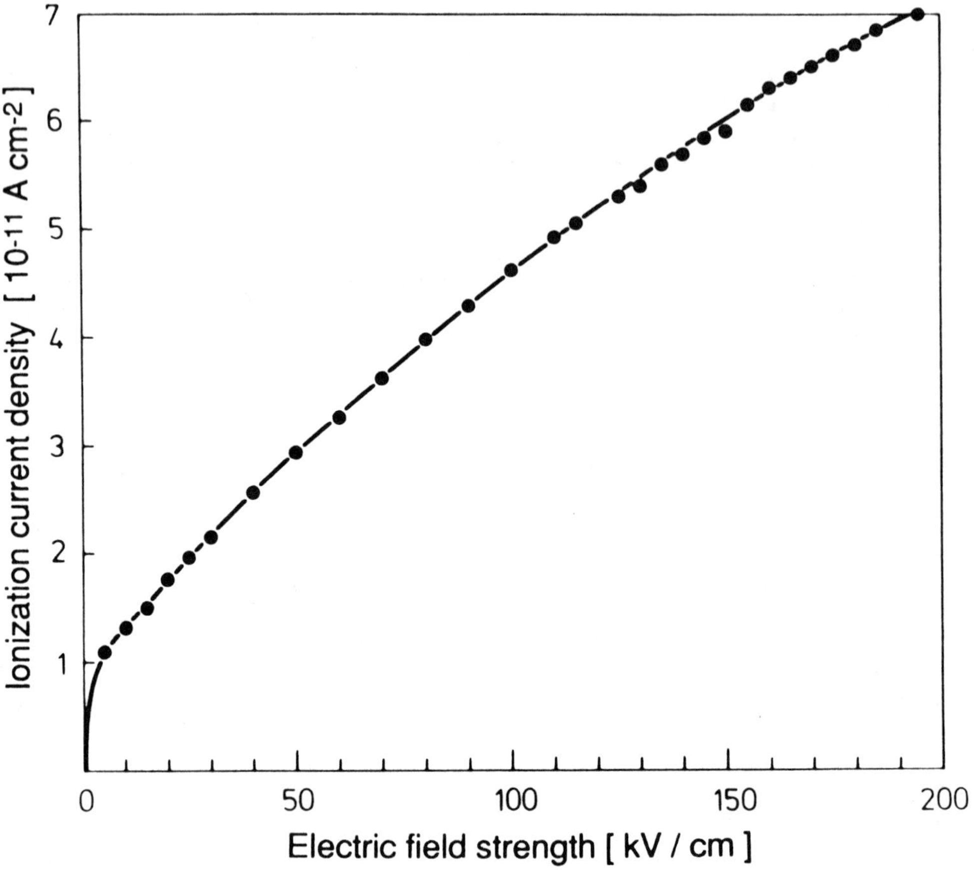

FIG. 2.56—*γ-radiation induced conductivity in n-hexane as a function of electric field strength (after Ref* 152).

where i is the ionization current with an applied voltage V. At low applied voltages charge carriers are produced with a rate g, and they are removed by recombination. The steady state is characterized by

$$dn/dt = g - k_r n^2 \tag{2.169}$$

where k_r is the rate constant for recombination either in L/mol s or in cm^3/s. The number of charge carrier pairs produced per cm^3 and s is proportional to the absorbed dose rate D_r in eV $cm^{-3}s^{-1}$ such that g is given as

$$g = D_r G_{fi} 10^{-2} \tag{2.170}$$

where G_{fi} is the free ion yield, the number of charge carrier pairs produced by the absorption of 100 eV of radiation energy. The radiation induced conductivity is given by

$$\sigma_{irr} = e_o n(\mu_+ + \mu_-) \tag{2.171}$$

Combining Eqs 2.169 and 2.170 yields

$$n = \sqrt{\frac{G_{fi}D_r}{100k_r}} \tag{2.172}$$

and with Eq 2.171 we obtain

$$\sigma_{\text{irr}} = e_o \sqrt{\frac{G_{fi}D_r}{100k_r}}(\mu_+ + \mu_-) \tag{2.173}$$

The radiation induced conductivity at low electric field strength is proportional to the square root of the dose rate; this has been verified experimentally in Ref *151*. The dose rate D_r must be high enough and the applied voltage must be small enough if the neutralization of charge carriers at the electrodes is to be negligible as compared to recombination. At higher values of the applied voltage (higher values of the electric field strength) volume recombination becomes increasingly less important, and, above a certain applied voltage, all free ion pairs are collected eventually at the electrodes. The ionization current then assumes the form

$$i(E) = e_o A\, dG_{fi}(E)\, 10^{-2} D_r \tag{2.174}$$

The ionization current is thus seen to be proportional to the dose rate D_r. The dependence of $G_{fi}(E)$ on the electric field strength can be explained by the Onsager model of geminate recombination [*130,49*] and experimentally, it was established that in many nonpolar liquids $i(E)$ is a linear function of the applied electric field strength [*49*]. At constant dose rate, D_r over a field strength range of several tens of kilovolts per centimeter the current can be expressed as

$$i(E) = f(E) + aE \tag{2.175}$$

here $f(E)$ represents the field dependence for the collection of the free ions while the term aE describes the influence of the electric field on the geminate recombination of the radiation induced charge carrier pairs.

The conductivity induced in hydrocarbons by other types of radiation (α-particles, high energy protons, fast neutrons) is determined by the $G_{fi}(E)$ term. At constant particle flux, the current-voltage dependence is contingent upon the type of radiation [*153–155*].

2.6.6 High Field Conductivity of Nonpolar Liquids

The conduction mechanisms, which were discussed in the Sections 2.6.1 to 2.6.5 exhibited a characteristic dependence on the electric field, E. Over a certain range of electric field strengths, the conductivity of a sample is dominated by the specific mechanism involved. As the current increases space charge effects become nonnegligible, and the original dependence on E is masked by the interference of the space charge field. In addition the Joule heat transferred to the liquid leads to changes of the thermophysical state. An important feature of this regime is the fact that the current density in a test cell with parallel plane electrodes is no longer independent of the spatial coordinates between the electrodes (Fig. 2.57). From optical investigations of the prebreakdown phenomena in stressed hydrocarbons it is known that shortly before breakdown occurs, the current develops as a streamer from one electrode to the opposite one (see Chapter 3 of this volume). Formation of filaments is the precursor of breakdown; filamentary currents have been observed experimentally and described theoretically in semiconductors [*156*]. Adaptation of such models to the field of liquid breakdown has not progressed very far.

An important step towards a better understanding of this conduction regime in nonpolar liquids was propounded by Kao [*157*]. The injection current between planar electrodes is at higher field

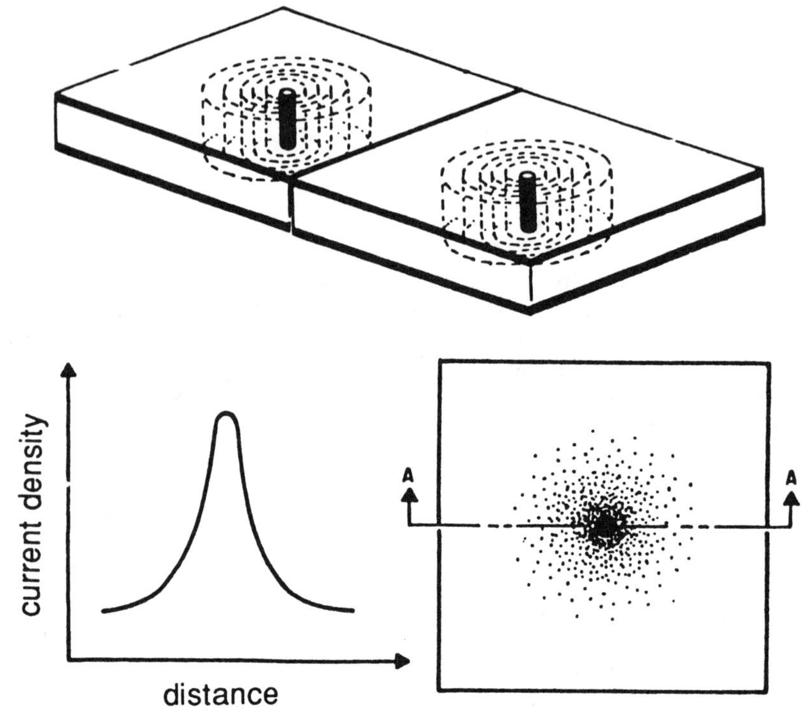

FIG. 2.57—*Schematic representation of the current density of a filament (after Ref* 156*).*

strength confined to microscopic asperities on the electrode and the conduction assumes a filamentary character. Although initially at lower electric field strength several filaments may be formed, above a certain value of the applied field strength, one filament becomes the dominating filament carrying almost all of the current which flows between the electrodes. The energy dissipated in this filament leads to rapid evaporation of the liquid and a gaseous discharge channel is formed. The conduction process is thus also determined by the external pressure on the liquid and on the temperature. The agreement between theory and the available experimental data is satisfactory, although such data are scarce.

Studies of high field conduction in cyclohexane using point-plane electrode geometry revealed that the conduction regime depends on the radius of curvature of the electrode tip, r_t [*123,158*]. For $r_t < 0.5$ μm, continuous current injection was observed for positive and negative polarity (see Section 2.6.2). At $r_t > 0.5$ μm, with the point negative, the current consisted of a series of pulses while the current with the point positive exhibited a steady d-c state upon which there were superimposed random current bursts [*123,158*]. Similar results were found in liquid argon and xenon [*159*]. The pulse regime was accompanied by light emission and erosion of the tip, and the regular emission of pulses with the tip negative was explained qualitatively as in the case of gases in the Trichel regime. Injected charges accumulate in the vicinity of the tip thus reducing the field strength at the tip, charge injection is stopped eventually until the negative space charge cloud has drifted away far enough from the tip. With the tip positive, the magnitude of the current reached before breakdown occurred is smaller than for the negative tip. No conclusive model has been developed for this case.

2.6.7 Streaming Liquids, Electrohydrodynamic Conduction

In liquids which flow through pipes or through the pores of solid insulation, triboelectric charges are generated. In liquids of low self-conductivity these charges can survive for long times, and under certain circumstances do give rise to undesirable, sometimes catastrophic phenomena. The well known electrostatic hazards arising with liquid fuel handling constitute a major problem [*160*]. In large power transformers, the circulation of the oil through the paper and pressboard insulation may lead to spurious charging and sparks, which can damage the insulation [*161,162*]. Controlled injection of electric charges into moving liquids might have certain applications in high voltage generation, heat transfer, and in electrostatic pumping (see Section 2.9.2).

If two different insulating materials or a metal and an insulator are brought into contact, charges flow from one material into the other and an electric double layer is set up. This double layer consists of fixed charges at the interface and of a diffuse layer of charges extending into the liquid. In the case of a metal in contact with a pure liquid hydrocarbon, electrons will flow into the liquid and a double layer is set up consisting of electrons in the fluid and positive charges in the metal (Fig. 2.58). The electron layer will extend into the liquid with a thickness given by the electron mobility of the liquid. In liquids with a high electron mobility the characteristic length is the mean free path while in low mobility liquids it is the jumping length (see Section 2.4.7). Liquid flow will sweep away these electrons into the bulk of the liquid and additional electrons will be injected until the electric field created between the opposite charges prevents further emission. In technically pure hydrocarbons, attachment of electrons to impurity molecules and formation of negative ions occurs. In addition, ionic charges due to dissociation of impurities exist already in the liquid and the nature of double layer formed depends on the type of ions (see Sections 2.3.5 and 2.3.6).

If a nonpolar liquid is brought into contact with a metal, an electrical double layer appears which consists of a compact or fixed region and of a diffusive region [*163,164*]. The diffusive layer can be affected by the flow of the liquid. The amount of charge, which can be swept away and detected at the exit of a metal pipe, depends on the process of charge formation in the diffusive layer. In the following, it is assumed that the space charge density in the pipe is small and that the formation of the space charge layer adjacent to the wall takes place on a time scale short compared to that

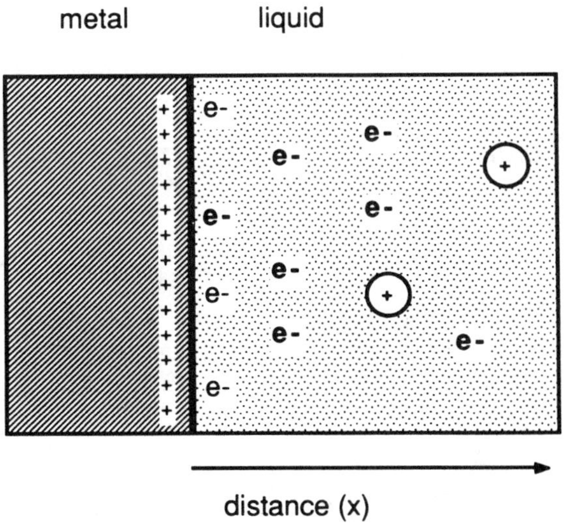

FIG. 2.58—*Double layer at the metal/insulating liquid interface.*

determined by the liquid flow. The diffusive space charge layer of singly charged unipolar charge carriers is governed by

$$\operatorname{grad} n_c + (n_c e_o/kT) \operatorname{grad} \psi = 0 \tag{2.176}$$

with n_c being the number density of charge carriers, e_o the elementary charge, and ψ the electrical potential. The conditions are more complex in the case of solid insulators in contact with the streaming liquid. The current, i_s, produced by the entràinment of charge in the flowing liquid in a tube of circular cross section can be expressed as

$$i_s = 2r_p\pi \int e_o n_c v \, dx \tag{2.177}$$

where r_p denotes the radius of the pipe, v is the liquid velocity in the diffusive part of the electric double layer, and x is the distance from the pipe wall (Fig. 2.59).

The motion of ions in a dielectric liquids involves momentum transfer to the liquid and, subsequently, generation of bulk movement. The liquid motion is influenced by the flow impedance of the experimental setup in which the process takes place [*165*]. If an electric field strength is applied to a liquid volume which contains ionic charge carriers, then initially the ions will move in a motionless liquid; but after a certain delay time, t_{EHD}, the liquid motion will set in. For the case of unipolar injection of ions from an electrode the onset of electrohydrodynamic (EHD) motion can be monitored either by the Schlieren method (see Section 2.8.2) or by recording the electrical transient current. Onset of EHD is characterized by a strong increase of the current above the level imposed by space charge limitation [*166*]. As an hypothesis, it was assumed in Ref *165* that liquid motion started when the potential difference between the front of the injected charge and the that of the injection electrode exceeded a certain critical value. This led to a dependence of the delay time upon the viscosity, η, of the liquid as

$$t_{\mathrm{EHD}} = \mathrm{const}\,\frac{\eta}{\epsilon_r \overline{E}^2} \tag{2.178}$$

where $E = V/d$ is the mean electric field strength.

2.6.8 Chemical Transformations and Aging

Passage of electrical current through molecular liquids is accompanied usually by chemical actions. At the electrodes, discharge of ions occurs, while, in the bulk, recombination of charge carriers or charge transfer takes place. One of the important processes is the polymerization which occurs when positive ions (M^+) accept an electron from a neutral molecule (M) and a chemical bond is formed between the ion and the neutral molecule

$$M + M^+ \longrightarrow M\text{—}M^+ \tag{2.179}$$

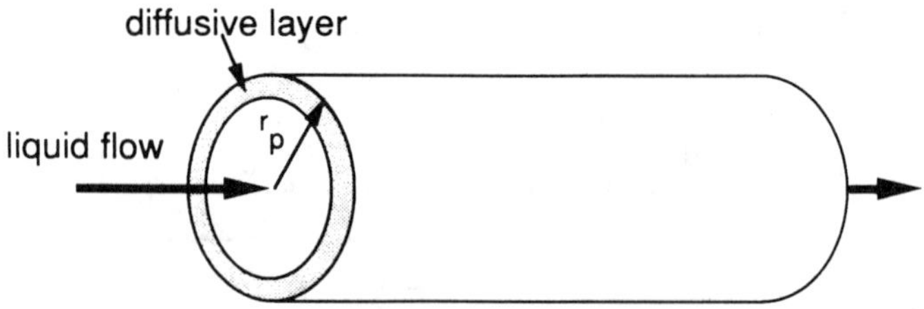

FIG. 2.59—*Streaming insulating liquid in a metal pipe.*

If this process continues a macromolecule is formed

$$M{-}M^+ + M \longrightarrow M{-}M{-}M^+ \tag{2.180}$$

$$M{-}M{-}M^+ + M \longrightarrow M{-}M{-}M{-}M^+ \tag{2.181}$$

The process of polymerization ends when the polymer-cation encounters an impurity molecule (S) with a lower ionization potential. An electron is transferred and the growth of the chain ceases

$$(M)_n{-}M^+ + S \longrightarrow (M)_n{-}M + S^+ \tag{2.182}$$

This polymerization process can be initiated at positive tips or edges where the high electric field enhancement promotes field ionization of the adjacent molecules. Many nonpolar liquids form polymers under these conditions. Wax formation in transformer oil under service conditions is probably related to these elementary processes. Recently, it was found that formation of polymers in hexane under repetitive discharges lead to a change of the breakdown voltage of a point/plane electrode gap. When the point electrode was at positive potential an increase in breakdown voltage was observed after 100 breakdown tests. Negative potential at the point electrode left the breakdown voltage unchanged [*168*]. This effect could be explained by the deposition of a polymer layer on the plane electrode, which blocked the current path.

Other chemical effects of current passage through liquids may involve a decomposition of molecules and formation of gases. This usually takes place in the prebreakdown and breakdown phase. As an example, the investigation of the decomposition of silicone oils under different types of discharges is described [*129*]. The amount of energy which was allowed to pass through a tip/plane gap was controlled by the magnitude of the capacitance connected to the gap. While there was an approximately proportional relationship between the total amount of gas produced and the charge which passed through the liquid (which is proportional to the discharge energy, since the breakdown voltage was nearly constant at a fixed electrode separation), the composition of the gas varied with the intensity of the discharge. In the Table 2.24 the composition of the gas produced in

TABLE 2.24—*Composition of gases produced by discharges of different intensities in silicone oils (after Ref* 129*).*

		Volume, %				
Discharge	Oil[a]	H_2	CH_4	C_2H_2	C_2H_4	C_2H_6
Partial	AK100	76.2	8.3	1.7	0.6	13.2
	AS100	75.3	7.3	5.8	1.7	9.9
a-c-breakdown	AK100	84.9	7.0	3.4	1.1	3.5
	AS100	83.4	5.6	6.5	0.8	3.6
d-c breakdown	AK100	86.8	6.6	3.6	0.3	2.7
$\approx 7\ 10^{-8}$ C/bd	AS100	86.2	6.2	5.4	0.5	1.7
	AP100	86.5	5.7	6.5	0.6	0.7
d-c-breakdown	AK100	89.4	4.0	5.1	0.4	1.1
$\approx 3\ 10^{-5}$ C/bd	AS100	86.9	4.2	7.6	0.6	0.6
	AP100	85.0	3.9	9.9	0.8	0.4

[a]AK100 dimethyl-siloxane, 100 cSt; AS100 methyl phenyl siloxane, 100 cSt; AP methyl phenyl siloxane, 100 cSt.

silicone oils by discharges of different intensities is given. With the intensity of the discharge increasing, a more complete breakdown of the oil molecules occurs and gases containing double and triple C—C-bonds increase, and, at the same time, more solid precipitate is produced. The influence of the type of discharge on the gas composition is brought out even more pronounced if the ratio of methane to hydrogen is plotted versus the ratio of ethylene to ethane produced (Fig. 2.60). Soft discharges (discharges with passage of a small amount of charge between the electrodes) produce hydrogen rich gases, while hard discharges (discharges with passage of large amount of charge) produce carbon rich gases. The analysis of the gases evolved in a transformer while in service can yield important information on the status of the insulation so that defects can be detected in an early stage [*169*].

Chemical changes in a transformer oil were demonstrated to have an influence on the conduction current [*171–174*]. Virgin oil was compared with aged oil from a transformer after 5 years of faultless service with the tests being carried out in measurement cells using tip-plane or tip-sphere electrode geometry. At a certain level of the applied d-c voltage current bursts were observed; in aged oil the rate and amplitude of these bursts was found to be higher than in the virgin oil. Although the explanation of this phenomenon is still lacking, further investigation of the observed effects may lead to the development of a diagnostic tool for ascertaining the condition of aged oils in transformers, thereby supplementing other more conventional monitoring methods. Complementary results on aged transformer oil were obtained when the electrification was investigated [*175,176*]. In these studies, the current was measured between a rotating disk immersed in the oil and the container wall. With increasing age of the oil the electrification currents were found to increase. Chemical transformations in liquids connected with the passage of current are of fundamental interest and may have important practical consequences. Except for aqueous solutions and

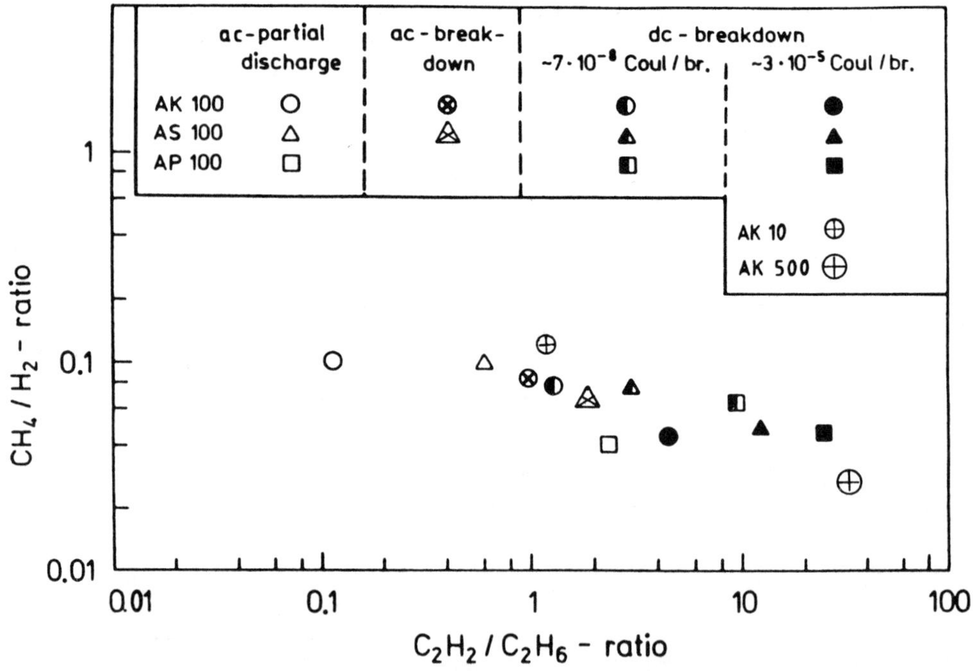

FIG. 2.60—*Dörnenburg diagram for silicone oils (after Ref* 129*).*

some other solvents little is known on these characteristics for most of the liquids used in electrotechnology.

2.7 Electrical Conduction in Polar Liquids

2.7.1 General Remarks

Polar liquids find applications in electrotechnology as dielectrics in pulsed power equipment (for example, water insulated capacitive discharge lines and electrostatic generators with propylene carbonate) or as electro-optic media in Kerr cells (for instance, nitrobenzene). Polar liquids usually show a higher conductivity than nonpolar liquids due to the effect of electrolytic dissociation. Electrolytic impurities dissolve in polar liquids much more readily than in nonpolar liquids. Pure polar liquid themselves show dissociation, too, with the consequence that a limiting conductivity exists, which cannot be diminished by purification. Two classes of polar liquids can be distinguished, aprotic liquids, which do not dissociate into protons and molecular anions (for example nitrobenzene) and liquids which dissociate into protons and anions (for example water). The dissociation is governed by the temperature and by the applied electric field strength (see Section 2.6.3).

2.7.2 Conductivity of Water and Aqueous Solutions

The specific conductivity σ of an aqueous solution is determined by the measurement of the current i through a conductivity cell of electrode area A and electrode separation d with an applied voltage V,

$$\sigma = \frac{i}{V}\frac{d}{A} \tag{2.183}$$

where σ is measured in Ω^{-1} cm^{-1} or S/cm (S = Ω^{-1}, Siemens) and depends on the concentration of ions in the solution. In order to compare the specific conductivity produced by a certain concentration c of one solute with that of another solute, the equivalent conductivity Λ is introduced as,

$$\Lambda = \sigma/c \tag{2.184}$$

It is measured in S/cm per mol/L ($\Omega^{-1}cm^{-1}$ L mol^{-1}) or in S/cm per mol/cm^3. If the solute dissociates into multiply charged ions, the equivalent conductivity, Λ is divided by the number of elementary charges z of these ions

$$\Lambda' = \Lambda/z \tag{2.185}$$

where Λ' is measured in S/cm^3 per val/cm^3. The number of val per cm^3 is equal to the number of moles per cm^3 divided by the charge of the ions measured in units of elementary charges. In many cases only a fraction of the solute dissolved in water dissociates into ions. The degree of dissociation α is defined as

$$\alpha = \Lambda/\Lambda_\infty \tag{2.186}$$

where Λ_∞ is the equivalent conductivity for an infinitely diluted solution. A compound which dissociates into singly charged ions (binary electrolyte) gives rise to the following equilibrium

$$\begin{array}{ccccc} AB & \longrightarrow & A^+ & + & B^- \\ c(1-\alpha) & & \alpha c & & \alpha c \end{array} \tag{2.187}$$

From the law of mass action the dissociation constant K_c is obtained as

$$K_c = \frac{[A^+][B^-]}{[AB]} \tag{2.188}$$

or

$$K_c = \frac{\alpha^2 c}{(1 - \alpha)} \tag{2.189}$$

where $[A^+]$ and $[B^-]$ are the concentrations of the reactants and $[AB]$ is the concentration of the product. With the definition of α in terms of Eq 2.186, we obtain

$$K_c = \frac{\Lambda^2 c}{\Lambda_\infty(\Lambda_\infty - \Lambda)} \tag{2.190}$$

which is called Ostwald's law of dilution.

Water undergoes a weak dissociation into hydronium ions and hydroxyl ions according to the following scheme

$$2\,H_2O \longrightarrow H_3O^+ + OH^- \tag{2.191}$$

The dissociation constant K_d is given by the law of mass action to

$$K_d = \frac{[H^+][OH^-]}{H_2O} \tag{2.192}$$

Here the quantities in brackets denote the concentrations of the species in moles per unit volume. Ultrapure water exhibits a small residual conductivity, which infers that the degree of dissociation is very small. The concentration of H_2O can be assumed to be constant and given by the number of moles of water per unit volume. Equation 2.192 can then be rewritten as

$$K_w = [H^+][OH^-] = \alpha^2[H_2O]^2 \tag{2.193}$$

K_w is called the ion product. The degree of dissociation can be calculated from conductivity measurements on ultrapure water. The equivalent conductivity, Λ, at 298 K was measured to be

$$\Lambda = 0.994\,10^{-6}\Omega^{-1}\text{cm}^2\text{mol}^{-1} \tag{2.194}$$

This equivalent conductivity corresponds to a specific conductivity of $\sigma = 5.497\,10^{-8}\Omega^{-1}\text{cm}^{-1}$ or to a specific resistivity of $\rho = 18.19$ MΩ cm. Modern water purifiers working with reverse osmosis and ion exchange produce water with resistivities close to this value (typically 18.16 MΩ cm). The equivalent conductivities of H^+ and OH^- are determined to be

$$\Lambda_\infty(H^+) = 349.8\,\Omega^{-1}\text{cm}^2\text{mol}^{-1} \tag{2.195}$$

and

$$\Lambda_\infty(OH^-) = 198.0\,\Omega^{-1}\text{cm}^2\text{mol}^{-1} \tag{2.196}$$

In terms of these data, α is calculated to be

$$\alpha = 1.815\,10^{-9} \tag{2.197}$$

and

$$K_w = 1.008\,10^{-14}[\text{mol/L}]^2 \tag{2.198}$$

The ion concentrations of ultrapure water are then given as

$$[H^+] = [OH^-] = 10^{-7}\ \text{mol/L at } 298\ \text{K} \tag{2.199}$$

The dissociation of water increases with temperature; at $T = 373$ K the ion product K_w is already

$$K_w = 7.4\ 10^{-13}\ [\text{mol/L}]^2 \tag{2.200}$$

The application of a high electric field strength increases the dissociation, too. From measurements with high voltage pulses the recombination constant for

$$H_3O^+ + OH^- \underset{k_d}{\overset{k_r}{\rightleftharpoons}} 2\,H_2O \tag{2.201}$$

was determined to be $k_r = 1.3\ 10^{11}$ L/(mol s) at 25°C while the dissociation constant was determined to be $k_d = 2.6\ 10^{-5}\ s^{-1}$ [*177*].

The specific conductivity, σ, of pure water or of mixtures of water and alcohols is an important quantity which determines the operation of capacitive discharge lines. The relaxation time τ is given by

$$\tau = \frac{\epsilon_o \epsilon_r}{\sigma} \tag{2.202}$$

In this relation τ determines the time it takes to counteract and nullify an externally applied electric field in the bulk of the liquid due to relaxation of the ionic charges. In practical applications, τ has to be longer than the time the discharge line is to be in the charged state. For ultrapure water the relaxation time is $\tau = 130\ \mu s$.

At higher values of the electric field strength enhanced dissociation increases the bulk concentration of charge carriers and a higher value of the conductivity results. In addition injection of ions at the electrodes may increase the conductivity. The polarity of the charge injected depends on the electrode material [*178*] and the experimental findings are summarized in Table 2.25.

2.7.3 Purification of Water

Common tap water is not pure. It is a solution of salts, gases, and sometimes organic substances. To remove these substances, multiple distillation can be employed. For the initial stages, the apparatus can be made out of glass; but, the final stage has to be constructed from quartz since the alkali ions are leached out of glass by the water. Nowadays, purification of water proceeds via ion exchange and reverse osmosis (Fig. 2.61). In the ion exchange process [*179*], the water is passed through two columns filled with ion exchange resins. In the first column the cations are replaced by protons (H^+), while in the second column the anions are replaced by hydroxyl ions (OH^-). The resins used in the ion exchange consist of macromolecules with a carbon backbone into which special groups are incorporated. In the case of the cation exchange resin, the incorporated group consist of acids. The protons of these acids exchange for the cations of the dissolved salts. In the anion exchange resin, bases are incorporated into the macromolecules which exchange their OH-groups for the anions. At the exit of the second column, the water contains H^+ and OH^- only, the concentration of which is determined by the dissociation constant (see Section 2.7.2). The ion exchange columns are usually at the end of the purifying line. An initial purification is performed by the process of reverse osmosis in which the water is forced through a filter consisting of a polymeric membrane. Pressures up to 80 bars are necessary. The reverse osmosis unit removes up to 95% of the initial ion concentration. In another stage, cells are deactivated and organic molecules are decomposed by irradiation of the water with ultraviolet light. Submicron filters retain the dead cells. Organic molecules are split into smaller fragments which form ultimately gases or electrolytic

TABLE 2.25—*Charge injecting electrodes in water (after Ref 178).*

Unipolar Positive Charge from the Anode	
Cathode	Anode
stainless steel	stainless steel
copper	copper
stainless steel	copper

Unipolar Negative Charge from the Cathode	
Cathode	Anode
aluminum	aluminum
brass	copper
brass	aluminum

Bipolar Homocharge Injection	
Cathode	Anode
brass	brass
copper	brass
aluminum	copper
aluminum	stainless steel
stainless steel	brass
copper	stainless steel
aluminum	brass
brass	stainless steel

compounds. The purifier is constructed usually with stainless steel containers and the connecting pipes are made out of polypropylene. With such a purifier, water with a specific resistivity of 18.2 MΩ cm can be produced, a value which is solely determined by the self-dissociation of water.

2.7.4 Aprotic Liquids

Some aprotic liquids, which have or might have applications in dielectric engineering, are propylene carbonate, acetonitrile, and nitrobenzene. Their important physical properties are compiled in Table 2.26. Their structure formulae are shown in Fig. 2.62.

2.7.4.1 Conduction in Propylene Carbonate—Propylene carbonate is a solvent used in electrochemical applications, as for instance in lithium batteries. Here, the propylene carbonate is doped with electrolytes to exhibit a certain required conductivity. For applications in electrostatic machines or as a dielectric in capacitors, its conductivity has to be reduced to the lowest values possible. Commercial propylene carbonate has a specific resistivity of approximately 10^6 Ω cm. By a combination of chemical and electrochemical purification procedures, propylene carbonate samples with a specific resistivity of 2 to 3 10^{13} Ω cm have been produced.

As production byproducts, water and propylene glycol are the main remnant impurities both of which dissociate in the propylene carbonate into cations and anions. The purification of propylene carbonate proceeds along the following steps: (*a*) treatment of the liquid with potassium permanganate in order to oxidize organic impurities; (*b*) fractionated distillation followed by filtration through activated alumina; and (*c*) circulation of the propylene carbonate through test cell and

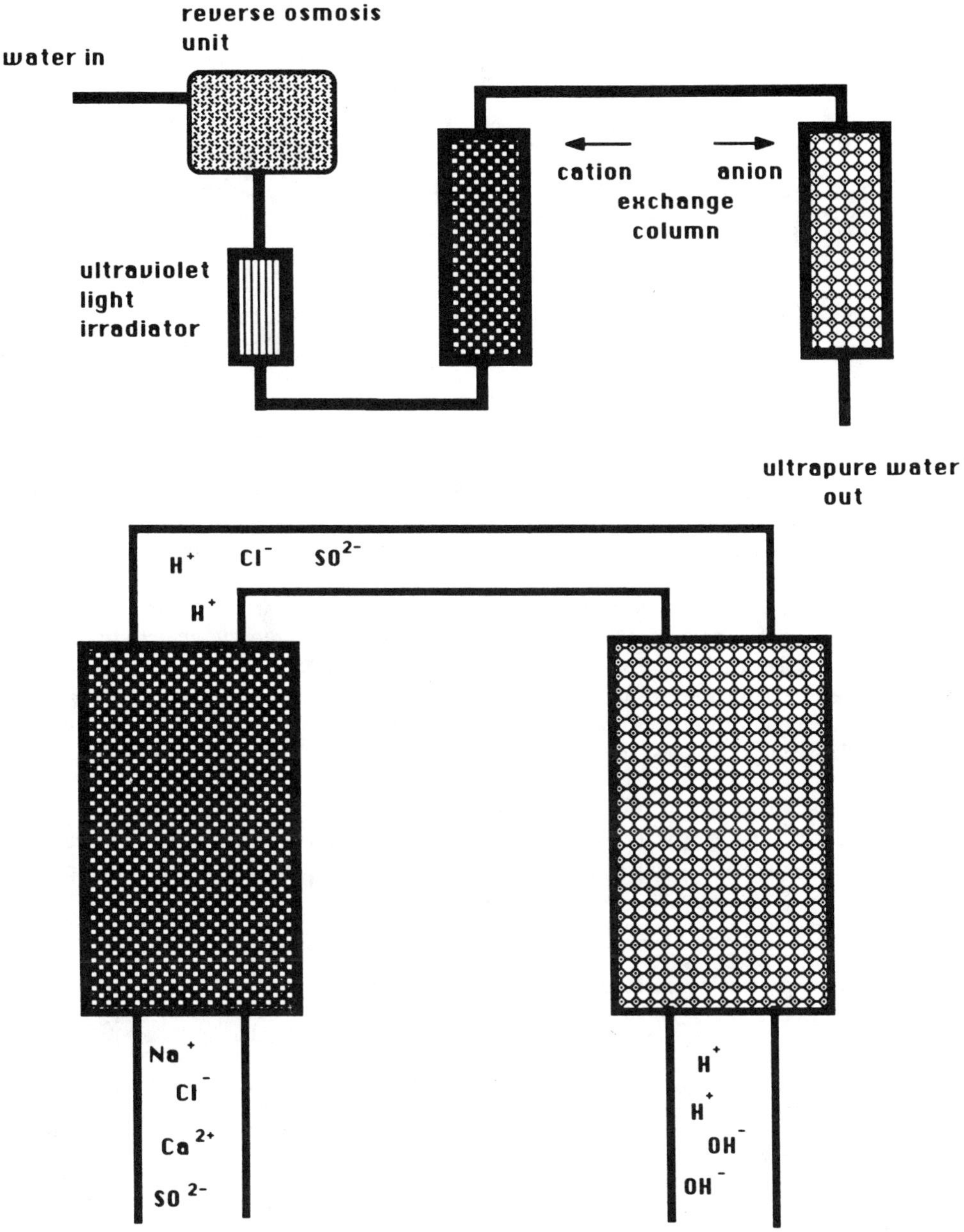

FIG. 2.61—*Water purification with reverse osmosis and ion exchange.*

electrodialysis cell in series (Fig. 2.63). Propylene carbonate belongs to the group of aprotic liquids, and, although it is a polar liquid with a high dielectric constant of $\epsilon_r = 65$ at 25°C, it exhibits a very low intrinsic conductivity provided it is purified in a special way. The commercial product, distilled in an apparatus with a fractionating column under vacuum, exhibits a specific resistivity of

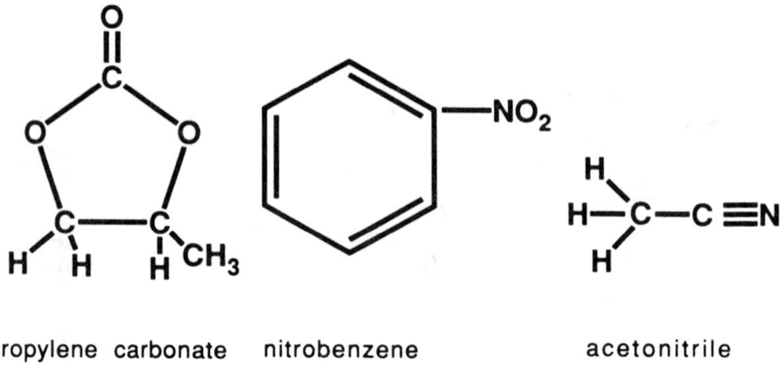

FIG. 2.62—*Structure formulae of some aprotic liquids.*

the order of 10^9 Ω cm. A further increase to values above 10^{12} Ω cm is achieved by means of the process of electrodialysis [*180*]. The principle of operation is illustrated in Fig. 2.64.

The liquid volume is partitioned into three compartments, which are separated by membranes. Then under the influence of the applied voltage the ions of different sign are segregated with the cations being enriched in the compartment adjacent to the cathode and the anions are enriched in the anode compartment. The liquid in the cathode and anode compartments is drained continuously into a storage container, while the purified liquid in the middle compartment is transferred to a clean storage container. In conduction experiments with propylene carbonate, the electrode compartments are formed by ion exchange membranes which are placed tightly over the electrodes. These membranes have a porous structure and contain enough storage capacity for the anions and cations trapped within their interior under the influence of the electric field. In order to monitor the progress of the purification, a test cell is connected in series with such an electrodialyzer (see Fig. 2.63).

The magnitude of the specific resistivity at the beginning of the purification procedure is due mainly to impurity ions, but, as the specific resistivity increases as a result of ion removal, the current-voltage characteristic exhibits a typical variation (Fig. 2.65). At high values of the specific resistivity it approaches the ideal characteristic described by Nikuradse (see Fig. 2.45*b*). At low field strengths an ohmic region is evinced (region I), which is followed by a more or less developed plateau (saturation region, region II). At high electric field strength, a steep increase of the current density becomes apparent which is attributed to charge injection from the electrodes. All current

TABLE 2.26—*Physical properties of some aprotic liquids (after Refs* 18, 46*).*[a]

Property	PC	AN	NB_z
Freezing point, °C	−49.2	−45.7	5.76
Boiling point, °C	241.7	81.6	210.8
Flash point, °C		7.2	88
Density at 25°C, g/cm^3	1.198	0.7766	1.1984
Viscosity at 25°C, centipoise	2.53	0.344	1.811
Dipole moment, Debye units	4.94	4.1	3.99
Relative dielectric constant at 25°C	64.9	36.0	34.82
Refractive index at 25°C	1.42	1.34	1.5506
Polarizability, Å^3/molecule		4.45	

[a]PC = propylene carbonate; AN = acetonitrile; NB_z = nitrobenzene.

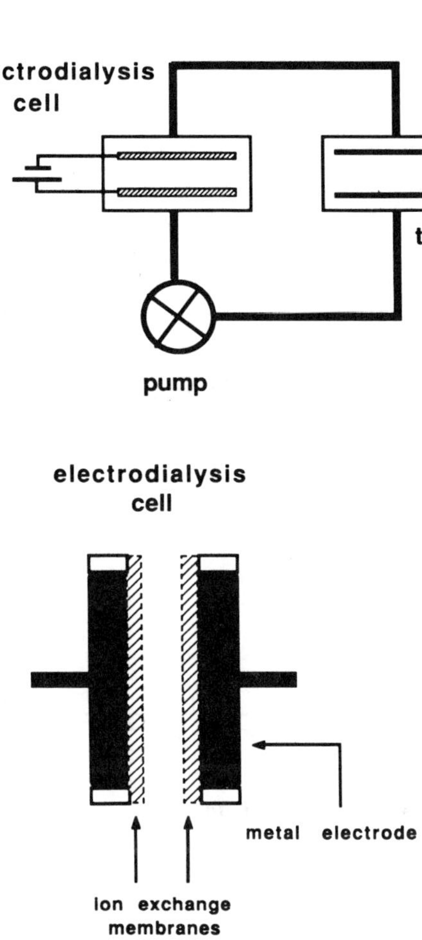

FIG. 2.63—*Conductivity measurement of propylene carbonate during purification by electrodialysis.*

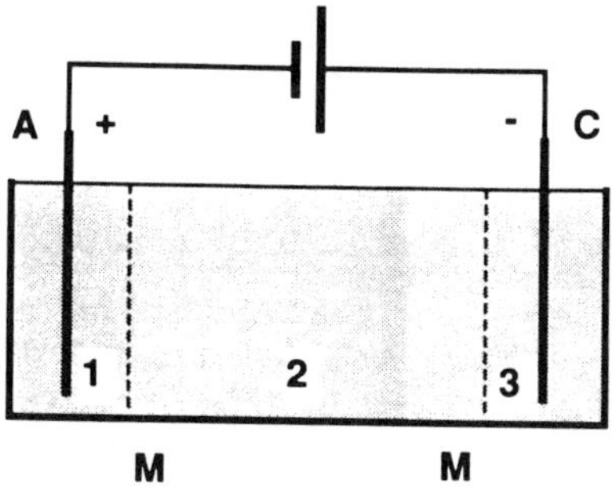

FIG. 2.64—*Principle of electrodialysis; M–membrane; 1–anode compartment; 3–cathode compartment; 2–liquid main compartment.*

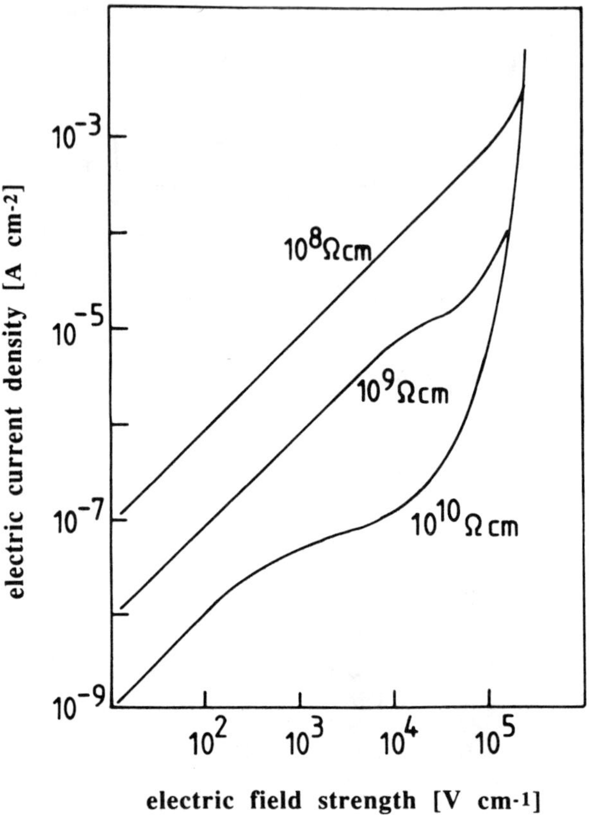

FIG. 2.65—*Current density as a function of the field strength for propylene carbonate for various values of the specific resistivity (after Ref* 181*).*

density versus field strength curves measured for different specific resistivities at low voltages converge to the same high field dependence, eventually.

2.7.4.2 Conduction in Nitrobenzene—Nitrobenzene is employed in Kerr cells for electro-optical switching where a low conductivity is required (see Section 2.3.6). The remnant conductivity after careful purification is conceivably due to ionic impurities at low electric field strengths and to ion injection at the electrode at higher electric field strengths as in the case of propylene carbonate (see Fig. 2.65). Since water in particular may become oxidized at the anode according to

$$H_2O - 2e \longrightarrow 2H^+ + \tfrac{1}{2}O_2 \qquad (2.203)$$

an injection of two positive charges will result. At the cathode, reduction of nitrobenzene is thought to occur with the formation of the radical anion

$$C_6H_5NO_2 + e \longrightarrow C_6H_5NO_2^- \qquad (2.204)$$

Very high specific resistivities in the range of 5 10^{12} Ω cm can be obtained by application of the method of electrodialysis [*180*] following the conventional chemical purification step. In such liquid specimen, at high electric field strength, cations and anions are injected by oxidation reduction reactions of the nitrobenzene itself at the electrodes.

2.7.4.3 Conduction in Acetonitrile—Acetonitrile is a solvent used widely in organic electrochemistry [*182*]. Its low field conductivity has been studied in detail, but the behavior of ultrapure specimens under high electric field conditions has not yet been investigated.

2.8 Electrical Breakdown and Prebreakdown

2.8.1 General Remarks

The electrical breakdown and prebreakdown phenomena in dielectric liquids have been investigated extensively. Since a variety of experimental conditions was employed in the various experiments in the past, a complex picture of the breakdown and prebreakdown processes emerged. The pertinent features have been summarized in a number of reviews [*182–194*], books [*2–5*], and in Chapter 3 of this volume. Other details of these phenomena in specific liquids are discussed in the various chapters of this volume, too. In this section, an attempt is made to present a brief simplified picture of the various stages of the electic breakdown process in nonpolar dielectric liquids. It is to guide the readers imagination rather than to present a detailed description. We assume that the electric breakdown process is initiated by a step function high voltage pulse. It is to proceed through various stages, which are depicted schematically in Fig. 2.66.

The *electronic stage* governs the injection of charge carriers at the electrodes. The important physical quantities are the effective work function of the electrode surface, the local field strength, and the ionization energy of the liquid. The mobilities of the charge carriers and the hydrodynamic impedance of the electrode arrangement are the decisive factors under which conditions this stage converts to the electrothermal stage. Purely electronic breakdown requires the duration of the high electric field strength to be short enough that the formation of the electrothermal stage is prevented. Pulse breakdown of liquefied rare gases and laser induced breakdown may represent this type of breakdown process. Given enough time for the injected charge to relax, the electrical energy will be imparted to the liquid by the frictional motion of the charge carriers.

The discharge goes into the *electrothermal stage* which is characterized by formation of current filaments and the rise of temperature along these filaments. At the same time liquid motion is induced, which enhances the formation of low density regions. Instabilities at the surface of the discharge channel occur. This stage is also referred to as the hydrodynamic stage. Once the gap between the electrodes is bridged the electric power expended in the discharge channel leads to an explosion like evaporation of the liquid and to the formation of a gaseous discharge channel. The *gas discharge stage* is connected with intense emission of light and with the decomposition of molecular liquids. During this stage the resistance of the electrode gap approaches the wave resistance of the connecting conductors. The voltage across the electrodes breaks down. After the power source has been discharged reformation of the insulating state takes place (*physico-chemical stage*). Recombination of the charge carriers reduces the conductivity of the discharge channel, the temperature decreases and radicals formed during the discharge recombine and give rise to a spectrum of molecular products. Each stage has its own time constants and it is possible, only to isolate a particular mechanisms and to study its kinetic behavior. In Fig. 2.66, in the left hand column the physical mechanisms are listed while in the right-hand column the measured quantities or the theoretical models are listed.

For the description of the electrical breakdown of polar liquids a simple scheme is lacking. Although much work has been performed on water, a rather complex picture emerged [*194*].

2.8.2 Nonpolar Liquids

A vast amount of experimental data on the electrical breakdown and prebreakdown of nonpolar liquids is available. Unfortunately, most of the results pertain to a specific type of experimental setup or to a specific sample of a given liquid, respectively. Generalization and development of a

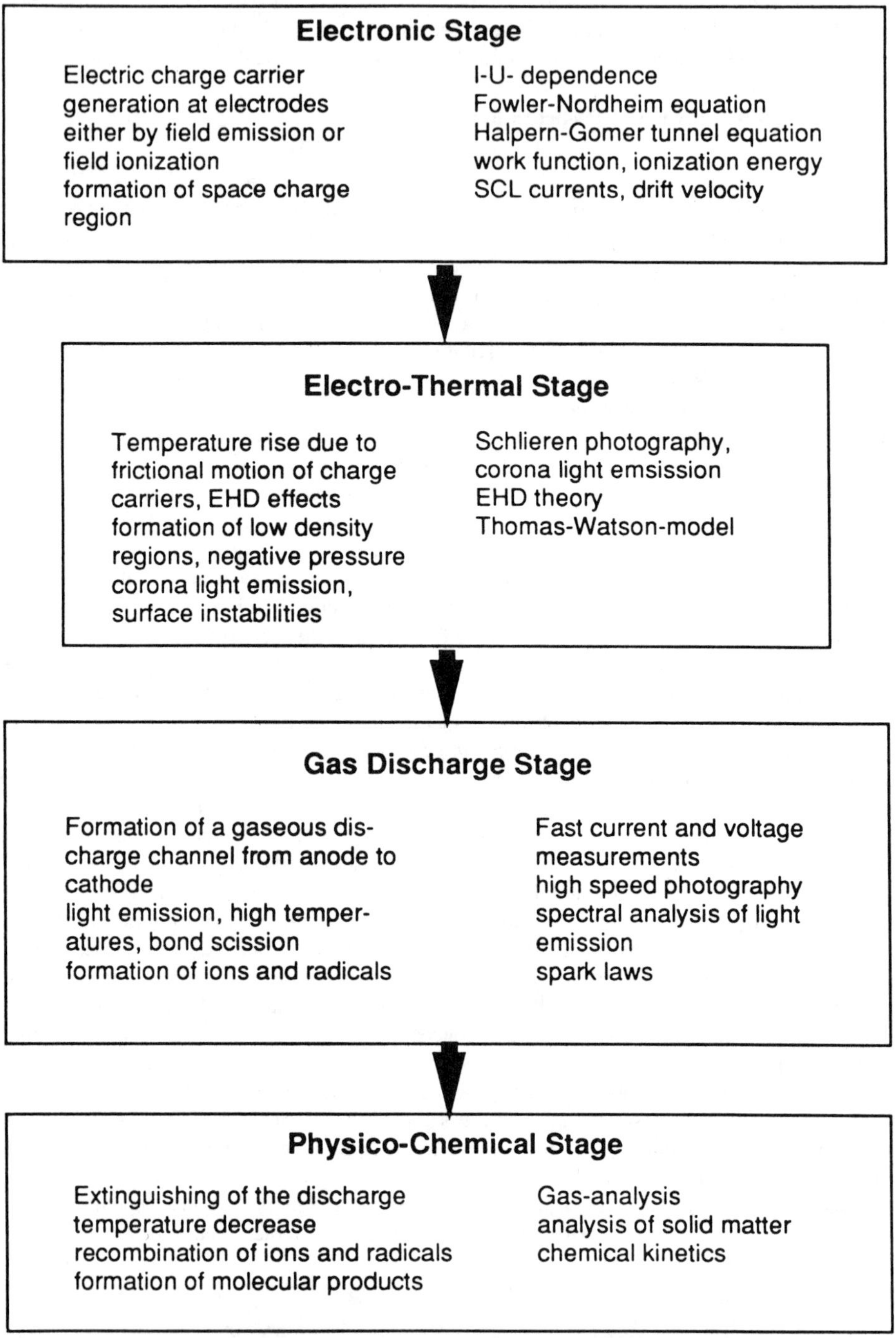

FIG. 2.66—*Stages of breakdown in nonpolar liquids (after Ref* 189*).*

theory of breakdown is hindered by this situation. In the following subsections, some theoretical treatments of electrical breakdown processes in hydrocarbons will be discussed. These examples are to elucidate some of the ideas which have to be incorporated into a general theory of liquid breakdown.

2.8.2.1 General Remarks—Most studies on prebreakdown and breakdown processes have been made with nonpolar liquids, as mineral oils, hydrocarbons and their derivatives, silicone oils, and cryogenic liquids. Prebreakdown processes were studied with optical, electrical, and acoustic methods which are summarized in Table 2.27. The shadow graph technique and the laser Schlieren method yielded results on the velocity of positive and negative streamers in various liquids. Generally, the positive streamer is faster than the negative streamer. Addition of an electron attaching compound, increased the speed of the negative streamer. The highly mobile electrons injected at the cathode were converted into low mobility negative ions (see Section 2.4.5). Addition of compounds with a low ionization potential enhanced the speed of the positive streamer [*195*]. By positive charge transfer the initially produced ions were converted into impurity ions (see Section 2.4.5). Ionic charge carriers in general, due to their lower mobility, also exhibit lower values of their diffusion coefficient. In this manner the repulsive action of the coulomb force leads to a slower relaxation of the space charge at the tip of the streamer, and higher electric fields can be maintained. Especially interesting are the measurements of the very high velocities of the positive and negative streamers in chlorinated hydrocarbons [*196,197*]. Once a streamer reaches the counter electrode the main-stroke channel is formed. The study of the temporal development of the main-spark discharge in liquid hydrocarbons requires fast current and voltage measurement techniques [*198–200*]. In addition, the test cell and the power source have to be integrated into the measurement setup. Generally, two situations can be distinguished, (*a*) the electrical energy stored is sufficient to generate a complete breakdown channel and (*b*) the electrical energy is sufficient for only an incomplete breakdown. In the first case, the spark resistance changes rapidly with time and approaches the wave resistance of the power source (Fig. 2.67). If the supply of electrical energy is limited, the breakdown proceeds at a slower rate and a much lower peak amplitude of the discharge current is reached (Figs. 2.68*a* and *b*). The theoretical description of the breakdown process in liquids is still confined to a few models. In the following subsections we give short summaries of the various ideas.

TABLE 2.27—*Detection methods for partial discharges (PD).*

Principle	Method	Quantity
Optical detection	Schlieren method, shadow graph method	streamer velocities
	light emission and spectral analysis	temperature of discharge channel decomposition products
	Kerr effect	spatial distribution of electric field
	Lichtenberg figures	streamer velocities along solid/liquid interface
Electrical detection	voltage and current measurements, partial discharge detection by a-c-bridge methods	energy or charge per PD event frequency of PDs
	d-c-currents bursts	ageing of liquid
Acoustic detection	transducers or microphones immersed in the liquid	detection of PD by sound and pressure

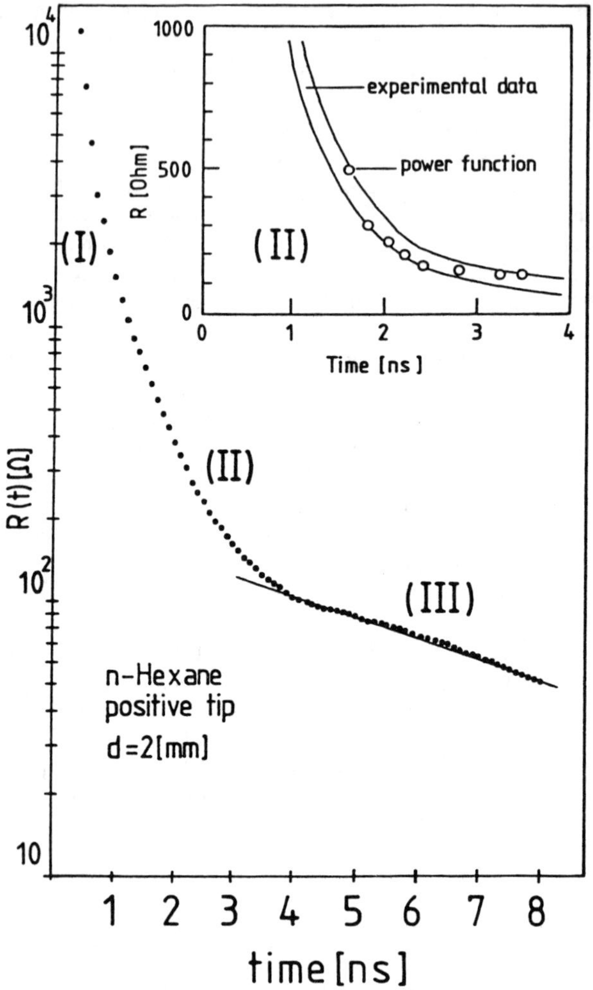

FIG. 2.67—*Spark resistance of n-hexane as a function of time (after Ref* 200*); test cell tip/plane geometry.*

2.8.2.2 Electronic Breakdown—Several early authors proposed breakdown models based on the formation of electron avalanches by the Townsend mechanism as observed in gases. The condition for the formation of an electron avalanche in a liquid is expressed as

$$eE_b\lambda = c_1 h\upsilon \tag{2.205}$$

where c_1 is a constant, E_b is the electric field strength at breakdown, λ is the electron mean free path, and $h\upsilon$ is an energy quantum lost by the electron to the atoms or molecules of the liquid in a collision. Lewis [*201*] pointed out that in hydrocarbons excitation of vibrations represent the major energy loss in electron/molecule collisions. The various groups CH, CH_2, and CH_3 of a paraffin

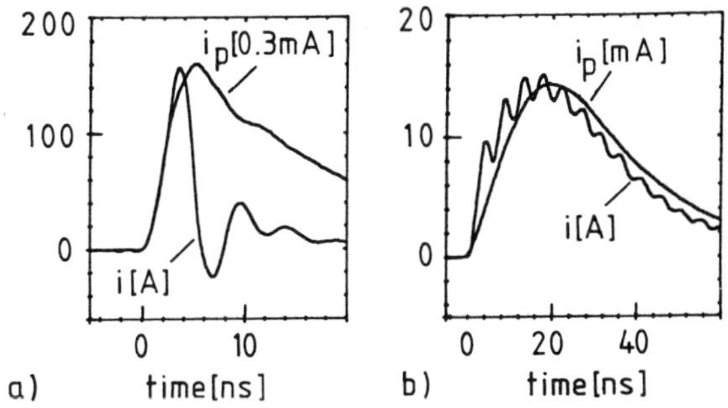

FIG. 2.68—*Complete* (a) *and incomplete* (b) *breakdown of n-hexane (after Ref* 199)*;* i_p-*photomultiplier current due to light emission.*

molecule were thought to act as independent scattering centers. The mean free path λ is then given as

$$\lambda^{-1} = \sum n_i Q_i \tag{2.206}$$

where n_i denotes the number density of the ith group and Q_i is the corresponding cross section. n_i is proportional to the number density N, of the liquid. It is obtained as

$$n_i = f_i N \tag{2.207}$$

where f_i denotes the number of groups of type i per molecule. It follows from Eqs 2.205, 2.206, and 2.207 for the breakdown field strength E_b

$$E_b = c_2 N \sum f_i Q_i \tag{2.208}$$

where c_2 denotes a constant. Such a dependence of the breakdown field on density and number of different CH_x groups was indeed observed for normal alkanes. Some deviations from Eq 2.208 occurred when branched hydrocarbons were included into the treatment, and the data obtained for alkyl benzenes required cross sections of the CH_x groups about a factor of 2 higher than for the alkanes [*188*]. The vibrational losses are in the energy range of 0.1 eV which is much smaller than the ionization energy in these liquids (see Section 2.3.3). However, they are comparable to values of the heat of vaporization (≈0.3 eV). It seems more likely that due to the collisions regions of low density are created in which electron avalanches are formed. This would also be in line with the experimental finding that the breakdown strength of the alkanes increases with hydrostatic pressure [*202*] (Fig. 2.69).

Another example for the decisive role of the electronic properties of the liquid on the breakdown behavior can be observed in 1 He [*25*]. An anomalous polarity effect was found in a point-plane test cell. Here, the positive tip gave a higher breakdown voltage than the negative tip. Furthermore, it could be demonstrated that the breakdown process always started at the cathode. A rational explanation for this phenomenon can be obtained by the consideration of the electronic energy levels at the tip/liquid interface (see also Fig. 2.32). Helium is characterized by a high gas phase ionization potential of I_g(He) = 24.5 eV. In the liquid phase, the ionization energy, I_{liq} is given by

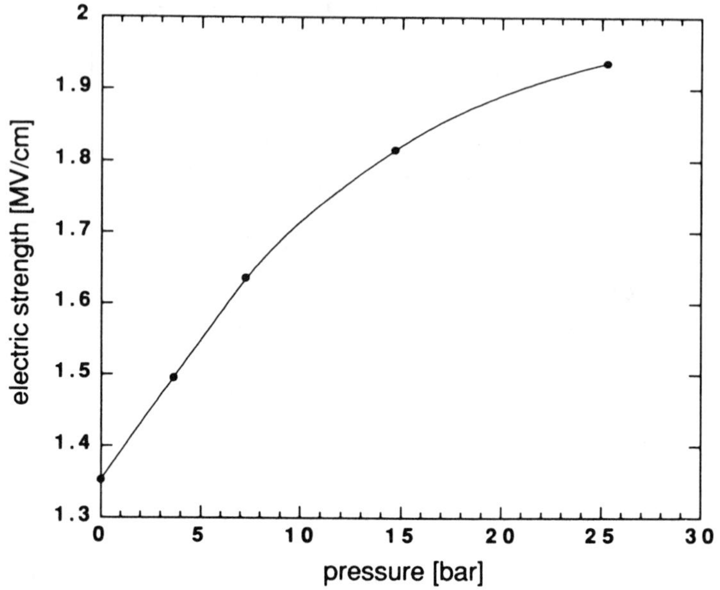

FIG. 2.69—*Pressure dependence of breakdown strength of n-hexane (after Ref* 188*); solid line–Eq 2.213 with* x = *1.5; data points–Kao and Higham* [202].

Eq 2.32. The relative dielectric constant of 1 He is low ($\epsilon_r = 1.048$), and $P_+ = -0.33$ eV is obtained from Eq 2.31. With $V_o = +1$ eV, an ionization energy in the liquid of I_{liq}(He) = 25.2 eV is estimated which is even higher than the gas phase value. The work function of the metal tip is approximately $\Phi_t = 4$ eV. The result is an extremely asymmetric situation with respect to the position of the Fermi level of the metal and the valence and conduction levels of the 1 He. From a negative tip, electron emission occurs since the barrier is approximately 5 eV, only. If the tip is positive, however, the barrier for valence electrons to enter the metal is $I_{liq} - V_o - \Phi_t = 20.2$ eV. Such a barrier would require much higher voltages in order to allow tunnelling of electrons. Before this condition is met, electron emission from asperities at the cathode probably sets in. Breakdown tests with different cathode materials exhibiting different work functions support this point of view [*25*].

2.8.2.3 Low Density or Bubble Formation—In Section 2.6.6 we discussed the filamentary character of the current at high electric field strength. This phenomenon suggests that near breakdown the energy input from the charge carriers moving in the electric field is confined to a small volume of the liquid. A steep rise of the temperature and vaporization would be the consequence. Watson and Sharbaugh [*203–204*] proposed a thermal mechanism for the liquid breakdown which involves the formation of a bubble or a region of lower density. Under typical conditions of pulse breakdown of alkanes, a breakdown field strength of about 1.6 MV/cm is measured. An enhancement of the electric field strength at the tips of asperities of 5 is assumed. The local current density is of the order of 10 Å cm^{-2}. The specific local power input W_{loc}, is then given as

$$W_{loc} = E_{loc} J_{loc} \tag{2.209}$$

and per microsecond an energy of 80 J cm^{-3} or 160 J g^{-1} is deposited. The heat ΔH, necessary to raise the temperature of a mass m, to the boiling point and to evaporate it is given by

$$\Delta H = m[c_p(T_b - T_a) + L_b] \tag{2.210}$$

where c_p is the specific heat at constant pressure, T_a is the ambient temperature, T_b is the boiling point, and L_b is the latent heat of vaporization. For n-hexane, $\Delta H = 400 \text{ J g}^{-1}$ is calculated, a value which is comparable to the energy input in the pulse breakdown experiment.

Watson and Sharbaugh derived a relationship between E_b and ΔH by making use of the observation that near breakdown the emission current from the cathode (or even the anode) is space charge limited (see Sections 2.3.7 and 2.6.7). Generally, SCL currents have a dependence on the applied voltage V, given as

$$i = c_3 V^x \tag{2.211}$$

where x is in the range of 1.5 to 3. c_3 is a constant. With Eq 2.209 it follows that the local energy input during pulse duration τ, can be expressed as

$$\Delta H = c_4 E^x \tau \tag{2.212}$$

with c_4 being a constant. Combining Eqs 2.210 and 2.212 yields the relationship between E and ΔH as

$$c_4 E^x \tau = m[c_p(T_b - T_a) + L_b] \tag{2.213}$$

The fidelity of Eq 2.213 was tested with respect to the data of Kao and Higham [*202*]. Some of the fits obtained are shown in Fig. 2.69 and Fig. 2.70.

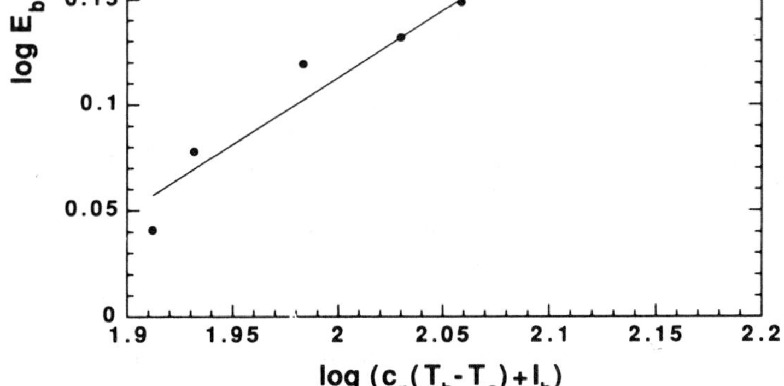

FIG. 2.70—*Temperature dependence of the breakdown strength of n-hexane (after Ref* 188*); solid line–Eq 2.213 with* x = *1.5; data points–Kao and Higham* [202].

A refinement of the Sharbaugh-Watson-theory was presented by Kao [*157*] who set up the energy balance as (cgs units)

$$J_{zo}E_{av} - c_p\rho\frac{dT}{dt} = \frac{9}{8}\epsilon\mu_n\frac{F\left[\frac{E_c}{E_{av}}\right]}{d}E_{av}^3 \tag{2.214}$$

where J_{zo} is the current density in the filament, μ_n is the mobility of the negative charge carriers, ϵ denotes the dielectric constant, F is a function of the ratio of the field at the cathode to the average field, and d is the electrode gap. The rhs of Eq 2.214 represents a SCL current with $x = 3$. Near the breakdown only one filament is strong enough to cause vaporization of the liquid. The mobility μ_n is taken as thermally activated with an activation energy E_{act}

$$\mu_n = \mu_o \exp\left[-\frac{E_{\text{act}}}{kT}\right] \tag{2.215}$$

Introduction of Eq 2.215 into Eq 2.214 and integration between T_a and T_b yields

$$\langle E_{av}\rangle = \frac{8}{9}\frac{c_p\rho kd}{E_{\text{act}}\epsilon\mu_o M}T_a^2\exp\left[\frac{E_{\text{act}}}{kT}\right] - T_b^2\exp\left[\frac{E_{\text{act}}}{kT}\right] \tag{2.216}$$

The fit of Eq 2.216 to the data of Kao and Higham for n-hexane [*202*] is also very good indicating that for this n-alkane the basic physical idea of the electrothermal breakdown seems to be sound. It should be pointed out that for some branched hydrocarbons (especially those with a quarternary C-atom) the electron mobility as a function of T is not described by Eq 2.215 (see Ref *8*).

2.8.2.4 Other Breakdown Phenomena—Investigations of the other various aspects of the breakdown process itself are too numerous to be even listed here. A variety of liquids and solutions sometimes in contact with solid materials have been studied. Physical conditions under which test have been performed include temperature, hydrostatic pressure pressure, electric field strength, impulse or d-c voltages, a-c voltages, homogeneous and inhomogeneous field distributions, liquid flow velocity, light and high energy radiation. Many studies report values of the breakdown strength and the influence of various parameters. This area of research becomes almost completely empirical when transformer oil is considered. Many brands and blends from different producers make it at present almost impossible to discover any underlying mechanism.

Moving liquids are connected with regions of lower density and increased defect concentration in the liquid which facilitate the initiation of electric breakdown. Two effects have to be discussed. First, the liquid motion leads to shear forces near the electrodes. A negative pressure is exerted on the liquid leading to lower density regions. Second, the moving liquid sweeps downstream charge carriers injected at one electrode such prolonging the current path to the counter electrode. The combined action of these two effects is that at low liquid velocity the breakdown voltage increases due to the second effect [*205*] while it decreases below the static value at high flow velocities. In Fig. 2.71 the variation of the breakdown voltage of transformer oil and n-hexane are shown as a function of the liquid velocity. Theodosiou et al. [*206*] confirmed the general dependence depicted in Fig. 2.71 but found another increase of the d-c breakdown voltage of transformer oil at velocities exceeding 50 cm/s.

Another group of experiments which have to be included in the development of a general picture of electric breakdown are time lag to breakdown measurements. Generally, time lag to breakdown measurements are characterized by a distribution which encompasses a statistical, τ_s and a formative time lag, τ_f. A thin liquid layer of 10 to 20 μm thickness is subjected to a step voltage. The

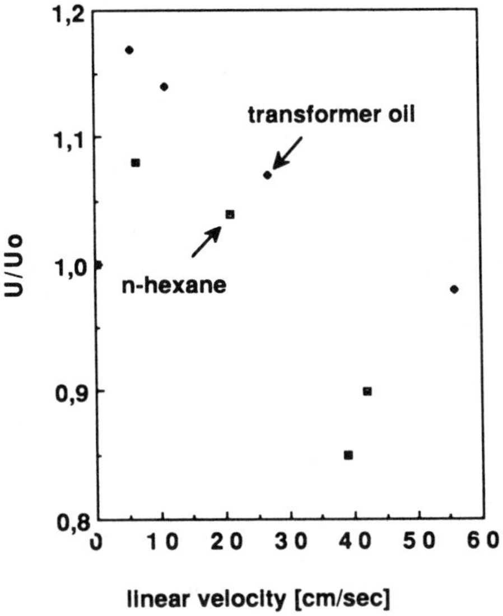

FIG. 2.71—*Breakdown of nonstationary transformer oil and n-hexane (after Ref* 205*).*

time lag from the application of the voltage step to the breakdown event is measured for a large number of trials. The result is a time lag distribution from which two characteristic time lags τ_f (fast time lag) and τ_n (normal time lag) can be obtained (Fig. 2.72). The fast time lag is observed as long as the electrical charges injected at the electrodes or present in the liquid have not yet induced electrohydrodynamical instabilities (see also Sections 2.6.7 and 2.4.4). Once EHD motion sets in the normal time lag is observed. The time delay t_{EHD} in seconds for the development of EHD instabilities has been calculated in Filippini et al. [*208*] as

$$t_{\text{EHD}} = B \frac{\eta}{\epsilon_o \epsilon_r} \frac{1}{\overline{E}^2} \tag{2.217}$$

where B is a dimensionless constant with a value between 160 and 320 depending on the injected charge density.

The influence of the magnitude of the injection current on the time lag to breakdown was demonstrated by changing the resistor in series with the discharge gap [*209*]. The higher the

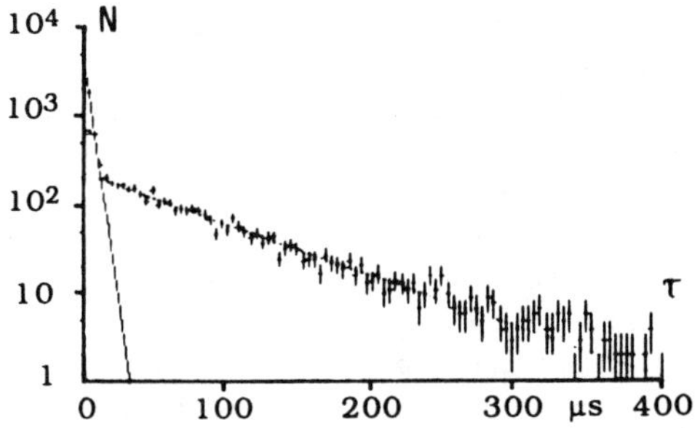

FIG. 2.72—*Distribution of time lag to breakdown (after Ref* 207*).*

injection current the shorter the time lag to breakdown. From this observation it was concluded that the time lag is determined by the time it takes to evaporate a certain liquid volume. Vapor bubbles which affect the electrical properties of a liquid can also be initiated at hot spots on the electrodes. This effect would be highly detrimental for the operation of superconducting coils immersed into liquid helium if partial quenching would occur at some part of the conductor. Hara et al. [*210*] carried out detailed investigations on the influence of vapor bubbles originating at an artificial hot spot on one electrode. Measurements were carried out in liquid nitrogen and liquid helium. Depending on the voltage wave form different mechanisms of breakdown were found. For slowly rising voltages in time (0.1 kV/ms) the discharge seemed to proceed with the growth of the bubble towards the counter electrode. For faster rising voltages (10 kV/ms) the breakdown seemed to originate in the bubble, but it propagated much faster than the expansion of the bubble. Fast rising voltages (standard lightning shape) were not influenced by the presence of thermally produced bubbles at the electrode surface.

2.8.3 Water and Aqueous Solutions

Due to its high dielectric constant, its ease of purification, handling, and disposal, water has become a very prominent insulating medium in pulse power capacitors, in pulse forming lines and transmission lines and in spark gap switches. In applications as pulse power capacitors or discharge lines, the water filled capacitors are charged by a fast rising voltage pulse to a given peak voltage. The electrical energy is then stored for a certain time until it is discharged via a triggered spark gap. The useful storage time is determined by the conductivity of the water (see Section 2.7.2). At higher values of the charging voltage, charge injection from the electrodes sets in and leads to a reduction of the useful storage time. Further increase of the charging voltage will eventually lead to electrical breakdown. All available experimental data seem to indicate that the electric strength attainable in a water insulated gap is limited by electrode processes.

Prebreakdown streamers in tap water and distilled water were observed by a Schlieren method and high speed photography with an image converter camera [*211*]. Mean positive streamer velocities between 1 and 4 10^6 cm/s were reported, which is about one order of magnitude faster than in liquid hydrocarbons.

The decisive influence of the electrodes on the breakdown of water was indicated already in the study of increased conductivity due to charge injection (see Section 2.7.2). By suitable choice of the electrode material an augmentation of the useful voltage range can be obtained; in this respect it is to be noted that Gehman et al. [*212*] found stainless steel and aluminum to form oxide layers, which in turn inhibit charge carrier injection. An increase of the breakdown strength of water/ethylene glycol mixtures was observed. Vorobev et al. [*213*] demonstrated in an ingenious experimental setup the influence of charged layers near the electrodes on the augmentation of the breakdown voltage of pure water; their experimental setup is shown schematically in Fig. 2.73. The test vessel was filled with purified water of 10 MΩ cm specific resistivity and the electrodes were made of stainless steel with a Rogowski profile with the flat portion being approximately 1 cm in diameter and consisting of porous stainless steel (pore size approximately 5 μm). An electrolytic solution slightly less dense than water was introduced into the gap via a porous upper electrode ($FeCl_3$ solution in ethyl alcohol): similarly, an electrolytic solution slightly more dense than water was introduced into the gap through the porous lower electrode ($CuSO_4$ in water). Diffusion layers at the surfaces of the electrodes of 0.3 to 0.4 mm thickness were formed. Photographic records indicated that in the majority of cases the breakdown originated at the anode, although for time lags to breakdown >2.5 μs, occasionally breakdowns originating from the cathode were observed. The diffusive layers had a dramatic effect on the breakdown field. With a rise time of the voltage pulse between 0.3 and 1 μs an approximately fourfold increase of the electric strength of pure water was achieved. Under these conditions the breakdown originated in the bulk of the liquid at the

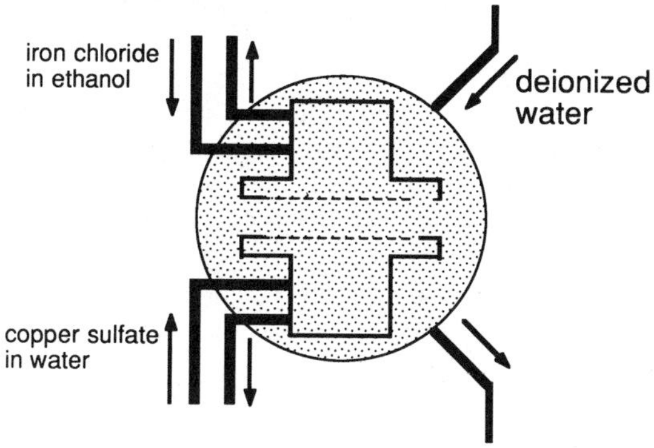

FIG. 2.73—*Diffusive electrolytic electrodes for measurement of breakdown in water (after Ref 212).*

electrolyte/pure water interface. These findings have been corroborated by the work of Zahn et al. [*178*] who found that in the case of bipolar homocharge injection which decreased the field at the electrode surface, generally higher voltages could be sustained without breakdown. With very fast rising pulses (rise time approximately 0.1 μs) and an electric field strength of 1 MV/cm, the breakdown occurred within 0.15 to 0.25 μs after the application of the voltage pulse and was initiated at the diffusive electrode with the lower conductivity solution regardless of the pulse polarity.

2.8.4 Laser-Induced Breakdown

High light intensity, as generated with a laser, is associated with a high electric field strength which may initiate an electric breakdown in the bulk of a liquid. Light from a pulsed laser is focussed into the liquid so that as the local laser power exceeds a certain threshold value, a bright spark is generated which is accompanied by shock waves and formation of gas bubbles. Here two elementary processes may play a role in the breakdown process: electron avalanches and multiphoton ionization. At low photon energies, avalanche formation is assumed to be the dominating mechanism while for higher photon energies multiphoton processes become important. Measurements on cryogenic liquids [*214*], on hydrocarbons and silicone oil [*215*] and on water have been reported [*216*].

Another effect observed with pulsed laser light is that of laser triggered switching. Electric breakdown is triggered in a liquid filled spark gap to which high voltage is applied. Moralda [*217*] triggered the breakdown of Shell Diala oil in a 0.356 cm gap with a 500 MW pulse of a ruby laser. Demidov et al. [*218*] triggered a spark gap with distilled water by a 20 ns pulse from a neodymium laser.

2.9 Applications

While the areas of application of dielectric liquids in high voltage apparatus are decreasing a better understanding of the physical and chemical processes occurring in dielectric liquids under electric stress bears a great potential for other novel applications. In this section we shall briefly summarize some pioneering developments, which involve the utilization of dielectric liquids.

Unipolar ion injection into dielectric fluids produces pressures and mechanical movement. The theoretical description of these processes has been developed thirty years ago [*219–225*]. The following conditions were assumed in the derivation of an equation for the pressure: (*a*) unipolar injection of one kind of ions, (*b*) the ion density is high enough to assume a space charge limited regime, (*c*) the electrodes do not interfere with the liquid flow, no turbulence, and (*d*) absence of surface charges on dielectric boundaries.

A space charge of density ρ exerts a force on the unit volume, which is given by ρE with E being the electric field strength. At equilibrium this force is balanced by a pressure gradient grad p

$$\text{grad}\, p = \rho E \tag{2.218}$$

Under space charge controlled conditions, the local potential V and ρ are related by the Poisson equation

$$\text{div grad}\, V = -\rho/\epsilon_r\epsilon_o \tag{2.219}$$

The current density J is determined by Ohm's law,

$$J = \rho\mu_{\text{ion}}E \tag{2.220}$$

where μ_{ion} is the mobility of the injected ions. In a parallel plate geometry, the emitter plane is assumed to be located at $x = 0$. In the layer δx between $x = 0$ and $x = x_0$ bipolar conduction is thought to take place. In the region between $x = x_0$ and $x = x_1$ unipolar conduction prevails. In other words, it is postulated that the generation of charge carriers occurs in the layer δx as a result of a corona discharge from which ions of one sign are extracted into the bulk of the liquid. Higher fluid velocity leads to a higher conversion efficiency of electrical energy into kinetic energy of the fluid. New applications of this principle in small pumps for liquids are emerging [*226*].

Transport of electrical charges by moving liquids can be employed for high voltage generation. The principle of the high voltage generation is the same as applied in a van de Graaf machine with the belt being replaced by a moving liquid. Highly insulating liquids are charged unipolarly by field emitter sources and the charge is pumped away from the source to a collector electrode. In laboratory models, voltages of several hundred kV could be generated, though limitations seem to exist due to the fact that the charges injected can move in the liquid flow column while on a conventional belt they are much more immobile. EHD effects begin to play a role at higher injection current levels [*227–228*].

Another line of developments of electrostatic HV machines employs polar liquids of high dielectric constant but low self conductivity as dielectric medium. Aprotic liquids are most suited. Propylene carbonate has been purified to very low levels of self conductivity and laboratory models of rotating electrostatic machines have been constructed [*229–231*].

Atomization of liquids in an electric fields can be applied in spraying of paint and generation of fuel aerosols for combustion. Charged droplets of liquids break up into smaller droplets when the electrostatic repulsive force of the electrical charges within the droplets exceeds the surface tension force. Hydrocarbon as well as water droplets of very small diameter can be generated this way. The liquid flows out of a nozzle and a plume of liquid droplets forms due to the electrostatic repulsion. For hydrocarbons, Kelley [*232*] developed a nozzle which incorporated a field emission source, consisting of a cathode tip and annular counter electrode with an orifice. The scheme is shown in Fig. 2.74. Rayleigh [*233*] analyzed the stability of charged droplets and found that a limited amount of charge could be placed onto the droplet only. Exceeding this limit leads to breakup of the droplet into smaller ones. The maximum charge is known as the Rayleigh limit and it is given as (cgs units for Eqs 2.221 and 2.222)

$$q_{\max} = 8\pi(\gamma)^{1/2}r_d^{3/2} \tag{2.221}$$

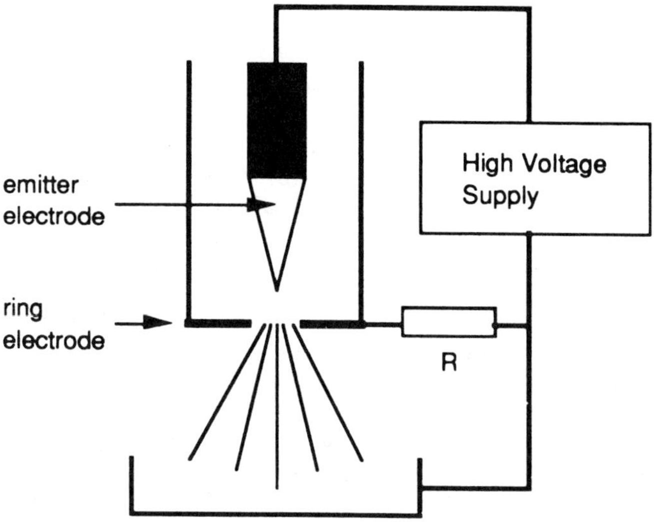

FIG. 2.74—*Electrode arrangement for electrical dispersion of hydrocarbons liquids (after Ref* 232*).*

where γ denotes the surface tension and r_d is the radius of the droplet. Dividing $q_{\max}$ by the mass m of the droplet yields

$$\left(\frac{q}{m}\right)_{\max} = \frac{6(\gamma)^{1/2}}{\rho r_d^{3/2}} \tag{2.222}$$

where ρ is the liquid density. Spraying of liquids of low conductivity can be achieved by charging the droplets. A sharp needle is placed near the end of the capillary. Field emission or field ionization can be used for the charge injection process (see Section 2.3.6).

Dielectrophoretic processes are the basis for a number of developments and applications where suspensions of particles in a nonconducting fluid are separated under the influence of an electric field. In a model due to Pohl [*234*] a dilute suspension of microscopic particles is assumed. The electric field strength E is constant in the whole volume, only in the vicinity of the particles distortions of the electric field occur. Particles on the same field line attract each other with a force F_E given as

$$F_E = \frac{d}{dr} U_E \tag{2.223}$$

where U_E is the work to bring together two particles separated by a distance r

$$U_E = \frac{A_1}{(r^3 - A_2)} \tag{2.224}$$

with

$$A_1 = 8\pi\epsilon_o\epsilon_f E_0^2 \beta^2 b^6 \tag{2.225a}$$

$$A_2 = 2b^3\beta \tag{2.225b}$$

and

$$\beta = \frac{(\epsilon_p - \epsilon_f)}{(\epsilon_p + 2\epsilon_f)} \tag{2.226}$$

where ϵ_f and ϵ_p are the relative dielectric constants of the fluid and the particle, respectively. Their motion in the electric field is characterized by two components, (*a*) translation with constant velocity in the direction of E and (*b*) relative motion towards their center of gravity. Shapiro et al. [*235*] solved the equation of motion of these particles and derived a flocculation time t_f. It turned out that this time is governed by the ratio of the dielectrophoretic force and the drag force. Separation of coal particles from hydrocarbon fluids is a problem in the liquefaction process of coal [*236*]. Other applications may develop in ore-dressing technology [*237*]. Water droplets immersed in a hydrocarbon can also be separated by dielectrophoresis. The potential of this process is recognized in the petroleum industry [*238*].

In liquid hydrocarbons and rare gases electronic conduction currents can be observed if the level of electron attaching impurities has been reduced sufficiently (see Section 2.4.4). The main application of such ultrapure liquids at the present is as detection media in liquid ionization chambers for high energy radiation and in liquid photocathodes for the detection of UV light [*239–241*]. The track of ionization produced by the passage of elementary particles or photons consists of electrons and positive ions. In an electric field, the electrons drift towards the anode and produce a fast current pulse which can be detected by a charge sensitive amplifier. The positive ions move much slower and they are ignored by the amplifier. Analysis of the electron current pulse yields information on the type of radiation and on the position of the track. In the liquid photo cathode ionization of an aromatic solute molecule of low ionization potential is affected by the absorption of a UV photon. An electron/positive ion pair is produced. The electron is extracted from the liquid by a strong electric field, and it is detected in the vapor space above the liquid by a proportional counting wire. Methods for the measurement of the electron lifetime in these liquids have been developed [*242–243*]. They provide a means for the detection of very low concentrations of electron attaching impurities. In liquid argon electron lifetimes of 10 ms were achieved which correspond to an equivalent concentration of oxygen of 30 ppt (parts per trillion [*239*]). These developments will certainly lead to new applications of ultrapure liquids in electrotechnology as well as in other fields of science [*244*].

Acknowledgment

The author wishes to express his sincere gratitude to Drs. R. Bartnikas, A. H. Sharbaugh, and W. Starr for reading the manuscript and offering constructive criticism and numerous suggestions which helped to improve the readability of the present chapter.

References

[*1*] Nikuradse, A., *Das Flüssige Dielektrikum,* Springer Verlag, Berlin, 1934.
[*2*] Zaky, A. A. and Hawley, R., *Conduction and Breakdown in Mineral Oil,* Peter Peregrinus Ltd., London, 1973.
[*3*] Adamczewski, I., *Ionization, Conductivity and Breakdown in Dielectric Liquids,* Taylor and Francis Ltd., London, 1969.
[*4*] Kok, J. A., *Der elektrische Durchschlag in flüssigen Isolierstoffen,* Philips Technische Bibliothek, Eindhoven, 1963.
[*5*] Gallagher, T. J., *Simple Dielectric Liquids,* Clarendon Press, Oxford, 1975.
[*6*] Gemant, A., *Liquid Dielectrics,* Wiley and Sons, New York, 1933.
[*7*] Gemant, A., *Ions in Hydrocarbons,* Interscience Publishers, New York, 1962.
[*8*] Schmidt, W. F., *Transaction on Electrical Insulation,* Institute of Electrical and Electronics Engineers, Vol. EI-19, 1984, pp. 389–418.

[*9*] Goldschvartz, J. M., ed., *Conduction and Breakdown in Dielectric Liquids, Proceedings,* Fifth ICDL, Delft, July 1975.

[*10*] *Proceedings,* Sixth International Conference on Conductance and Breakdown of Dielectric Liquids, Mont Saint Aignan, July 1978, Editions Frontieres, Dreuy, France and *Journal of Electrostatics,* Vol. 7, 1979.

[*11*] *Proceedings,* Seventh International Conference on Conductance and Breakdown of Dielectric Liquids, Berlin-W.-Germany, July 1981, IEEE Publication No. 81CH1594-1 and *Journal of Electrostatics,* Vol. 12, 1982.

[*12*] *Proceedings,* Eighth International Conference on Conductance and Breakdown of Dielectric Liquids, Pavia, Italy, July 1984, and *Transaction on Electrical Insulation,* Institute of Electrical and Electronics Engineers, Vol. EI-20, 1985, No. 2.

[*13*] *Proceedings,* Ninth International Conference on Conductance and Breakdown of Dielectric Liquids, Salford, England, July 1987 and *Transaction on Electrical Insulation,* Institute of Electrical and Electronics Engineers, Vol. EI-23, 1988, No. 4.

[*14*] Pauling, L., *General Chemistry,* W. H. Freeman & Co., San Francisco, 1970.

[*15*] Stuart, H. A., *Molekülstruktur,* Springer-Verlag, Berlin, 1967.

[*16*] Moore, W. J., *Physical Chemistry,* Fourth Edition, Prentice Hall Inc., Englewood Cliffs, NJ, 1972.

[*17*] Malmberg, C. G. and Maryott, A. A., *Journal of Research,* National Bureau of Standards, Vol. 56, 1956, pp. 1–8.

[*18*] Landolt-Börnstein, *Zahlenwerte und Funktionen,* Vol. I, Part 3, 1951 and Vol. II, Part 6, 1959, Springer Verlag, Berlin.

[*19*] Smyth, C. P., *Dielectric Behaviour and Structure,* McGraw-Hill, New York, 1955.

[*20*] Tabor, D., *Gases, Liquids, and Solids,* Cambridge University Press, Cambridge, 2nd edition 1979.

[*21*] Frenkel, J., *Kinetic Theory of Liquids,* Dover Publications, New York, 1955.

[*22*] Gubanov, A. I., *Quantum Electron Theory of Amorphous Semiconductors,* Consultants Bureau, New York, 1965.

[*23*] Born, M., *Zeitschrift für Physik,* Vol. 1, 1920, pp. 45–48.

[*24*] Holroyd, R. A. and Allen, M., *Journal of Chemical Physics,* Vol. 54, 1971, pp. 5014–5021.

[*25*] Yoshino, K., Ohseko, K., Shiraishi, M., Terauchi, M., and Inuishi, Y., *Journal of Electrostatics,* Vol. 12, 1982, pp. 305–314.

[*26*] Tauchert, W. and Schmidt, W. F., *Zeitschrift für Naturforschung,* Vol. 30a, 1975, pp. 1085–1086.

[*27*] Tauchert, W., Jungblut, H., and Schmidt, W. F., *Canadian Journal of Chemistry,* Vol. 55, 1977, pp. 1860–1866.

[*28*] Casanovas, J. and Guelfucci, J. P., *Transaction on Electrical Insulation,* Institute of Electrical and Electronics Engineers, Vol. 23, 1988, pp. 515–528.

[*29*] Nakagawa, K., Itoh, K., and Nishikawa, M., *Transaction on Electrical Insulation,* Institute of Electrical and Electronics Engineers, Vol. 23, 1988, pp. 509–514.

[*30*] Buschick, K. and Schmidt, W. F., *Transaction on Electrical Insulation,* Institute of Electrical and Electronics Engineers, Vol. 24, 1989, pp. 353–356.

[*31*] Casanovas, J., Grob, R., Delacroix, D., Guelfucci, J. P., and Blanc, D., *Journal of Chemical Physics,* Vol. 75, 1981, pp. 4661–4668.

[*32*] Casanovas, J., Guelfucci, J. P., and Terrissol, M., *Radiation Physics and Chemistry,* Vol. 32, 1988, pp. 361–373.

[*33*] Baron, P. L., Casanovas, J., Guelfucci, J. P., and Laou Sio Hoi R., *Transaction on Electrical Insulation,* Institute of Electrical and Electronics Engineers, Vol. 23, 1988, pp. 563–570.

[*34*] Reininger, R., Asaf, U., and Steinberger, I. T., *Physical Review,* Vol. B 28, 1983, pp. 3193–3199.

[*35*] Böttcher, E. H. and Schmidt, W. F., *Journal of Chemical Physics,* Vol. 80, 1984, pp. 1353–1359.

[*36*] Yasufuku, S., Umemura, T., and Tanii, T., *Transaction on Electrical Insulation,* Institute of Electrical and Electronics Engineers, Vol. EI-14, 1979, pp. 28–35.

[*37*] Stumper, U. in *Molecular Motions in Liquids,* Jean Lascombe, Ed., Reidel Publishing Co., Dordrecht, NL, 1974, pp. 654–660.

[*38*] Coelho, R., *Physics of Dielectrics for the Engineer,* Elsevier Scientific Publishing Co., Amsterdam, 1979.

[*39*] "Techniques of Chemistry," *Organic Solvents,* Weissberger, A., Ed., Wiley & Sons, New York, 1986.

[*40*] *Handbook of Chemistry and Physics,* CRC Press, Boca Raton, FL.

[*41*] Landolt-Börnstein, *Zahlenwerte und Funktionen,* Springer Verlag, Berlin, 1959.

[*42*] Winslow, W. M., *Journal of Applied Physics,* Vol. 20, 1949, pp. 1137–1140.

[*43*] Deynega, Yu. F., Popko, K. K., and Kovganich, N. Ya., *Heat Transfer-Soviet Research,* Vol. 10, 1978, pp. 50–56.

[*44*] Atten, P., McCluskey, F. M. J., and Perez, A. T., *Transactions on Electrical Insulation,* Institute of Electrical and Electronics Engineers, Vol. EI-23, 1988, pp. 659–667.

[*45*] Lipshtein, R. A. and Shakhnovich, M. I., "Transformer Oil," Israel Program for Scientific Translation, Jerusalem, 1970, U.S. Department of Commerce, Clearinghouse.

[46] Fujinaga, T. and Izutsu, K., "Propylene Carbonate," *Recommended Methods for Purification of Solvents and Tests for Impurities,* J. F. Coetzee, Ed., IUPAC Publication, Pergamon Press, Oxford, 1981, pp. 19–24.
[47] Martin, T. L. and Leonard, W. F., *Electrons and Crystals,* Brooks/Cole Publishing Co., Belmont, 1970.
[48] Mozumder, A. and Magee, J. L., *Radiation Research,* (a) Vol. 28, 1966, pp. 203–214; (b) Vol. 28, 1966, pp. 215–231.
[49] Hummel, A. and Schmidt, W. F., *Radiation Research Review,* Vol. 5, 1974, pp. 199–300.
[50] *Kinetics of Nonhomogeneous Processes,* Freeman, G. R., Ed., Wiley, New York, 1987.
[51] Randriamalala, Z., Denat, A., Gosse, J. P., and Gosse, B., *Transactions on Electrical Insulation,* Institute of Electrical and Electronics Engineers, Vol. EI-20, 1985, pp. 167–176.
[52] Denat, A., Gosse, B., and Gosse, J. P., *Journal of Electrostatics,* Vol. 12, 1982, pp. 197–205.
[53] Denat, A., Goose, B., and Gosse, J. P., *Journal of Electrostatics,* Vol. 7, 1979, pp. 205–225.
[54] Hellemans, L. and DeMaeyer, L., *Journal of Chemical Physics,* Vol. 63, 1975, pp. 3490–3498.
[55] Nauwelaers, F., Hellemans, L., and Persoons, A., *Journal of Physical Chemistry,* Vol. 80, 1976, pp. 767–775.
[56] Theoleyre, S. and Tobazeon, R., *Proceedings,* International Symposium on Electrical Insulation, Institute of Electrical and Electronics Engineers, Philadelphia, June 1982.
[57] Briére, G. and Goose, J. P., *Journal of Chemical Physics,* Vol. 65, 1968, pp. 1341–1348.
[58] Kleinheins, G., *Berichte der Bunsengesellschaft für Physikolischie Chemie,* Vol. 73, 1969, pp. 1011–1014.
[59] Wollers, F., *Berichte der Bunsengesellschaft für Physikolischie Chemie,* Vol. 77, 1973, pp. 1133–1139.
[60] Wollers, F. and Kleinheins, G., *Zeitschrift für Naturforschung,* Vol. 23a, 1968, pp. 1865–1866.
[61] Proud, J. M. and Auborn, J., *Journal of Applied Physics Letters,* Vol. 27, 1975, pp. 265–268.
[62] Zahn, M., Ohki, Y., Rhoads K., LaGasse, M., and Matsuzawa, H., *Transactions on Electrical Insulation,* Institute of Electrical and Electronics Engineers, Vol. EI-20, 1985, pp. 199–211.
[63] Zahn, M., "Space Charge Effects in Dielectric Liquids," *The Liquid State and Its Electrical Properties,* E. E. Kunhardt, L. G. Christophorou, and L. H. Luessen, Eds., Plenum Press, New York, 1987, pp. 367–430.
[64] Maeno, T., Nonaka, Y., and Takada, T., *Transactions on Electrical Insulation,* Institute of Electrical and Electronics Engineers, Vol. EI-25, 1990, pp. 475–480.
[65] Jakst, A. and Gäfvert, U., *Transactions on Electrical Insulation,* Institute of Electrical and Electronics Engineers, Vol. EI-23, 1988, pp. 723–727.
[66] Maxwell, J. C., *A Treatise on Electricity and Magnetism,* Vol. 1, 3rd edition, Clarendon Press, Oxford, 1892, p. 276.
[67] Derenzo, S. E., Mast, T. S., Zaklad, H., and Miller, R. A., *Physical Review,* Vol. A9, 1974, pp. 2582–2591.
[68] Arii, K., Kitani, I., and Kawamura, M., *Journal of Physics D: Applied Physics,* Vol. 12, 1979, pp. 787–796.
[69] Haidara, M. and Denat, A., (a) Conference Record Tenth International Conference on Conductance and Breakdown of Dielectric Liquids, Grenoble, Sept. 1990, pp. 397–401, IEEE Publication No. 90CH2812-6; (b) *Transactions on Electrical Insulation,* Institute of Electrical and Electronics Engineers, Vol. EI-26, 1991, pp. 592–597.
[70] Schmidt, W. F., "Nonpolar Liquids," Chapter 9, *Photoconductivity and Related Phenomena,* J. Mort and D. M. Pai, Eds., Elsevier Publishing Co., Amsterdam, 1976.
[71] Loeb, L. B., *Basic Processes of Gaseous Electronics,* University of California Press, Berkeley and Los Angeles, 1955.
[72] Dolezalek, F. K., "Experimental Techniques," Chapter 2, *Photoconductivity and Related Phenomena,* J. Mort and D. M. Pai, Eds., Elsevier Publishing Co., Amsterdam, 1976.
[73] Romanets, R., Bolshakov, V., Dikarev, B., and Karasev, G., Conference Record 10. International Conference on Conductance and Breakdown of Dielectric Liquids, Grenoble, Sept. 1990, IEEE Publication No. 90CH2812-6, pp. 196–200.
[74] Reiss, K. H., *Annalen der Physik,* Leipzig, Vol. 28, 1937, pp. 325–352.
[75] Zessoules, N., Brinkerhoff, J., and Thomas, A., *Journal of Applied Physics,* Vol. 34, 1963, pp. 2010–2014.
[76] Schmidt, W. F., *Canadian Journal of Chemistry,* Vol. 55, 1977, pp. 2197–2209.
[77] Holroyd, R. A. and Schmidt, W. F., *Annual Review of Physical Chemistry,* Vol. 40, 1989, pp. 439–468.
[78] Munoz, R. C., Holroyd, R. A., and Nishikawa, M., *Journal of Physical Chemistry,* Vol. 89, 1985, pp. 2969–2972.
[79] Munoz, R. C. and Holroyd, R. A., *Journal of Chemical Physics,* Vol. 84, 1986, pp. 5810–5815.
[80] Munoz, R. C. and Ascarelli, G., *Journal of Physical Chemistry,* Vol. 88, 1984, pp. 3712–3715.

[81] Bakale, G., Tauchert, W., and Schmidt, W. F., *Journal of Chemical Physics,* Vol. 63, 1975, pp. 4470–4473.
[82] Munoz, R. C., *Radiation Physics and Chemistry,* Vol. 32, 1988, pp. 169–176.
[83] Allen, A. O., "Drift Mobilities and Conduction Band Energies of Excess Electrons in Dielectric Liquids," NSRDS-NBS 58, National Bureau of Standards, 1976.
[84] Schmidt, W. F., Chapter 7 in *Electron-Solvent and Anion Solvent Interactions,* L. Kevan and B. S. Webster, Eds., Elsevier Science Publishing Co., Amsterdam, 1976.
[85] Kalinowski, I., Rabe, J. G., Schmidt, W. F., and Uth, J., 1976 Annual Report, Conference on Electrical Insulation and Dielectric Phenomena, National Academy of Sciences, Washington, DC, 1978, pp. 159–166.
[86] Sowada, U., Bakale, G., Yoshino, K., and Schmidt, W. F., pp. 1–4, of Ref *9.*
[87] Schmidt, W. F., Bakale, G., and Sowada, U., *Journal of Chemical Physics,* Vol. 61, 1974, pp. 5275–5278.
[88] Gzowski, O., *Zeitschrift für Physik und Chemie,* Vol. 221, 1962, pp. 288–296.
[89] Chong, P. and Inuishi, Y., Technical Reports, Osaka University, Vol. 10, 1960, pp. 545–547.
[90] Casanovas, J., Grob, R., Chemin, A., Guelfucci, J. P., and Crine, J. P., *Transactions on Electrical Insulation,* Institute of Electrical and Electronics Engineers, Vol. EI-20, 1985, pp. 143–147.
[91] Robinson, R. A. and Stokes, R. H., *Electrolyte Solutions,* Butterworths, London, 1959.
[92] Birks, J. B., "Electrophoretic Deposition of Insulating Materials," in Vol. 1 of *Progress in Dielectrics,* J. B. Birks and J. H. Schulman, Eds., Heywood Co., London, 1959.
[93] Matsuda, Y., Nakashima, H., Morita, M., and Takasu, Y., *Journal of the Electrochemical Society,* Vol. 128, 1981, pp. 2552–2556.
[94] Harned, H. S. and Owen, B. B., *The Physical Chemistry of Electrolytic Solutions,* Reinhold Publishing Corp., New York, 1958.
[95] Felici, N., *Revue Génerale d' Electricité,* Vol. 78, 1969, pp. 730–741.
[96] Wada, T., Shinsaka, K., Namba, H., and Hatano, Y., *Canadian Journal of Chemistry,* Vol. 55, 1977, pp. 2144–2155.
[97] Shinsaka, K., Codama, M., Srithanratana, T., Yamamoto, M., and Hatano, Y., *Journal of Chemical Physics,* Vol. 88, 1988, pp. 7529–7536.
[98] Warman, J. M., deHaas, M. P., and Hummel, A., pp. 70–73 of Ref *9.*
[99] Sowada, U. and Holroyd, R. A., *Journal of Chemical Physics,* Vol. 70, 1979, pp. 3586–3591.
[100] Mehnert, R., Brede, O., Bös, J., and Naumann, W., *Journal of Electrostatics,* Vol. 12, 1982, pp. 107–114.
[101] de Haas, M. P., Hummel, A., Infelta, P. P., and Warman, J. M., *Journal of Chemical Physics,* Vol. 65, 1976, pp. 5019–5020.
[102] Umehara, M. in *Vacuum Ultraviolet Radiation Physics,* Proceedings of the Seventh International Conference, Jerusalem, August 1983, pp. 279–281.
[103] Cohen, M. H. and Lekner, J., *Physical Review,* Vol. 158, 1967, pp. 305–309.
[104] Davis, H. T., Schmidt, L. D., and Minday, R. M., *Physical Review,* Vol. A3, 1971, pp. 1027–1037.
[105] Davis, H. T. and Brown, R. G. in *Advances in Chemical Physics,* Vol. 31, 1975, pp. 329–464.
[106] Bagley, B. G., *Solid State Communications,* Vol. 8, 1970, pp. 345–348.
[107] Berlin, Yu. A., Nyikos, L., and Schiller, R., *Journal of Chemical Physics,* Vol. 69, 1978, pp. 2401–2406.
[108] Yamada, N., Tanaka, T., and Shiomi, H., *Denki Gakkai Ronbunshi,* Vol. 96A, 1976, No. 3, pp. 95–102 (in Japanese).
[109] Sharbaugh, A. H. and Barker, R. E., "Ionic Impurity Conduction in Organic Liquids," *Colloques Internationaux du CNRS,* No. 179, Grenoble, 1968, pp. 349–361.
[110] Cottrell, G. A., Hatto, C. E., Reed, C., and Rose-Innes, A. C., *Journal of Physics D: Applied Physics,* Vol. 17, 1984, pp. 989–1005.
[111] Holroyd, R. A., Dietrich, B. K., and Schwarz, H. A., *Journal of Physical Chemistry,* Vol. 76, 1972, pp. 3794–3800.
[112] Blossey, D. F., *Physical Review,* Vol. B9, 1974, pp. 5183–5187.
[113] Michel-Beyerle, M. E. and Bube, W., *Zeitschrift für Naturforschung,* Vol. 30a, 1975, pp. 506–515.
[114] Baker, E. B. and Boltz, H. A., *Physical Review,* Vol. 51, 1937, pp. 275–282.
[115] Cox, B. M., *Journal of Physics D: Applied Physics,* Vol. 8, 1975, pp. 2065–2070.
[116] Halpern, B. and Gomer, R., *Journal of Chemical Physics,* Vol. 51, 1969, pp. 1031–1047.
[117] Halpern, B. and Gomer, R., *Journal of Chemical Physics,* Vol. 51, 1969, pp. 1048–1056.
[118] Gomer, R., *Accounts of Chemical Research,* Vol. 5, 1972, pp. 41–48.
[119] Schmidt, W. F. and Schnabel, W., *Zeitschrift für Naturforschung,* Vol. 26a, 1971, pp. 169–170.
[120] Dotoku, K., Yamada, H., Sakamoto, S., Noda, S., and Yoshida, H., *Journal of Chemical Physics,* Vol. 69, 1978, pp. 1121–1125.

[*121*] Arii, K. and Schmidt, W. F., *Transactions on Electrical Insulation,* Institute of Electrical and Electronics Engineers, Vol. EI-19, 1984, pp. 16–23.
[*122*] Denat, A., Gosse, J. P., and Gosse, B., *Revue de Physique Appliqué,* Vol. 22, 1987, pp. 1103–1111.
[*123*] Denat, A., Gosse, J. P., and Gosse, B., *Transactions on Electrical Insulation,* Institute of Electrical and Electronics Engineers, Vol. 23, 1988, pp. 545–554.
[*124*] McClintock, P. V. E., *Physics Letters,* Vol. 29A, 1969, pp. 453–454.
[*125*] McClintock, P. V. E., *Physics Letters,* Vol. 35A, 1971, pp. 211–212.
[*126*] McClintock, P. V. E. and Read-Forrest, H., *Cryogenics,* Vol. 13, 1973, pp. 363–371.
[*127*] Hickson, A. and McClintock, P. V. E., *Physics Letters,* Vol. 34A, 1971, pp. 424–425.
[*128*] Schnabel, W. and Schmidt, W. F., *Journal of Polymer Science,* Symposium No. 42, 1973, pp. 273–280.
[*129*] Rabe, J. G., Schmidt, W. F., and Klein, W., *Journal of Electrostatics,* Vol. 7, 1979, pp. 253–266.
[*130*] Onsager, L., *Journal of Chemical Physics,* Vol. 2, 1934, pp. 599–615.
[*131*] Staudhammer, P. and Seyer, W. F., *Journal of Applied Physics,* Vol. 28, 1957, pp. 405–410.
[*132*] Blanc, D., Mathieu, J., and Boyer, J., *Nuovo Cimento,* Vol. 19, 1961, pp. 929–938.
[*133*] Shimokawa, H., Ohashi, A., and Ueda, M., *Journal of Electrostatics,* Vol. 7, 1979, pp. 187–197.
[*134*] Shimokawa, H., Ohashi, A., and Ueda, M., *Journal of Electrostatics,* Vol. 12, 1982, pp. 179–187.
[*135*] Kahlstatt, P. and Wollers, F., *Physics Letters,* Vol. 38A, 1972, pp. 301–302.
[*136*] Asano, K. and Krämer, H., *Journal of Electrostatics,* Vol. 2, 1976, pp. 269–276.
[*137*] Bachmann, K. C. and Munday, J. C., Technical Report No. AFWL-TR-72-90, 1973.
[*138*] Takashima, T., Hanaoka, R., Ishibashi, R., and Ohtsubo, A., *Transactions on Electrical Insulation,* Institute of Electrical and Electronics Engineers, Vol. 23, 1988, pp. 645–658.
[*139*] Krasucki, Z., *Colloques Internationaux du CNRS,* No. 179, 1970, pp. 311–323.
[*140*] Denegri, G. B., Liberti, G., Molinari, G., and Viviani, A., *Transactions on Electrical Insulation,* Institute of Electrical and Electronics Engineers, Vol. EI-12, 1977, pp. 114–127.
[*141*] Molinari, G. and Viviani, A., *Journal of Electrostatics,* Vol. 5, 1978, pp. 343–354.
[*142*] Molinari, G. and Viviani, A., *Journal of Electrostatics,* Vol. 7, 1979, pp. 27–32.
[*143*] Bozzo, R., Coletti, G., Molfino, P., and Molinari, G., *Transactions on Electrical Insulation,* Institute of Electrical and Electronics Engineers, Vol. EI-20, 1985, pp. 343–348.
[*144*] Matuszewski, T., Terlecki, J., and Sulocki, J., *Acta Physica Polonica,* Vol. A48, 1975, pp. 861–869.
[*145*] Nelson, D. A. and Ruoff, A. L., *Review Letters,* Vol. 42, 1979, pp. 383–386.
[*146*] Mellis, W. J., Hamilton, D. C., Trainor, R. J., Radousky, H. B., Mitchell, A. C., and Holmes, H. C., *Proceedings,* Tenth AIRAPT High Pressure Conference on Research in High Pressure Science and Technology, Amsterdam, July 1985, UCRL-92924 Preprint. Physica B + C, No. 1–3, 1986, pp. 565–567.
[*147*] Scaife, W. G., *Journal of Physics D: Applied Physics,* Vol. 7, 1974, pp. 647–652.
[*148*] B. Garben, *Physics Letters,* Vol. A47, 1974, pp. 1281–1288.
[*149*] Vij, J. K. and Scaife, W. G., unpublished work, cited in Gallagher, Ref *5*.
[*150*] Brazier, D. W. and Freeman, G. R., *Canadian Journal of Chemistry,* Vol. 47, 1969, pp. 885–889.
[*151*] Hummel, A. and Allen, A. O., *Journal of Chemical Physics,* Vol. 44, 1966, pp. 3426–3431.
[*152*] Casanovas, J., Grob, R., Blanc, D., Brunet, G., and Mathieu, J., *Journal of Chemical Physics,* Vol. 63, 1975, pp. 3673–3675.
[*153*] Munoz, R. C., Cummings, J. B., and Holroyd, R. A., *Journal of Chemical Physics,* Vol. 85, 1986, pp. 1104–1115.
[*154*] Blanc, D., Mathieu, J., and Torres, L., *Nuclear Instruments and Methods,* Vol. 27, 1964, pp. 353–354.
[*155*] Aubert, B., Ghez, P., Lacotte, J. C., Colas, J., Dobrzynski, L., Kryn, D., Mendiburu, J. P., Salin, P., Mansoulie, B., and Teiger, J., *Nuclear Instruments and Methods,* Vol. A286, 1990, pp. 147–154.
[*156*] Barnett, A. M., "Current Filament Formation," Chapter 3, *Semiconductors and Semimetals,* R. K. Willardson and A. C. Beer, Eds., Vol. 6, *Injection Phenomena,* Academic Press, New York, 1970.
[*157*] Kao, K. C., *Transactions on Electrical Insulation,* Institute of Electrical and Electronics Engineers, Vol. EI-11, 1976, pp. 121–128.
[*158*] Lesaint, O., Gournay, P., and Tobazeon, R., *Transactions on Electrical Insulation,* Institute of Electrical and Electronics Engineers, Vol. EI-26, 1991, pp. 699–707.
[*159*] Arii, K. and Schmidt, W. F., *Transactions on Electrical Insulation,* Institute of Electrical and Electronics Engineers, Vol. EI-19, 1984, pp. 16–23.
[*160*] Klinkenberg, A. and van der Minne, J. L., *Electrostatics in the Petroleum Industry,* Elsevier Publishing Co., Amsterdam, 1958.
[*161*] Okubo, H., Ikeda, M., Honda, M., and Menju, S., Paper 23.14, Third International Symposium on High Voltage Engineering, Milan, 1979.
[*162*] Lee, M. J. and Nelson, J. K., *Transactions on Electrical Insulation,* Institute of Electrical and Electronics Engineers, Vol. 26, 1991, pp. 739–748.
[*163*] Felici, N. J., Gosse, J. P., and Solofomboahangy, A., *Journal of Electrostatics,* Vol. 12, 1982, pp. 369–376.

[*164*] Touchard, G. and Romat, H., *Journal of Electrostatics,* Vol. 12, 1982, pp. 377–382.
[*165*] McCluskey, F. M. J. and Atten, P., *Transactions on Electrical Insulation,* Institute of Electrical and Electronics Engineers, Vol. EI-20, 1985, pp. 405–412.
[*166*] Atten, P., Malraison, B., and Ali Kani, S., *Journal of Electrostatics,* Vol. 12, 1982, pp. 477–488.
[*167*] Filippini, J. C., Lacroix, J. C., and Tobazeon, R., *C.R. Academy of Science,* Paris, Vol. 271B, 1970, pp. 73–76.
[*168*] Fuhr, J. and Schmidt, W. F., *Transactions on Electrical Insulation,* Institute of Electrical and Electronics Engineers, Vol. 23, 1988, pp. 485–487.
[*169*] Dörnenburg, E. and Strittmatter, W., *Brown Boverie Mitteilungen,* Vol. 61, 1974, pp. 238–242.
[*170*] El-Sulaiman, A. A., Ahmed, A. S., and Qureshi, M. I., *Journal of Electrostatics,* Vol. 12, 1982, pp. 621–627.
[*171*] El-Sulaiman, A. A., Ahmed, A. S., Qureshi, M. I., and Alrashood, K. A., *Pakistan Journal of Industrial Research,* Vol. 32, 1989, pp. 590–594.
[*172*] El-Sulaiman, A. A., Ahmed, A. S., Qureshi, M. I., and Hassan, M. M. A., *Arabian Gulf Journal of Scientific Research,* Vol. 4, 1986, pp. 385–400.
[*173*] El-Sulaiman, A. A., Ahmed, A. S., and Qureshi, M. I., *Canadian Electrical Engineering,* Vol. 11, 1986, pp. 35–39.
[*174*] El-Sulaiman, A. A., Ahmed, A. S., and Qureshi, M. I., *Transactions on Electrical Insulation,* Institute of Electrical and Electronics Engineers, Vol. EI-16, 1981, pp. 453–457.
[*175*] Brzostek, E. and Kedzia, J., *Transactions on Electrical Insulation,* Institute of Electrical and Electronics Engineers, Vol. 21, 1986, pp. 609–612.
[*176*] Kedzia, J., *Transactions on Electrical Insulation,* Institute of Electrical and Electronics Engineers, Vol. 24, 1989, pp. 175–178.
[*177*] Eigen, M. and DeMayer, L., *Zeitschrift für Elektrochemie,* Vol. 59, 1955, pp. 986–994.
[*178*] Zahn, M., Ohki, Y., Rhoads, K., LaGasse, M., and Matsuzawa, H., *Transactions on Electrical Insulation,* Institute of Electrical and Electronics Engineers, Vol. 20, 1985, pp. 199–211.
[*179*] Taylor, T. L. and Gorski, J. J., ''Ultra High Purity Water-New Frontiers,'' Chapter 6, *Handbook of Contamination Control in Microelectronics,* D. L. Tolliver, Ed., 1988, Noyes Publishing, Park Ridge, NJ.
[*180*] Felici, N. J. and Tobazeon, R. E., *Journal of Electrostatics,* Vol. 11, 1981, pp. 135–161.
[*181*] Theoleyre, S. ''Etude de la conduction electrique dans un liquide fortement polaire et de grande resistivite, le carbonate de propylene,'' PhD thesis, Institut National Polytechnique de Grenoble, May 1983.
[*182*] *Recommended Methods for Purification of Solvents and Tests for Impurities,* Coetzee, J. F. Ed., Pergamon Press, Oxford, New York, Toronto, 1982.
[*183*] Inuishi, Y., *Journal of Electrostatics,* Vol. 7, 1979, pp. 1–11.
[*184*] Yoshino, K., *Transactions on Electrical Insulation,* Institute of Electrical and Electronics Engineers, Vol. 21, 1986, pp. 847–853.
[*185*] Forster, E. O., *Transactions on Electrical Insulation,* Institute of Electrical and Electronics Engineers, Vol. 20, 1985, pp. 905–912.
[*186*] Forster, E. O., *Journal of Electrostatics,* Vol. 12, 1982, pp. 1–12.
[*187*] Rzad, S. J., Devins, J. C., and Schwabe, R. J., *Transactions on Electrical Insulation,* Institute of Electrical and Electronics Engineers, Vol. 18, 1983, pp. 1–10.
[*188*] Sharbaugh, A. H., Devins, J. C., and Rzad, S. J., *Transactions on Electrical Insulation,* Institute of Electrical and Electronics Engineers, Vol. EI-13, 1978, pp. 249–276.
[*189*] Schmidt, W. F., *Transactions on Electrical Insulation,* Institute of Electrical and Electronics Engineers, Vol. EI-17, 1982, pp. 478–483.
[*190*] Lewis, T. J., ''An Overview of Electrical Processes Leading to Dielectric Breakdown of Liquids,'' *The Liquid State and Its Electrical Properties,* E. E. Kunhardt, L. G. Christophorou, and L. H. Luessen, Eds., NATO ASI Series, B 193, Plenum Press, New York, 1988, pp. 431–453.
[*191*] Tobazeon, R. E., ''Streamers in Liquids,'' *The Liquid State and Its Electrical Properties,* E. E. Kunhardt, L. G. Christophorou, and L. H. Luessen, Eds., NATO ASI Series, B 193, Plenum Press, New York, 1988, pp. 465–501.
[*192*] Hebner, R. E., ''Measurement of Electrical Breakdown in Liquids,'' *The Liquid State and Its Electrical Properties,* E. E. Kunhardt, L. G. Christophorou, and L. H. Luessen, Eds., NATO ASI Series, B 193, Plenum Press, New York, 1988, pp. 519–537.
[*193*] Zahn, M., ''Space Charge Effects in Dielectric Liquids,'' *The Liquid State and Its Electrical Properties,* E. E. Kunhardt, L. G. Christophorou, and L. H. Luessen, Eds., NATO ASI Series, B 193, Plenum Press, New York, 1988, pp. 367–430.
[*194*] Naugolnych, K. A. and Roi, N. A., *Electrical Discharges in Water (Hydrodynamic Description),* in Russian, Nauka Publisher, Moscow, 1971.

[*195*] Devins, J. C., Rzad, S. J., and Schwabe, R. J., *Journal of Physics D, Applied Physics,* Vol. 9, 1976, pp. L87–L91.
[*196*] Yamada, H. and Sakamoto, S., *Journal of Electrostatics,* Vol. 7, 1979, pp. 155–168.
[*197*] Sakamoto, S. and Yamada, H., *Transactions on Electrical Insulation,* Institute of Electrical and Electronics Engineers, Vol. 15, 1980, pp. 171–181.
[*198*] Fuhr, J. and Schmidt, W. F., *Transactions on Electrical Insulation,* Institute of Electrical and Electronics Engineers, Vol. 20, 1985, pp. 321–325.
[*199*] Fuhr, J., Schmidt, W. F., and Sato, S., *Journal of Applied Physics,* Vol. 59, 1986, pp. 3694–3701.
[*200*] Fuhr, J. and Schmidt, W. F., *Journal of Applied Physics,* Vol. 59, 1986, pp. 3702–3708.
[*201*] Lewis, T. J., *Journal of the Electrochemical Society,* Vol. 107, 1960, pp. 185–191.
[*202*] Kao, K. C. and Higham, J. B., *Journal of the Electrochemical Society,* Vol. 108, 1951, pp. 522–526.
[*203*] Sharbaugh, A. H. and Watson, P. K., *Progress in Dielectrics,* J. B. Birks and J. Hart, Eds., Vol. 4, 1962, Heywood, London.
[*204*] Watson, P. K. and Sharbaugh, A. H., *Journal of the Electrochemical Society,* Vol. 107, 1960, pp. 106–112.
[*205*] Schmidt, W. F., *Transactions on Electrical Insulation,* Institute of Electrical and Electronics Engineers, Vol. EI-24, 1989, pp. 179–183.
[*206*] Theodossiou, G., Nelson, J. K., Lee, M. J., and Odell, G. M., *Journal of Physics D: Applied Physics,* Vol. 21, 1988, pp. 45–50.
[*207*] Frei, C. J., He, L., and Müller, K., *Transactions on Electrical Insulation,* Institute of Electrical and Electronics Engineers, Vol. 24, 1989, pp. 169–174.
[*208*] Filippini, J. C., Lacroix, J. C., and Tobazeon, R., *C.R. Academy of Science,* Paris, Vol. 271B, 1970, pp. 73–76.
[*209*] Goto, Y. and Aso, T., *Record of Electrical and Communication Engineering Conversation* (in Japanese), Vol. 45, No. 1, 1976, pp. 40–44.
[*210*] Hara, M., Koishihara, H., and Saita, K., *Transactions on Electrical Insulation,* Institute of Electrical and Electronics Engineers, Vol. EI-26, 1991, pp. 685–691.
[*211*] Klimkin, V. F. and Ponomarenko, A. G., *Zhurnal Tekhnicheskoi Fiziki,* USSR, Vol. 49, 1979, pp. 1896–1904.
[*212*] Gehman, V. H., Fenneman, D. B., and Gripshover, R. J., ''Effects of Alloys and Surface Treatments on Electrical Breakdown Strength of Water and Water Mixtures,'' *Digest of Technical Papers,* Fifth IEEE Pulsed Power Conference, 10–12 June 1985, Arlington, VA, pp. 335–344.
[*213*] Vorobev, V. V., Kapitonov, V. A., Kruglyakov, E. P., and Tsidulko, Yu. A., *Soviet Physics Technical Physics,* Vol. 25, 1980, pp. 598–602.
[*214*] Brueck, S. R. J. and Kildal, H., *Journal of Applied Physics,* Vol. 52, 1981, pp. 1004–1010.
[*215*] Fujii, H., Yoshino, K., and Inuishi, Y., *Journal of Electrostatics,* Vol. 7, 1979, pp. 93–101.
[*216*] Smith, W. Lee, Liu, P., and Bloembergen, N., *Physical Review,* Vol. A15, 1977, pp. 2396–2403.
[*217*] Marolda, A. J., *Journal of Quantum Electronics,* Institute of Electrical and Electronics Engineers, Vol. 4, 1969, pp. 503–505.
[*218*] Demidov, B. A., Ivkin, M. V., Petrov, V. A., and Fanchenko, S. D., *Pribory i Tekhnika Eksperimenta,* No. 1, 1974, pp. 120–122.
[*219*] Pickard, W., *Journal of Applied Physics,* Vol. 34, 1963, pp. 246–250.
[*220*] Pickard, W., *Journal of Applied Physics,* Vol. 34, 1963, pp. 251–258.
[*221*] Jorgensen, G. V. and Will, E., *Review of Scientific Instruments,* Vol. 33, 1962, pp. 55–56.
[*222*] Stuetzer, O. M., *Physics of Fluids,* Vol. 2, 1959, pp. 642–648.
[*223*] Stuetzer, O. M., *Journal of Applied Physics,* Vol. 31, 1960, pp. 136–146.
[*224*] Stuetzer, O. M., *Journal of Applied Physics,* Vol. 30, 1959, pp. 984–994.
[*225*] Stuetzer, O. M., *Review of Scientific Instruments,* Vol. 32, 1961, pp. 16–22.
[*226*] Sharbaugh, A. H. and Walker, G. W., *Transactions on Industrial Applications,* Institute of Electrical and Electronics Engineers, Vol. 21, 1985, pp. 950–955.
[*227*] Dittrich, W., ''Hochspannungserzeugung mittels einer strömenden elektrisch geladenen Isolierflüssigkeit,'' *Fortschritt-Berichte der VDI Zeitschriften,* Reihe 9, No. 26, VDI Verlag, Düsseldorf, 1981.
[*228*] Dittrich, W., ''Über die elektrische Aufladung einer strömenden Isolierflüssigkeit durch ein elektrisches Spitzenfeld,'' *Fortschritt-Berichte der VDI Zeitschriften,* Reihe 9, No. 22, VDI Verlag, Düsseldorf, 1979.
[*229*] Dittrich, W., *Elektrotechnische Zeitschrift (ETZ) Archiv,* Vol. 1, 1979, pp. 127–137.
[*230*] Dittrich, W., *Elektrotechnische Zeitschrift (ETZ) Archiv,* Vol. 6, 1984, pp. 261–265.
[*231*] Dittrich, W., *Elektrotechnische Zeitschrift (ETZ) Archiv,* Vol. 3, 1981, pp. 397–398.
[*232*] Kelley, A. J., *Journal of Applied Physics,* Vol. 49, 1978, pp. 2621–2628.
[*233*] Lord Rayleigh, *Philosophical Magazime,* Vol. 14, 1882, pp. 184–186.
[*234*] Pohl, H. A., *Dielectrophoresis,* Cambridge University Press, Cambridge, 1978.
[*235*] Shapiro, M., Shalom, A. L., and Lin, I. J., *Journal of Applied Physics,* Vol. 58, 1985, pp. 1028–1031.

[*236*] Lockhart, N. C., *Journal of Applied Physics,* Vol. 51, 1980, pp. 2085–2092.
[*237*] Andres, U. Ts., *Powder Technology,* Vol. 23, 1979, pp. 85–97.
[*238*] Joos, F. M., Snaddon, R. W. L., and Johnson, N. A., *Transactions on Industrial Applications,* Institute of Electrical and Electronics Engineers, Vol. 21, 1985, pp. 1366–1372.
[*239*] Buckley, E., Campanella, M., Carugno, G., Cattadori, C., Gonidec, A., et al., *Nuclear Instrument Methods,* Vol. A275, 1989, pp. 364–372.
[*240*] Albrow, M. G., Apsimon, R., Aubert, B., Bacci, C., Bezaguet, A., et al., *Nuclear Instrument Methods,* Vol. A265, 1988, pp. 303–318.
[*241*] Peskov, V., Charpak, G., Miné, P., Sauli, D., Scigocki, D., et al., *Nuclear Instrument Methods,* Vol. A269, 1988, pp. 149–160.
[*242*] Ochsenbein, S., Schinzel, D., Gonidec, A., and Schmidt, W. F., *Nuclear Instrument Methods,* Vol. A273, 1988, pp. 654–656.
[*243*] Lopes, I. and Schmidt, W. F., *Transactions on Electrical Insulation,* Institute of Electrical and Electronics Engineers, Vol. 23, 1988, pp. 937–940.
[*244*] Schmidt, W. F., Key-Note Lecture, International Conference on Material Engineering for Resources, Akita, Japan, Nov. 1991, pp. 196–202.

E. O. Forster[1]

Chapter 3

Electrical Breakdown in Dielectric Fluids

3.1 Introduction

To start the discussion of the subject matter, it appears desirable to ascertain what is meant by *dielectric fluids.* Obviously there are many ways in which to classify materials which depend usually on their end use. Of interest here is the one which considers a material's resistance to the flow of electricity as a guide. For example, materials having a specific resistance of less than one microohm per centimeter are called conductors, those having less than one gigaohm per centimeter are called semiconductors, and those having higher resistivities are referred to as insulators or dielectric materials. It is this last group, that comprises in addition to fluids, solids, and gases, that is of interest here.

Having defined the meaning of dielectric fluids, it is advisable to define also the meaning of electrical breakdown. When a dielectric material loses its insulation ability, that is, when the barriers to the flow of electricity are suddenly removed by some process, one speaks of electrical breakdown. For this event to occur the material has to be placed between two conductors of different electrical potential. The process leading to breakdown actually involves not only the dielectric but also the whole system formed by the conductors and the insulating material. If the insulating material is a solid, such as polyethylene or an epoxy compound, then the electrical breakdown will lead to permanent failure of the system. If, on the other hand, the dielectric is a fluid such as transformer or silicone oil, then electrical breakdown will lead only to a short interruption of the functioning of the system since the fluid will recover rapidly and resume its insulating function. This ability to recover from failure renders dielectric fluids very useful in many practical applications. It should be pointed out here that such temporary failures naturally cause chemical changes in the fluid which will lead to a gradual deterioration of the fluid and thus shorten its useful life. Obviously their ability of recovery make liquids quite desirable as insulating materials except for the need for their containment.

The preceding comments point to an interesting fact. Electrical breakdown always involves the development of a conducting channel in the insulating material. Such channels are of a permanent nature in solids but only of short existence in fluids and gases. These channels involve only a very small fraction of the insulator, and their existence suggests that the failure is of a local nature although their presence renders the total system temporarily or permanently inoperable. Since society depends on the reliable delivery of electric power, the occurrence of such interruptions due to electric breakdown is highly objectionable. The question then arises as to why these failures occur and what one could do to prevent them. To provide answers to these and similar more

[1]Department of Physics, Rutgers, the State University of New Jersey at Newark, NJ: formerly with the Corporate Research Laboratories of Exxon Research and Engineering Company, Clinton, NJ.

detailed questions, scientists have studied for many decades the breakdown process in a variety of dielectric fluids. In the following, the principal results of these studies will be discussed, and a qualitative model of the breakdown process in dielectric liquids will be presented. The general physico-chemical characteristics of such fluids are reviewed in Section 3.2. In Section 3.3 a classification of electrical breakdown according to types of potential difference, electrode geometry, and other extraneous factors is given. This is followed in Section 3.4 by a historical review of the most important studies performed in the last fifty years. The advantages and disadvantages of various experimental procedures are reviewed in Section 3.5 to facilitate the interpretation of experimental results reported in the literature which are summarized in Section 3.6. Based on historic and more recent findings, a model of electrical breakdown is presented in Section 3.7. Conclusions and recommendations for future research, based on today's knowledge, are given in Section 3.8. References to the literature are summarized in Section 3.9.

3.2 Physico-Chemical Characteristics of Dielectric Fluids

For the purpose of the following discussion it is convenient to redefine dielectric fluids as liquids in which self-dissociation, if any, is extremely small. This implies that under carefully controlled conditions, which shall be reviewed later on in this section, few, if any, free ions will be present in the fluid unless generated by external means such as high applied voltages or electromagnetic radiation (light or X-ray pulses). The preceding statement infers that the dielectric constant of dielectric fluids is generally below ten and in most cases below five. Under these conditions ionic species have a very low probability of forming by internal or external means.

Furthermore, it is convenient to classify these dielectric fluids in terms of their intermolecular interactions into polar and nonpolar ones. As the name implies polar liquids are made up of molecules having a permanent dipole moment due to the asymmetric distribution of electrical charges within the molecule. The presence of the permanent dipoles leads to intermolecular interactions that impart some local structure to the fluid. The molecules and atoms forming nonpolar fluids have symmetric charge distributions within their structure which means that in the absence of any external perturbation the center of mass of the molecule and the centers of positive and negative charges coincide. Under these conditions these molecules and atoms are said to have no measurable dipole. They interact with each other via weak van der Waal forces, and there is very little local structuring in these fluids. Those fluids consisting of atoms are referred to as cryogenic fluids (with the exception of mercury) because they exist as fluids only at very low temperatures. In addition to these two classes of compounds there exists another group of molecules in which dipolar forces can be induced. These molecules are characterized by a high degree of symmetry such as methane (CH_4), benzene (C_6H_6), and carbon tetrachloride (CCl_4). In these molecules intermolecular collisions cause deformation of the colliding molecules and thus the collisions create asymmetric charge distributions that last for short periods. During these periods the molecules behave as if they had a dipole and cause the formation of some structural units which are often referred to as clusters.

The preceding remarks have alluded to the existence of structural units within the dielectric fluids. The existence of such units is important in understanding some of the physical properties which will be discussed in the next paragraphs. Before starting this discussion it is necessary to point out that there exist, in addition to the aforementioned electrostatic (polar) and van der Waal forces, other means for causing formation of structural units. Here, reference is made to the mechanical entanglement of molecules that can rotate more or less freely around the bonds connecting the constituting atoms. For example *n*-hexane is known to be able to form such entanglements. This tendency of mechanical entanglement increases with molecular complexity, and it is particularly pronounced in polymeric materials such as polydimethyl siloxane (silicone oil). This type of cluster formation, unlike the collision induced dipole or permanent dipole clusters, do not affect to any measurable degree the electrical properties of the fluid. This fact is illustrated in Table 3.1 in which are shown the differences between the static dielectric constant and that measured near infinite frequency (defined as the square of the refractive index).

The fluids containing either permanent or induced dipoles show significant differences between the two dielectric constant which reflect the structural peculiarities caused by these interactions. The fluids made up of atoms rather than molecules follow closely the behavior of simple, nonpolar liquids and their two dielectric constants are the same within experimental error.

With this background information the stage is set for a discussion of the properties important in the use of dielectric fluids in the electrical industry.

3.2.1 Important Physical Properties

Dielectric fluids serve not only as insulators but also as heat transfer media so that their transport properties are important. These include in addition to viscosity, thermal and electrical conductivity, melting, as well as boiling temperatures, and density.

The molecular structure and the degree of intermolecular interaction influence the viscosity as does temperature and pressure. It is convenient to express the viscosity-temperature relation in terms of two or more viscosities determined at 20 and 100°C or higher. Also of importance is the temperature at which the fluid stops flowing, the pour point. This property is particularly important when the fluid is to be used in outdoor applications at extremely low temperatures such as are encountered in northern Canada or Alaska. The thermal conductivity of a liquid is a property that requires great skill in its determination. As a general rule, it can be stated that the denser the fluid the better is its thermal conductivity.

One of the main requirements of a dielectric fluid is its low level of electrical conductivity whether it is determined under a-c or d-c conditions. In principle these fluids should have conductivities of less than 10^{-12} S/cm as determined under a-c conditions as discussed later on. Acceptable d-c values range between 10^{-12} and 10^{-14} S/cm. In general, the lower this value, the better the fluid and the more desirable it is as an insulating medium.

Melting and boiling points are determined by well defined methods, and they reflect the composition and purity of a fluid, particularly if the latter is a single compound such as *n*-hexane or toluene (methyl benzene). In the case of organic liquids these properties define the temperature range over which the fluid can be used safely, and they are often used to provide the design engineer with a guideline to fire safety. For this class of liquids, it is generally true that the higher their boiling is, the safer is their use in critical applications such as within buildings. For liquids containing chloride, silicone, or similar inorganic materials, these rules do not apply.

TABLE 3.1—*Experimentally observed difference between the static dielectric (ϵ_s) constant and that at infinite frequency (ϵ_∞) (25.00 ± 0.01)°C.*

Fluid	Mol Interaction	Observed Difference,[a] $\epsilon_s - \epsilon_\infty$
n-hexane	entanglement	0.0011 ± 0.0005
Cyclohexane	van der Waal	0.0031 ± 0.0011
Benzene	induced dipole	0.030 ± 0.002
Chlorobenzene dioctyl	permanent dipole	3.39 ± 0.15
Dioctyl phthalate	permanent dipole	3.48 ± 0.15

[a]The values represent the average of seven independent measurements and the standard deviation from these averages at the 75% confidence level. The measurements were carried out at 1000 Hz and at the sodium D-line for the refractive index.

The density of a fluid, as mentioned previously, reflects not only the ability to conduct heat but also acts as an indicator of its ability to prevent electrical failure. Again the same empirical rule applies here as for thermal conductivity. The higher the density, the higher, in principle, is the voltage at which failure will occur. Today the densities of dielectric fluids cover the range from about 0.6 to 1.4 g/cm^3, and their boiling points range from below 70 to nearly 300°C. Their melting temperatures range from -30 to -60°C and lower. These broad variations reflect differences in their chemical composition.

3.2.2 Important Chemical Characteristics

Dielectric fluids comprise several large generic families of materials. There are the hydrocarbon fluids which include straight and branched chain paraffins as well as cyclic saturated and unsaturated (aromatic) hydrocarbons and combinations of the various types, that is, combinations of cyclic with branched or linear chain hydrocarbons (naphthenes and alkyl or aryl aromatics). Mixtures of all three classes are found to constitute the products derived from the distillation of crude oil (petroleum) such as transformer oils.

Then there are the variations of these compounds in which one, several, or all hydrogen atoms are replaced by either chlorine or fluorine or both. In the first group belong the *freons* which have been associated recently with deliterious effects on the ozone layer. The latter group, in which all hydrogens have been replaced producing the perfluoro hydrocarbons, has very unusual electric properties which are believed to be due to the high affinity of fluorine for electrons. In this class belong also the polychlorinated aromatic hydrocarbons (PCB) which, because of their possible toxicity have been removed from use in industry.

A third group includes hydrocarbon molecules in which one or more carbon atoms are replaced by either nitrogen, phosphor, or sulfur. These substitutions are found particularly in aromatic molecules. When these unsaturated, cyclic compounds contain one of these elements mentioned previously, they are referred to as heterocyclic molecules. Some of them are believed to be less desirable because their ability to react with oxygen is much greater than that of the homocyclic molecules, and, thus, they contribute to the early degradation of the fluid.

Then there are those molecules made up of oxygen in addition to carbon and hydrogen. These include, according to their functionality and structure, the organic acids, aldehydes, ketones, and alcohols. Of interest here are combinations of acids and alcohols which are called esters. Examples of these molecules are the phthalate esters which are derived from the combination of phthalic acid (1,2, benzene dicarboxylic acid and alcohols such as octyl (eight carbon atoms) or nonyl (nine carbon atoms) alcohols. Oils derived from natural products such as from the castor bean also belong in this classification.

Another group of fluids is based on combinations of silicone, oxygen, carbon, and hydrogen. These are the polydimethyl siloxane, polymethyl phenyl siloxane, and similar combinations. Depending on the number of monomers in the chain, these materials cover the range from thin fluids to 1 cSt to syrup like liquids of 10 000 cSt and higher. These silicone oils are of particular interest in those applications requiring maximum protection from fire hazards.

Finally mention should be made here of the family of cryogenic fluids such as liquified nitrogen, helium, argon, neon, and xenon. Of these the first two are of some practical importance since they find use as insulating fluids in such superconducting structures as power cables and rotating machinery. The others are rarely used, and they attracted attention mainly for scientific purposes. These cryogenic fluids have a draw back in terms of their use for prebreakdown and breakdown studies since it is extremely difficult to carry out *in situ* observations at these low temperatures. For these reasons they have not received much attention in recent years, and no further reference will be made to them in subsequent sections.

The selection of a given type of material will be governed by several considerations which include economic as well as technical aspects. For example silicone fluids and perfluorinated hydrocarbons are very expensive compared to petroleum derived oils, but they have a much lower flamability level than the latter. Another important consideration is the reactivity of these fluids with oxygen. The slower a fluid reacts with oxygen, oxidizes, and the more stable it is, the longer it will last in a given type of system. Therefore, the aging characteristics of these oils have to be considered too, since the equipment in which they are used is to be kept operative for several decades. Finally, the electrical characteristics have to be taken into account, as will be shown in the next paragraph.

3.2.3 Electrical Properties

The dielectric fluids mentioned in the previous paragraphs cover a broad range of electrical properties. These properties include dielectric constant, conductivity, and ability to resist electrical breakdown. Of interest here is conductivity and its bearing on the breakdown process. It has been observed empirically that conductivity of an oil can be reduced by purification. This process can involve several conventional or cryogenic distillations combined with filtration. It can be replaced or supplemented with percolation techniques employing molecular sieves or ion exchange columns to produce dry, pure fluids of very low conductivity. The purity of a fluid can be determined by several means such as mass spectrography, gas chromatography, and electrical conductivity measurements. The latter can be performed under a-c or d-c conditions as will be discussed in Section 3.3. While the d-c conductivity measurements are simpler to perform than the a-c ones, they are also less precise than the latter. It is for this reason that a-c conductivity measurements are preferred. Such a-c measurements are particularly useful when studying the level of moisture in dielectric fluids. These measurements supplement the conventional moisture determinations using the Karl Fischer method. Studies have shown that very pure, dry fluids require higher voltages to produce breakdown than do impure ones. This is illustrated in Table 3.2 in which the breakdown voltages of four samples on *n*-hexane of different degree of purity are compared under nonuniform field conditions. The significance of the use of nonuniform fields over uniform ones will be discussed in a subsequent section. The above data clearly indicate the detrimental influence of certain impurities on both conductivity and breakdown voltage. Hence, it is important in all investigations to be aware of the purity level of the fluids used and to handle the fluids with great care. It appears advisable, also, to characterize the fluid prior to an investigation not only in terms of its physical properties but also in terms of its electrical ones.

With this information in mind, it is now possible to go into the actual study of the breakdown process. As a first step, the various types of breakdown will be discussed in the next section.

TABLE 3.2—*The effect of impurities on the breakdown voltage of n-hexane, (25.0 ± 0.1)°C, 0.3 cm gap.*

		Breakdown Voltage, kV	
	Conductivity, S/cm	Point Negative	Point Positive
Ultrapure	$<1 \times 10^{-16}$	>203	40
Chemical grade	1.6×10^{-15}	148	37
10 ppm ASA 3[a]	1.9×10^{-10}	100	37
0.1 *M* DMA[b]	1.5×10^{-15}	>200	40

[a]A mixture of 1/3 chromium salicilate, 1/3 calcium succinate and 1/3 vinyl copolymer in kerosene (50/50 mixture by volume).

[b]DMA = N,N′ dimethylaniline.

3.3 Classification of Electrical Breakdown

As mentioned in Section 3.1, electrical breakdown is the result of a system's failure to prevent the flow of electrons from one conductor, the cathode, to another conductor, the anode. Naturally there has to exist a potential difference between the two conductors, and, in the following discussions, it will be always the cathode that is at a considerably higher potential than the anode. This approach is somewhat different from that customarily used by electrical engineers and avoids the confusion introduced by the concept of the electric current flowing in the opposite direction to that of the electrons.

The flow of electrons or charge carriers in a dielectric is governed by the electrical field which in turn is determined by the potential difference and the distance of the conductors from each other as well as their respective shapes. Thus, one distinguishes various physical and temporal ways of applying an electrical stress to a system. They include, in addition to the type of stress, the geometric shapes of the conductors or electrodes, their separation, and the environmental conditions under which the materials are to be stressed, that is, temperature, atmosphere, pressure, radiation, and combinations of these. In the next paragraphs these parameters will be reviewed briefly.

3.3.1 Types of Electrical Stresses

To produce a potential difference between two conductors or electrodes, one needs a generator (in the field) or a power supply (in the laboratory). These power supplies can be classified into two classes, those producing direct current (d-c) and those yielding alternating current (a-c). In the d-c case the two electrodes maintain their specific functions or polarities with time. One is the cathode, which injects electrons, and the other, the anode, which helps remove the charge carriers from the dielectric. In the second case the polarities of the two electrodes change systematically with time, that is, with the frequency of the applied field. These two conditions are illustrated in Fig. 3.1. For electric power transmission either 50 of 60 Hz are used. In the laboratory other frequencies are often employed to study a material's response to a-c stresses. In addition to the preceding difference, it is important to realize that the potential cannot be applied instantaneously but that it takes a finite time to reach a given potential difference. One has to distinguish, therefore, among various ways of raising the potential. There are power supplies that allow raising the potential at variable rates to produce single pulses such as square or lightning like pulses, saw tooth pulses of variable rise time, repetition rate, and pulse width, as well as constant potentials, as illustrated in Fig. 3.2. It

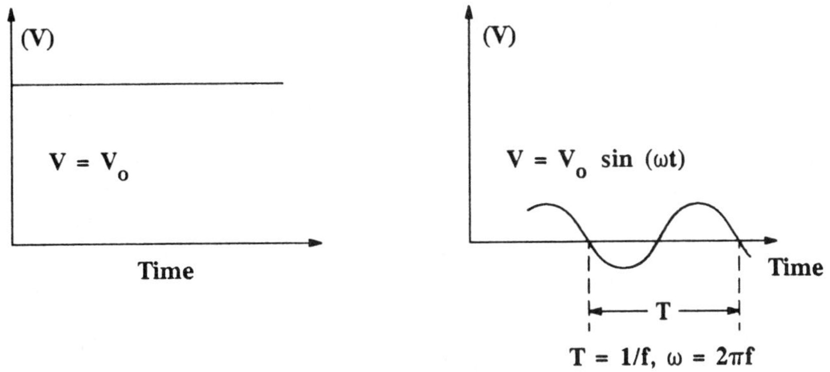

a) direct current b) alternating current

FIG. 3.1—*Schematic representation of potentials.*

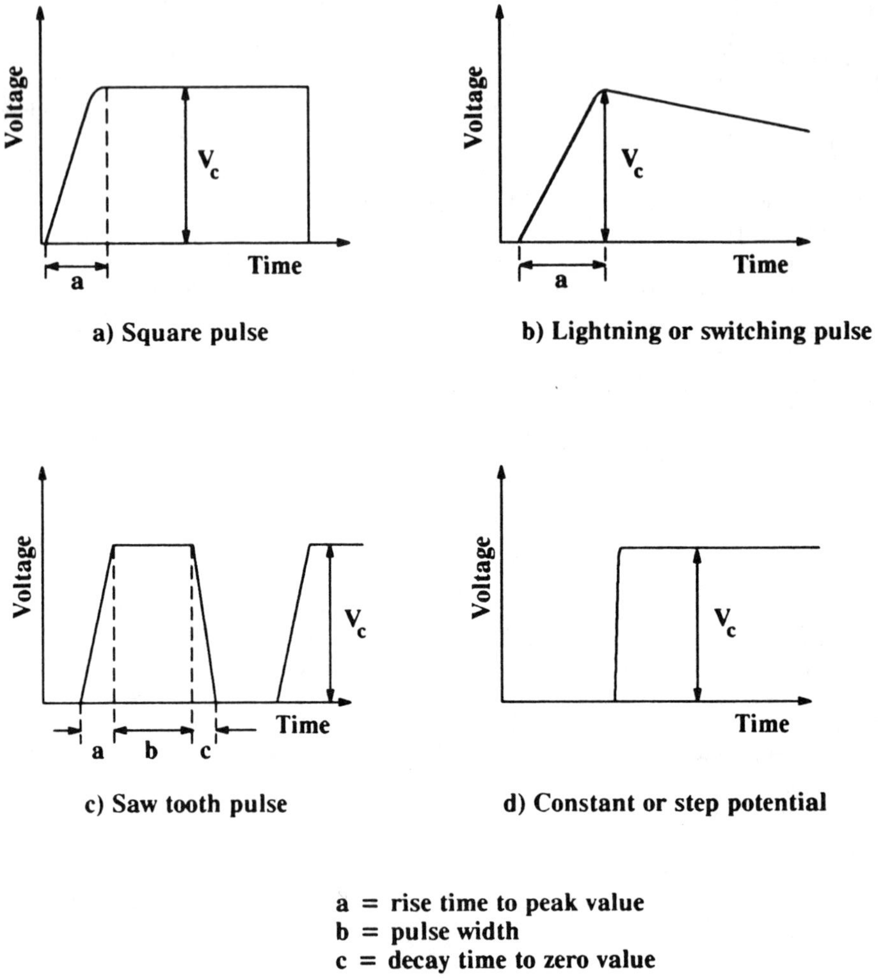

FIG. 3.2—*Various types of d-c potentials used in testing dielectrics.*

should be mentioned here that the various instruments used as power supplies do not necessarily deliver the same energy to the test system. For example there are a great variety of pulse generators particularly those based on the Marx type that use capacitor banks to deliver the power to the system. Depending on the number and size of the capacitors used, the power stored in them can range from a fraction of a joule to several 1000 J. The degree of damage produced by such test equipment will reflect the energy stored in the power supply. Another type of pulse generator uses a combination of inductor coils and a pulse transformer. A schematic diagram of this type of pulse generator is shown in Fig. 3.3. The pulse produced by this arrangement is usually in the form of a trapezoid; hence, the instrument is referred to generally as a trapezoidal pulse generator. Its main advantage is that it generates considerably less electromagnetic radiation than do the Marx type generators. The influence of electrode geometry is discussed in the next paragraph.

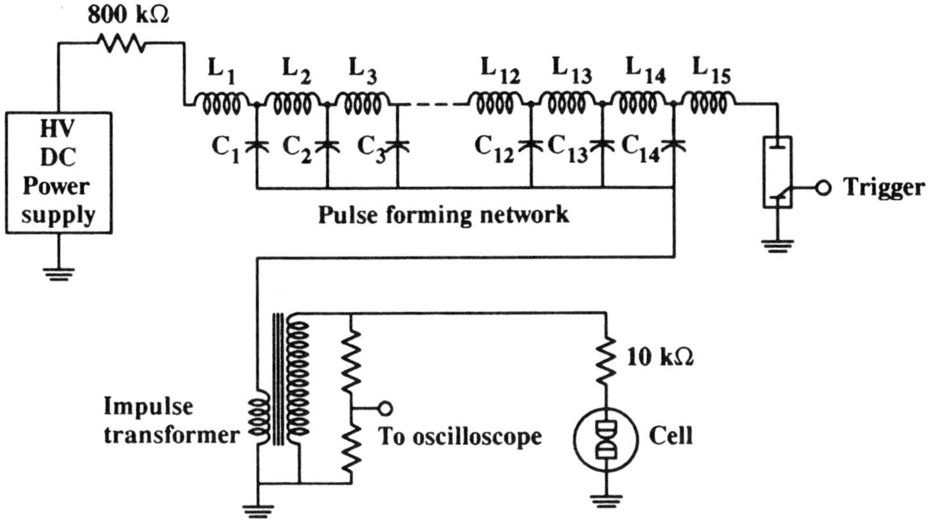

FIG. 3.3—*Schematic circuit diagram of high voltage pulse generator.*

3.3.2 Electrode Geometry

In many practical applications, electrical insulation is used in coaxial geometries such as in cables where the electrodes consist basically of two concentrically arranged cylinders. In the laboratory, where one is interested in studying a material's response to electrical stressing under more severe conditions, other geometries are used in addition to the coaxial one mentioned previously as shown in Fig. 3.4*a*. In particular one often attempts to achieve uniform field conditions by arranging two flat, cylindrical metal surfaces to be closely parallel (Fig. 3.4*b*). Under these conditions the material between them is said to be exposed to a uniform field which is given by the ratio of the applied potential difference and the separation of the metal plates. To assure the uniformity of the field the sample is exposed to, a guard ring surrounds the measuring electrode and prevents exposure of the sample to edge effects. To simulate imperfections occurring in actual service, use is made of point/plane, sphere/sphere configurations or combinations of the two (Fig. 3.4*c*, *d*, and *e*). In these cases the electrical field produced by the external application of a potential difference is nonuniform, and the results obtained with such asymmetric geometries depend on the polarity of the electrodes. If instead of d-c, a-c potentials are used in combination with asymmetric geometries, the observations made are very difficult to interpret since both polarities make contributions that are interdependent on the events occurring in the preceding cycles. For this reason many fundamental studies have been carried out using only d-c conditions. In all these cases the gap between the two electrodes has a definite influence on the advent of electrical breakdown. In addition to the role of potential difference and electrode geometry, the breakdown process is influenced also by environmental conditions as will be shown in the next paragraph.

3.3.3 Environmental Effects

It has been known for a long time that the potential difference necessary to create electrical breakdown at a given electrode geometry and gap is a function of temperature, pressure, humidity, and radiation.

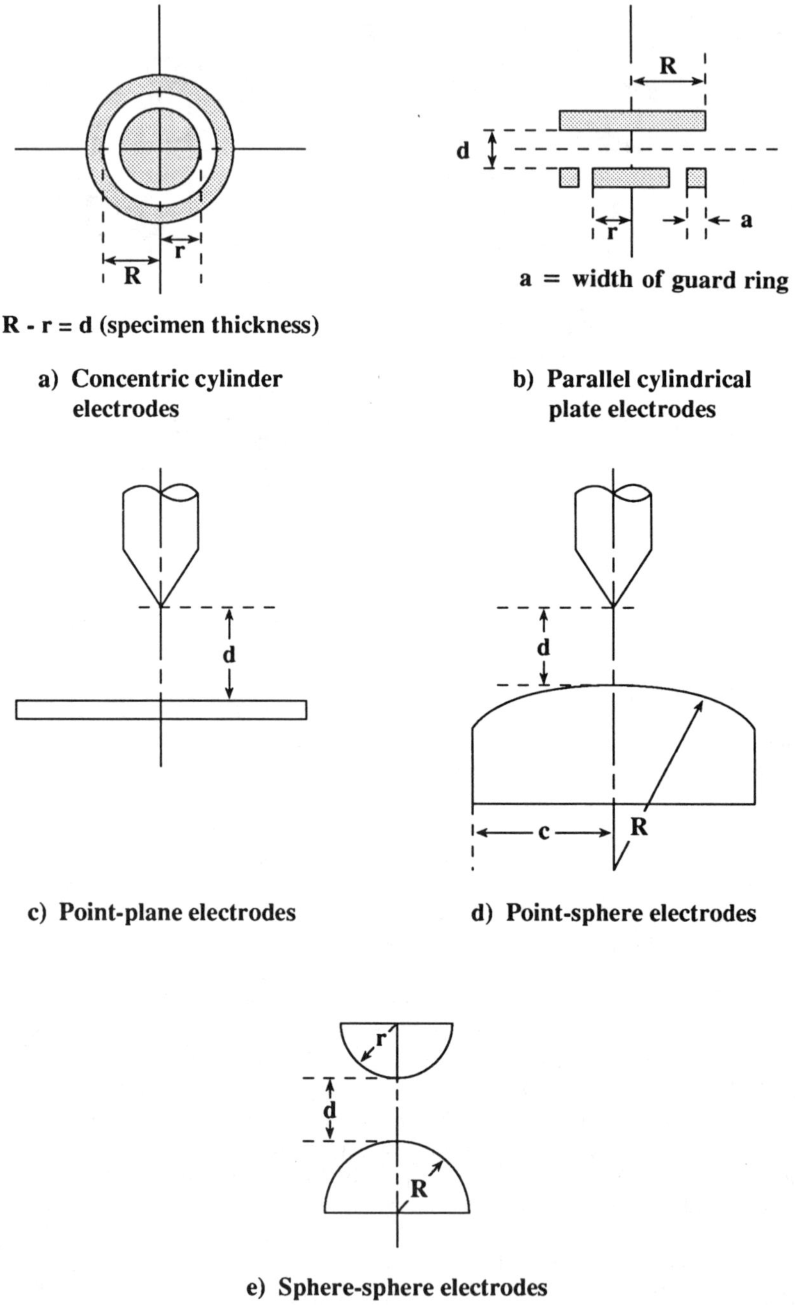

FIG. 3.4—*Commonly used electrode geometries.*

3.3.3.1 Role of Temperature—In general the voltage required for electrical breakdown decreases with increasing temperature. This behavior, in general, is due to a reduction in the dielectric liquid's density and, hence, its decrease in viscosity leading to greater ease of charge transport as reflected in an increase in conductivity.

3.3.3.2 Role of Pressure—The effect of increasing pressure is opposite to that produced by increasing temperature. As the pressure is increased the dielectric increases in density and, as recent studies have shown, a decrease in conductivity in most instances. As a consequence, more energetic electrons are needed to cause breakdown. These energetic electrons are produced at considerably higher applied potentials leading to higher breakdown voltages. Conversely, a reduction in pressure below atmospheric conditions leads to a lowering of the breakdown voltage.

3.3.3.3 Role of Atmosphere—Here, one is concerned with the effect of moisture and of oxygen. The presence of both increases the dielectric fluid's conductivity which in turn causes a lowering of the breakdown voltage. Experience has shown that when the external potential required to produce electrical failure in a given electrode system at a fixed gap is higher, the dielectric's conductivity is lower. The presence of water or oxygen or both is believed to cause the creation of more mobile charge carriers. Neither nitrogen nor hydrogen show this effect. Also, the presence of oxygen, as mentioned earlier, leads to chemical changes that affect adversely the useful life of the fluid.

3.3.3.4 Role of Radiation—The effect of electromagnetic radiation depends on its ability to interact with the dielectric liquid and its intensity. Within the electromagnetic radiation spectrum there are regions which allow such interactions. For example, in the infrared region certain molecular motions can absorb this radiation and thereby increase their energy content as witnessed by increased intramolecular motion. This effect is obviously similar to that of heating the material. Interaction with radiation in the ultraviolet and vacuum ultraviolet region can lead to more serious electronic excitation and eventual ionization of molecules. X-rays and nuclear radiation are known to cause ionization of dielectric molecules. These processes, in addition to chemical changes, can lead to charge carrier production that, in turn, causes an increase in the fluid's conductivity and, hence, a lowering of the breakdown voltage. Recently, it has been shown that the use of coherent radiation produced by lasers in the infrared and visible spectrum can cause electrical breakdown when focused on a volume element within the electrode gap. Here, the high electric field associated with the focused beam apparently leads to ionization of molecules contained in this volume element that triggers the overall process at much lower applied voltages.

3.3.4 Intrinsic and Thermal Breakdown

No discussion of the classification of electrical breakdown would be complete without mention being made of the terms intrinsic and thermal breakdown. The former term, *intrinsic,* refers to the development of a local field that literally tears apart a molecule or rather the bonds holding atoms together in a molecule. It has been determined theoretically that such a field would have to be of the order of 10^8 V/cm (10^{10} V/m) to achieve this ionization. The latter term, *thermal,* has been coined to refer to breakdown caused by thermal effects such as are produced by local heating. Unfortunately, all breakdowns involve thermal effects and the use of the term thermal has caused considerable confusion. Thermal breakdown may occur more readily under a-c rather than d-c conditions because the alternating electric field acts on the fluid's molecules and causes local heating that can influence the development of breakdown channels. Under d-c conditions one is only concerned with joule heating that is very localized as will be shown later.

The preceding review of factors controlling electrical breakdown was presented so as to facilitate the understanding of the experiments reported in the literature and their interpretation. In the next section a historical overview is presented of what, in the opinion of this writer, appears to be the major contributions made to the field of electrical breakdown in the past fifty years.

3.4 Historical Overview

Ever since Benjamin Franklin and even before him, man has tried to understand the origin of sparks. As man's knowledge of electricity progressed this desire to understand electrical breakdown of air and other materials intensified. Rather than go into the early history, this review starts arbitrarily with the contribution made by von Hippel in his paper "Electrical Breakdown of Solid and Liquid Insulators" [*1*] in which he suggests that one apply the ideas and concepts developed for gaseous breakdown to condensed matter and modify them so as to reflect the decrease in the distance electrons can travel between collisions in solids and liquids (their mean free path).

3.4.1 Electron Movement in Condensed Matter

Early studies of electrical breakdown in gases and vacuum had shown that electrons were injected from the cathode into the gas and, depending on the concentration of gas molecules, the electrons could travel some distance in the direction of the externally applied field and thereby gain energy from the field. On collision with gas molecules the latter could be ionized if the electrons were energetic enough. According to Townsend [*2*] this process could lead to electron avalanches and, hence, electrical breakdown. Using these basic ideas and those of Paschen [*3*], von Hippel concluded that in condensed matter electrons could not readily gain sufficient energy to cause impact ionization. However, the electrons could collide inelastically with molecules of the dielectric causing transfer of energy and, hence, excitation of vibrations in the molecules. The transfer of electron energy to the lattice of solid halides seemed to explain their breakdown behavior very well. However, this concept was less successful in predicting breakdown behavior in amorphous solids and in liquids. Some twenty years later, Lewis [*4*] and Adamczewski [*5*] took up these ideas and tried to modify the Townsend avalanche process for use in liquid hydrocarbons. They visualized that electrons injected from the metal electrode into the liquid would exchange their energy with molecules making up the liquid, thus increasing the latter's energy to the point where the molecules could ionize. Working along similar lines, Swan [*6*] concluded that electron emission and charge multiplication were essential for electrical breakdown to occur in dielectric liquids. These models appear quite reasonable, but they fail to explain the pressure dependence of the breakdown process. This pressure dependence suggests that a phase change is involved in the overall process.

3.4.2 The Bubble Concept

Another concept was introduced by Watson and Sharbaugh [*7*] who suggested that the electron injection process would cause intense heating of the region adjacent to the injection site. This heating process could then lead to the vaporization of the liquid and the formation of a bubble at that site on the cathode. Similar bubble mechanisms were also proposed by Kao [*8*], Krasucki [*9*], and later on by Thomas and Forster [*10*]. Verification of this bubble concept required appropriate optical techniques that allowed ultrahigh speed photography. Among the first attempts in this direction were those of Hakim and Higham [*11*], as well as Chadband and Wright [*12*]. Further support for the concept of a two phase (liquid/vapor) system came from more elaborate optical studies.

3.4.3 Streamers, Bushes, and Trees

Using schlieren or shadowgraph techniques, it was possible to produce single pictures of streamers that resembled in structure a bush or a tree which grew out of the cathode into the liquid. Actual pictures of breakdown were also obtained that suggested the existence of a plasma column [*13–19*]. The major drawback of this approach was that one obtained only one or two photographs during the breakdown event taken at random in a hit and miss fashion. This limitation was imposed by the inability to control the camera's shutter fast enough which in turn required the use of a short laser [*10*] or flash lamp pulse [*11,12*] to illuminate the gap between the test electrodes while the camera's shutter was left open. This situation is illustrated in Fig. 3.5, which shows two views of the gap between two parallel cylindrical disks illuminated by a 15 ns wide laser pulse that was split into two beams with one delayed 40 ns with respect to the other. The two beams entered the test cell at right angle to each other. A schematic arrangement of test equipment is shown in Fig. 3.6. A photograph of the actual arrangement in the author's laboratory as it existed around 1980 is shown in Fig. 3.7.

Alternatively, one could study very viscous liquids or larger gaps or both in which the streamer development was sufficiently slow to be observable by more conventional means [*9*].

Based on thc information gathered by conventional voltage-current experiments and the limited optical studies, a more detailed knowledge of the electrical breakdown process evolved. As mentioned in the preceding paragraph, the early pictures revealed shadowgraphs of structures that resembled bushes or trees. Indeed, in some pictures taken at low resolution, the shadows resembled the outline of a bubble as shown in Fig. 3.8, under uniform field conditions. Later, when using nonuniform fields and better resolution more details of these structures became evident as shown in Fig. 3.9. It became apparent that these structures were made up of many branches and streamers, growing in all directions so that their overall appearance resembled that of a bush or a tree. Sharbaugh et al. [*13*] reviewed the state of knowledge in 1978, and many of the conclusions reached by these authors are still valid today. This review article also provides an excellent list of pertinent references to earlier studies.

3.4.4 Latest Developments

In the late 1970s more sophisticated electro-optical devices became available, such as, high speed framing cameras and image intensifying devices which were capable of taking up to 2×10^7 frames per second. With the help of these devices it was possible to gain more detailed knowledge of the processes leading to breakdown. It became apparent that initiation of bush or tree-like structures involved the formation of a streamer in the low density, two phase region in the vicinity of the injection site. The growth of this region, leading to the development of branches, was a function of the polarity of the electrode and of the local field conditions. Actual breakdown was preceded by the formation of secondary streamers growing much faster than the primary ones as can be seen in Fig. 3.9. It was also shown that the growth of negative streamers emanating from a cathode was caused by a different mechanism than those growing from an anode. The negative streamers were found to be bigger and to have more branches than do the positive ones (Fig. 3.10). The continued growth of negative streamers was found to be associated with the availability of electrons coming from the cathode whereas the positive streamer growth was connected with the supply of electrons in the fluid [*30*]. These prebreakdown and the actual breakdown events were studied by many investigators [*14–33*], and the experimental techniques used by them are presented in the next section.

3.5 Experimental Procedures

There are basically four procedures available for the study of electrical breakdown of a system. The oldest and most commonly used one deals with measurements of the externally applied voltage which causes breakdown. Next come experiments in which the current is determined as a function of voltage, temperature, and time. More recently, shadow and schlieren photography have been used to record the development of streamers in liquid dielectrics. Then there are the measurements involving light and acoustical emission associated with partial and full breakdown. The utility of these various techniques is discussed in the following paragraphs.

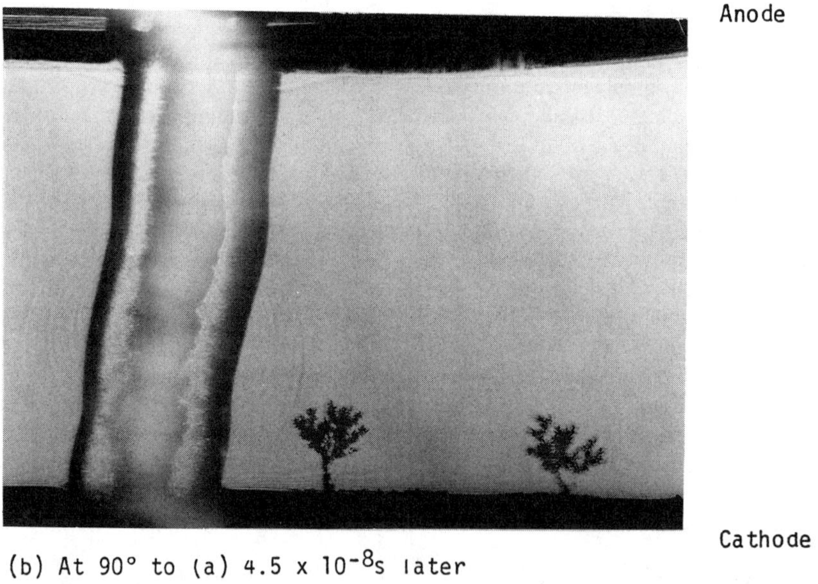

FIG. 3.5—*Breakdown regime in n-hexane under uniform field conditions (after Wong and Forster* [16]*).* (a) *direct beam and* (b) *at 90° to* (a) *and 45 ns later.*

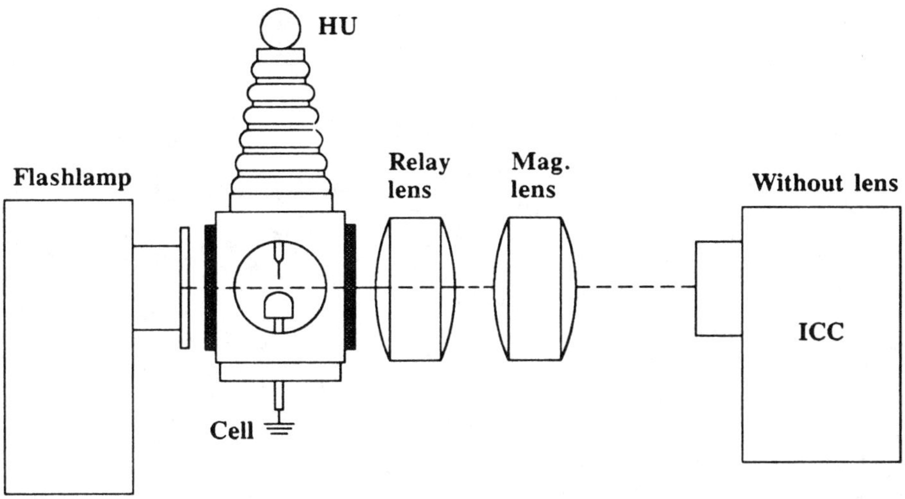

FIG. 3.6—*Experimental arrangement for the observation of breakdown phenomena in dielectric liquids (courtesy of E. F. Kelley, National Bureau of Standards).*

3.5.1 Determination of the Breakdown Voltage

Obviously there exists great interest in determining the highest externally applied voltage that an insulation system can safely withstand. Surprisingly, this determination is very difficult to carry out and the interpretation of the phrase "... the voltage a system can safely withstand" has caused great confusion among electrical engineers. As will be shown later, electrical breakdown of an insulation system is a statistical process that depends among other factors on the reproducibility of the system under test. Equally difficult is the measurement of the voltage at which breakdown occurs 100% of the time. Further complicating the picture is the desire of engineers to consider the ratio of this breakdown voltage and the gap between the electrodes, the breakdown field. The electric field at which breakdown occurs is often called the dielectric strength or the breakdown strength of a material. In fact, this ratio of voltage to gap is not a material's property since it depends nonlinearly on the gap. This ratio defines the electric stress under uniform field conditions for a given electrode separation. Under nonuniform field conditions this ratio defines only the average electrical stress since the actual one varies continuously across the gap. In an effort to provide a uniform measuring technique to the industry, the American Society for Testing and Materials (ASTM) and the International Electrotechnical Commission (IEC) have developed various methods for such determinations under both alternating and direct current conditions. These test methods are good for making comparisons among dielectric fluids, but they are of no help in the study of the events leading to breakdown. In general, there are a variety of conditions one may select to carry out the determination of a system's breakdown voltage. These reflect the ultimate use of the equipment and the information one is interested in obtaining. They include various forms of potential applications and electrode geometries, as well as environmental conditions.

The basic requirements for performing these and similar tests include a power supply capable of producing a stable high potential, or voltage pulse in the desired time interval, or an instrument, which allows the potential to be raised at a predetermined rate or to be kept at a constant level for a long time. Such power supplies could be either of the a-c or d-c type. The determination of the actual voltage to which the test fluid is subjected is carried out usually with a capacitive or resistive voltage divider. The latter is more desirable since it has a faster response time, it does not load

FIG. 3.7—*General view of author's laboratory showing on left the optical delay line which the operator is adjusting; at right foreground is the test cell with high voltage feed through. In the background is the ruby laser which produced the pulses (15 ns width) to illuminate the electrode gap and various mirrors* (ca. *1980).*

unduly the test system, and it limits the energy flow through the test cell. The output of this divider is displayed on a properly calibrated oscilloscope or a voltmeter. A common practice has been to rely on the voltmeter reading provided by the power supply which does not necessarily correspond to the voltage one measures with the voltage divider near the test cell. This disagreement is particularly noticeable if the applied potential is time dependent. The wires connecting the various components should be of the coaxial type, and their internal impedance should be matched to that of the cell. It is possible to include also current measurements in this setup. This aspect is discussed in the next paragraph.

3.5.2 Current Measurements

Dielectric materials in general and fluids in particular are known for their high resistivity and their inability to conduct electricity. Thus, the current flowing through the material once a potential difference is applied across the electrodes is very small and of the order of nanoamps (10^{-9} A) or less. As the potential difference is increased the current level increases only slightly until potentials close to those causing breakdown are reached when the current suddenly increases rapidly to milliamps or higher, as shown in Fig. 3.11. At breakdown, current levels of thousand amps or higher have supposedly been measured [*34*]. This change in current level occurs in about one to two microseconds. At the end of this period the current reverts back to the nanoamp or lower level depending on the experimental conditions used for control of the applied potential. None of the

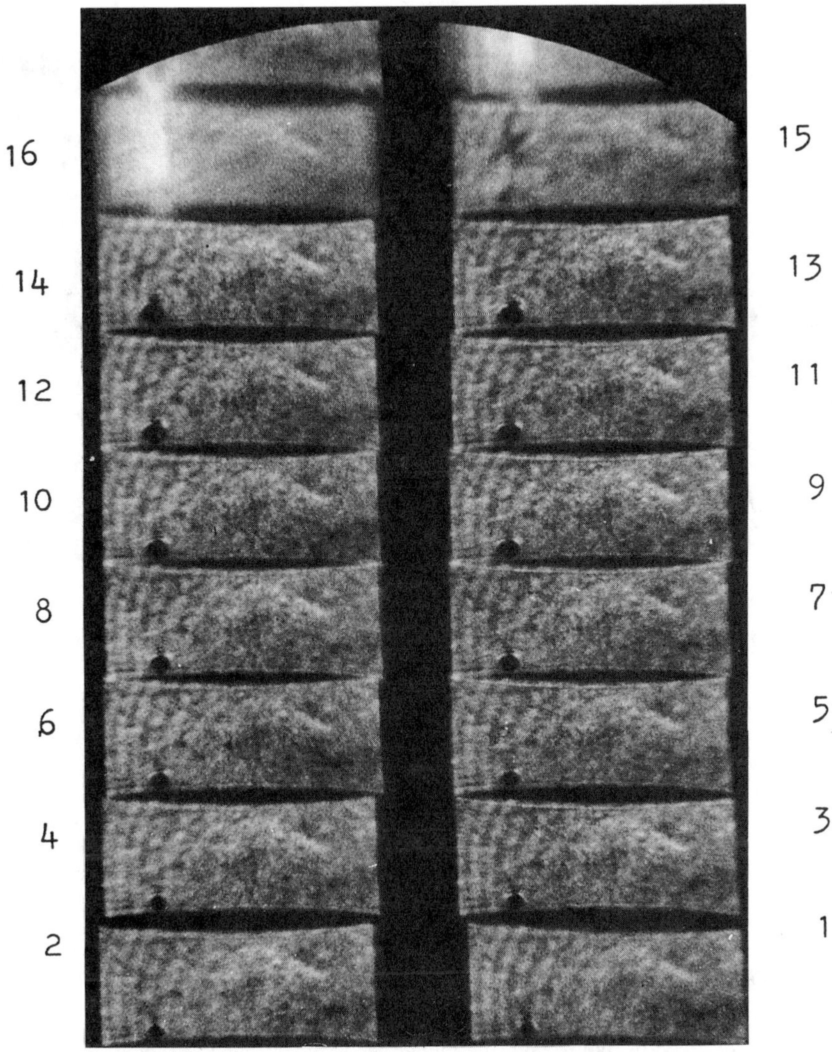

FIG. 3.8—*Streamer development in n-hexane under uniform field conditions; bottom electrode is the cathode. Trapezoidal pulse, 220 kV crest, 2 mm gap, 500 ns between frames.*

existing current measuring devices have a dynamic range of 10^{11} or bigger in this short time interval so that a complete set of measurements of the whole event is at present not possible. Another problem associated with this type of measurement is the definition of the electrode area that participates in the transport of charges. The actual area involved in charge injection and extraction is a small fraction of the total area. Thus when reporting current densities based on the actual electrode area, these quantities are underestimates [*35*]. The actual current densities will be much higher than expected. As a result of this mistake, erroneous predictions have been made (see, for example, Ref *7*).

To carry out these current measurements it is necessary to select, in advance, the instrument most

FIG. 3.9—*Negative streamer development in n-hexane. Needle cathode. Trapezoidal pulse, 60 kV crest, 3 mm gap, 500 ns between frames.*

suitable to the expected level of charge carrier flow to assure proper sensitivity. Since breakdown is such an unpredictable event it will be always advisable to protect the measuring device from voltage transients associated with breakdown. Special diodes and current limiting resistors are used frequently for this purpose. While voltage and current measuring techniques are relatively simple and well documented, the electro-optical techniques, which were developed in the last two decades, are much more elaborate as will be shown in the next paragraph.

3.5.3 Electro-Optical Techniques

Early attempts at making optical observations involved single frame shadow and schlieren photography. More recently multiframe, ultrahigh speed cameras have become available, which transform photons into electrons and, after proper displacement, convert them back into photons so that they can be recorded on photographic film. In either case a light source such as a Xenon flash lamp or a pulsed laser is used to illuminate the gap between the two electrodes. The electrodes are mounted in a cell provided with at least two windows positioned at 180° of each other. To guide the light beam from the flash source through the cell to the detection device, several lenses and a recording device are mounted on one or two optical benches. A typical arrangement is shown in Fig. 3.6. For single frame operation use is made of regular cameras with shutters operated electromechanically to permit their remote activation. For multiframing, image converter cameras (ICC) are used with or without image intensifiers. The cells used in these experiments are usually of cylindrical design, manufactured from some inert, stable polymer such as polytetrafluoro ethylene, and they have typically four diametrically opposed windows as well as two high voltage feed throughs. The latter are aligned at 90° to the optical plane defined by the four windows. In addition some cells are equipped with an external circulation system to remove particulates created during

(a) needle cathode, crest
voltage 50 kV, no breakdown

(b) needle anode, crest
voltage 30 kV breakdown occurred
1.3×10^{-6}s after picture was taken

FIG. 3.10—*Laser-schlieren photographs of streamer development under nonuniform field conditions in isooctane (1.5 × 50 μs impulse wave):* (a) *needle cathode, 50 kV crest, no breakdown and* (b) *needle anode, 30 kV crest, breakdown occurred 1.3 μs after photograph was recorded.*

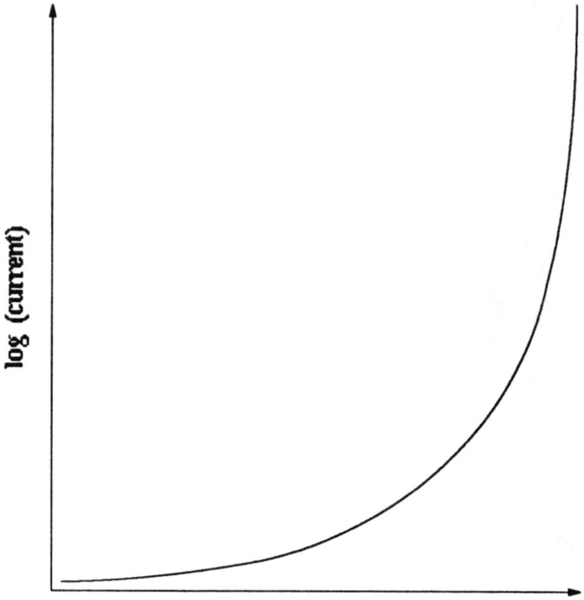

FIG. 3.11—*Typical current-voltage relationship in a dielectric liquid.*

breakdown. A typical arrangement is shown schematically in Fig. 3.12. In some cases provisions are also made to control temperature during the photographic observations.

Usually, the electrical stress is supplied by a pulsing unit such as a Marx generator or a pulse transformer. The reason for the choice of rather fast rising pulses for these types of observation is based on the assumption that under pulse conditions the fluid remains relatively free of excess charges (space charges) which would cause distortion of the local electrical field. Such space charge effects are known to complicate studies under a-c conditions, as mentioned in Section 3.3. These effects, however, are also present in long term d-c studies in which the time to failure of the system is measured under constant electrical stress. Another concern is the amount of electromagnetic noise generated upon triggering of a generator such as one of Marx type. This noise was found to interfere seriously with the operation of a framing camera. A pulse transformer usually generates a tolerable level of such electromagnetic interference.

The interpretation of information generated by these electro-optical techniques relies heavily on the assumption that the observed change in refractive index in the liquid near the injection site is due to either the development of a low density region or the localized change in polarizability of the dielectric or both due to the presence of thermalized electrons. This aspect will be discussed in more detail in Section 3.6. In any case the photographs so obtained give a two dimensional representation of a three dimensional event. Attempts to gain better spatial resolution were only of limited success [*21*].

With presently available commercial equipment, temporal resolutions of the order of 50 ns are possible. This is the shortest interval between consecutive frames achievable with existing framing cameras. In other words, 10 consecutive pictures can be obtained in 0.5 μs. Similarly, streak camera pictures are obtainable today with spatial resolution of 1 mm in 10 ns. This latter technique records continuously the light transmission via an angularly moving mirror. Neither of these two electro-optical methods allows the study of the fast events, which occur in time frames of less than 10 ns

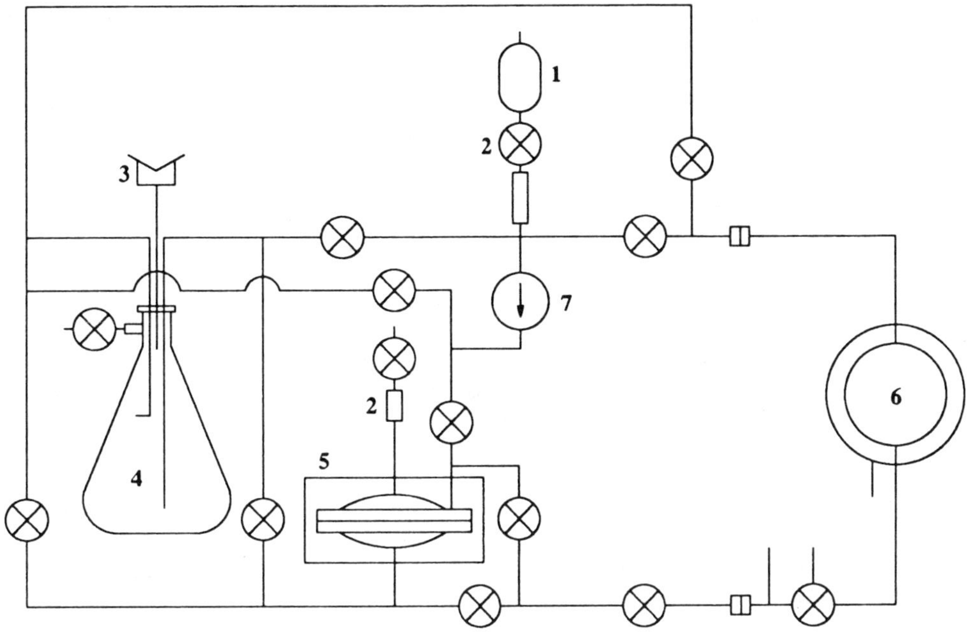

1 Evacuation chamber
2 Sight tube
3 Pressure relief
4 Fluid reservoir
5 Filter
6 Circulating pump
7 Dielectric cell

FIG. 3.12—*Filtering system for dielectric fluids (courtesy of E. F. Kelley, National Bureau of Standards).*

just prior to breakdown. The presently available cameras, however, have given valuable insight into the events leading to breakdown and thus have led to an improved understanding of this process. The information obtained from light and acoustic measurements certainly supports the electro-optical observations as will be shown in the next paragraph.

3.5.4 Light and Acoustical Measurements

Both light and acoustical techniques have been used for some time in detecting prebreakdown events which are also known as partial or corona discharges. This subject is treated in detail in an ASTM special technical publication [*36*]. It shall be summarized here briefly to provide the reader a complete picture of the various detection techniques used in the study of electrical breakdown.

The detection of light emission involves the use of photomultiplier tubes (PMT) that detect the photons given off by the prebreakdown or breakdown events or both. These PMT devices are characterized by their spectral response, in terms of their sensitivity to light, to a given wavelength (photon energy). Since it is well known that the light emitted covers, in addition to the infrared, the visible and ultraviolet region [*21,24,25*], PMTs with broad spectral response and fast risetime generally are used. The latter requirement stems from the fact that prebreakdown light emission involves short light pulses that last from 1 to 10 ns, and their detection requires that the PMT have a risetime of 1 ns or faster. Also, the windows through which these observations are made should be of quartz or similar material transparent to both visible and ultraviolet radiation to permit transmission of as broad a spectral range of radiation as possible. Even under the most favorable conditions only a fraction of the emitted light will be detected by the PMT because of some absorption and scattering by the fluid. Also, the radiation is emitted in all directions, whereas it is only detected over a solid angle of usually less than 45°.

The use of PMT provides information on the total, integrated amount of photons given off without distinction of their energy distribution. To obtain knowledge about the latter requires the use of spectrum analyzers [*24,25,32*]. These analyzers usually combine the classical prism technique of spectral analysis with electro-optical methods involving solid-state diodes with multichannel analyzers to determine the spectral distribution of the incident radiation. With the help of such devices it is possible to determine the major atomic and molecular events contributing to the light emission processes accompanying partial discharges and total breakdown.

The techniques used in the detection of acoustical signals are discussed in detail in Ref *37*. A distinction should be made here between the acoustical signals generated by pressure gradients created by the partial or total breakdown of the dielectric resulting in formation of a plasma and consequent rapid local heating of the dielectric (shock waves) and the electromagnetic radiation produced by this plasma which is detected in the audio frequency range using conventional microphones which transform this radiation into audible signals. The former, truly acoustical signals are very difficult to detect since they are transient in nature, of nanosecond duration and of low amplitude. The response time of most detectors (microphones or piezo electric transducers) is in the microsecond range. At present the use of acoustic detection devices appears most useful as a screening device to detect a certain level of repetitive partial discharges in actual service. Because of the previously mentioned limitations, these techniques are not as helpful in the mechanistic study of partial discharges as are the light detection systems mentioned earlier. The results obtained by the various methods mentioned here are summarized in the next section.

3.6 Experimental Results

As will be shown later, electrical breakdown is in many respects similar to the advent of a natural catastrophy such as a tornado that touches ground. It lasts for a very short time, and it has devastating effects. Depending on the energy stored in the system, it will produce more or less sever degradation of the insulation system. The mechanism of this process will be discussed in detail in this section.

Results of electric breakdown studies of liquids can be found in the Institute of Electrical and Electronics Engineers (IEEE) *Transactions on Electrical Insulation,* the *Journal of Physics D;* the *Annual Reports of the Conference on Electrical Insulation and Dielectric Phenomena,* CEIDP (published till 1980 by the National Research Council in Washington, DC, thereafter by IEEE); the *Proceedings of the International Conference on Conduction and Breakdown in Dielectric Liquids,* ICDL (published triannually since 1981 by IEEE, before that by the individual sponsoring units); and also the *Proceedings of the IEEE International Symposium on Electrical Insulation,* ISEI (published biannually by IEEE). For a summary of early works the book by Gallagher [*38*] is recommended as a most useful resource as is the article by Sharbaugh et al. [*13*].

In discussing experimental results it is convenient to distinguish between those that are merely concerned with the determination of the breakdown voltage of a liquid under certain experimental conditions and those that deal with an inquiry into the events leading to final breakdown, the so-called prebreakdown studies, which require much more sophisticated equipment. Accordingly these two aspects will be discussed separately in the following paragraphs.

3.6.1 Determination of Electrical Breakdown Voltages

In looking over the published data one cannot help but note the variety of experimental conditions used by the various investigators. Starting with the electrode configuration that includes both uniform and nonuniform field conditions, gaps that range from microns to centimeters, voltages that cover the gamut from fast rising square pulse, to lightning impulses, as well as stepwise or linearly rising d-c and a-c voltages. The liquids tested include a long list of ill-defined commercial fluids under the label of transformer oils and silicone oils, and the typical hydrocarbons such as n-hexane, as well as more exotic materials such as fluorinated hydrocarbons or polychlorinated aromatic hydrocarbons (PCB). In most cases the purity of the hydrocarbons and their composition are poorly described, thus making the interpretation of the experimental results and their comparison with others very difficult if not impossible. Such comparisons of published results are further complicated by the different experimental techniques used to determine the voltage at which breakdown occurs. In the majority of cases this voltage is read from the meter provided by the instrument manufacturer and not from the sample itself. When using other than square pulsed voltages, the actual voltage at the time of breakdown might differ considerably from the *nominal* crest voltage and the *charging* voltage read from the meter referred to previously. Finally, and most importantly, the breakdown voltage is not a material constant as is the dielectric constant. As was mentioned earlier, the breakdown process is a statistical phenomenon; thus, the associated breakdown voltage is a reflection of this statistical behavior. To determine a breakdown level for a given set of experimental conditions requires repetitive testing involving at least seven or more identical samples. Again, there is a lack of information in the literature as to the number of tests that were carried out to obtain the breakdown voltage reported. Attempts to provide a standard test to determine this breakdown voltage under other than a-c conditions have not been very successful.

The ASTM D 3300 procedure is one example of such an attempt of determining the breakdown voltage under impulse conditions. It uses a point-sphere electrode geometry and a 2.54 cm (1 in.) gap. The electrodes are mounted in a transparent plastic container that allows visual observation. Using a Marx type impulse generator to produce the desired 1.25×50 μs lightning pulse and a voltage divider in conjunction with a digital oscilloscope, one can determine the crest voltage, the actual breakdown voltage, and the time to breakdown. With the point (needle) cathode and using the same generator, voltage divider, and oscilloscope, the breakdown voltage can be determined with precision of about 5%. Slightly less scattering is observed when the needle is an anode [*39*]. If data from different laboratories using the same liquid are compared, considerably higher scatter is found which is due in part to differences in the calibration of the equipment used and in part to the lack of strict adherence to the experimental procedure. It should be pointed out here that another contributing factor is the availability of the proper equipment such as oscilloscopes that can translate analogue signals into numbers (digitizing) or store data or both to allow their subsequent reproduction and analysis at a much slower speed. The energy stored in the impulse generator system has also been found to influence, to some extent, the end results, particularly if proper current limiting devices are omitted. The preceding comments indicate the most obvious causes of variations in the experimental results reported in the literature. A somewhat different picture prevails in the area of prebreakdown studies where the need for complex instrumentation limits the variations in experimental procedures. This aspect is discussed in the next paragraphs.

3.6.2 Determination of the Prebreakdown Events

The early students of electrical breakdown were hindered in their efforts to unravel the events leading to the actual breakdown because their experimental means were limited [*4–7*]. They used narrow gaps ranging from 100 μm to 2 or 3 mm because the high voltage output of available power supplies was insufficient to cause breakdown of larger gaps. Also the temporal resolution of the electronic equipment in usage then was limited to microsecond whereas today nanosecond resolution is considered a standard requirement. In spite of these limitations, Lewis [*4*] and Watson and Sharbaugh [*7*] made correct deductions as to the events taking place prior to breakdown. Lewis suggested that electrons, injected from the cathode into the liquid, exchanged energy with the molecules making up the liquid. Watson and Sharbaugh suggested that the flow of electrons from the metal into the liquid caused local heating which lead to the vaporization of the liquid and the formation of a vapor bubble. As will be shown in the next paragraphs both concepts have been found to be indeed operative [*40*].

3.6.2.1 The Streamer, Bush, and Tree Concepts—In the last twenty years techniques have been perfected for making photographic observations of these prebreakdown events. Early results suggested that structures resembling bushes and trees formed in front of the needle electrode [*11–16*]. With the help of high magnification techniques, FitzPatrick et al. [*41*] established that pencil shaped streamers emanate from needle cathodes to form negative streamers that developed into bush-like structures as shown in Fig. 3.13. These bushes grow at rates ranging from 10 to 100 m/s and eventually stop growing until an as yet unexplained event creates an electrodynamic instability that leads to the development of a secondary, negative streamer [*42*]. These secondary streamers grow with speeds exceeding 10 km/s until they nearly reach the anode. Then, a very fast event takes place that establishes a short circuit. Whereas the slow growing primary streamer or bush has usually a highly branched structure, the secondary, fast growing one has little if any branches and has in general a smaller cross section. Figure 3.14 illustrates the growth of primary and secondary streamers in various carefully purified hydrocarbons. These negative streamers are quite different from those formed when the needle is an anode. Positive streamers appear to be formed by a different mechanism than are the negative ones. They have a smaller cross section, are less branched, and appear to form more tree-like structures. The photographs shown in Fig. 3.15 illustrate such structures. They suggest that there is a single channel connecting the delta-shaped structure to the anode. Again one can observe, under favorable conditions such as large gaps, the occurrence of a secondary fast streamer that connects the primary one to the cathode. Examples of such a fast event are shown in Fig. 3.16. The growth rate of the primary positive streamer was found to be of the order of 1000 m/s which is at least an order of magnitude faster than that of the negative one. Estimates of the fast secondary streamer suggest that it moves with velocities of the order of 10 to 100 km/s [*30*]. These observations were made on purified liquids. It became of interest to establish the role of various impurities on these growth rates. The results of these studies are summarized in the next paragraph.

3.6.2.2 The Role of Impurities—Among various impurities, oxygen and water appear to be the most important ones. As was shown in Section 3.2, these two exert measurable influences on a liquid's conductivity and breakdown voltage. Other chemical compounds can produce similar or even more significant effects, such as the antistatic additive ASA 3 also mentioned in Section 3.2, together with dimethylaniline, a low ionization potential compound that can act as a source of both positive and negative ions. FitzPatrick et al. [*43*] and Hebner et al. [*44*] showed that when the needle was a cathode, the presence of impurities caused a lowering of the breakdown voltage. These impurities appeared also to speed up the onset of the electrodynamic instability so that the primary bush development was stopped earlier than in the pure case. This is illustrated in Fig. 3.17. In the case of the needle being an anode the breakdown voltage was not affected by these impurities when using gaps of less than 2 cm [*43*]. At wider gaps of 2.54 cm it became clear that the presence of dimethylaniline caused an increase in the breakdown voltage. Photographs shown in Fig. 3.18 suggest that this increase was prompted by the greater availability of negative charge carriers that modified the overall shape of the tree from one with sharp edges to one resembling an umbrella so that the electric field resembled more like those formed by a plane and a hemisphere than like a point-plane configuration. The former is known to require a higher breakdown voltage than the latter. These results indicate that the role of impurities need not necessarily be a detrimental one as

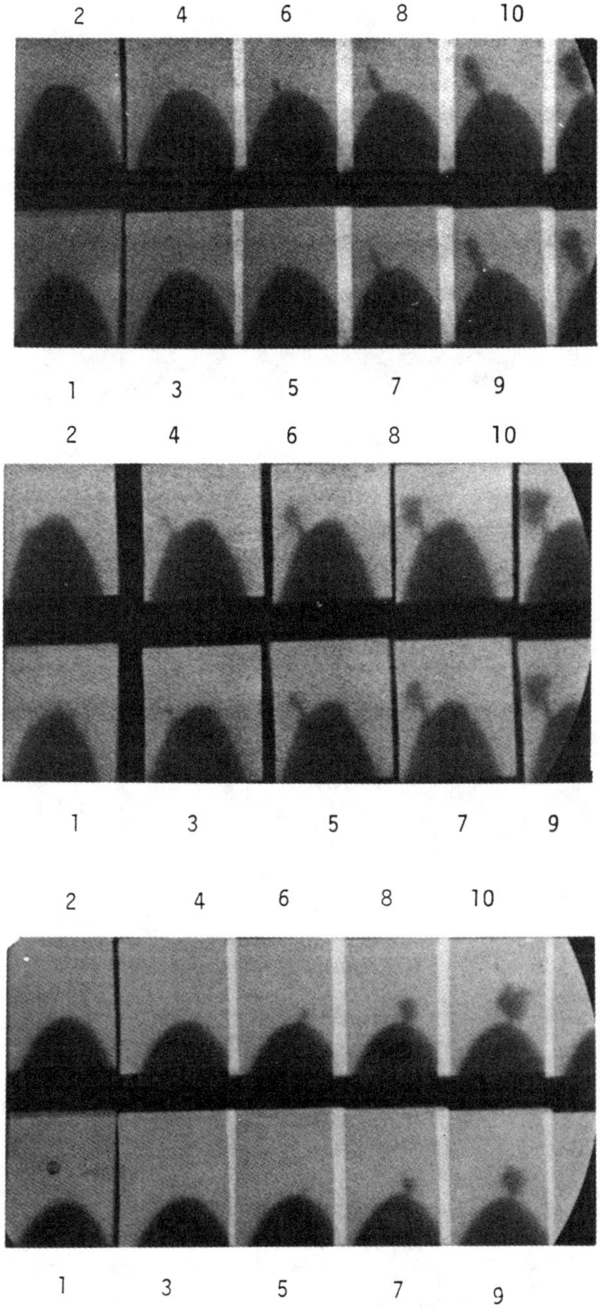

FIG. 3.13—*Negative streamer initiation (after FitzPatrick et al.* [41]*). Trapezoidal pulse, magnification X100, 50 ns between frames. From top to bottom: 1 cm gap, 41 kV crest voltage, toluene; 2 cm gap, 47 kV crest isooctane; and 1 cm gap, 53 kV crest, Marcol 70 (an aromatic-free white oil of viscosity similar to transformer oil).*

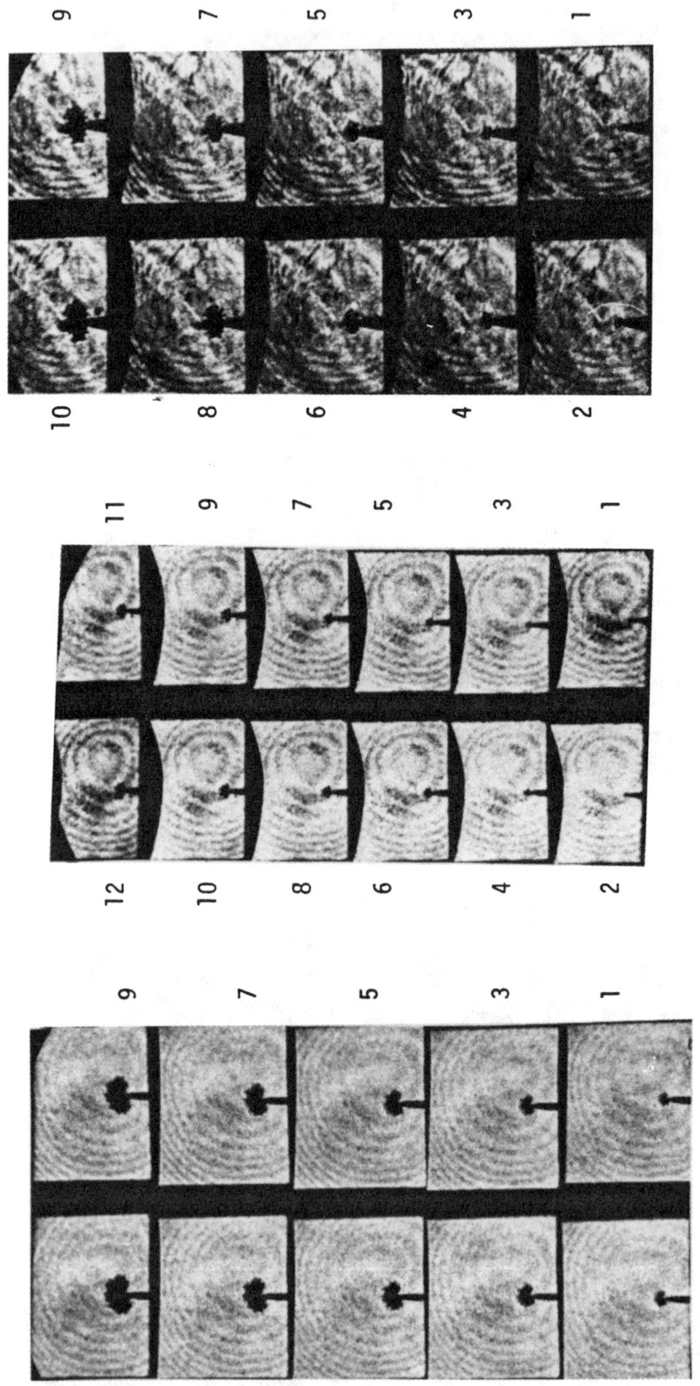

FIG. 3.14—*Streamer growth in pure hydrocarbons (after Hebner et al. [30]). Trapezoidal pulse, 500 ns between frames. From left to right: 6.4 mm gap, 156 kV crest, n-hexane; 5 mm gap, 120 kV crest, cyclohexane; and 3 mm gap, 168 kV crest, toluene.*

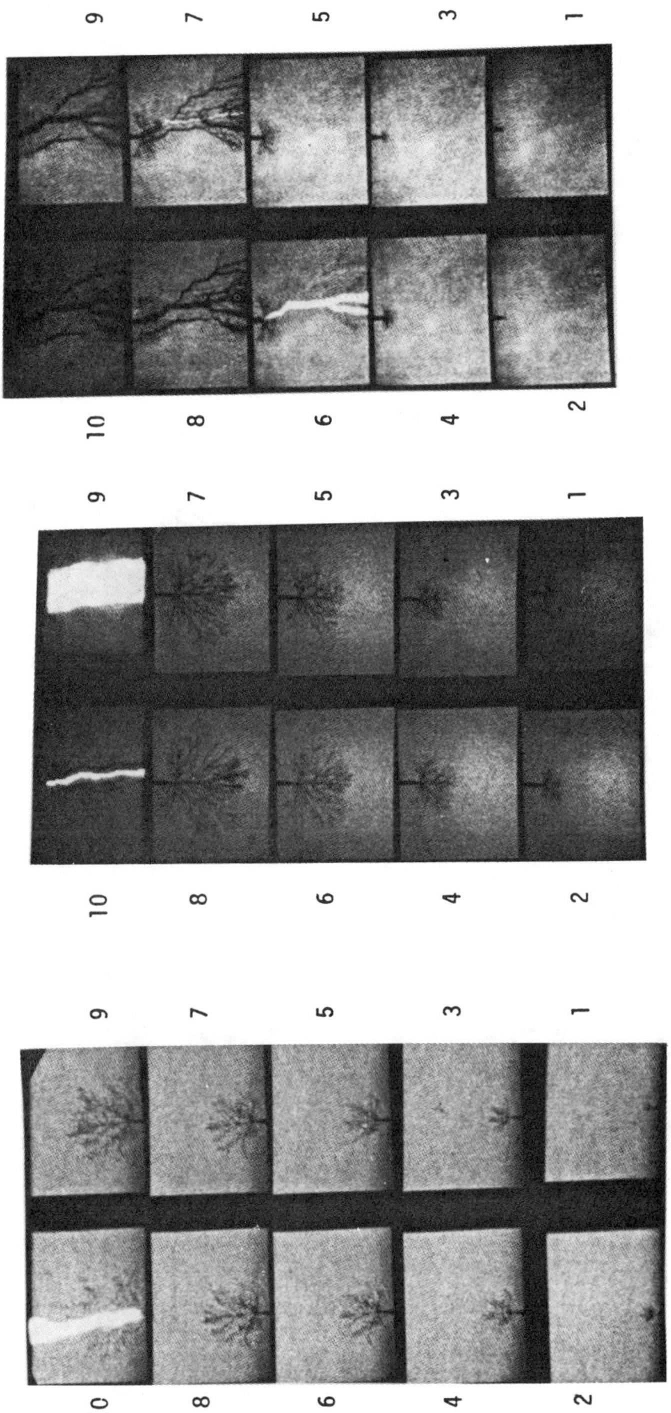

FIG. 3.15—*Positive streamer growth in liquid hydrocarbons (after Hebner et al. [30]). Needle anode, trapezoidal pulse, 96 kV crest, 1.27 cm gap. From left to right: n-hexane, 200 ns between frames: Marcol 70, 500 ns between frames; and toluene, 500 ns between frames.*

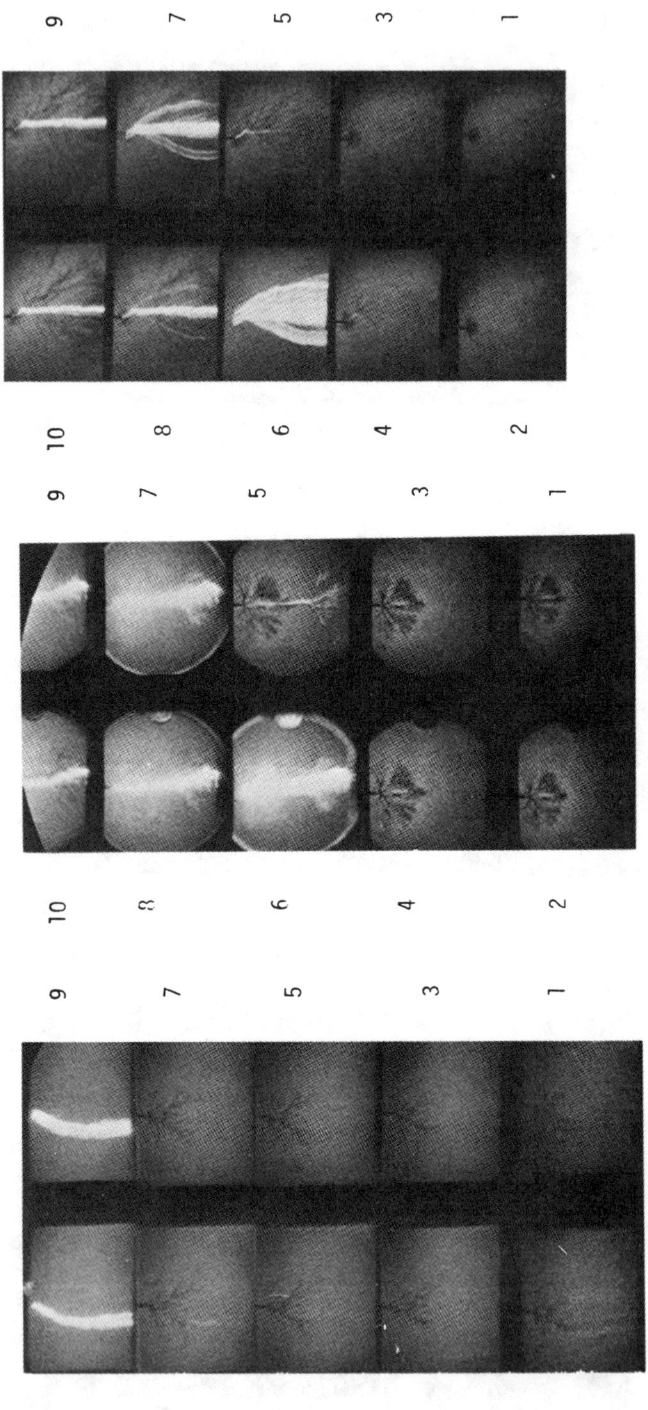

FIG. 3.16—*Secondary streamer development in liquid hydrocarbons (after Hebner et al.* [30]*). Needle anode, trapezoidal pulse, 2.5 cm gap. From left to right: pure n-hexane, 120 kV crest, 50 ns between frames; Marcol 70, 240 kV crest, 200 ns between frame; and toluene, 216 kV crest, 50 ns between frames.*

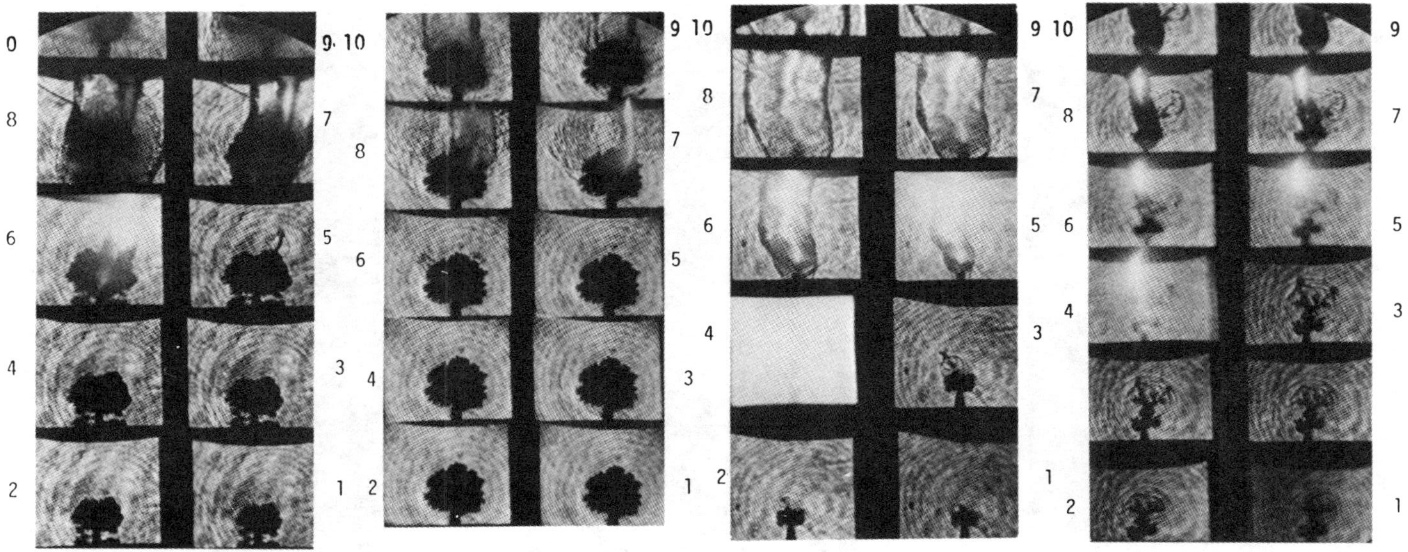

FIG. 3.17—*Effect of contaminant on secondary streamer development (after Hebner et al.* [30]*). Needle cathode, trapezoidal pulse, 3 mm gap. From left to right: pure n-hexane, 192 kV crest, 500 ns between frames; n-hexane + 0.15 M DMA, 180 kV crest, 500 ns between frames; chemical grade n-hexane saturated with O_2 and H_2O, 148 kV crest, 200 ns between frames; and toluene + 20 ppm ASA 3, 108 kV crest, 50 ns between frames.*

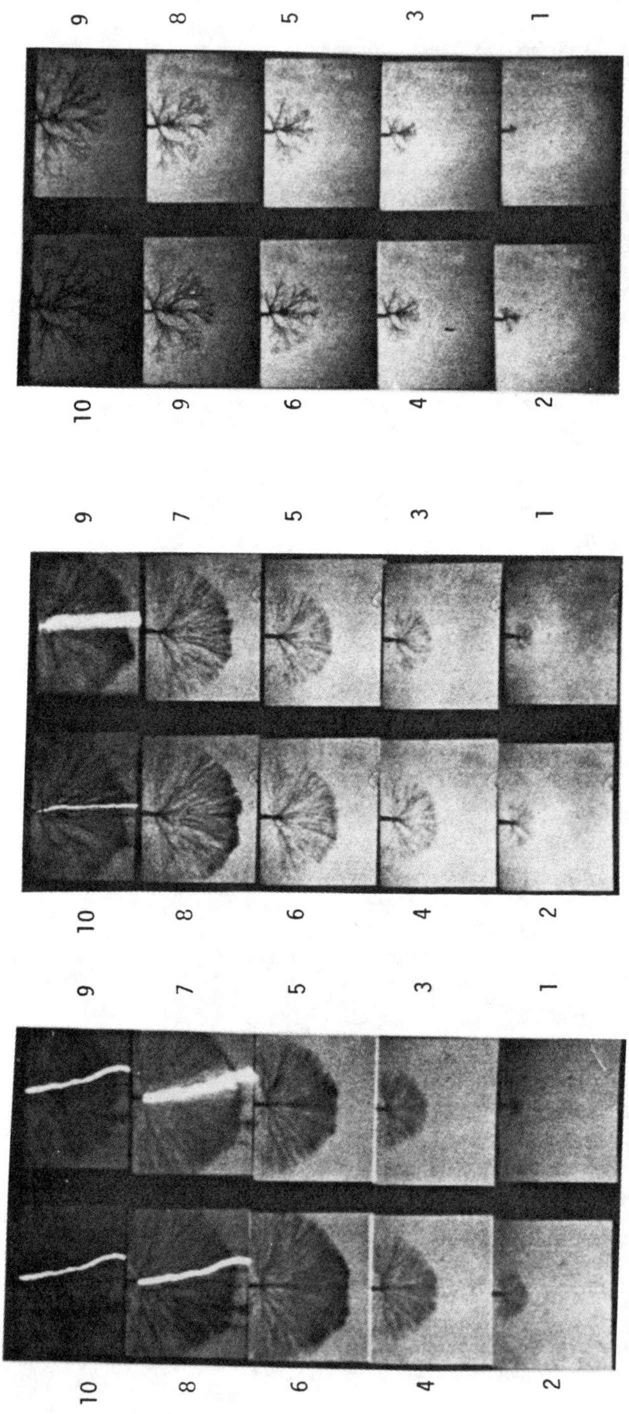

FIG. 3.18—Effect of 0.2 M dimethyl aniline (DMA) on positive streamer structure (after Hebner et al. [30]). Needle anode, trapezoidal pulse, 1.27 cm gap, 500 ns between frames. From left to right: n-hexane, 120 kV crest; Marcol 70, 96 kV crest; and toluene, 96 kV crest.

suggested earlier by Devins et al. [*45*]. They also provide further evidence that the mechanisms involved in the growth of positive streamers are most likely quite different from those causing negative streamers to grow. Possible mechanisms explaining these observations will be discussed in Section 3.7. Before this can be done, it is necessary to inquire into what effect the molecular structure of the fluid has on the preceding observations.

3.6.2.3 The Role of the Molecular Composition of Fluids—In the preceding paragraph comparisons were presented on fluids of different degrees of contamination. They included saturated, straight chain hydrocarbons such as *n*-hexane and cyclic, and unsaturated ones such as toluene. Other examples studied include white oils, that is, petroleum derived oils containing only saturated linear and cyclic compounds as well as combinations of these two [*46*], transformer oils that contain in addition to the components of white oils, aromatic molecules [*47*], silicone oils (polydimethyl siloxane) and perfluorinated hydrocarbons. All these types of fluids showed the same basic behavior, and the streamers, bushes, and trees growing in them under the influence of externally applied electrical voltage pulses had similar structures. Of particular interest was the role of viscosity in these examples which ranged from a low of about 1 cSt to as high as 10 000 cSt for a high molecular weight silicone fluid. The growth rate of the positive streamers and of the breakdown voltage was not affected significantly by this variation in viscosity, while the negative streamers showed only a slight effect for the samples having the highest viscosity of 10 000 cSt [*48*] as shown in Table 3.3. The time to breakdown for the 10 000 cSt oil at −168 kV was of the same order as that for 1000 cSt oil at −144 kV. It is interesting to point out that in the case of the 10 000 cSt sample no bush was formed and the streamer showed little if any branching. Also, higher breakdown voltages were required for this fluid than for the lower viscosity ones. This is illustrated in Fig. 6.19. Similar observations were made when simple hydrocarbons were subjected to significant hydrostatic pressures as shown in the next paragraph.

3.6.2.4 The Effect of Pressure—The photographs of the prebreakdown events taken under atmospheric conditions suggested that these streamers represented low density regions, the perimeter of which were populated by thermalized electrons [*41*]. To test this idea of the existence of a two phase, low density region, Hebner et al. [*49*] undertook a study of the effect of pressure on the streamer growth. From their observations they concluded that up to a pressure of about 1.0 MPa, there existed indeed a low density region within the streamer which most likely contained vaporized fluid. Above that pressure the streamer structure assumed a shape similar to that observed

TABLE 3.3—*Effect of viscosity on breakdown voltage and time to breakdown in polydimethyl siloxane*[a] *(25°C, 3 mm gap between needle point and plane).*

Applied Voltage, kV	Viscosity, cSt 10 000 V_b[b]	t_b[c]	1000 V_b	t_b	100 V_b	t_b	10 V_b	t_b
				NEEDLE CATHODE				
120	110 ± 2	6 ± 0.5	105 ± 5	4.0 ± 0.3	115 ± 3	4.6 ± 0.3	97 ± 3	4 ± 0.2
144	132 ± 5	4.5 ± 0.3	109 ± 4	3.3 ± 0.2	120 ± 7	3.9 ± 0.4	105 ± 4	3 ± 0.2
168	139 ± 4	3.4 ± 0.2	117 ± 3	2.8 ± 0.1	127 ± 4	3.3 ± 2	111 ± 2	3 ± 0.2
				NEEDLE ANODE				
36	33 ± 1	4.4 ± 0.4	31 ± 2	3.9 ± 0.3	30 ± 2	3.9 ± 0.2	32 ± 3	4 ± 0.5
60	39 ± 0.5	2.6 ± 0.1	40 ± 1	2.7 ± 0.1	40 ± 1	2.7 ± 0.1	39 ± 1	2.7 ± 0.1

[a]Data taken from Ref *48*.
[b]V_b = Breakdown voltage in kV.
[c]t_b = time to breakdown in μs.

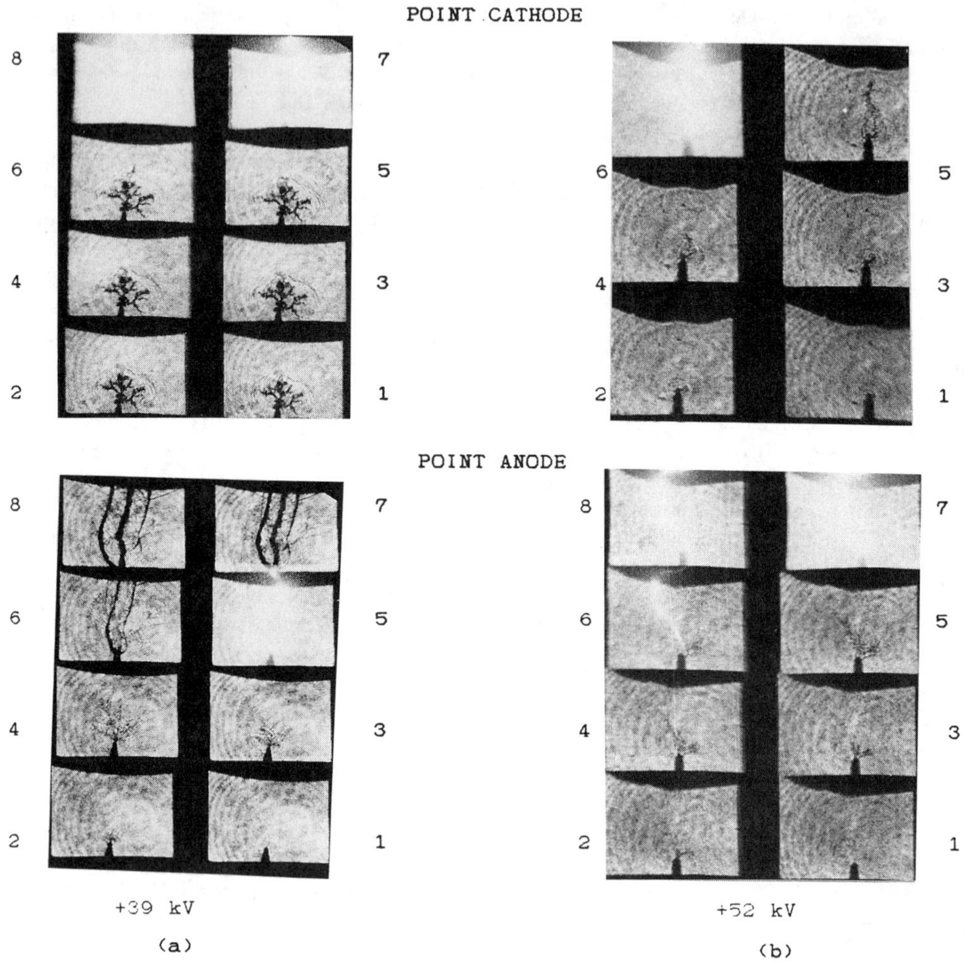

FIG. 3.19—*Effect of viscosity on streamer shape and growth in dimethyl siloxane, DC 200, (after Kelley et al.* [48]*). Trapezoidal pulse, 5 mm gap,* (a) *1000 cSt, crest voltages −104 kV and +39 kV, respectively, 500 ns between frames, and* (b) *10 000 cSt, crest voltages −169 kV and +52 kV, respectively, 200 ns between frames.*

for positive streamers, and considerably higher external voltages were required to cause a breakdown. At these higher voltages the negative streamers propagated at speeds similar to those observed for the fast event under both polarities. The shape and propagation rates of positive streamers appeared not affected adversely by pressure. Figures 3.20 and 3.21 illustrate these differences. Apparently in water, which is occasionally used as dielectric liquid, the negative streamer pattern does not change with pressure, exhibiting the low density region even at pressures as high as 10 MPa [*49*].

The results of this pressure study appear to confirm the existence, at atmospheric pressures, of a low density region within the structures observed by shadowgraph and schlieren techniques when the needle point is a cathode. It became of interest to establish the growth mechanism of these structures. For this purpose studies were undertaken to measure their growth simultaneously, as

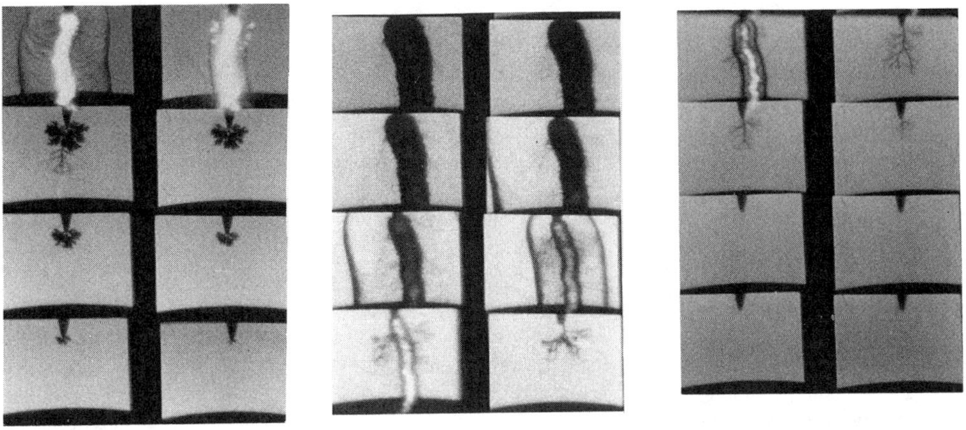

FIG. 3.20—*Effect of pressure on negative streamer shape and growth (after Hebner et al.* [49]*). Needle cathode, trapezoidal pulse, 2.3 mm gap, n-hexane. From left to right: 1 atm −65 kV crest, 500 ns between frames; 12.5 atm −78 kV crest, 500 ns between frames; and 24 atm −91 kV crest, 200 ns between frames.*

well as the current and light pulses that had been observed in the same time frame. The results of these investigations are discussed in the next paragraph.

3.6.2.5 Effect of Partial Discharges on Growth of Streamers—Although many investigators had studied the growth rate of streamers in dielectric fluids, no detailed study of the growth mechanism had been undertaken. It usually was assumed that partial discharges were involved in that mechanism. There were, however, no data available to substantiate this belief. If this assumption were correct, then the growth could well occur in steps which coincided with the occurrence of a partial discharge. The photographic records available consisted mainly of sequences obtained at intervals

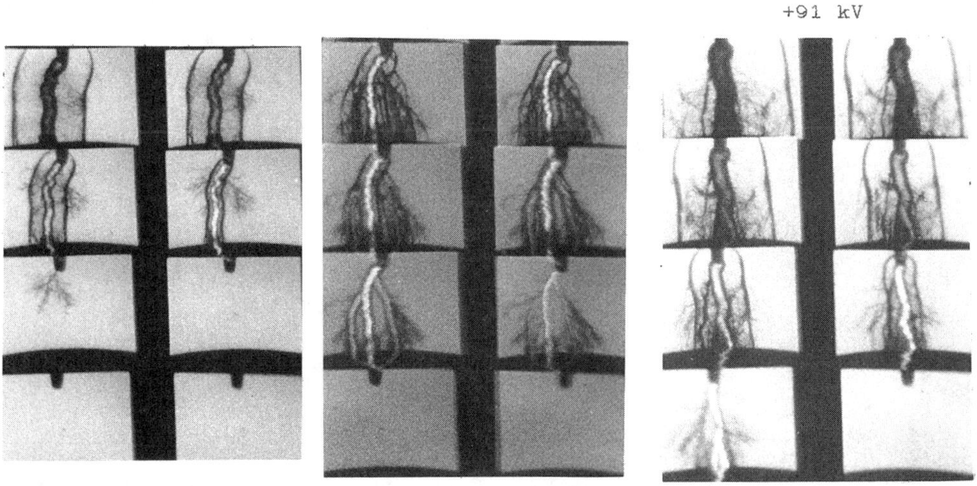

FIG. 3.21—*Effect of pressure on positive streamer shape and growth (after Hebner et al.* [49]*). Needle anode, trapezoidal pulse, 2.3 mm gap, n-hexane. From left to right: 1 atm +52 kV crest, 200 ns between frames; 6.25 atm, +78 kV crest, 50 ns between frames; and 12.5 atm +91 kV crest, 200 ns between frames.*

of 50 ns or longer. These sequences did not permit one to make a decision regarding the growth of the streamers. To provide a more detailed understanding of this growth process, Kelley et al. carried out simultaneous measurements of light emission, current pulses, and growth of prebreakdown streamers in *n*-hexane [*50*]. Similar but less comprehensive studies were also conducted by Winterberg et al. [*51*]. The results of these studies are illustrated in Figs. 3.22 and 3.23 taken from Ref *50*. From these data Kelley et al. concluded that the streamer growth is continuous and does not depend on the energy released from partial discharges. The light and current pulses occur simultaneously and thus appear to be related to each other. The authors also noted that in the absence of streamers no light and current pulses were observed. Furthermore the light and current pulses appeared randomly and more frequently during growth periods of negative streamers than during stagnation which usually occurred prior to the growth of the secondary streamer. These observations lead the authors to conclude that the streamer growth was not related to the occurrence of partial discharges but had to be caused by a process that allowed for continuous growth. Such a process, for example, could involve the ionization of the liquid at the tip of the streamer brought about by the local electric field.

Calculations performed by Girdinio et al. [*52*] and of FitzPatrick et al. [*53*] of the electric field in front of a streamer have indeed suggested that such ionization might be feasible since, in front of a streamer having a sharp point, the local field may approach values corresponding to the intrinsic breakdown field of 10^{10} V/m. The growth of the streamer would slow down or stop completely when either the tip of the streamer would broaden or several tips would be sufficiently close to interact with each other, thus reducing the local fields. The latter condition seems to occur as the numerous streamers mature into a tree- or bush-like structure. Hence, the growth slows down and stops until a disturbance causes the development of a secondary streamer, the tip of which is sharp enough to cause field enhanced ionization. As has been noted earlier, these secondary streamers

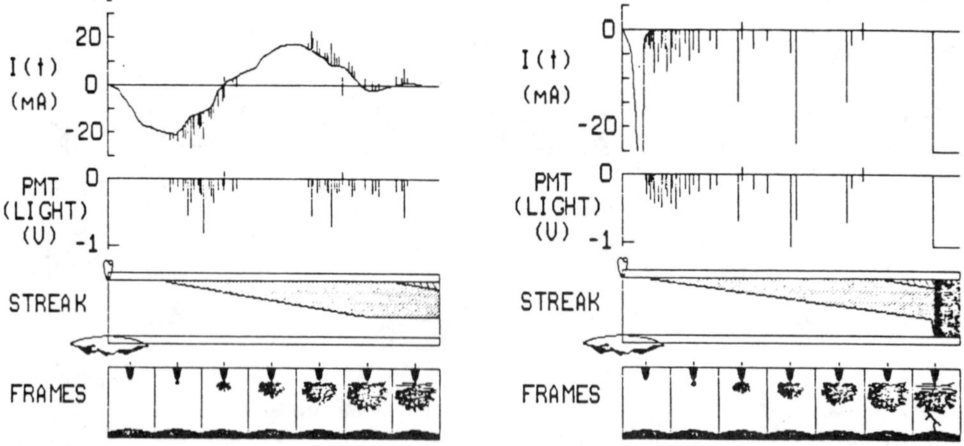

FIG. 3.22—*Simultaneous measurement of negative streamer growth, electrical and light pulses in n-hexane (after Kelley et al.* [50]*).* (a) *Gaussian shaped pulse and* (b) *square voltage pulse.*

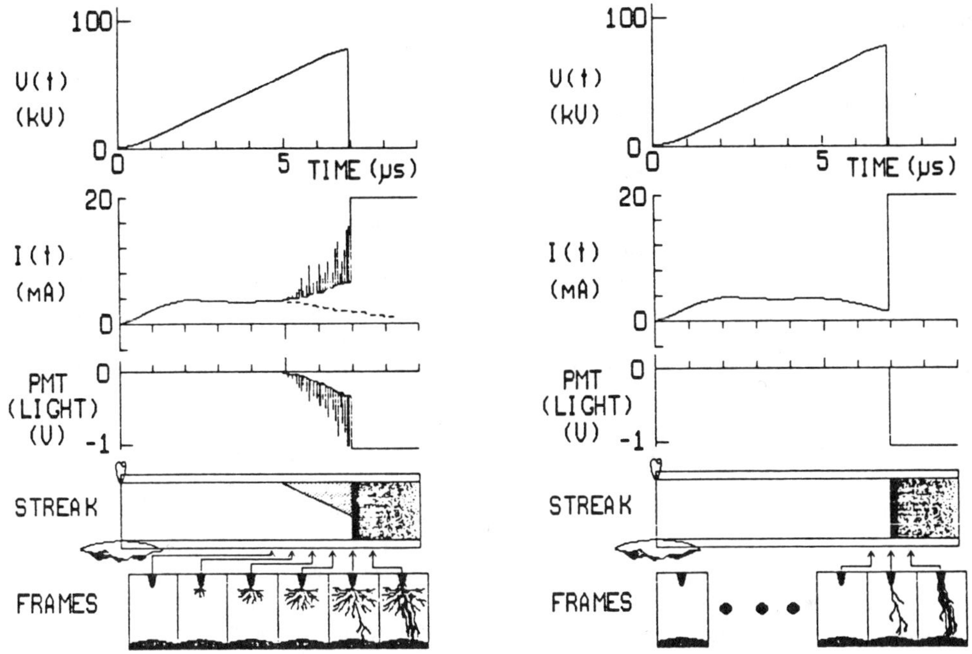

FIG. 3.23—*Simultaneous measurement of positive streamer growth, electrical and light pulses in n-hexane (after Kelley et al.* [50]*).* (a) *Streamer development before breakdown and* (b) *no streamer before breakdown.*

grow much faster than the primary ones. This is probably due to the fact that the local field conditions are more favorable for the growth of the secondary than they are for the primary. The bush- or tree-like structure represents an extension of the needle electrode into the liquid [*54,55*] so that the average electric field is considerably higher at the start of the secondary streamer than it is at the start of the primary streamer. These findings seem to explain well the observed streamer growth pattern, but they do not explain either the role of partial discharges and the associated light emission. Obviously more detailed studies are required to unravel these events.

Based on the preceding results and observations it has become possible to propose some qualitative models for the breakdown process under nonuniform field conditions. While there is good agreement among the various groups regarding the negative streamer initiation and growth, there exists still some doubt on the model describing positive streamer growth. These models will be discussed in the next section.

3.7 Breakdown Models

As mentioned in Section 3.3, there exist several types of electrical stresses and electrode geometries that find usage in the determination of the breakdown voltage of an insulating system (the word system refers to the fluid and both electrodes). Thus, as mentioned in Section 3.2, one does distinguish customarily between breakdown of systems employing parallel plates (uniform field) and those using point-plane, point-point, rod-plane, sphere-plane, or similar asymmetric arrangements (nonuniform fields). In addition each of these systems can make use of a variety of electrical stresses ranging from very fast rising square pulses to constant voltages under d-c or a-c conditions. As a consequence one cannot assign a value to the breakdown voltage of a fluid unless one identifies the system at the same time including the electrode separation, the type of applied electric stress, and temperature. Given all these variations one may raise the question regarding the effect of the system and the type of electrical stress on the breakdown mechanism. It has been observed that, in all cases, breakdown is accompanied by a spark resulting from the formation of a short circuit across the insulating fluid and the resulting heating effects, thermal degradation of the liquid, and associated acoustical effects. Preceding this breakdown one finds that in almost all cases one observes partial discharges or coronas, and, depending on the temporal nature of the voltage stress and its amplitude, one can measure these partial discharges more or less readily.

From what is known today about the breakdown processes occurring under those varied conditions, one concludes that the only major difference among them is the time scale of the appearance of these events. In all cases the development of a conducting channel has to take place. The speed with which this stage is reached depends, in principle, on the time required for primary and secondary streamers to cross the gap between the electrodes. This growth will in turn depend on the local field conditions and the latter depend among other factors on the applied potential difference, its amplitude as well as its time dependence. While it is easier to study and understand the development of breakdown under d-c conditions, one has learned from comparison of the polarity effects under nonuniform field conditions, that breakdown will occur most of the time on the positive cycle. It should be recalled here that under point-plane or similar electrode geometries the breakdown voltage for identical gaps is generally lower when the needle is an anode than when it is cathode. This difference is maintained even when using slowly rising or constant voltages. Under uniform field conditions these differences do not exist, and it has been suggested that in this case the surface conditions at the electrode/dielectric interface become significant in controlling breakdown [*35*]. In view of these considerations it appears feasible to describe all the various breakdowns in terms of a single model, the time scale of which can be modified to meet the requirements of the particular case. In the next paragraph this model will be developed, and subsequently a couple of cases will be discussed to show how these modifications make the model adaptable to those circumstances.

3.7.1 The Basic D-C Breakdown Model

To start with it will be assumed that the dielectric liquid is free of any charges except those that are unavoidably transferred from the metal electrodes to the dielectric in the absence of any externally applied potential. For the purpose of the model it is not essential, in the beginning, to stipulate a certain electrode geometry since it has been shown by Latham [*56*] that local surface conditions rather than overall electrode geometry control electron emission from the metal into the dielectric. Also, later on the movement of the charge carriers is controlled by the local field rather than the externally applied one. In this respect the model is a general one although as yet only a qualitative one. Polarity effects come into play only when the streamer development and their propagation are considered. At that point it is necessary to discuss the development of negative and positive streamers separately. The model is divided into four steps each dealing with a given phase of the overall breakdown process. Step 1 describes the events occurring at the metal dielectric interface and gives an insight as to how electrons get from one medium into the other. In Step 2 the conditions for streamer development are spelled out, while in Step 3 primary and secondary streamer growth are discussed. Step 4 deals with the very fast final step that leads to the formation of the conducting channel that bridges the electrode gap and leads to the collapse of the applied potential difference, the actual breakdown step.

3.7.1.1 The Charge Injection Process—The metallic cathode acts as a source of electrons that need to be transported across the metal/dielectric boundary through the influence of an externally applied potential difference. This difference is produced by a power supply that is capable of providing energetic electrons. At low applied fields, the electrons coming from the conduction band of the metal have to reach the surface and than overcome the metal's work function, which describes the energy required to remove an electron from the metal's surface into vacuum. This energy barrier usually is defined to be the energy difference between the Fermi level of the metal and its ionization potential. In the presence of an external field the energy barrier will be assuming a shape of a potential energy barrier which becomes narrower with increasing magnitude of the field. At the same time the Fermi level is raised and the distance the electrons have to travel across the energy barrier decreases..Figure 3.24 illustrates this situation. This means that the probability of electrons tunneling through the energy barrier increases proportionally so that, with increasing applied potentials, the flow of electrons into the dielectric will increase. Since this increased flux of electrons involves electrons having higher energies than those injected at lower potential differences, they will be able to interact more vehemently with molecules of the dielectric. According to Lantham [*56*] the most favorable sites for electron injection from a metal surface into a dielectric are grain boundaries, dislocations or asperities of molecular dimension, and dust particles having diameters of the order 5 to 10 nm. These sites are characterized by a much lower than average work function (<1 eV) and a local field which is nonuniform and which can range from 100 to 300 fold the average applied field. The cross section area of such injection sites is of the order of 10 to 100

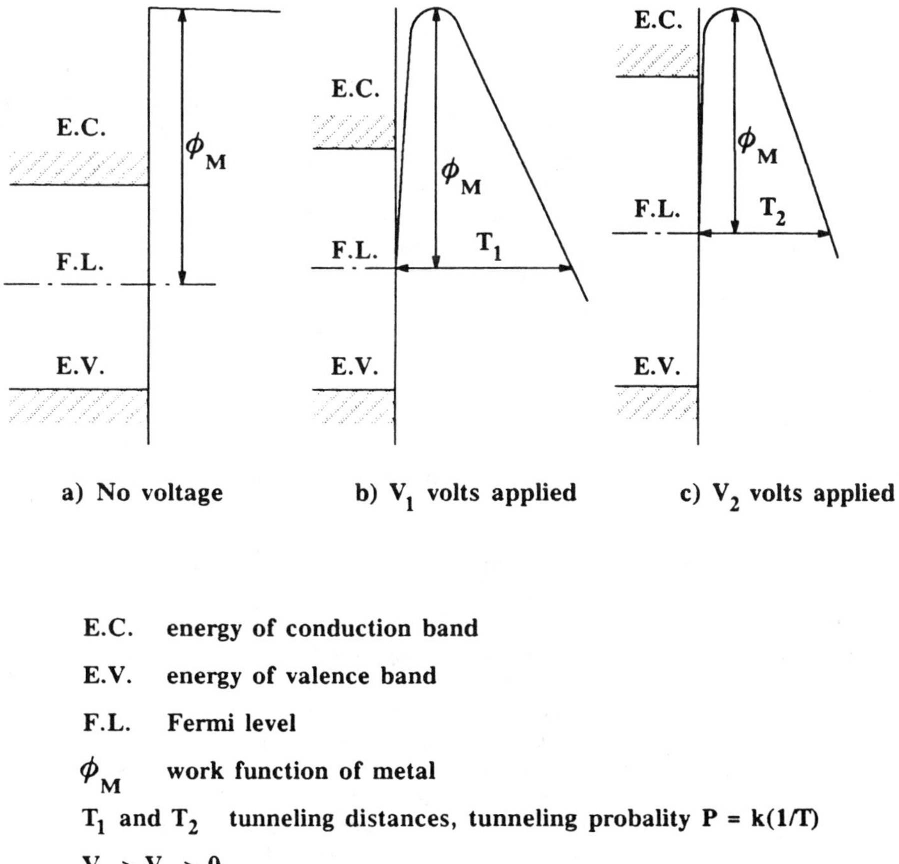

FIG. 3.24—*Potential barrier at metal-dielectric interface: the effect of increasing applied potential.*

nm^2, and there are as few as 5 to 200 such sites per square centimeter of metal surface. Lewis [*57*] suggested further that the presence of oxide films and polymerized organic molecules might further facilitate charge injection.

The preceding picture predicts the existence of a critical potential difference below which the electrons have not enough energy to initiate a streamer. These are the "conduction" electrons responsible for the frequently observed leakage current [*51*], which ranges from as low as fractions of a picoampere to nanoamperes. Above this critical voltage the electrons contribute not only to the conduction but they also can cause streamer development as described in the next paragraph.

3.7.1.2 Streamer Development—Once the electrons have left the metal and entered the dielectric they move under the influence of the local field. In the course of this movement they collide with molecules making up the dielectric fluid. These collisions can involve simple energy transfer leading to an increase in vibrational energy of the molecules and hence to local expansion. If the electrons have energies in excess of 3 to 4 eV they can cause breakage of carbon-hydrogen or carbon-carbon bonds [*58*] and thereby release energy to the environment causing further local expansion. A third possibility was suggested by Kao [*58*] which involves energy exchange between electrons leading to the formation of a "hot" electron capable of causing impact ionization. This Auger-type process and the other two processes mentioned earlier all cause heating of the liquid near the injection site, thus helping in the formation of the low density region in that area. It may be recalled that Watson and Sharbaugh [*7*] had postulated local heating and bubble formation as the initial steps in the formation of a streamer. The explanation offered here is thus an elaboration of ideas advanced nearly three decades earlier. At the end of the energy exchange the electrons are thermalized, and their average kinetic energy is equal to that of the surrounding molecules. At this stage they interact with their next neighbors by polarizing them and forming clusters that drift under the influence of the prevailing local field for short periods, then open up and reform again leading to the classical picture of the hopping process in amorphous materials. These clusters can also be viewed as electron traps, the depth of which depends on the degree of molecular interaction as determined by the structure of the molecules. For example highly symmetric spherical molecules such as methane (CH_4), carbon tetrachloride (CCl_4), or neopentane [$C(CH_3)_4$] form very shallow traps because they do not interact strongly with each other as discussed in Section 3.2. A molecule of low symmetry such as *n*-hexane can interact with its neighbors, in addition to the normal van der Waal forces, by mechanical entanglement and thus form a more stable, deep trap. Consequently, it can be stated that the charge carriers, the thermalized electrons, will move more freely in the case of symmetric molecules than in that of molecules capable of undergoing mechanical entanglement.

At this stage of the discussion, it is necessary to point out that the above ideas help greatly to understand the formation of a low density region when electrode injection takes place from a point cathode. These ideas do not explain what one observes under opposite polarity conditions. As was mentioned earlier, the structure of positive streamers is considerably different from that of negative ones. The preceding mechanism no longer seems to apply, and a different concept needs to be introduced. Before doing so it should be recalled that in the case of negative streamers electron injection takes place in the high field region. By analogy such injection would occur from the plane cathode in the low field region. This implies that the electrons have to move from the low field to the high field region, the exact opposite of what happens in the negative cathode case. One can recognize easily that the movement from low to high field is the more natural one. The positive anode thus seems to act as a focusing device for all electrons present in the fluid. As charge carriers congregate near the anode they modify the local field by increasing it considerably over the average applied one. This local field increases as the charges approach the surface of the anode, and suddenly the field equals or exceeds the intrinsic breakdown field. The result is the ionization of the liquid volume element contained between the charge cloud and the anode surface. The ionization leads to the formation of a conducting channel which can be photographed after it has been formed. This process is so fast that it cannot be analyzed in detail with presently available electro-optical devices. This inability to study it in detail has to be blamed for the existing disagreement as to its exact nature. The steps just proposed are supported to some extent by photographic evidence such as shown in Fig. 3.25 in which the approach of a dark spot, identified as a charge cloud, is shown to lead to a streamer formation.

The proposed process thus predicts that electrons are removed from the liquid through a discharge channel that rarely is branched and apparently grows from the anode into the liquid, while in actuality it forms in a very short interval of probably less than nanoseconds between the focused charges and the anode. Once the first channel has formed, it acts, like the negative streamer, as an extension of the electrode since its walls are formed by thermalized electrons or fragments of the ionization process or both. Once formed the channel's tip will cause a repetition of the process and thus the positive streamer pattern develops. Since the mechanism does not change, the rate of growth is uniform until the positive streamer approaches the vicinity of the cathode, and the fast event sets in which is probably very similar to the one mentioned in the case of the negative streamer. In this respect the model correctly predicts the uniform growth of the streamer and its dependence on the availability of electrons in the liquid. The data obtained by Devins et al. [45] and

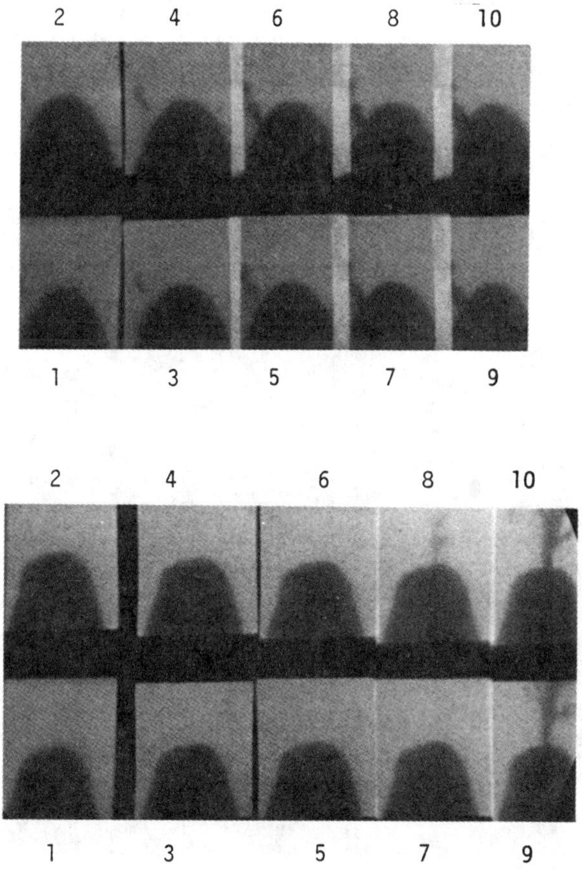

FIG. 3.25—*Positive streamer initiation (after FitzPatrick et al.* [41]*). Needle anode, trapezoidal pulse, 50 ns between frames. From top to bottom: toluene, 1 cm gap, +78 kV crest voltage; and isooctane, 5 mm gap, +53 kV crest.*

Hebner et al. [*30*] have been explained by Forster [*40*] in terms of a model that is discussed in detail in the next paragraph.

3.7.1.3 Primary and Secondary Streamer Growth—As mentioned earlier, when considering streamer development it is necessary to include the effect of electrode geometry. With the point a cathode, electrons are injected from the negative needle into the liquid while the opposite appears to happen when the needle point is an anode. Accordingly the two cases will be discussed separately.

3.7.1.3.1 Negative streamers—In the case of negative streamers it is agreed generally that the growth proceeds in at least four steps. The first one is the appearance of a pencil shaped streamer that grows at an average speed of about 100 m/s (see Fig. 3.13). Within 100 to 400 ns it develops several branches that create the appearance of a tree- or bush-like structure. This tree growth, Step 2, is somewhat slower at about 10 m/s both in the direction of the anode and perpendicular to that direction. Then, depending on the purity of the liquid and the local field conditions, growth may stop temporarily. Step 3 then starts with a pertubation of some kind which causes the onset of an electrodynamic instability [*42*] that leads to the development of the secondary streamer that grows at rates ranging from 1 to 10 km/s. Finally the fast event, Step 4, involves growth rates in excess of 100 km/s and is the one least understood. It should be pointed out that this sequence of steps is dependent on the continuous supply of electrons from the injection site and the local field conditions. If the supply of electrons decreases or ceases completely or both, the trees or bushes will gradually dissolve into a flurry of bubbles which, in turn, will also disappear. This condition occurs also whenever several sites are injecting charges simultaneously. Eventually only one site will cause the streamer to grow across the gap and produce breakdown as shown in Fig. 3.26. The other injection sites will stop injecting and the streamers associated with them will disintegrate.

3.7.1.3.2 Positive streamers—The progress of positive streamers is in some ways simpler than that of the negative one. It grows at more uniform speed, and it grows in a simpler, less branched pattern as shown in Fig. 3.15. Its overall shape is more like that of a river delta with all branches leading into the principal one that makes contact with the anode. One observes very rarely two such channels reaching the electrode. With positive streamers one can observe also their cessation of

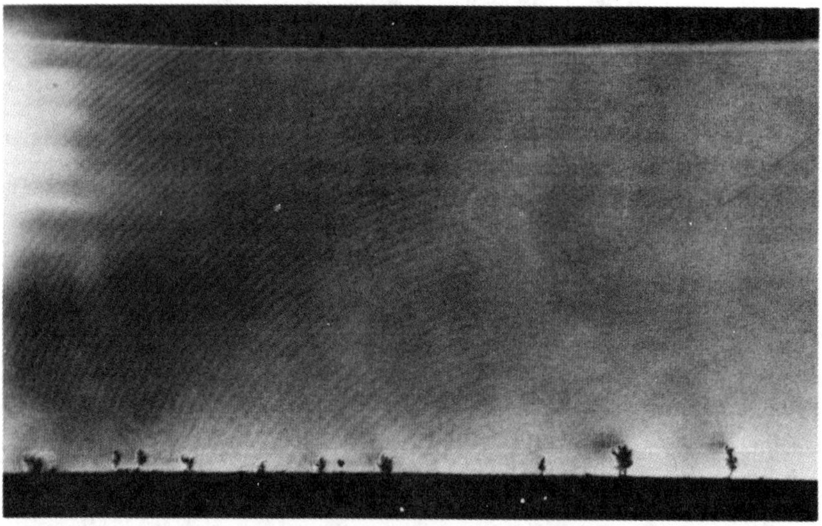

FIG. 3.26—*Laser-schlieren photograph of streamer development under uniform field conditions in toluene (after Forster* [35]*). 1.5 × 50 μs lightning pulse, 6 mm gap, −260 kV crest, using a ruby laser pulse (15 ns width) for illumination. Laser fired 2 μs before breakdown.*

growth, but, unlike the negative streamers, they seem to disappear without or with only a few bubbles left behind. Their growth thus can be described as occurring in two steps. Step 1 involves the primary streamer growth at an average speed of about 1 km/s which is followed by Step 2, the fast event which propagates at speeds in excess of 100 km/s. This fast event is not noticeable in gaps of less than 1 cm under presently available means of observation. Evidence of its existence can be obtained under favorable conditions in large gaps of 2 or 2.5 cm [*38*].

In the presence of electron donating molecules such as dimethylaniline, one observes an increase in the number of discharge channels formed, and the overall pattern produced resembles an umbrella. This geometric pattern also represents an approach to the hemisphere/plane configuration of electrodes. This latter configuration approaches that of a uniform field which requires higher voltages to cause breakdown. To produce such higher fields the positive streamers have to grow closer to the cathode which takes more time. Thus, breakdown takes place later and possibly at higher voltage as was observed experimentally by Hebner et al. [*38*] who reported an increase in the time to breakdown in the presence of dimethylaniline over the pure hydrocarbon in gaps bigger than 1 cm.

In addition to the structural differences mentioned here, the two types of streamers exhibit other dissimilarities which appear to reflect on the former ones. These involve the emission of electrical and light pulses. The former are often referred to as partial discharges or corona. Their origin and development are discussed in the next paragraph.

3.7.1.3.3 Streamers and partial discharges—The occurrence of partial discharges has been studied extensively [*36*], and they have been attributed to electrical discharges occurring in voids within the dielectric without causing a complete collapse of the applied voltage. The model of breakdown proposed here associates these partial discharges with the presence of streamers and explains their occurrence in the following manner. As mentioned earlier, a low density region develops at the injection site. Within this region the pressure is lower than that of the surrounding medium and decreases as this region growth. At a given instant the internal pressure will drop below the Paschen limit [*3*], and electrical breakdown will occur within this region provided that there exists an electric field in it. The existence of such a field is also necessary to sustain electron injection from the electrode into the dielectric.

Before discussing the existence of such a field it is necessary to explore the envelop surrounding this low density region. Krasucki [*9*] had postulated the existence of bubbles in dielectric liquids, and he had developed a criterion for the stability of such bubbles which involved surface tension, coulomb interactions, and other electrostrictive forces. Based on the model developed here, the thermalized electrons are forming the outermost layer of the envelop. Their mutual repulsion is overcome by surface tension and hydrostatic pressure. Inside that low density region positive charges coming from the earlier mentioned impact ionization (Auger process) constitute the lining of that region thus forming a sort of double layer at the far end of that region facing the anode. It is this collection of positive charges that provides, together with the negatively charged injection site, the source of the field in the low density region. As the field collapses these charges are neutralized, and no field exists for a very short time until the plasma associated with the collapse cools down and charges left from it can collect at the end of the region and restart the emission process. If the growth of the streamer were depending on the energy release from this partial discharge, it would show only intermittent growth that could be correlated to the electrical pulses one detects in conjunction with the partial discharges. As mentioned in Section 3.6 no such observations were made. On the contrary, the streamer growth was found to be continuous on a time scale of nanosecond, the shortest time electro-optically resolvable at this writing [*50*]. This absence of any dependence of growth on partial discharges was further confirmed by observations reported earlier in Section 3.6, that, in the absence of optically observable streamers, no partial discharges were detected. These observations support the conclusions reached on the basis of the present model, that partial discharges emanate from the streamer and that streamer growth is not controlled by the energy released by the discharges.

Two questions then arise. The first is concerned with what causes the continuous growth of the streamer and the second one deals with the disposition of the energy released by partial discharge. The answer to the first question is rather simple. As shown by Giordinio et al. [*52*] the local field in front of the sharp tip of the streamer is sufficiently high to cause field enhanced ionization which leads to the streamer growth. The answer to the second question is more speculative and not supported by either experimental or theoretical results. FitzPatrick and Forster [*59,60*] suggested that the energy released by partial discharges is used to create instabilities along the sides of the low density region, and these instabilities then lead to the growth of branches. Indeed, comparisons of the magnitude of partial discharges occurring with negative streamers and those observed with positive ones show that the former are much more energetic than the latter. This appears to be consistent with the high degree of branching observed with negative streamers and the low degree found with positive ones.

The observation of partial discharges usually involves the measurement of some voltage or light pulses [*36*]. There exist many different methods for carrying out the former as shown by Bartnikas [*60*]. The detection of emitted light usually involves the use of photomultiplier tubes and appropriate amplifiers as shown by Wong and Forster [*39*]. The studies by Kelley et al. [*50*] suggest that electrical and optical pulses have the same origin, namely, the collapse of the electric field within the low density region caused by the development of a plasma. Thus, partial discharges can now be identified with the processes going on within the streamer itself. The similarity of the discharge within the streamer to the actual total breakdown of the whole system was shown by Wong and Forster [*39*] in their spectroscopic study of the light emitted during partial and complete discharges.

The model developed here for negative streamers appears to be quite consistent with diverse experimental observations. The proposed mechanism of positive streamer growth leads to the prediction that both light emission and electrical pulses would be fewer and weaker than those observed with negative ones. Indeed, the experimental observations again support these predictions. Both light and electric pulse activity, in general, is weaker for positive than for negative pulses. It should be kept in mind here that the detection limits, for both types of pulses, restrict detection to times when the streamers have grown sufficiently to become visible under the prevailing optical conditions. This means that the streamers have grown to be at least 1 μm long and above the shadow of the needle tip. The observer cannot tell from these photographs where on the needle surface the streamer originated and how long it took it to grow to the detectable size. Similar uncertainties do limit present knowledge of the last phase of streamer development as will be shown in the next paragraph.

3.7.1.4 The Fast Event—As mentioned earlier there suddenly appears to develop a condition that leads to supersonic growth of the streamer. At present one can only guess at its origin. One is tempted to model this event in the following manner. As the streamer approaches the anode, the field in front of it increases rapidly. If one then assumes that the tip is sharp as it was in the beginning for the primary streamer, this local field will be high enough to cause ionization of the liquid molecules in its immediate vicinity, thus leading to the development of a narrow conducting channel that represents the forward growth of the streamer resulting in a pressure drop that can lead, subsequently, to a partial discharge. After this phase is complete within fractions of a nanosecond the whole cycle repeats itself, and each time a larger segment of liquid molecules is affected, thus accounting for the accelerated growth. This model predicts an ever increasing pulse activity involving both electrical and light pulses. Such increased pulse activity was indeed observed just prior to breakdown by Wong and Forster [*39*] and Kelley et al. [*49*]. As a matter of fact just prior to breakdown the pulses came so fast that they were no longer resolved by the system used. To permit detailed observation of these events requires much faster means of observation including framing cameras allowing framing rates as high as 10^9 and streak cameras capable of operating a speeds of 1 mm per 10 ps. Naturally, to carry out observations in these short times will require light sources of sufficient intensity yet weak enough so as not to affect the breakdown process.

3.7.2 The Practical D-C Model

The preceding model was developed with the assumption that a rapidly rising, fast electric pulse is applied to a system free of excess charges. Obviously this is a situation which can only be produced under carefully controlled conditions in a laboratory. Most of the time one is interested in understanding and explaining electric failure under actual service conditions. The most realistic would involve a-c conditions. Before these are discussed, it appears helpful to review the d-c situation particularly the case where one tries to determine either the breakdown voltage for a given dielectric fluid system or the time to failure under a given set of stress conditions. In the first case, one deals with a relatively slow rising voltage which is stopped when a certain level of leakage current is reached. In the second case, one applies a preselected voltage and measures the time to failure. Frequently one uses this test at elevated temperatures to accelerate failure. In the present discussion, no comments will be made regarding the merits of such accelerated testing since one is only interested in understanding the cause for failure. A few remarks, however, will be made later on in reference to thermal effects on the breakdown mechanism.

3.7.2.1 Breakdown Under Slowly Rising D-C Potentials—The gradual increase of the applied potential will cause a corresponding slow increase in the injection of charge carriers which will be reflected in a slight increase in the level of the leakage current and a gradual build up of charges near the cathode. These space charges will drift slowly away from the cathode under the influence of the gradually increasing field and spread out in the liquid. At a critical voltage, the injection process will become more efficient, as mentioned in Section 3.7.1, and a low density region will start to form. Little is known of the overall concentration of these space charges and still less is known of their effect on the local field. In principle the same steps are operative here than in the basic model except that they occur at a much slower rate and under less well defined conditions. Eventually the low density region will develop into tree- or bush-like structures, and, finally, the fast event will produce the conducting channel that leads to breakdown. In many instances one uses a current limiting device which will shut off the voltage when a given current level is reached. The response time of these devices is of the order of microseconds. As has been shown earlier, the fast event is much faster than this, and thus, in most instances, one finds that breakdown did occur in spite of this device. Unless high-speed detecting devices are used, it is very difficult to determine precisely the voltage at which the breakdown took place. Some power supplies have a memory meter that literally freezes the voltage indicating needle at the moment the cut off device is activated. Again the response time of these meters is much slower than the actual event, and considerable uncertainty exists in the values recorded with the help of these meters. A similar condition prevails in the case of a constant voltage as will be shown in the next paragraph.

3.7.2.2 Breakdown Under a Constant D-C Potential—One can understand the electron injection process in this case as consisting of a more or less constant stream of electrons entering the dielectric liquid at several favorable sites. These electrons, after exchanging their energies with the molecules constituting the liquid, are thermalized; they form clusters that slowly drift and hop across the electrode gap. Some of the more energetic electrons will form hot electrons by the aforementioned Auger process and cause some chemical changes due to impact ionization. Also the constant stream of electrons will gradually raise the local temperature of both the metal and the fluid in the immediate vicinity of the injection sites. Eventually the combined effect of the thermalization, the Auger process, and the Joule heating, a low density region will form at one or several of the injection sites as shown in Fig. 3.26. Depending on the local conditions and the space charge distribution, one or more of these regions will grow faster than the others and it will develop into a tree or bush. As it grows towards the anode the local field in front of it will increase and eventually the conditions for the fast event will be encountered leading to breakdown. The main difference between this and the preceding model is the time scale on which these events take place. It can be appreciated readily that the events described in this paragraph take much longer to happen than in the case of a continuously increasing potential, since in the latter case the local field conditions change both because of the growth and because of the increase in the externally applied potential.

Based on the proposed mechanisms one would expect that partial discharge activity will start in either case once the low density region starts to develop. At present, one has many reports of the voltage at which one begins observing these discharges, but there are no corresponding optical observations available to verify this correlation. This is due in part to experimental difficulties of making observations at the "right" time and in part to the sensitivity of the circuits used to measure the electrical signals. It is interesting to note that many investigators attribute the onset of these partial discharges to the physico-chemical properties of the dielectric material. In actuality the onset of these discharges is essentially a reflection of events taking place at the metal dielectric interface. The extensive data available on the injection process indicate that the metal and its Fermi level, as well as the nature of the applied potential and the dielectric, are all involved in determining the formation of the low density region that leads to the development of partial discharges.

Before leaving d-c potentials it is noteworthy that the discussion of the two last cases did not distinguish between positive and negative streamers because their initial development is affected similarly by the applied potential. Their growth is delayed, but, once it starts, its progress is determined by local field conditions, by the electron supply at the injecting electrode, and, in the case of the positive streamer, by the availability of excess charges in the liquid. As mentioned earlier, in case of a point anode, breakdown will occur at a lower voltage than in the case of a point cathode, no matter what type of d-c potential is applied, except in those cases involving fluids made up of molecules with high electron affinity such as perfluorocarbons.

3.7.3 The A-C Breakdown

Most of the following remarks will relate to investigations in which the frequency of the a-c potential is 50 or 60 Hz. These two frequencies are also referred to as the power frequencies. The majority of breakdown studies has been carried out at these frequencies in view of their practical importance. The studies can be divided into those in which the a-c potential is gradually raised until breakdown occurs, those in which the a-c potential is applied at a preselected level, and the time to failure is measured as a function of temperature or other environmental stresses or both. The latter type of studies are also referred to as "life" or "voltage endurance" tests. The former is intended to indicate the voltage at which breakdown occurs 100, 80, or 50% of the time. These values appear to be useful in the design of electrical insulation systems.

Electrical breakdown under the influence of an a-c potential, be it constant or steadily increasing, is a much more complex process then the one observed with corresponding d-c potentials. The main reason for these complications is the buildup of space charges in the dielectric material and their influence on the local fields. This buildup comes about in the following manner. During one half of the cycle one electrode is injecting electrons under conditions similar to the case of a slowly rising d-c potential, while the other electrode acts as an anode and extracts some charges from the dielectric. In the next half cycle the roles of the electrodes are reversed. The discharge of injected charges into what was before a cathode will depend on the local field under which these charges will drift towards the electrode. This local field is much lower for discharge purposes than it is for injection, particularly if one deals with asymmetric, nonuniform electrode geometries. Even under uniform field conditions there will be some difference in the rate of charge injection and charge extraction due to the fact that energetic electrons are injected on one side and thermalized, trapped electrons are extracted on the other side. The net difference between the concentration of injected and extracted charges is then referred to as an excess, space charge. This scheme is based on the detailed analysis of the electronic processes taking place at the two electrodes by Lewis [*57*] in the light of present knowledge of amorphous phase electronics and of the electrochemistry of electrode/dielectric interfaces. Based on ample experimental evidence Lewis concludes that there exist indeed quasi-free electrons in dielectric liquids and possibly equally mobile positive hole states that allow the development of models for extremely fast cathode and anode initiated prebreakdown events. As a result of these considerations and the ones discussed in Sections 3.6 and 3.7, there will always be a build up of excess space charges during the first few cycles until steady-state conditions are reached. Once these steady-state conditions are established, the alternating field will merely cause the gradual growth of the streamers under the influence of the space charge modified local field. Obviously the end result will be the same for the two types of applied a-c potentials mentioned previously. The rate with which breakdown is reached will depend on the level of the applied potential.

A further complication arises from the application of an a-c potential. It stems from the interaction of the excess free charges in the dielectric fluid with the a-c field. The alternating field will cause them to move, and the resulting motion will complicate the study and analysis of the actual breakdown events [*62,63*]. These electrodynamic effects become particularly noticeable near breakdown when the liquid motion becomes so strong that cavitation can set in [*64–67*].

Unlike the d-c cases mentioned earlier, much less detailed information is available on the initiation of the a-c breakdown process. Most prebreakdown a-c studies concern themselves with the determination of the voltage at which the first partial discharges can be detected electrically. This voltage is commonly referred to as the partial discharge or corona inception voltage (PDIV or CIV). A survey of the techniques used for measuring the intensity of partial discharges is given in Ref *61*, and it shows that the detection limits are determined by the sensitivity of the various instrument circuits used. It is, therefore, not surprising that different methods produce PDIV's that are quite dissimilar. The idea that the PDIV is a characteristic of the insulating material is erroneous too because, as was shown in Section 3.7.2, the partial discharges originate at the metal/dielectric interface, and they are the result of the effect of electron injection at that interface. It would be more correct to assign the responsibility for these partial discharges to the conditions prevailing at that interface. These statements are applicable to both types of applied a-c potentials as will be shown in the following paragraphs.

3.7.3.1 Breakdown Under Slowly Rising A-C Potentials—The electrode configurations used with this type of potential include both parallel plate and point-plane arrangements. In either case the PDIV determination is the prime objective [*68*] using one of the techniques discussed in Ref *61*. Similar to the remarks made in conjunction with the effect of electrode gap on the observation of transient phenomena under d-c stresses, it is advisable to utilize larger gaps with a-c stresses. For example Mallet and Fallou [*68*] used a 5 cm needle/sphere gap to study PDIV, and they noted interesting differences in the behavior of transformer oil and simpler organic fluids such as phenylxylene ethane, a new type of capacitor fluid. In their studies they were able to discriminate partial discharges having charges as low as 10 pC from those having higher ones. These authors also studied the gas evolution accompanying these discharges and concluded that the molecular structure of the liquid had a definite influence on this process.

To explain this conclusion in the absence of more detailed studies, one can only assume that the same mechanism as that observed under d-c stressing was operative here. This assumption appears reasonable if one recalls that under d-c stress linear saturated hydrocarbons produce considerably more gas (particularly hydrogen) than do cyclic hydrocarbons containing high levels of unsaturation (aromatic hydrocarbons) such as benzene or toluene. The phenylxylene ethane (PXE) is such a highly unsaturated hydrocarbon which contains at least twice as many aromatic carbon atoms as does transformer oil. The d-c breakdown model predicted the breakage of carbon-hydrogen and carbon-carbon bonds. The former lead to the formation of atomic and molecular hydrogen gas which can be absorbed by the aforementioned aromatic structures to form more or less saturated hydrocarbons, thereby avoiding the evolution of hydrogen gas. Saturated hydrocarbons have no such ability; hence, in their presence, gas evolution is considerably higher than in dielectric fluids containing aromatic molecules. These observations thus suggest that in the case of slow rising a-c voltages the same prebreakdown processes will take place, albeit at a slower pace since the voltage rises usually slower than in the d-c case and the potential changes cause further delay. The final breakdown process will take place at a lower potential than that occurring under identical d-c conditions, except in the presence of asymmetric fields when the point is an anode resulting in a close similarity of both. As mentioned earlier more detailed studies under a-c conditions are necessary.

3.7.3.2 Breakdown Under Constant A-C Potentials—Similar comments can be made in the case of the application of a steady a-c potential to either uniform or nonuniform electrode systems than have been made in the case of a rising a-c potential. The majority of studies reported in the literature deals merely with the determination of the time to failure which information is obviously of significant interest to the design engineer. In this case one often carries out a series of tests with increasingly higher a-c voltages and reports the times to failure. With the help of statistical methods (such as Weibull statistics) one can then predict the life of the system at a much lower a-c stress level which could correspond to the one the designer is contemplating for a practical application. Similar tests can also be conducted at different temperatures or environmental conditions or both to carry out more realistic predictions about the useful life of a system.

3.7.4 Other Potentials

As mentioned earlier in Section 3.3, other potentials are often used. Of interest here are those that consist of repeating d-c pulses, the so-called saw-tooth pulses. They distinguish themselves from the a-c potentials in that there is no polarity reversal. They resemble the a-c potential because the applied potential is cyclic in nature, increasing from zero to a crest value, residing there for a limited time, and returning again to zero. The repetition rate as well as the rate of rise and the time at crest stress can be varied depending on the contemplated applications. This form of potential is used mainly in pulsed power applications. Here, again, the crest voltage is selected to be below the breakdown voltage, but the accumulation of space charges contributes to the deterioration of the insulating material as pointed out in the preceding paragraphs. The effect of space charges on the local field distribution in the dielectric fluid has not been studied. There are also few reports in the literature on the effect of repetition rate, amplitude, and pulse duration on dielectric systems. Most of the published studies have been conducted on composite systems including special paper or polymer films impregnated with dielectric fluids, such as castor oil or fluorinated hydrocarbons [*69,70*]. Little is known about the behavior of dielectric fluids alone under these pulsed conditions. Also, the effect of rapid discharge, prior to recharging, has not received sufficient attention. Apparently, the discharge rate is such that there remains always a residual charge in the system so that it can influence the next pulse. The exact nature of this influence, however, has not been investigated as yet. In any case, based on the limited amount of observations available, it appears that the same mechanisms are operative under steady a-c potential as are functioning with rising a-c potentials, the corresponding d-c potentials, and the repeated d-c pulse potentials. The only significant differences are the time scale on which these events occur and the effect of thermal energy liberated during the application of the potential. In the case of a-c potentials there is also dielectric heating to be considered which can influence, to some extent, the time scale on which electric breakdown occurs. The final outcome is the same no matter what potential is being used. The dielectric fluid loses temporarily its insulating capabilities as the conducting channel is established and a plasma develops within its confines.

The preceding discussion dealt with the development of a model that described the events leading to the actual breakdown of the dielectric fluid. It did not mention the consequences of this development and the possibilities of avoiding it. These aspects are discussed in the next section.

3.8 The Breakdown Event and Its Implications

It has been known for some time that electrical breakdown of dielectric materials, no matter what their physical state, is a statistical event. It is always destructive and only in gases and fluids are the consequences of a temporary nature. Dielectric fluids together with gases have the ability to heal to become again a viable insulating system. The magnitude of the damage caused by breakdown depends, among other factors, on the amount of energy stored in the system, the rate of drainage of this stored energy during breakdown, and environmental conditions. Although the dielectric liquid can overcome the electric effect of a breakdown, it experiences a certain degree of degradation; thus, it is sensitive to repeated breakdowns that causes a gradual loss of its insulating ability. To a lesser extent the partial discharges occurring prior to breakdown contribute to this deterioration so that after some time it is prudent to change the liquid by replacing the old, used fluid with fresh one.

In addition to the physico-chemical changes produced by the partial discharges and the actual breakdown, the electrodes or metallic conductors in contact with the fluid are subject to erosion. Each partial discharge and certainly each breakdown leaves its imprint on the metal surface. Since plasma conditions are created in either case, it is not surprising that the metal at the injection site of the event is subject to temperatures in excess of 2000 K so that melting of the metal is occurring [*34*]. This erosion is particularly noticeable when point-plane electrode geometries are used and one observes the tip of the needle which changes with each application of a high voltage pulse [*52,53*]. On a practical scale this erosion might be less noticeable, but it nevertheless will occur and contribute to the contamination of the dielectric fluid.

In Section 3.6 it was mentioned that the present facilities used in the study of electric breakdown are limited by the time resolution they offer. Indeed the ultimate solution of the puzzle presented by the breakdown of a dielectric will come from electro-optical observations in the picosecond range. One hopes that with these ultrafast techniques the fast events will become accessible to the researchers and bear out the correctness of what is presently a mere speculation. With these new optical techniques one will have to use also larger electrode gaps to obtain additional time to carry out the desired observations. Once these objectives are achieved, it will become necessary to carry out studies of the environmental effects, including in addition to humidity and temperature, pressure and radiation. Clearly, the future research activities will be much more complex and costly, demanding a significant team effort to achieve the desired goals.

Finally, after all that is known has been reviewed, one has to raise the question regarding the prevention of breakdown. Based on today's knowledge, breakdown of a dielectric system appears unavoidable. One can perhaps postpone its occurrence by building saveguards into the system, such as overdesigning the insulating system by limiting the load factor and hence operating temperatures, as well as by proper maintenance. In designing the system one could, for example, make allowance for twice the maximum voltage stress the system will actually be exposed to. Another approach would involve the selection of the proper materials without regard for cost. Finally, it would be prudent to incorporate in the system sufficient protection from the environment. Since most of these steps are too costly to be incorporated, the actual system will be more prone to failure and, hence, have a shorter useful life. The introduction of mixed dielectrics such as solid/liquid, gas/liquid, and the like is a step in the right direction provided the dielectric properties of the two dielectrics are not too dissimilar. Boundaries between dissimilar dielectrics can produce space charge effects, distortion of the local field, and lower the partial discharge inception voltage. In the end, the economic factors will have to dictate which solution is most viable from all points of view.

3.9 Acknowledgments

The author wishes to express his gratitude to his colaborators and colleagues at the National Institute of Standards and Technology (formerly the National Bureau of Standards), Dr. R. E. Hebner, Dr. E. F. Kelley, and Dr. G. J. FitzPatrick with whom he was associated for nearly a decade. Many thanks go also to Mr. P. P. Wong, who collaborated with the author over a dozen years while at the Exxon Research and Engineering Company's Corporate Research Laboratories.

References

[*1*] von Hippel, A., *Journal of Applied Physics,* Vol. 8, 1937, pp. 815–832.
[2] Townsend, J. S., *Philosophical Magazine,* Vol. 1, 1901, pp. 198–227 and Vol. 6, 1903, pp. 598–618.

[3] Paschen, F., *Annual der Physik,* Vol. 37, 1889, pp. 69–96.
[4] Lewis, T. J., *Journal of Applied Physics,* Vol. 27, 1956, pp. 645–650.
[5] Adamczewski, I., *Zeszyty Naukowe Politechniki Gdanskie,* Vol. 2, 1958, pp. 13–24.
[6] Swan, D. W., *Proceedings of the Physical Society,* Vol. 78, 1961, pp. 423–432.
[7] Watson, P. K. and Sharbaugh, A. H., *Journal of the Electrochemical Society,* Vol. 107, 1960, pp. 516–531.
[8] Kao, K. C., Conference Paper 60-84, Winter Meeting, American Institute of Electrical Engineers, New York, 1960.
[9] Krasucki, Z., *Proceedings of the Royal Society,* Vol. A294, 1966, pp. 393–404.
[10] Thomas, W. R. L. and Forster, E. O., *Conduction and Breakdown in Dielectric Liquids,* Delft University Press, Delft, 1975, pp. 49–54.
[11] Hakim, S. S. and Higham, J. B., *Nature,* Vol. 189, 1961, p. 996.
[12] Chadband, W. G. and Wright, G. T., *British Journal of Applied Physics,* Vol. 16, 1965, pp. 305–313.
[13] Sharbaugh, A. H., Devins, J. C., and Rzad, S. J., *Transactions on Electrical Insulation,* Institute of Electrical and Electronics Engineers, Vol. EI-13, 1978, pp. 249–276.
[14] Yamashita, H., Amano, H., and Mori, A., *Journal of Physics D.,* Vol. 10, 1977, pp. 1753–1760.
[15] McGrath, P. B. and Nelson, J. K., *Proceedings,* Institution of Electrical Engineers, Vol. 124, 1977, pp. 183–187.
[16] Wong, P. P. and Forster, E. O., *Canadian Journal of Chemistry,* Vol. 55, 1977, pp. 1980–1985.
[17] Sueda, H. and Kao, K. C., *Applied Optics,* Vol. 18, 1977, pp. 3990–3994.
[18] Chadband, W. G. and Calderwood, J. H., *Journal of Electrostatics,* Vol. 7, 1979, pp. 75–91.
[19] Devins, J. C., Rzad, S. J., and Schwabe, R. J., *Journal of Applied Physics,* Vol. 52, 1981, pp. 4531–4545.
[20] Kelley, E. F. and Hebner, R. E., *Applied Physics Letters,* Vol. 38, 1981, pp. 231–233.
[21] Forster, E. O. and Wong, P. P., *Transactions on Electrical Insulation,* Institute of Electrical and Electronics Engineers, Vol. EI-12, 1977, pp. 183–187.
[22] Hebner, R. E., Kelley, E. F., Forster, E. O., and FitzPatrick, G. J., *Journal of Electrostatics,* Vol. 12, 1982, pp. 265–283.
[23] McGrath, P. B. and McKenny, P. J., *Journal of Electrostatics,* Vol. 12, 1982, pp. 241–246.
[24] Arii, K. and Kitani, I., *Journal of Electrostatics,* Vol. 12, 1982, pp. 247–252.
[25] Yamashita, H. and Amano, H., *Journal of Electrostatics,* Vol. 12, 1982, pp. 253–263.
[26] Hebner, R. E., Kelley, E. R., FitzPatrick, G. J., and Forster, E. O., *1983 Annual Report, Conference on Electrical Insulation and Dielectric Phenomena,* IEEE 83CH1902-6, 1983, pp. 26–34.
[27] Sufian, T. M. and Chadband, W. G., *1983 Annual Report, Conference on Electrical Insulation and Dielectric Phenomena,* IEEE 83CH1902-6, 1983, pp. 20–25.
[28] Hanna, M. C., Thompson, J. E., and Sudarshan, T. S., *1983 Annual Report, Conference on Electrical Insulation and Dielectric Phenomena,* IEEE 83CH1902-6, 1983, pp. 245–250.
[29] Sufian, T. M. and Chadband, W. G., Eighth International Conference on Conduction and Breakdown in Dielectric Liquids, Conference Record, IEEE 84CH2055-2, 1984, pp. 153–158.
[30] Hebner, R. E., Kelley, E. F., Forster, E. O., and FitzPatrick, G. J., *Transactions on Electrical Insulation,* Institute of Electrical and Electronics Engineers, Vol. EI-20, 1985, pp. 281–292.
[31] Xie, H. K. and Kao, K. C., *Transactions on Electrical Insulation,* Institute of Electrical and Electronics Engineers, Vol. EI-20, 1985, pp. 293–297.
[32] Yamada, H. and Sato, T., *Transactions on Electrical Insulation,* Institute of Electrical and Electronics Engineers, Vol. EI-20, 1985, pp. 261–266.
[33] Chadband, W. G. and Sufian, T. M., *Transactions on Electrical Insulation,* Institute of Electrical and Electronics Engineers, Vol. EI-20, 1985, pp. 239–246.
[34] Wong, P. P. and Forster, E. O., *Transactions on Electrical Insulation,* Institute of Electrical and Electronics Engineers, Vol. EI-17, 1982, pp. 203–220.
[35] Forster, E. O., *Transactions on Electrical Insulation,* Institute of Electrical and Electronics Engineers, Vol. EI-19, 1984, pp. 524–528.
[36] R. Bartnikas and E. J. McMahon, Eds., *Engineering Dielectrics, Corona Measurement and Interpretation,* Vol. 1, STP 669, American Society for Testing and Materials, Philadelphia, 1979.
[37] Harrold, R. T., *Transactions on Electrical Insulation,* Institute of Electrical and Electronics Engineers, Vol. EI-21, 1986, pp. 781–792, see also Chapter 10 in Ref *36.*
[38] Gallagher, T. J., *Simple Dielectric Liquids,* Clarendon Press, Oxford, UK, 1975.
[39] Forster, E. O., *Conference Publication 239,* Institution of Electrical Engineers, 1984, pp. 1–5.
[40] Forster, E. O., *Transactions on Electrical Insulation,* Institute of Electrical and Electronics Engineers, Vol. EI-20, 1985, pp. 891–896.
[41] FitzPatrick, G. J., Forster, E. O., Hebner, R. E., and Kelley, E. F., *1985 Annual Report, Conference on Electrical Insulation and Dielectric Phenomena,* IEEE 85CH2165-9, 1985, pp. 26–32.

[42] Watson, P. K., *1981 Annual Report, Conference on Electrical Insulation and Dielectric Phenomena,* IEEE 81CH1668-3, 1981, pp. 370–376.
[43] FitzPatrick, G. J., Forster, E. O., Hebner, R. E., and Kelley, E. F., *1982 Annual Report, Conference on Electrical Insulation and Dielectric Phenomena,* IEEE 82CH1773-1, 1982, pp. 464–472.
[44] Hebner, R. E., Kelley, E. F., FitzPatrick, G. J., and Forster, E., *1983 Annual Report, Conference on Electrical Insulation and Dielectric Phenomena,* IEEE 83CH1902-6, 1983, pp. 26–34.
[45] Devins, J. C., Rzad, S. J., and Schwabe, R. J., *1976 Annual Report, Conference on Electrical Insulation,* National Academy of Sciences, Washington, DC, 1976, pp. 182–192.
[46] Kelley, E. F., Hebner, R. E., FitzPatrick, G. J., and Forster, E. O., *1984 International Symposium on Electrical Insulation,* IEEE Conference Record 84CH1964-6-EI, 1984, pp. 284–287.
[47] Yamashita, H. and Amano, H., *Transactions on Electrical Insulation,* Institute of Electrical and Electronics Engineers, Vol. EI-20, 1985, pp. 247–255.
[48] Kelley, E. F., Hebner, R. E., Forster, E. O., and FitzPatrick, G. J., *1982 International Symposium on Electrical Insulation,* IEEE Conference Record 82CH1780-6-EI, 1982, pp. 255–258.
[49] Hebner, R. E., Kelley, E. F., FitzPatrick, G. J., and Forster, E. O., *Ninth International Conference on Conduction and Breakdown in Dielectric Liquids,* IEEE Conference Record 87CH2403-4, 1987, pp. 26/1–26/5.
[50] Kelley, E. F., Nehmadi, M., Hebner, R. E., McKenny, P. J., and Forster, E. O., *1987 Annual Report, Conference on Electrical Insulation and Dielectric Phenomena,* IEEE 87CH2462-0, 1987, pp. 132–137.
[51] Winterberget, A. L., Pace, M. O., Blalock, T. V., Foust, J. V., and Alexeff, I., *1987 Annual Report, Conference on Electrical Insulation and Dielectric Phenomena,* IEEE 87CH2462-0, 1987, pp. 95–100.
[52] Girdinio, P., Malfino, P., Molinari, G., Viviani, A., FitzPatrick, G. J., and Forster, E. O., Ninth International Conference on Conduction and Breakdown in Liquid Dielectrics, IEEE Conference Record 87CH2403-4, 1987, pp. 311–317.
[53] FitzPatrick, G. J., Forster, E. O., Hebner, R. E., and Kelley, E. F., *Transactions on Electrical Insulation,* Institute of Electrical and Electronics Engineers, Vol. EI-2, 1987, pp. 453–458.
[54] Kelley, E. F. and Hebner, R. E., *Journal of Applied Physics,* Vol. 52, 1981, pp. 191–195.
[55] Kelley, E. F. and Hebner, R. E., *1986 Annual Report, Conference on Electrical Insulation and Dielectric Phenomena,* IEEE 86CH2315-0, 1986, pp. 272–277.
[56] Latham, R. V., *Vacuum,* Vol. 32, 1982, pp. 137–140.
[57] Lewis, T. J., *Transactions on Electrical Insulation,* Institute of Electrical and Electronics Engineers, Vol. EI-20, 1985, pp. 125–132.
[58] Kao, K. C., *Journal of Applied Physics,* Vol. 55, 1985, pp. 752–755.
[59] FitzPatrick, G. J. and Forster, E. O., *1988 International Symposium on Electrical Insulation,* IEEE Conference Record 88CH2594-0-DEZI, 1988, pp. 8–11.
[60] FitzPatrick, G. J. and Forster, E. O., *Conference Publication 289,* Institution of Electrical Engineers, 1988, pp. 13–16.
[61] Bartnikas, R., *Transactions on Electrical Insulation,* Institute of Electrical and Electronics Engineers, Vol. EI-22, 1987, pp. 629–653.
[62] Cross, J. D., Nakano, M., and Savannis, S., *Journal of Electrostatics,* Vol. 7, 1979, pp. 361–372.
[63] Haidari, M. and Atten, P., *Transactions of the Industry Applied Society 1983,* Institute of Electrical and Electronics Engineers, pp. 1125–1129.
[64] Atten, P., Palraison, N., and Ali Kani, S., *Journal of Electrostatics,* Vol. 12, 1982, pp. 477–488.
[65] Nakano, M. and Cross, J. D., *Transactions on Electrical Insulation,* Institute of Electrical and Electronics Engineers, Vol. EI-20, 1985, pp. 381–383.
[66] Watson, P. K., *Transactions on Electrical Insulation,* Institute of Electrical and Electronics Engineers, Vol. EI-20, 1985, pp. 395–399.
[67] LeSaint, G. and Tobazeon, R., *Transactions on Electrical Insulation,* Institute of Electrical and Electronics Engineers, Vol. EI-20, 1985, pp. 269–273.
[68] Mallet, Ph. and Fallou, B., *Ninth International Conference on Conduction and Breakdown in Dielectric Liquids,* IEEE Conference Record 87CH2403-4, 1987, pp. 387–392.
[69] Galperin, I. and White, W., *1982 Annual Report, Conference on Electrical Insulation and Dielectric Phenomena,* IEEE 82CH1773-1, 1982, pp. 563–571.
[70] Mauldin, G. H., Nunnally, W. C., and Sarjeant, W. J., *1982 Annual Report, Conference on Electrical Insulation and Dielectric Phenomena,* IEEE 82CH1773-1, 1982, pp. 557–562.

M. Duval[1] *and T. O. Rouse*[2]

Chapter 4

Physical and Chemical Properties of Mineral Insulating Oils

4.1 Introduction

Dielectric liquids are a major part of the electrical insulation system in many types of electrical equipment—transformers, circuit breakers, bushings, cables, and capacitors. Often the insulation liquid provides additional functions. It is also the cooling medium in transformers and cables, the arc extinguishing phase in switchgear, or even the lubricant in tap changers and circulating pumps. Each application then demands an insulating liquid with somewhat different electrical, chemical, and physical characteristics. In transformers, for example, the dielectric liquid must possess high specific heat and thermal conductivity along with low viscosity and pour point in order to provide effective heat transfer. Electrical stresses are not extremely high. However, these stresses are high in capacitors, and the dielectric liquid must provide exceptional resistance to partial electrical discharges (high discharge inception voltage, high gas adsorption). Less heat is generated in normal operation of a capacitor, and the physical characteristics of the liquid related to heat transfer are of lesser concern. Still another set of characteristics is required in circuit breakers. The electrical arc associated with operation must die abruptly as the breaker is opened or closed. The initial degradation in an arc in the liquid must not produce a situation in which the arc is sustained as the voltage across the liquid falls.

The electrical equipment used in the transmission and distribution of electrical energy is expected to have a service life of years or decades. In addition to possessing suitable properties initially, the fluid should maintain them throughout the long useful life of the equipment. Metals are used as elements of construction, as conductors and, often, as magnetic materials. Other nonconducting or insulating materials are present in most cases. The insulating liquid used in each type of equipment must have demonstrated compatibility with the other materials present in that equipment. The requisite properties of the particular insulating liquid used must neither rapidly degrade these materials nor be rapidly degraded by them.

The requirements that the several kinds of electrical equipment place on liquid insulation are often in conflict, and no single class of liquids serves equally well in all applications. Liquids have been developed for specific applications; synthetic fluids with high permittivity are preferred in capacitors and synthetic hydrocarbon liquids are now widely used in cables. However, mineral oils refined from petroleum crude stocks have been, and will continue to be, the most widely used electrical insulating liquids. The largest single use of electrical insulating liquids, by far, has been in transformers (and circuit breakers), and the bulk of the liquid used in transformers around the world

[1]Institut de Recherche d'Hydro-Quebec, Varennes, Quebec, Canada J3X 1S1.
[2]General Electric Company, Pittsfield, MA 01201.

has been mineral oil. This situation is not likely to change. In the United States, for example, it has been estimated that as much as two billion gallons of mineral oil are in service in transformers at present. Fifty to one hundred million gallons of transformer mineral oils are produced annually by oil refiners in the United States today for use in new units and as makeup in transformers already in service. An economically attractive alternate produced in these quantities and compatible with the oil in service does not appear likely.

The first liquid-insulated transformers were produced commercially nearly one hundred years ago, and the insulating liquid used in them was a mineral oil. Despite the long- and wide-spread use of mineral oil as dielectric liquid, the underlying mechanisms of electrical breakdown and conduction and of oil degradation are not understood fully. Often the effects of the other materials in the insulation system and body of the equipment are not known in detail. Some properties of mineral insulating oils are, therefore, characterized by their phenomenological response in specified test environments. Others, particularly those related to mass and heat transfer, are measured in directly functional ways.

For purposes of discussion, the properties and characteristics of insulating liquids are often categorized as chemical, physical, and electrical. The distinctions are not simple and clearcut in many cases, and a particular classification is sometimes arbitrarily assigned. For instance, both dissolved water and particles are often included in the chemical category. However, both are actually contaminants, one chemical and one physical, and their primary effect is on an electrical property—the dielectric strength of the oil. A different grouping is used in the present chapter. The chemical composition of these mineral oils and its characterization are considered together, then the several properties related primarily to heat transfer. Characteristics relating to safe use and to health and environmental effects of mineral insulating oils are considered separately.[3] Finally, the electrical insulation properties and the factors affecting them are grouped together. In outline form, (*a*) chemical composition is described in terms of refractive index, molecular weight, carbon-type analysis, aniline point, chromatographic absorption, spectrometric properties, and paraffinic, aromatic, naphthetic, hydrocarbons and nitrogen and sulfur contents; (*b*) cooling properties are considered in terms of the thermal conductivity, specific heat, coefficient of thermal expansion, as well as density, relative density, viscosity, pour point, and cloud point; (*c*) safety and environmental effects are discussed in terms of the flash and fire points, autoignition temperature, volatile content, heat of combustion, and PCB content; (*d*) insulating properties are considered in terms of oxidation, voltage stress resistance and contaminant effects. In-service oxidation resistance is characterized in terms of oxidation resistance in standard tests, neutralization number, peroxide number, sludge content, antioxidant content, copper content, polar oxidation products and color; stress resistance is related to the dissolved-gas content, gas solubility, gassing tendency and impulse strength. Contamination levels are considered in terms of the water and particle contents, dielectric breakdown strength, inorganic chlorides and sulfates and metal contents, compatibility with other insulating fluids, dissipation factor, a-c loss characteristics and resistivity.

General information on the physical and chemical properties of mineral oils may be found in Refs *1–12*, and *84* while more specific data have been published on composition [*13–34*], cooling [*35–62*], safety and environmental effects [*63–66*], and insulation-related properties [*67–234*], as well as on standard test methods and specifications (the interested reader should consult other chapters of this book for discussions of electrical properties of the liquids, that is, dielectric strength or a-c loss characteristics).

In many countries, a number of these physical and chemical properties are the subject of

[3]The intention here is to draw attention to several liquid characteristics which relate to safety and to health and environmental effects. The actual requirements placed on these liquids are defined frequently by local and national codes, regulations, and legislation; the interested reader is urged to consult the official documents directly.

specifications issued by the national standards setting groups. In Canada, transformer oils are governed by Standard C-50, issued by the Canadian Standards Association. In the United States, the ASTM has published a Standard Specification for Mineral Insulating Oil Used in Electrical Apparatus (D 3487) and similar ones for cable oils (D 1818 and D 1819). A standard Guide for High Fire-Point Electrical Insulating oils of Petroleum Origin has recently been issued. At the international level, the International Electrotechnical Commission (IEC) has issued a ''Specification for Unused Mineral Insulating Oils for Transformers and Circuit Breakers'' (IEC Publication 296-1982), which is used widely in Europe. However, despite concerted efforts, differences between national and international specifications have not all been eliminated.

The primary standard test methods and specifications for the physical and chemical properties of new mineral oils are listed in Table 4.1, together with typical values for different types of oil. Many standard methods of testing specific to electrical insulating fluids and several specifications based on the results of these test methods have been developed for mineral insulating oils by ASTM Committees. These are contained in Volume 10.03 of the Annual Book of ASTM Standards. Standard Guide <u>D 117</u> provides a brief summary of each of these standardized test methods; these have been underlined in Table 4.1. Those methods whose results are included in Standard Specification D 3487 are underlined twice. A number of standard test methods, developed for other petroleum-based materials and occasionally useful in evaluating insulating oils, are also listed in Table 4.1. Other, nonstandardized, tests are available in the literature and will be discussed later.

Not all these standard tests and methods are used by the same individuals or serve the same purposes. Utilities and other operators of transformers and circuit breakers who already have accepted a given type of commercial oil will use only a relatively short list of basic tests (acceptance tests) to check the continuity of production of newly received batches of oil, to assure the absence of contamination and to follow the deterioration of properties in service. The qualification of a new commercial oil usually will require additional tests from the oil supplier to ensure satisfactory performance. Investigations of failures or malfunctions in service may call for still other tests, while special information concerning heat transfer and the dielectric properties of the oil will be required by equipment design engineers. Finally, in order to advance the knowledge of insulating oils, to identify important components or characteristics not considered before, to help to improve refining processes and to provide new tools in equipment maintenance, novel techniques resulting from basic or applied research will have to be developed. A summary of the applications of these tests and techniques is shown in Table 4.2.

Standard D 3487 specifies the qualities of new unused insulating oils. However, the transfer of new oil from its original container and the contact of new oil with equipment construction materials will affect several of the properties given in Table 4.1. Acceptable values for the oil properties inside electrical equipment before energizing and in service vary widely among users, depending on the stresses imposed. After several months or years of in-service aging, the oil properties, particularly those related to oxidation and to degradation resulting from thermal or electrical stresses, will further deteriorate, and the differences in acceptable limits will vary even more. Surveys of industrial practices are made regularly, however, and recommendations are issued by institutions such as Institute of Electrical and Electronics Engineers (Guide No. C57.106 for Acceptance and Maintenance of Insulating Oil in Equipment) concerning the acceptable values of oil properties after filling and after in-service aging (Table 4.3).

4.2 Chemical Composition

Mineral insulating oils are refined from petroleum crude stocks. Like the crude oils themselves, insulating oils are mixtures of many hydrocarbon molecules having a variety of structures and a distribution of molecular sizes and weights. Most of these hydrocarbons fall into three broad classes: alkanes (paraffins), cycloalkanes (also called naphthenes, naphthenics or cycloparaffins),

TABLE 4.1—*Physical and chemical properties of new insulating oils. Test methods, specifications, and typical values (oils from North America and other parts of the world). ASTM tests specifically recommended for transformer insulating oils in* D3487 *(property requirements) and* D117 *(test descriptions) are underlined twice and once, respectively.*

	Characteristic	ASTM Test Method	IEC Publication	ASTM Specification (D 3487) Types I, II	IEC Specification (Publ. 296) Types I, II, III	Typical Values[a]
Chemical composition	refractive index	D 1807	588-2			1.46 to 1.50
		D 1218				
	molecular weight	D 2224				200 to 400
		D 2878				
		D 2502				
		D 2503				
	aniline point	D 611	588-6	63 to 84°C		68 to 110°C
	carbon type	D 2140				5 to 20% C_A
		D 3238				
	chromatographic absorption	D 2007				
		D 1319				
		D 2549				
	infrared absorption	D 2144	590			
	ultraviolet absorption	D 2008				
	mass spectrometry	D 3239				
		D 2786				
	nitrogen content	D 3228				50 to 500 ppm
		D 3431				
	sulfur content	D 1275		noncorrosive		0.1 to 1%
		D 2622	296A			
		D 3120				
		D 129				
Heat transfer	thermal conductivity	D 2717				0.11 to 0.16 W/m·K
	specific heat	D 2766				1000 to 2300 J/kg·K
	coefficient of expansion	D 1903				7 to $9.10^{-4}K^{-1}$
	density, specific gravity	D 1298	296	<0.91 (15°C)	≤ 895 kg/m^3	825 to 890 kg/m^3
		D 1481	588-2			
		D 287				
		D 941				
		D 1217				

TABLE 4.1—*Continued.*

	Characteristic	ASTM Test Method	IEC Publication	ASTM Specification (D 3487) Types I, II	IEC Specification (Publ. 296) Types I, II, III	Typical Values[a]
Heat transfer (con't.)	viscosity	D 445 D 88 D 2161 D 2983	296A	<12 (40°C) <6000 (−40°C) <1800 (−30°C) (cSt)	≤ 16.5 ≤ 11 ≤ 3.5 (40°C) ≤ 800 (−15°C) ≤ 1800 (−30°C) ≤ 150 (−40°C) (cSt)	3 to 16 cSt (+40°C) 2 to 50 Pa·s (−40°C)
	pour point cloud point	D 97 D 2500	296A	< −40°C	< −30°C ≤ −45°C ≤ −60°C	−30 to −60°C −35 to −8°C
Environmental effects	flash point	D 92 D 56 D 93	588-2	>145°C	≥ 140°C ≥ 130°C ≥ 95°C	100 to 170°C
	fire point	D 92	296-2			110 to 185°C
	auto-ignition temperature	E 659	79-4			230 to 410°C
	vapor pressure	D 2551 D 86 D 2878 D 2887				
	heat of combustion	D 240				30 kJ/g
	PCB content	D 4059		nondetected		
Oxidation	oxidation resistance	D 2440 D 943	74 474 813	acidity (mg/g) sludge (%) 72 H <0.5 to <0.3 mg/g <0.15 to <0.1% 164 H <0.6 to <0.4 mg/g <0.3 to <0.2% <195 min		
	neutralization number	D 2112 D 974 D 664 D 3339	296	≤ 0.03 mg/g	≤ 0.03 mg/g	0 to 0.5 mg/g (aged oil)
	peroxide number	D 1563				0 to 15 ppm (aged oil)

Oxidation (con't.)	sludge content	D 1698				0 to 0.5%
		D 4055				(aged oil)
	antioxidant content	D 2668	666	≥ 0.08% ≥ 0.3%		0 to 0.3%
		D 4768				
	copper content	D 2608				0.1 to 10 ppm
		D 2675				
		D 3635				
	color	D 1500	588-2	<0.5		
Stress resistance	Dissolved-gas content	D 3612				
		D 831	567			
		D 1827	599			
		D 2945				
	gas solubility	D 2780				
		D 2779				
	gassing	D 2300 A,B	628 A,B	D2300 A + 15 μL/min		−35 to +30 mm^3/min
		D 2298	815	B + 30		
	impulse strength	D 3300	897	145 kV		
Contamination	water content	D 1533	733	<35 ppm	<30 ppm	0 to 80 ppm
	particle content	F 661	814			
	inorganic chlorides and sulfates	D 878	588-2			
	compatibility	D 3455	588-5			
	dielectric strength	D 877	156	≥ 30 kV	≥ 30 kV	
		D 1816		≥ 28 kV (1.02 mm) ≥ 56 kV (2.03 mm)		
	interfacial tension	D 971		>40 dynes/cm (25°C)	≥ 40.10^{-3} N/m (25°C)	40 to 60 mN/m
		D 2285				
	power factor	D 924	247	≤ 0.05% (25°C)	≤ 0.5% (90°C)	0.1 to 0.5% (90°C)
	resistivity	D 1169	247			

[a]The typical values indicated in the last column are extracted from general references [*1–12*].

TABLE 4.2—*Applications of test methods and techniques.*

Application	Physical and Chemical Tests
Basic quality (continuity) control tests for new and used oils	Viscosity, density, flash point, refractive index; interfacial tension, color, neutralization number; dissolved gases, water, PCB, antioxidant, sludge contents; pour point; dielectric strength, power factor.
Additional tests for the qualification of new types of oil	Aniline point, IR, chromatography absorption; carbon-type, molecular weight, compatibility, oxidation resistance; n-paraffin content; cloud, fire point, heat of combustion; gassing tendency, impulse strength; inorganic chlorides, sulfates, nitrogen, sulphur contents; mass spectrography.
Additional tests for failure investigation and equipment design	Copper, metal, particle contents, ferrography; polar oxidation and furanic compounds, X-wax; flow-improver content; auto-ignition temperature, vapor pressure; coefficient of expansion, thermal conductivity, specific heat.
Additional tests for applied R & D	MW distribution (GPC); RMN, DSC analysis; gas solubility.

TABLE 4.3—*Physical and chemical properties of aged insulating oils. Acceptable limits proposed in ASTM* D3487 *and IEEE Guide 64 (1977).*

	New Oil As-Received	New Oil After Filling Electrical Equipment, 345-kV class	Service-Aged Oils, 325-kV class	Oil After Reclaiming
Interfacial tension, dynes/cm	40	40	30.7	30
Neutralization number, mg KOH/g	0.03	0.03	0.28	0.15
Color	0.5	1.0	...	...
Water content, ppm	35	15	15	19

and aromatics. The alkanes have chain-like backbones of singly bonded carbon atoms. The chains may be linear or, more frequently, may have one or several side branches. Cycloalkanes contain one or more ring-like structures of singly bonded carbon atoms, usually five or six. Two or more of these ring structures can be "fused" together and naphthenes may have one or more linear or branched alkane chains (alkyl) attached. Aromatic hydrocarbons as well contain one or more ring-like structures of carbon atoms, most commonly six, but with carbon-carbon double bonds. Two or more of these rings also may be fused together, and both mono- and polyaromatic hydrocarbons may have alkyl or naphthenic substituents attached. Several examples of these types of structures, with molecular weights likely to be found in insulating oils (see Table II of Ref *28*), are shown in Fig. 4.1.

The three classes of hydrocarbon molecules each contribute to the behavior of the ensemble. Aromatic hydrocarbons are in general more reactive than saturated hydrocarbons; much of the chemical behavior of a mineral insulating oil results from the aromatic species present. Alkanes and cycloalkanes together are present in larger amounts; many of the physical characteristics of an

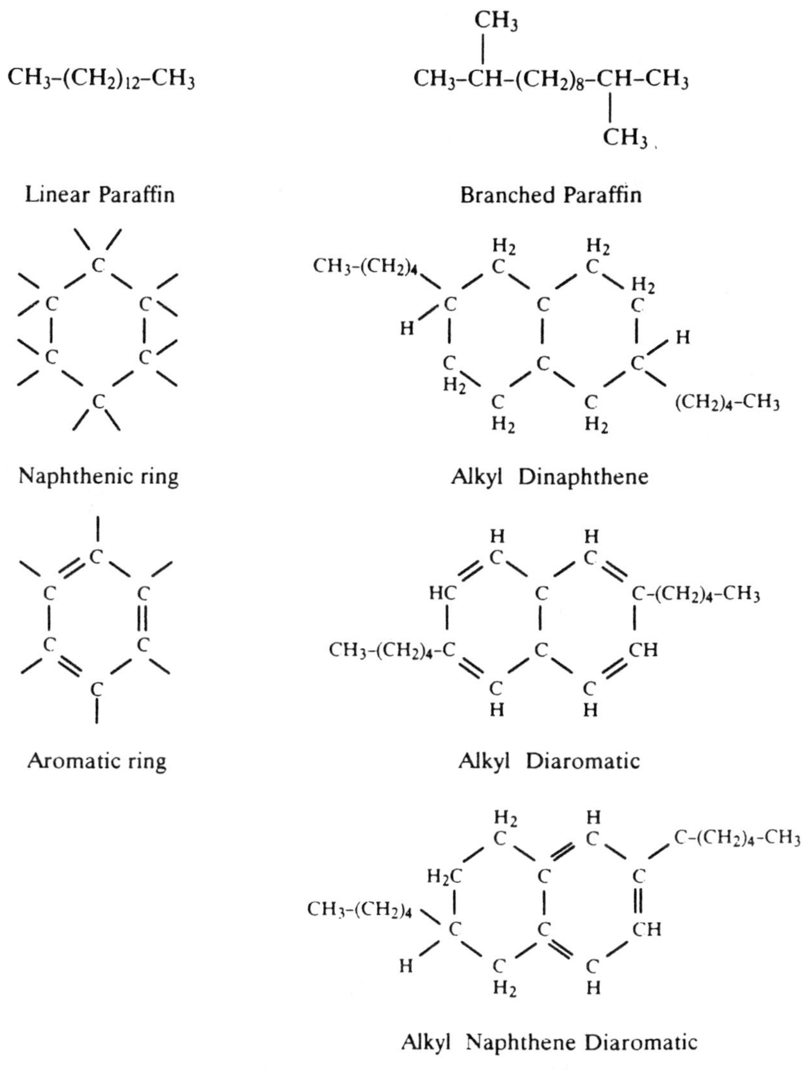

FIG. 4.1—*Structures of hydrocarbon molecules in oil.*

insulating oil reflect their presence, although these characteristics are also influenced greatly by the aromatic content, as indicated in Table 4.4 and Fig. 4.2.

Crude oils can vary widely from field to field, from well to well within a field, and even with depth in a given pool. Some are rich in aromatics, others in naphthenes, still others in paraffins. Some are rich in large high molecular weight molecules, others in lighter molecules. Traditionally, insulating oils have been refined from crudes containing large fractions of naphthenes and small quantities of paraffins, particularly of the straight chain ''normal'' paraffins which form wax on cooling. The resulting insulating oils inherently have low pour points, and do not usually require special processing to provide satisfactory performance at low temperatures. Satisfactory insulating oils can also be made from wax containing paraffinic or mixed base crudes when the additional processing to remove wax can be justified, or when used in warm climates.

TABLE 4.4—*Influence of the chemical character of insulating oils on some of their physical properties* [7,10].

Chemical composition						
% CA	46	21	14	8	6	10.2
% CN	45	75	36	33	28	81
% CP	9	5	50	59	66	9
Specific gravity at 20/20°C	0.931	0.906	0.915	0.901	0.882	0.870
Viscosity, cSt at 37.8°C	730	581	159	154	85	62
at 98.9°C	35	33	10.9	11.7	9.5	8.5
Pour point, °C	+54	−18	...	...	...	−18

Electrical insulating mineral oils are produced in oil refineries by the technology commonly used in the production of petroleum lubricating oils [*13*]. A crude oil is first separated into fractions boiling over selected temperature ranges by distillation under atmospheric pressure. The flash and fire points, the viscosity characteristics, and the density of the fraction of a given crude oil are determined largely by the molecular composition of the hydrocarbons boiling in the temperature range over which that fraction is collected. The flash and fire points of a separated fraction will increase as the boiling temperature of the initial portion collected is increased. The viscosity of a fraction will increase as the average boiling point of the collected fraction is increased. For example, the boiling range of a typical naphthenic transformer oil is 283 to 429°C and is 239 to

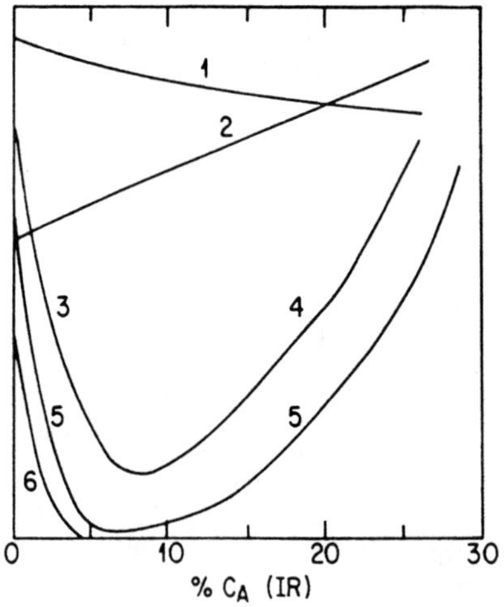

1. Specific heat, thermal conductivity
2. Viscosity, refractive index, density
3. Gas evolution
4. Gas absorption
5. Oxidation
6. Inhibitor response

Arbitrary relative ordinate units.

FIG. 4.2—*Influence of aromatic content on physical and chemical properties of oil* [7,10].

531°C for a paraffinic transformer oil. The average molecular weights of the two fractions are 250 and 300, respectively. The flash points of the oils resulting from these fractions exceed the minimum value of 145°C specified for transformer oils in the United States, and the viscosities are lower than the specification value of 12 cSt at 40°C. The temperature ranges over which the individual fractions are collected are then chosen to give the physical characteristics required of specific lubricating, turbine, hydraulic, and insulating oils. The initial and average boiling temperatures are higher for the higher fire point oils used in special transformer applications and for the more viscous pipe and impregnating oils used in cables. Fractions collected from the initial atmospheric pressure distillation are often distilled a second time at reduced pressure to tailor the boiling range and molecular composition more closely to the application, although atmospheric distillation alone may give a sufficiently cleancut separation of desirable cuts from some crude oils.

Quantities of olefinic and other reactive hydrocarbons, and sulfur, oxygen, and nitrogen containing compounds which also carry over in distillation fractions must be eliminated to produce satisfactory insulating oils. Traditionally this was done by treatment with sulfuric acid. The resulting oil was neutralized by caustic, washed, and filtered through clay to remove residual ionic and polar impurities. Over the last 25 years, acid refining has been replaced largely by a combination of mild treatment with hydrogen in the presence of catalysts, and light extraction with solvents such as *n*-methylpyrolidone, sulfur dioxide, furfural, or phenol. The undesirable impurities can be eliminated by either severe hydrogen treatment or thorough extraction; however, both the oxidation resistance and the aromatic content of the refined oil is reduced substantially by conditions harsh enough to eliminate the undesirable materials by either process alone. Also, in the United States, Occupational Safety and Health Administration regulations now require that naphthenic oils produced by mild hydrorefining, mild solvent extraction or acid/clay treatment be labeled as potential carcinogens. The proper balance of properties can be obtained in refining insulating oils by a combination of modest solvent extraction and mild hydrogen treatment. An optimum amount of aromatics is needed, in particular, to maintain good oxidation resistance and inhibitor response (see Fig. 4.2). Wax is formed in oils by the ordering of long straight chain normal paraffins (and to a lesser degree, of lightly branched iso-paraffins). The presence of wax can result in higher non-Newtonian viscosities which may slow the movement of mechanical parts through the oil to unacceptable levels. Cracks can form as the stiffened waxy liquid is further cooled. In most naphthenic insulating oils, the concentration of n-alkanes is low enough that wax formation does not occur at temperatures above −40°C. It may be necessary to remove the wax-forming hydrocarbons from paraffinic oils or to disrupt the ordering process by the addition of viscosity modifying additives (pour point depressants). Wax forming molecules have been removed by chilling and filtering or by reaction with solvents. Over the last decade, catalytic processes involving zeolites have been introduced which shorten the wax-forming hydrocarbons.

4.2.1 Molecular Structure

A mineral insulating oil fraction is a complex mixture containing thousands of individual molecules which differ in both molecular weight and chemical structure.

Figure 4.3 shows the molecular size distribution (MSD) of three commercial naphthenic oils, obtained by gel permeation chromatography (GPC) [*28*]. The upper curves, obtained with the refractometer detector, represent the MSD of all the molecules present in the oils, and the lower curves the MSD of the ultraviolet (UV) absorbing aromatic molecules only. The MSD can be translated to some extent into a molecular weight distribution (MWD) or a distribution according to the number of carbon atoms in the molecules, using molecular model sets (see Section 4.2.3.1 for a description of GPC principles and the interpretation of GPC curves). Such oil properties as viscosity or flash point may be affected by the MSD, the MWD, or the average molecular weight.

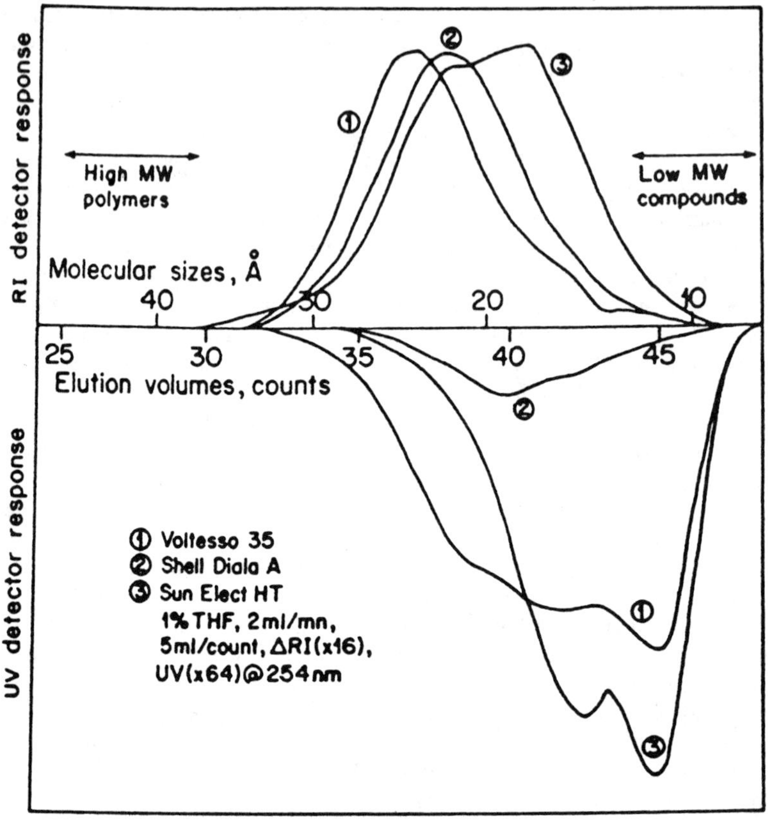

FIG. 4.3—*Molecular size distribution of three commercial naphthenic oils* [28].

The presence of low-molecular-weight material (with low boiling points) in particular may lead to an unacceptably flammable oil (see Section 4.4.1).

Because of the complexity of the hydrocarbon mixture, mineral insulating oils are characterized usually in terms of molecular classes and subclasses present rather than by individual chemical species. One approach is based on empirically determined relationships between physical properties of oils (viscosity, specific gravity and density, refractive index) and the percentage of the carbon atoms in the oil present in each of the three major structural organizations of hydrocarbons, that is, aromatic, paraffinic and naphthenic. These relationships have been expressed algebraically and numerical values (usually as a percentage) of C_A, the fraction of aromatic, C_P, of paraffinic, and C_N, of naphthenic carbons can be calculated. In modern insulating oils, C_A is typically 10 to 15%, C_P is around 40% and C_N is around 50%. Oils can be characterized easily and quickly with this approach, and the results can provide the basis for useful correlations with physical or chemical properties of oils. However, it should be kept in mind that no distinction is made between paraffinic carbons actually present in paraffinic molecules and those present in alkyl substituents on naphthene or aromatic molecules, or between naphthenic carbons in naphthene molecules and those in naphthenic substituents on aromatic molecules. For example, consider hexylbenzene on the one hand and an equimolar mixture of benzene and hexane on the other. Characterized in this way, C_A is 50% in both mixtures as is C_P, but the molecular weight, boiling point, etc. is quite different for the two. Correlation (or the lack of correlation) between a physical property or chemical behavior and carbon type determined in this way should be viewed cautiously.

Mass spectrometry (MS) can provide a more detailed characterization of the composition of mineral insulating oils in two ways. First, it distinguishes the aromatic, naphthenic, and paraffinic components on a molecular, rather than a structural, basis. (The mixture of benzene and hexane is quite distinct from hexylbenzene). The molecular paraffin and cycloparaffin contents of an oil can be determined individually. Secondly, MS analysis provides a more detailed description of the subclasses of the aromatic and naphthenic fractions. An example of the results obtained for a naphthenic oil and a catalytically dewaxed paraffinic oil is given in Table 4.5, with the relative amounts of iso (branched) paraffins, 1- to 6-ring cycloparaffins, 1- to 5-ring aromatics, sulfur-containing aromatics and oxygen- and nitrogen-containing polars, as well as the types of substitution groups on the aromatic rings (alkylated or cycloparaffinic).

The "naphthenic" oil (see Table 4.5) was a widely used commercial oil; the paraffinic oil met all the requirements of Specification D 3487. The composition of the two oils described here does not vary by much. Cycloparaffins predominate in both examples, and, although the paraffinic hydrocarbon content of the "paraffinic" oil is greater than that of the naphthenic oil, the alkanes are minority components in both oils. The concentration of *n*-alkanes is negligible in both naturally wax-free and dewaxed oils. The aromatic content of the two oils is quite different, both in total and by subclass. This must reflect differences in the compositions of the original crudes and, perhaps even more, in the refining processes needed to produce acceptable finished oils from these crudes.

TABLE 4.5—*Mass spectrometer analyses of naphthenic and paraffinic oils.*

Hydrocarbons (% by weight)	Paraffinic Oil			Naphthenic Oil		
Total paraffins	80.3			70.6		
Iso-		23.0			11.4	
Cyclo- (naphthenes)		57.8			59.2	
1-ring			23.3			16.3
2-ring			16.0			16.8
3-ring			11.7			12.2
4-ring			6.8			10.0
5-ring			0.0			3.9
6-ring			0.0			0.0
Total aromatics	19.2			29.4		
Mono-		12.6			18.3	
alkylbenzenes			4.5			4.9
naphthene benzenes			4.0			6.0
di-naphthene benzenes			4.0			7.4
Di-		4.8			8.7	
naphthalenes			2.2			4.1
acenaphthenes, dibenzofurans			1.5			2.5
fluorenes			1.1			2.1
Tri-		0.4			0.7	
phenanthrenes			0.4			0.6
naphthene phenanthrenes			0.0			0.0
Tetra-		0.4			0.3	
pyrenes			0.3			0.1
chrysenes			0.2			0.2
Penta-		0.0			0.0	
Sulfur heterocyclics		0.8			1.2	
benzothiophenes			0.6			0.6
di-benzothiophenes			0.1			0.6
naphthene benzothiophenes			0.0			0.0
Unidentified		0.2			0.2	
Polar compounds	0.0			0.0		

Nuclear magnetic resonance (NMR) spectrometry can also be used to determine the total aromatic carbon content of a mineral insulating oil [*84*]. Carbons in monoaromatic moieties (for example, benzenes, biphenyls) can be distinguished from those in condensed polyaromatic structures (for example, naphthalenes, phenanthrenes) by NMR; both the monoaromatic and polyaromatic carbon content of an oil can be determined by NMR. The latter can also provide information on the average number of methylene groups bonded to monoaromatic and polyaromatic carbons and, hence, to the average levels of alkyl or naphthenic substituents (and benzylic carbons) on the mono- and polyaromatic molecules. A ratio of methylene groups to methyl groups ($—CH_2—/—CH_3$) can also be obtained to characterize the degree of branching in the alkane structure; the ratio increases as the linearity of the alkane increases.

Total aromatic carbon content of an oil can be measured by infrared (IR) spectrometry. It is difficult to obtain additional definitive information from conventional IR measurements; the mixture of hydrocarbon structures is too complex. Fourier transform (FTIR) methods do not seem to have been applied to the study of insulating mineral oils as of yet.

Taken together, the results of the measurements of the molecular size and weight distributions, molecular and structural compositions, and structures of substituents and saturated hydrocarbons themselves can present rather detailed descriptions of many of the typical molecules making up mineral insulating oils. These measurements can be combined with gel and high-pressure liquid chromatographic separations to give still more detailed descriptions.

Much remains to be done, however, to evaluate the specific influence of each chemical subclass on the chemical and physical properties of mineral insulating oils. An analysis of the relationship between paraffin content and wax formation, the roles of aromatic hydrocarbons and sulfur and nitrogen-containing materials in oxidation, and the influence of the amounts and structure of the naphthenic molecules are cases in point. The increasingly sophisticated analytical techniques now available provide better tools towards understanding, optimizing and controlling the behavior of these electrical insulating materials, and present interesting opportunities for future research.

4.2.2 Test Methods

In the text that follows, the test methods specified in ASTM D 3487 and D 117 are indicated by double and single underlining, respectively.

4.2.2.1 Refractive Index—The refractive index is defined as the ratio of the velocity of light in vacuum to the velocity of light of the same wavelength in the substance. Its value varies with the wavelength of the light used and with temperature, and both must be quoted. The "*D* line" of the sodium spectrum (wavelengths of 5890 and 5896 Å) is commonly used for refractive-index measurements with hydrocarbons.

Specific optical dispersion is the difference between the refractive indices of light of two different wavelengths, with both indices measured at the same temperature, the difference being divided by the specific gravity, also measured at the test temperature. Wavelengths of 6563 and 4861 Å of the spectral lines of hydrogen are commonly used.

The refractive index can be used as a quick means of checking for continuity of production and contamination, and only indirectly for composition analysis, as part of D 2140 or D 3238 (see Section 4.2.3.4).

Two ASTM test methods are available for measuring the refractive index and specific optical dispersion. D 1218 and D 1807 differ mainly in the accuracy of the refractometer used, D 1218 being a precision method for determining a refractive index accurate to 0.00006. The oils must be transparent and light-colored and have a refractive index between 1.33 and 1.50.

Measurement is based on the critical angle method (Fig. 4.4): a ray of light at a grazing angle to the interface between a prism and the liquid sample will be refracted into the more refractive medium (the prism) at the critical angle. Someone looking into the field of refracted rays through a

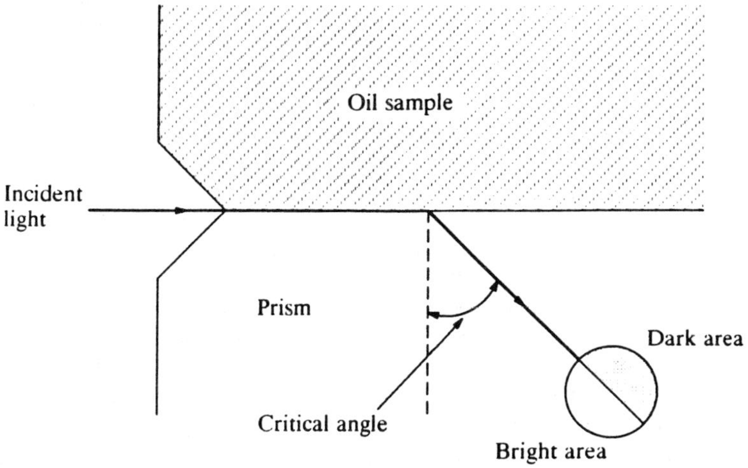

FIG. 4.4—*Principle of refractive index measurement.*

telescope would see a bright area and a dark area, with a sharply divided line between them. Precise measurement of the angle this line makes with the interface yields the refractive index.

4.2.2.2 Molecular Weight—Determination of the number average molecular weight of a mineral insulating oil is useful when characterizing it by hydrocarbon-type analysis (see Section 4.2.3.4). It can be evaluated according to ASTM Test Method D 2224, a cryoscopic method of measuring the freezing-point depression caused by a known weight of oil in a solvent, applying Raoult's law. Other methods are also available, such as ASTM D 2878, which yields the apparent vapor pressure and molecular weight of oil by comparing its evaporation rate to that of pure *m*-terphenyl, or ASTM D 2502, where molecular weight is evaluated from correlations with the kinematic viscosity of oil at 37.7 and 98.8°C (100 and 210°F), and ASTM D 2503, where oil is dissolved in a solvent, and the evaporation rate of the solution compared to that of the pure solvent by means of temperature measurements.

The number average molecular weight is based usually on a much broader molecular weight distribution, as discussed in Section 4.2.2 and illustrated in Fig. 4.3. Gel permeation chromatography (GPC), which is used to determine the MWD, is depicted schematically in Fig. 4.5. Basically, a solvent is pumped at high pressure through a chromatographic column into a detector. The

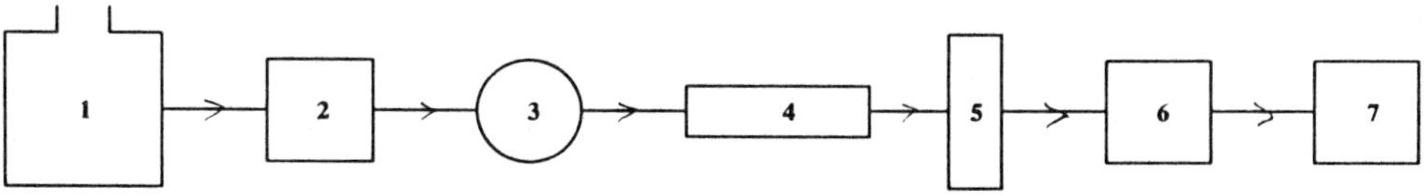

1. Solvent reservoir
2. High pressure pump
3. Sample injector
4. Column
5. Detector
6. Recorder
7. Data processing

FIG. 4.5—*Schematic representation of a GPC/HPLC chromatograph.*

specimen to be analyzed is injected between the pump and the column, in pure form or dissolved in the solvent. The high molecular weight components of the specimen cannot enter the small porous sites of the GPC column and are the first to appear (elute) in the detector. The low molecular weight components are retained longer and are eluted last. Thus, elution volumes can be translated into molecular sizes or molecular weights by calibration with model compounds, and the GPC curve can be used as a representation of the molecular size (or weight) distribution. In Figs. 4.3 and 4.6 the measurement of elution volumes is in siphon count units of 5 mL each. Detection is by refractive index, ultraviolet, or other detectors [*23*]. Low molecular weight material resulting from the cracking of oil molecules under arcing conditions and X-wax formed in corona discharges have thus been detected in oils (Fig. 4.6) [*28*].

4.2.2.3 Aniline Point—The aniline point can give an indication of the solvent power of insulating oils with respect to such materials as rubber gaskets or the rosin products used in some cable impregnates. Therefore, it may complement the results of compatibility tests and is also commonly related to the aromatic content of the oil. The aniline point can be determined according to ASTM Test Method D 611 by mixing specified volumes of aniline and oil samples, heating them at a controlled rate until the two phases become miscible, and observing the temperature at which they separate when cooled at a controlled rate.

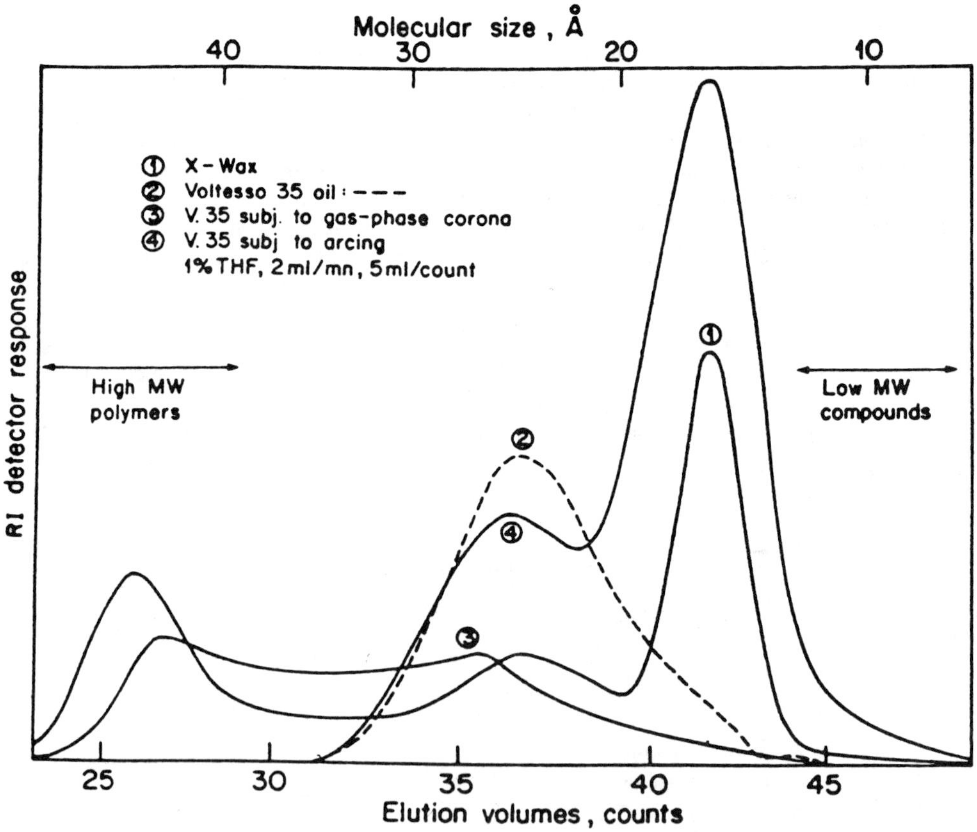

FIG. 4.6—*GPC chromatographs of X-wax and LMW products formed by gas-phase corona and arcing in oil, respectively* [28].

4.2.2.4 Carbon-Type Composition—These methods are used to determine the carbon-type composition of mineral oils by correlation with basic physical properties. Carbon-type composition, expressed as the percentage of aromatic, naphthenic (or cycloparaffinic) and paraffinic carbons, is used to relate the chemical nature of the oil to some of its other physical and chemical properties. Again, it must be remembered that the values of carbon-type composition obtained may be affected by the nature of the chemical subclasses, for instance the relative proportion of polyaromatics in the aromatic fraction.

Although total aromatic content now is more commonly evaluated by infrared or chromatographic absorption, carbon-type composition methods are still useful to differentiate between paraffinics and cycloparaffinics when more sophisticated methods like mass spectrometry are not available.

Two ASTM methods are available: D 2140 and D 3238. The former requires measurement of the refractive index (D 1218), density and specific gravity (D 1481) and viscosity (D 445), with % C_A, % C_N, and % C_P determined from a triangular representation as depicted in Fig. 4.7. The viscosity-gravity constant (VGC) is first calculated from the measured values of the viscosity ν (in cSt at 37.8°C) and specific gravity G (at 15.6°C), and the refractivity intercept (τ_i) is computed from the measured values of the density d (at 25°) and the refractive index n_D^{20} (at 20° for the D line sodium), as follows

$$VGC = \frac{G + 0.0887 - 0.776 \log \log (10\nu - 4)}{1.082 - 0.72 \log \log (10\nu - 4)}$$

$$\tau_i = n_D^{20} - (d/2)$$

The values of VGC and τ_i correspond to one point in the correlation chart of Fig. 4.7, and this point corresponds to one set of values of % C_A, % C_N, and % C_P in the same chart. A simple computer program can be used to obtain these calculations and correlations.

ASTM D 3238 is the so-called *n-d-M* method, requiring measurement of the refractive index and density at 20°C, as above, and the molecular weight. These data are used to calculate the carbon distribution (% C_A, % C_N, and % C_P) using appropriate sets of equations.

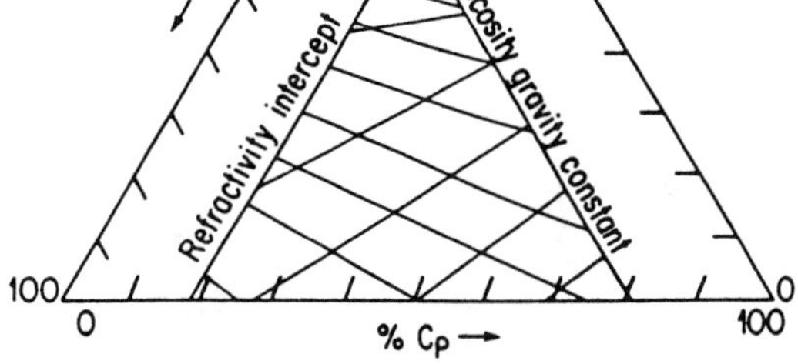

FIG. 4.7—*Correlation chart for determining* % C_A, % C_N, *and* % C_P *(ASTM test method D 2140).*

For the determination of refractive index, molecular weight, density, and viscosity, see Section 4.2.3.1, Section 4.2.3.2 and Section 4.3.3.4 and Section 4.3.3.5, respectively.

4.2.2.5 Chromatographic Absorption—Oil components can be separated into three categories (polar compounds, aromatics, and saturates) based on their tendency to be absorbed on clay or silica gel columns and selectively eluted by an n-pentane solution. ASTM D 1319, D 2007, or D 2549 may be used for this purpose.

Other elution chromatography methods, more accurate albeit more complex, allow the characterization of mono-, di-, and polyaromatics [*30,31*], using various combinations of absorbants and solvents. The advent of modern high performance liquid chromatography (HPLC), capable of further separation depending on the polarity of the oil molecules, brought a wealth of additional information on the various types of aromatic and polar materials contained in insulating oils [*24–29*]. Figure 4.8 provides an example of the differences in aromatic composition that HPLC has revealed in three commercial oils [*28*]. The equipment used is basically as in Fig. 4.5, with differences in the column packings and eluents. In absorption chromatography, a nonpolar solvent such as heptane is used with a strongly polar column packing (pure silica, or silica bonded with polar groups). In the reversed phase mode, a polar solvent mixture such as water and methanol is used with a nonpolar, C_{18} or C_8 bonded silica packing, to selectively separate the components of an oil sample [*23*]. X-wax (Section 4.5.4.7) may also be detected using these techniques [*28*].

4.2.2.6 Spectrometric Methods—Structural information can be obtained from the spectrometric properties of an oil, using methods such as infrared, ultraviolet, NMR, and mass spectrometry.

In conventional infrared spectrometry (Fig. 4.9*a*), a beam of continuous infrared radiation is sent through the sample to be analyzed, and through a second reference optical path (not shown in the figure) to eliminate experimental variations. This beam is then passed through a monochromator (typically, a grating or prism/driving system combination), where it is dispersed as a function of infrared frequency. The intensity of the dispersed rays is measured by a detector and plotted as a function of frequency on a recorder, to provide the infrared spectrum. A decrease in the intensity of some regions of the spectrum indicates that the corresponding infrared frequencies have been absorbed by the sample. These frequencies may be associated with specific chemical structures present in the sample and used for identification and quantitation. For mineral insulating oils, infrared absorption at 1605 cm^{-1} (6.23 μm) and 725 cm^{-1} (13.8 μm) is used to determine the percentage of aromatic and paraffinic carbons respectively (IEC Publication 590), while absorptions at 1754 to 1667 cm^{-1} (5.7 to 6.0 μm) and 3571 cm^{-1} (2.8 μm) are useful for determining oxidation products and certain phenol-type oxidation inhibitors (ASTM D 2144).

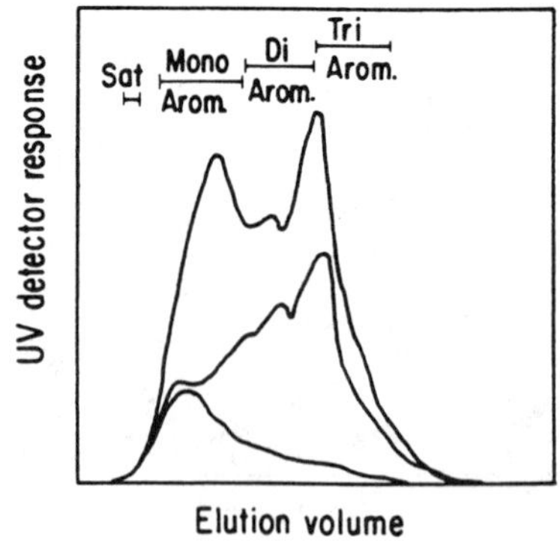

FIG. 4.8—*Aromatic composition of three commercial oils* [28].

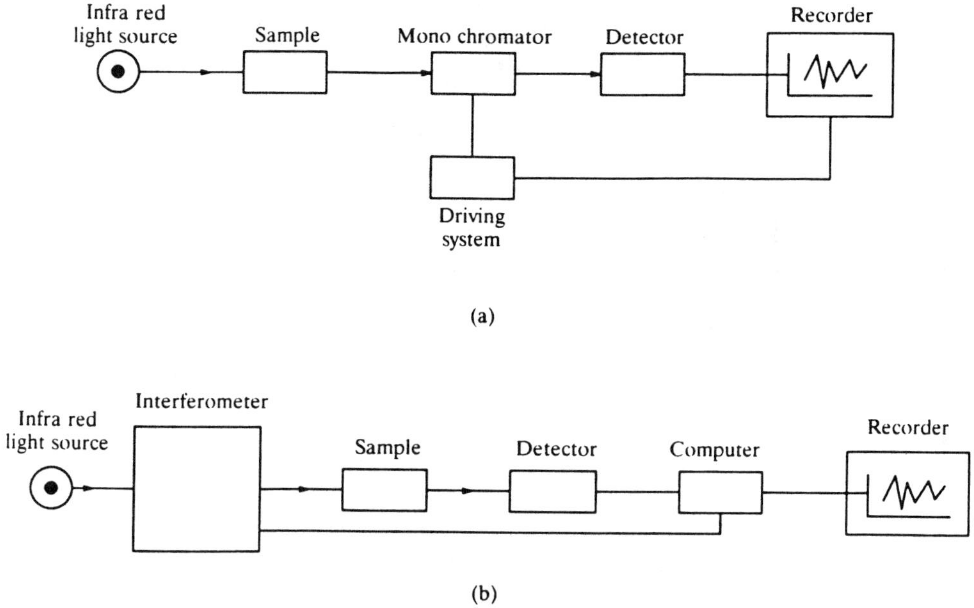

FIG. 4.9—*Schematic diagram of* (a) *conventional infrared spectrometry and* (b) *Fourier-transform infrared spectrometry.*

New insulating oils can be analyzed by transmission spectrometry, that is, by passing the infrared beam through the oil sample contained in a quartz cell. Dark, contaminated oils and solids may be examined by reflection spectrometry, that is, with the infrared beam being reflected several times inside a crystal plate and along the surface of the sample in contact with the plate.

Spectacular improvements in frequency resolution and signal-to-noise ratios are achieved by modern Fourier transform spectrometers. Instead of using a monochromator to disperse the infrared beam coming from the sample, the beam is passed through an interferometer (producing an output signal containing all the frequencies and varying as a function of optical path difference in the interferometer), before being sent to the sample (Fig. 4.9*b*). A mathematical procedure known as a Fourier transform, achieved by a digital computer attached to the spectrometer, yields the spectrum as a function of frequency.

Ultraviolet absorption can be used to qualitatively detect mono-, di-, and polynuclear aromatics in the oil (ASTM D 2008) but calibration and quantitation are difficult because they strongly depend on the relative proportions of the specific aromatic species present in the oil.

Mass spectrometry has been used to probe further into the types and configurations of aromatic and naphthenic rings, as illustrated in Table 4.5. In a mass spectrometer, an electron beam is made to collide with the molecules of the sample, breaking them into ion fragments (Fig. 4.10). These ions (electrically charged atoms or molecules) are sent through a combination of electric and magnetic fields, where they are deflected and separated according to mass. An electric detector placed behind a slit records the current produced by the impinging ions. Correlations can be made between the ion fragments observed and the issuing molecules, thus allowing their precise identification and quantitation.

Nuclear magnetic resonance (NMR) can be used to evaluate the number of CH_2 and CH_3 groups

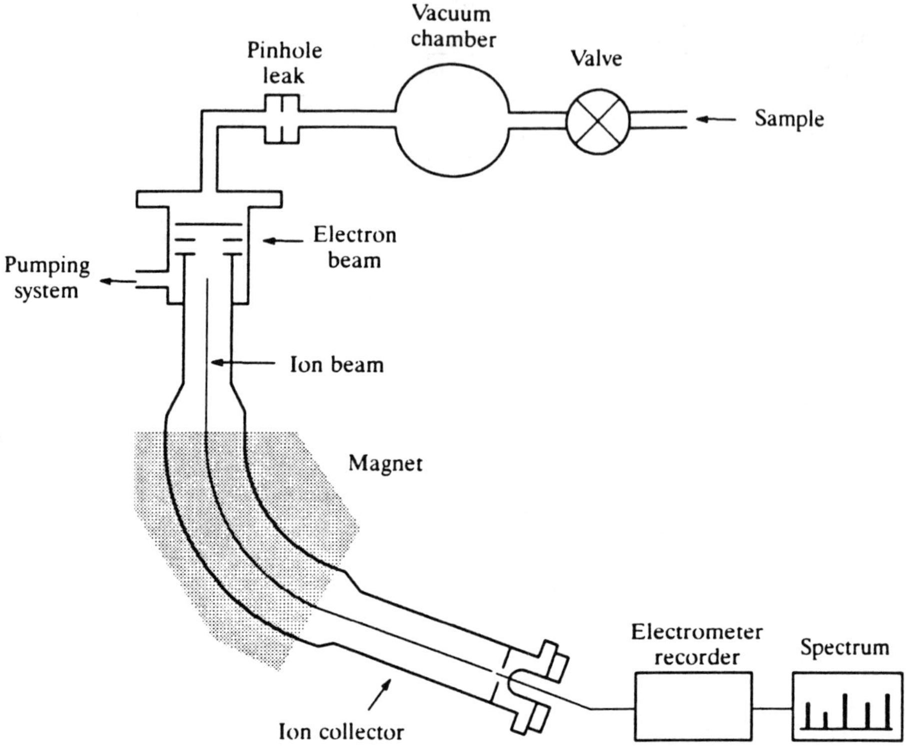

FIG. 4.10—*Schematic diagram of mass spectrometer.*

in insulating oils, as well as other chemical structures (see Section 4.2.1). NMR is based on the presence of spinning nuclei in atoms, which behave as tiny magnets, assuming certain allowed (quantized) orientations when placed in a strong external magnetic field. They can be reoriented only by adding energy of the exact transition (radio) frequency. A difference in the chemical environment (neighboring atoms) of a nucleus can modify its local magnetic field and, therefore, its resonance frequency. The small shift is called "chemical shift" and it is used to obtain information about the molecular structures (the various functional groups) present in a sample. NMR analysis based on the magnetic resonance signals of the hydrogen nuclei (^{1}H) is most used because it is easy to do and is the most sensitive, but analyses for (^{13}C) and others are also useful.

The values for the aromatic content obtained for the same oil by the infrared absorption, chromatographic elution, carbon-type composition and mass spectrometric methods are not the same: they include varying amounts of the saturated carbons surrounding the aromatic rings, from none with the infrared method to all with the chromatographic method. An example of this variation is shown in Table 4.6.

4.2.2.7 Paraffinic Content—Free straight-chain paraffins are undesirable compounds as they increase the viscosity at low temperatures and the pour point. Their paraffinic content can be determined either by urea-adduction methods [*19–22*] or differential scanning calorimetry [*14–18*]. In concentrated solutions of urea in water, urea molecules form cage-like structures into which only linear paraffins can enter because of steric considerations. The linear molecules are trapped, and

TABLE 4.6—*Aromatic content of insulating oils as measured by different methods.*

Hydrocarbon Type/ Analytical Method	Aromatic	Paraffinic	Naphthenic
Infrared	6	25	69
Carbon-type composition (n-d-M)	6	32	62
Chromatographic separation	16	84[a]	
Mass spectrometry	22	21	57

[a]Paraffinic and naphthenic carbons not differentiated.

precipitated out of the solutions, then weighed. The paraffinic content is calculated from results of calibration tests. Differential scanning calorimetry makes use of the fact that linear paraffins are crystalline materials and relates the size of the endothermic peak of fusion of these crystallites to their content in oil. The structure of other fully saturated paraffins can be understood through the saturate fraction separated by elution chromatography (see Section 4.2.3.5), followed by mass spectrometry (see Section 4.2.3.6) or the refractivity intercept method (D 2159), which requires measurement of the refractive index and density of the saturate fraction. The refractivity intercept, calculated from these data according to Section 4.2.3.6, is plotted as a function of density in a correlation chart (given in D 2159) providing the total volume percentage of naphthenes in the sample. Mass spectrometry goes a step further by quantifying the naphthenes as a function of the cycloparaffinic ring number (see Table 4.5).

4.2.2.8 Nitrogen and Sulfur Contents—ASTM Methods D 3228 and D 3431 can be used for nitrogen and D 1275, D 2622, D 3120, and D 129 for sulfur. More precise, specific measurements can also be obtained by mass spectrometry and neutron activation analysis [*72*].

4.3 Cooling Properties

4.3.1 Heat Transfer Properties

Dissipation of the heat generated by joule and other electrical losses in electrical equipment is a major function of electrical insulating oils. For example, efficient heat transfer by the oil reduces the temperature in transformer windings during both normal and emergency overload conditions [*35–40*]. This in turn prolongs the life of the transformer by minimizing the rate of degradation of the cellulosic paper and press board used as the primary electrical insulation in the winding. Several devices or methods have been proposed in recent years to evaluate the temperature of hot spots resulting from unproper heat dissipation in the equipment [*41–47*].

Heat dissipation in a transformer is effected mainly by the natural or forced convection of oil through cooling ducts incorporated in the transformer windings. Knowledge of the thermal conductivity, specific heat, viscosity, coefficient of thermal expansion, and density of the oils used in a transformer, at the various operating temperatures, is essential for transformer design engineers to decide: (*a*) the type of oil circulation (natural or forced convection, with the addition of oil circulation pumps); (*b*) the size and configuration of the cooling ducts that will provide the adequate oil flow to dissipate the heat generated; and (*c*) the thermal time constants of the equipment, under normal and overloading conditions and in the case of accidental forced-pumping failures. The coefficient of thermal expansion must also be known in order to select oil conservator volumes adequate to accommodate expansion and contraction during in-service temperature variations.

The thermal conductivity, specific heat, and coefficient of thermal expansion of mineral oils vary with density (that is, naphthenic or aromatic content or both) and with temperature (Fig. 4.11) according to the following expressions

$$C_P = (0.403 + 0.008t)d^{-1/2}$$
$$\lambda = 0.000\,28(1 - 0.0054t)d^{-1}$$

where d is the density of oil at 20°C, t the temperature in °C, C_P the specific heat in cal/g/°C, and λ the thermal conductivity in cal/s/°C/cm. For a given transformer design or in places where oil is allowed to flow freely, the cooling efficiency will not be significantly affected by C_P or K, but will depend mainly on the flow properties and, hence, the viscosity of the oil, whatever the type of mineral oil used.

Specific heat and thermal conductivity values are useful for evaluating temperature gradients and possible hot-spot formation in places where the oil is trapped. Inside oil-impregnated paper insulation for instance, a temperature gradient of 15°C has been reported across a layer of only 0.4 mm of oil-impregnated paper covering a copper conductor [*44*].

Density is used to calculate the mass of a given volume of oil and in subsequent calculations. Knowledge of density is also a guide to chemical composition, but only as part of D 2140 or D 3238 (see Section 4.2.3.4). In general, paraffinic, naphthenic, and aromatic hydrocarbons have comparatively low, intermediate, and high densities, respectively.

Knowledge of density may be important in specific applications, such as electrical equipment in cold climates where ice may form in de-energized transformers exposed to temperatures below 0°C. The maximum oil density in such equipment should ensure ice will not float at any temperature the oil attains (see Fig. 4.12). Also, the density of insulating oil added to apparatus in service should not be so different that the oil forms a separate layer, since this could cause serious overheating of self-cooled apparatus.

4.3.2 Flow Properties

The flow of the cooling liquid is the principal factor in the transfer of heat by convection, and the resistance to flow of the liquid, the dynamic viscosity (the ratio of the force required to maintain

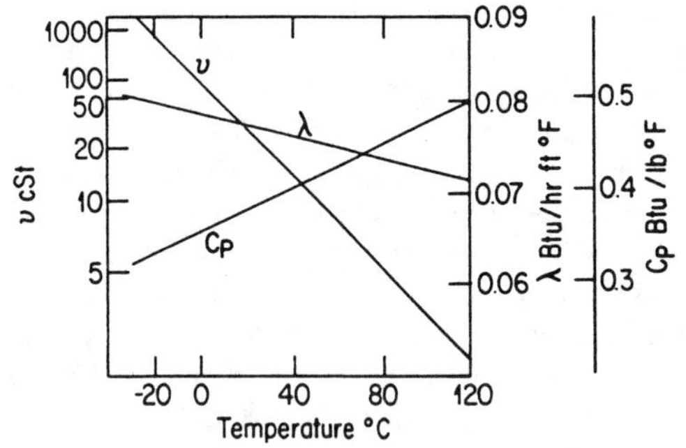

C_P = specific heat
λ = thermal conductivity
ν = viscosity

FIG. 4.11—*Variation of thermal properties with temperature* [5].

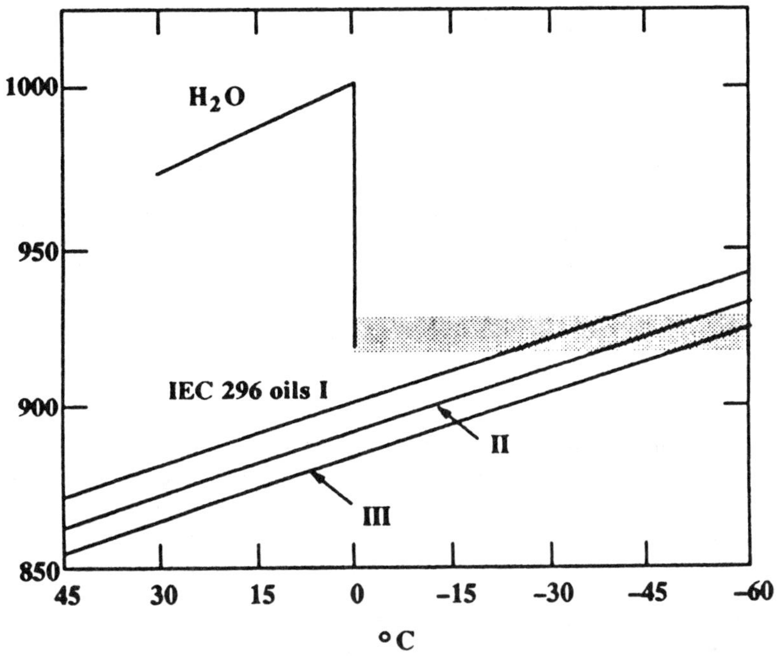

FIG. 4.12—*Density of IEC 296 oils and ice, in kg · m^{-3}, versus temperature* [7].

unit flow velocity of the liquid) is the primary parameter determining the efficiency of heat transfer systems in electrical equipment. Viscosity is dependent strongly on molecular composition, increasing with molecular weight, size, and rigidity. So long as the molecular organization is random and there is little long range ordering in the liquid, its viscosity is independent of velocity of flow, and the liquid is said to be "Newtonian." The viscosity of a Newtonian liquid increases markedly as the temperature is lowered according to the empirical Walther equation

$$\log \log (\nu + K) = A + B \log T$$

where ν is the kinematic viscosity (the dynamic velocity divided by the density of the oil at absolute temperature T (see Section 4.3.2.5), A and B are constants for a given liquid, and K is a universal constant for values of $\nu \geq 1.5$. Heat transfer requirements differ from one type of electrical equipment to another, and the temperature dependent viscosity of the insulating oils used is individually specified for each type of equipment. As an example, the temperature dependence of the viscosity of transformer and switchgear oils specified in IEC Standard 296 is shown in Fig. 4.13. The low temperature viscosity of insulating oils is also specified in another indirect way. On cooling (under specified test conditions), an oil will reach a temperature at which it will no longer flow under its own weight; the temperature at which this occurs is termed the pour point. The maximum pour point of transformer oils specified in ASTM D 3487 is −40°C, for example. The preceding relationship is valid from approximately 5°C below the flash point to 5°C above the pour point for mineral oils which contain few long chain linear paraffins, for example, naphthenic and dewaxed paraffinic oils.

As the temperature of a mineral insulating oils is lowered, any long chain linear paraffin molecules present begin to order themselves [*49,84*]. If the concentration of these molecules is sufficiently high, microcrystals can form and precipitate, and ultimately grow together into a solid

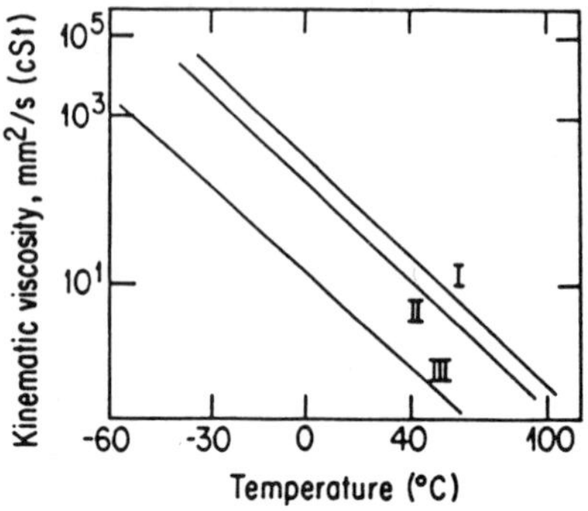

FIG. 4.13—*Viscosity of IEC 296 oils versus temperature* [7].

wax phase as the temperature is lowered still further. As this ordering occurs, the force required to maintain unit flow velocity increases with velocity and the observed viscosity is found to increase as the shear rate on the oil decreases. The effect of temperature on the viscosities of a paraffinic base oil containing 8% linear paraffins and of a naphthenic base oil containing 1% are shown in Fig. 4.14. In rheological terms, the paraffinic base oil has become a non-Newtonian "pseudoplastic" fluid. A "yield stress," a minimum force, may be required to even initiate flow in the waxy phase. The standard measurement of viscosity is made at low and internally variable flow rates. As the temperature is lowered and the degree of non-Newtonian behavior increases in an oil, viscosities measured by the standard methods are found to be increasingly higher than those predicted by the Walther equation from higher temperature measurements on the same oil. Viscosities measured at high shear rates, however, approach the predicted values at low temperatures.

The cloud point is the temperature at which precipitation of wax crystallites can be observed visually as a haze (under the same conditions specified for the measurement of pour point). Undewaxed paraffinic transformer oils with acceptable room and operating temperature viscosities were found to have cloud points as high as −10 to −15°C and pour points in the −20 to −30°C range [*84*]. Obviously the temperatures in an operating transformer are much too high for wax formation to occur, and it is not entirely clear that this behavior presents a problem with heat transfer during start up in a transformer even at very low ambient temperature. The poorer heat

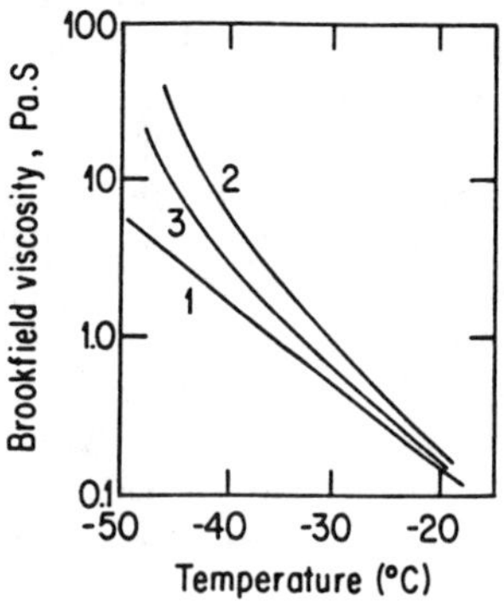

1. Naphthenic oil
2. Paraffinic oil
3. Paraffinic oil + 0.5 to 2% flow-improving additive.

FIG. 4.14—*Non-Newtonian behavior of paraffinic oils at low temperatures* [49].

transfer due to higher viscosities simply hastens the local rise in temperature which in turn hastens the melting of the wax, and the reversal of the ordering returns the heat transfer to its expected level. Unacceptably high local temperatures should not be reached. (Two other potential effects of this low temperature behavior, however, should not be overlooked. The increased drag on moving parts such as switches may result in improper operation. Secondly, the formation of wax and its contraction with decreasing temperature can lead to cracks and fissures in the oil in out-of-service equipment in cold climates. This could result in electrical breakdown on voltage application during startup.)

Flow-improving additives enhance the viscosity of such oils at low temperatures, without eliminating their non-Newtonian character (see Fig. 4.14). Some additives are complex aromatic hydrocarbons which contain an alkyl side chain of similar length to the linear paraffins present in the oil [*59,61–62*]. Other efficient flow-improving additives, alkyl-methacrylate polymers, and alkylated-polystyrenes, act as surface active agents, preventing the paraffinic crystals from growing into long interlocking chains and dispersing them rather into clusters around the additive molecules. Additive efficiency is optimum at concentrations around 0.5 to 1% for the paraffinic oil of Fig. 4.15. Flow-improving additives also act as pour point depressants in the case of oils containing free linear paraffins. The incorporation of an additive is usually equivalent to removing all the linear paraffins from the oil at the refining level (see Fig. 4.15).

However, the beneficial effect of the additive disappears when the amount of linear paraffins is too high. Figure 4.15, for instance, depicts what can be observed in the case of a paraffinic oil containing 8% of linear paraffins and 0.45% of pour point depressant additive. This oil has a pour point temperature of −49°C and a cloud point temperature of −21°C. When this oil is cooled under the relatively rapid conditions of ASTM D 97, it solidifies only on reaching −49°C (pour point temperature). However, if it is cooled down to −45°, and kept there for several days, it will remain fluid and suddenly solidify after approximately ten days. After the ten-day soaking period, the pour point temperature of the oil is found to be −45°C. If it is cooled down to −22°C and kept at this temperature, it will again first stay fluid and suddenly solidify, this time after only two days (see Fig. 4.16). Its new pour point temperature will therefore be −22°C after the two-day soaking period. Similarly, when kept at −20 and −29°C, the oil will take ten days to solidify (see Fig. 4.16). It thus clearly appears

1. and that the pour point temperature of this paraffinic oil is not constant;
2. and that it depends on the period of time the oil has been kept at a low temperature (soaking effect); and

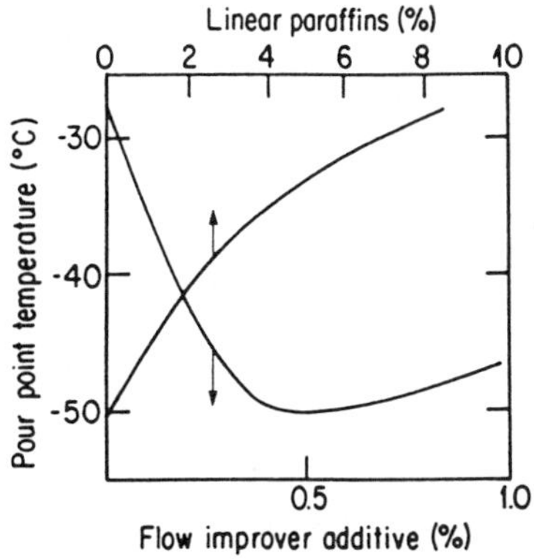

FIG. 4.15—*Pour-point depression of paraffinic oil by dewaxing versus addition of a flow improver* [51].

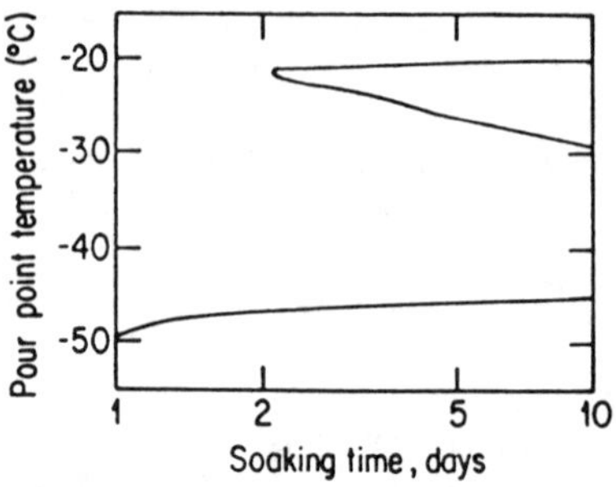

FIG. 4.16—*Variation of the pour points of a paraffinic oil versus soaking time at various low temperatures* [51]. *(Paraffinic oil with 8% linear paraffins and 0.45% flow-improver additive).*

3. that the pour point depressant additive, as mentioned above, is not active anymore under such conditions.

A similar soaking effect can be observed with other types of paraffinic oils, and has been related to the amount of linear paraffins present in these oils [*51*]. In particular, it has been shown that the amount of linear paraffins in oil must be under a critical value in order to avoid such a soaking effect and to keep the pour point depressant additive fully efficient. For the oil of Fig. 4.16, for example, this critical value has been evaluated experimentally as 4% (by fully dewaxing the oil, then adding increasing amounts of its own linear paraffins to the oil, until the soaking effect is observed again). For other types of oils, the linear paraffin critical content may be quite different (Fig. 4.17): for example 2% for paraffinic oil P_3 (as compared to the 4% of paraffinic oil P_1 the oil used previously for Fig. 4.16, and to the 4% of naphthenic oil N). Further investigations have shown that these critical contents are related to the average melting temperatures or carbon atom numbers of the linear paraffins present in the oils. In Fig. 4.17, for example, the linear paraffins in oils P_1 and N were C_{15} waxes melting at 13°C (C_{18} waxes melting at 27°C in P_3, and C_{14} waxes melting at 6°C in P_2), and it can be seen that the higher the melting temperature of the linear

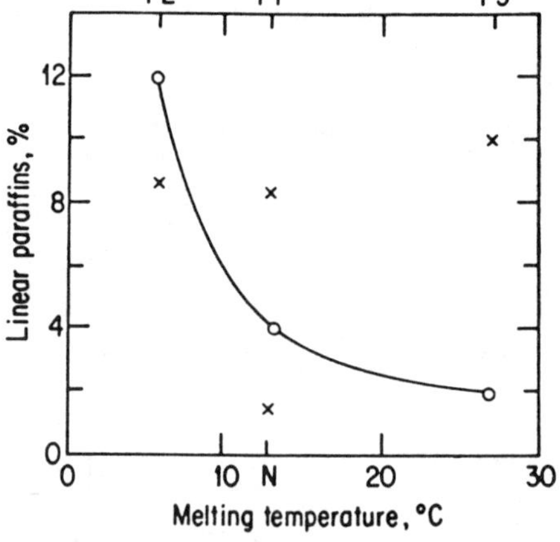

FIG. 4.17—*Critical content of linear paraffins in oils versus their average melting temperature:* ○ *critical content;* × *actual content in four commercial oils (three paraffinic oils P_1, P_2, P_3, and one naphthenic N)* [51].

paraffins, the lower their critical content in the oil. The practical implications of Fig. 4.17 are the following: in order to be acceptable for use at low temperatures:

1. Oils P_1 and P_3 should be dewaxed at least down to 4 and 2%, respectively;
2. Oils P_2 and N can be used without treatment; and
3. New paraffin oils proposed on the market should be analyzed for linear paraffinic content and melting temperature in order to determine the extent of any further dewaxing necessary.

The presence of a pour point depressant and of free linear paraffins in oil can be detected from the difference between the pour point and cloud point temperatures. Methods developed to analyze quantitatively the additive content in mineral oils [*32*] have shown the additive to be quite stable with respect to oxidation and arcing but degraded by gas-phase partial discharges, like the other aromatic compounds of oil [*52*].

The efficiency of oil flow and cooling is affected mainly by the oil viscosity, but it may also be impaired by sludge accumulation in the cooling ducts of very old equipment, resulting from oil aging in service. Desludging with hot new oil is possible in some cases to regain some of the original efficiency.

4.3.3 Test Methods

4.3.3.1 Thermal Conductivity—The thermal conductivity of an oil is a measure of its ability to transfer energy as heat in the absence of mass transport phenomena. In the case of mineral oils, the thermal conductivity can be determined according to ASTM Test Method D 2717 by measuring the temperature gradient produced across the oil by a known amount of energy introduced into a glass cell by electrically heating a platinum resistance element. It is expressed in such units as W/m·K or cal/s·°C, or Btu·in.h·ft^2·°F.

4.3.3.2 Specific Heat—The specific heat or heat capacity of oil is the amount of energy required to produce a given temperature change within a unit quantity of oil (cal·K^{-1}·g^{-1}). It can be measured according to ASTM Test Method D 2766 by transferring a container with the oil sample from a heated furnace into a water-filled adiabatic calorimeter.

4.3.3.3 Coefficient of Thermal Expansion—The coefficient of thermal expansion of oil is the change in volume per unit volume per degree of temperature. It can be calculated according to ASTM Test Method D 1903 by determining the specific gravities of oil (using Test Method D 1298) at any two temperatures below 90°C but not less than 5°C or more than 14°C apart. The average coefficient of expansion (commonly expressed in 10^{-4}K^{-1}) over the given temperature range is

$$K_{exp} = \frac{S - S_1}{S(T_1 - T)}$$

where S is the specific gravity at the lower temperature T and S_1 is the specific gravity at higher temperature T_1.

4.3.3.4 Density—Density is defined as the mass (weight *in vacuo*) of liquid per unit volume at 15.6°C (60°F). Specific gravity is the ratio of the mass of a given volume of liquid at 15.6°C to the mass of an equal volume of pure water at the same temperature. In some cases, API gravities are also found, derived from the specific gravity by the relationship

$$\text{degrees API} = 141.5/\text{sp. g } 60/60°\text{F} - 131.5$$

The density (or specific gravity, or API gravity) varies with temperature, as shown in the comprehensive tables published by the Institute of Petroleum (see also ASTM D 1250).

Density and specific gravity can be measured according to ASTM Test Methods D 1298 (using a hydrometer) or D 1481 (using a pycnometer). For oils less viscous than 15 cSt at 20°C, ASTM D 941 and D 1217 are available. API gravity can be measured according to ASTM methods D 287 or D 1298, using a hydrometer.

4.3.3.5 Viscosity—The viscosity of an oil is a measure of its resistance to flow. It can be thought of as a type of internal friction which prevents the oil from moving as freely as it would normally. There are several ways of measuring and reporting viscosities, the most commonly used being dynamic viscosity and kinematic viscosity. The dynamic viscosity (η) is defined as the force per unit area required to maintain a unit velocity gradient in the oil. If the ratio of the force to the velocity gradient remains constant with increasing velocity (rate of shear), the liquid is said to be Newtonian. The SI unit of dynamic viscosity is 1 $N{\cdot}s/m^2$ or Pa·s, which is equivalent to 10 poises (P). (The centipoise is used frequently, 1 cP = 10^{-2}P). Since viscosity is determined normally by the time of flow of a liquid under its own weight in a capillary of known dimensions, its value is also reported in terms of kinematic viscosity (ν), which is defined as the dynamic viscosity divided by the density. The SI unit of kinematic viscosity is 1 m^2/s, which is equivalent to 10^4 stokes (St). Frequently the centistoke (cSt) is used (1 cSt = 10^{-2} St).

Dynamic viscosities less than 15 mPa·s can be measured according to ASTM Test Method D 445 using U-tube capillary glass viscometers of the Oswald, suspended-level or reverse-flow types, with insulating oils having a Newtonian flow behavior. Viscometer calibration and operating instructions are detailed in ASTM Test Methods D 446 and D 2162. A viscometer basically consists of a calibrated capillary glass tube through which a fixed volume of oil is allowed to flow under gravity at a closely controlled temperature (Fig. 4.18). The measured flow time (in seconds), multiplied by the calibration constant of the viscometer (determined with viscosity standards), provides the viscosity of the sample.

Viscosity data from the literature are sometimes referred to in terms of empirical units of no real physical significance, such as Saybolt seconds, which correspond to the efflux time of a 60-mL oil sample flowing through a calibrated orifice, measured under carefully controlled conditions, according to ASTM Test Method D 88 (now considered valid only for asphalts and heavy fuel oils). Conversion of these units into dynamic or kinematic viscosities is only possible by complicated and inaccurate correlations, which are described in ASTM Test Method D 2161.

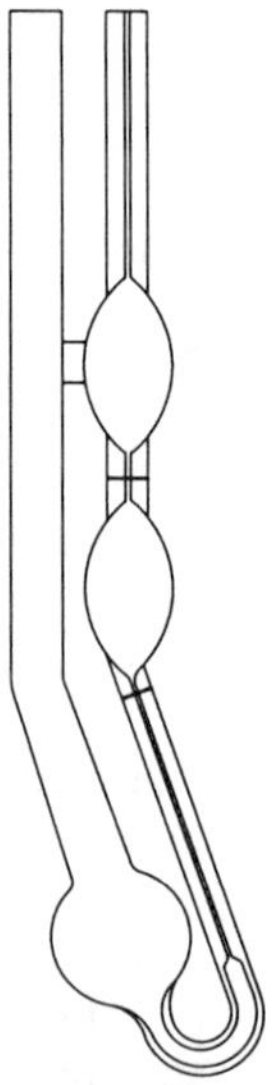

FIG. 4.18—*Typical viscometer for transparent liquids (Cannon Fenske Routine type).*

When the oil has an excessively high viscosity or does not show Newtonian behavior (as often occurs at very low temperatures), only relative values of apparent viscosity can be obtained. These depend on the configuration of the measurement setup or on the shear rate during the measurements. Measurement of the pressure drop across a continuous flow of oil through a pipe at low temperatures has been used, for instance, to simulate real conditions in transformer cooling ducts and radiators [*49*].

The Brookfield viscometer method, although not officially recommended in ASTM D 117, is probably the most widely used under such conditions. As described in ASTM Test Methods D 2699 and D 2983, it consists in measuring the resistance of a spindle placed in the oil sample to stirring. The apparent viscosity must be given as a function of spindle speed (shear rate) and temperature, the viscosity range being 1 to 1000 Pa·s. With non-Newtonian paraffinic oils especially, the cooling rate or soaking time of the oil sample prior to measurement of its viscosity is an important factor and should also be controlled carefully.

4.3.3.6 Pour Point Temperature—The pour point temperature of oils is the lowest temperature at which the oil is observed to flow when cooled and examined under the prescribed conditions of ASTM Test Method D 97. A special setup (Fig. 4.19) allowing smaller volumes of oil to be used and left for long periods of time (several hours to several days) at a low temperature before measuring their pour point has been described [*51*]. The setup was designed to avoid ice condensation from the atmosphere on the oil container during the soaking period. Cold nitrogen gas from a liquid-nitrogen carboy is circulated in a copper coil installed in the fixed insulated jacket to refrigerate the mobile cylindrical chamber and the capped glass test tubes containing the oil samples. A temperature programmer, connected to an electrovalve and a thermocouple placed in one reference tube, adjusts the flow of cold nitrogen to obtain the desired temperature ramp. Two other thermocouples are used to verify temperature uniformity in the chamber. Triple-glazed windows attached with rubber gaskets to each end of the cylindrical chamber prevent the condensation of ambient atmospheric moisture and allow for checks of the onset of oil cloudiness (cloud point). The chamber is rotated to observe the absence of oil flow (pour point).

4.3.3.7 Cloud Point Temperature—The cloud point temperature is the temperature at which haziness is first observed in a sample of oil cooled at a specified rate according to ASTM Test Method D 2500. A cylindrical glass jar containing the oil sample (approximately 50 mL) and a

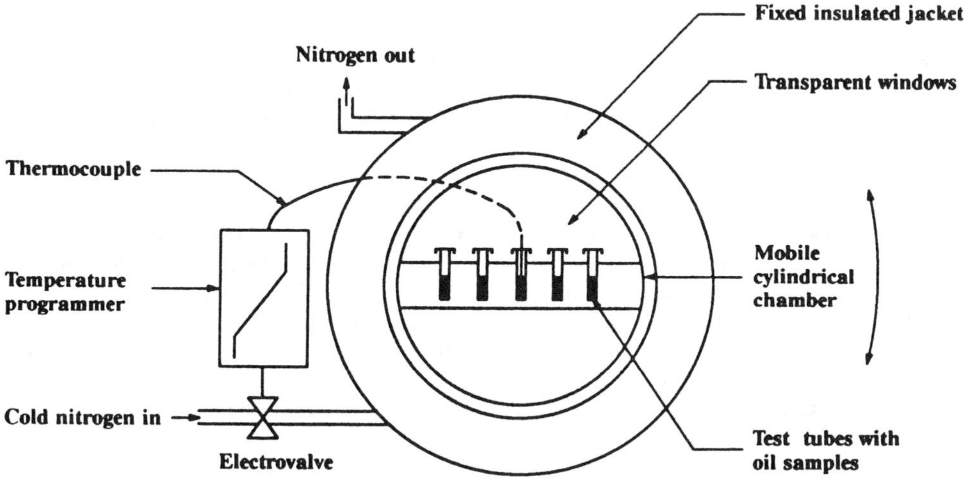

FIG. 4.19—*Experimental setup to measure the pour point and cloud point* [51].

thermometer are placed in a tube immersed in a refrigerating bath. Each time the oil temperature drops by 1°C, the jar is removed and examined until haziness is observed. Several baths may be necessary to reach the cloud point temperature. The special setup described in Fig. 4.19 has been found faster and more convenient to use.

The measurement of cloud point temperature, although not officially recommended in ASTM D 117, is very useful for evaluating paraffinic oils (see Section 4.3.2).

4.4 Flammability, Handling, and Disposal

4.4.1 Flammability

A mineral insulating oil is a mixture of hydrocarbons and, as such, exhibits the flammability characteristics typical of any petroleum product of similar boiling range and molecular composition. Fire hazard classifications and requirements for proper storage, installation, and use of these liquids are established by fire code making panels, government regulatory agencies and insuring companies. These bodies define the conditions under which the fire hazard of these liquids is judged to be acceptable, and their requirements should be adhered to strictly.

The usual basis for classification of the flammability of mineral insulating liquids is flash point (Cleveland Open Cup) and fire point. The flash point is the lowest temperature at which, under precisely controlled test conditions, the vapor forming above a heating pool of liquid ignites in air at a pressure of one atmosphere. The physical arrangement of the test apparatus, the rate of heating, and the characteristics of the ignition source are specified. The fire point is the lowest temperature at which, on further heating under the same conditions, the vapor formed over the pool will support combustion for 5 s. While the conditions specified for this test may never occur in actual use, experience has shown that the results of the test can be used to establish the requirements placed on mineral insulating oil in many applications. The minimum accepted value of the flash point of the oil used in transformers installed outdoors, the most common of the mineral insulating liquids, is 145°C (in some locales 135°C).

The use of mineral oils in new or different applications may require more intensive consideration of their flammability characteristics. A primary case in point today is the use of liquid filled transformers installed in confined spaces such as buildings. The accepted approaches to fire protection in enclosed spaces have been to install the transformer in a fire proof vault, to provide an adequate sprinkler system, or to use an insulating liquid with high fire resistance. Until the middle and late 1970s, nonflammable askarels, all of which then contained polychlorinated biphenyls (PCBs), could be used as this fire resistant liquid. Bans on production of PCBs eliminated that alternative in the manufacture of transformers throughout most of the world. The search for askarel replacements has resulted in the evaluation of the flammability characteristics of a number of candidate liquids and, in consideration of the relevance of the evaluation procedures themselves to actual fires involving transformers. Measurements have been made of oxygen indices, of net calorific values and heats of combustion, of rates of heat release and flame propagation and spread, of opacity of smoke, and of corrosiveness and toxicity of combustion products, as well as flash and fire points. New tests continue to be introduced and standardized. While it appears that none of these controlled laboratory tests alone can fully assess the fire hazards resulting from the actual use of liquid insulated electrical equipment, there is growing acceptance that a requirement for a fire point greater than 300°C is basic to acceptance of an insulating liquid in transformers installed indoors. Mineral insulating oils meeting this requirement have been in this service in the United States for at least a decade, as have polydimethyl siloxanes (''silicones'') and polyol esters.

The flash and fire point of a mineral oil are the result of the ignition and combustion of its most easily vaporized components. An increase in average molecular weight and an attendant increase in viscosity of a mineral oil is almost inevitably associated with an increased flash fire point. The heat transfer characteristics of the high fire point mineral transformer oil discussed above then differ from those of the more commonly used 145°C flash point oil, as does the design practice of transformers for its use.

4.4.2 Handling and Disposal

The effects on the natural environment of the spilling or improper disposal of petroleum crude oils and refined products have been examined widely in recent years. As a result, the contact of these materials with the environment is controlled in many countries by regulating transport, storage, and disposal. Procedures are often in place to minimize the effect of new spills and emissions and to clean up old sites. Mineral insulating oils are refined petroleum products, albeit highly refined, and normally contain only a small amount of an environmentally innocuous oxidation inhibitor as the single additive. As such, they must be handled and used with the same attention given to any mineral oil.

Special attention must be given to mineral insulating oils in transformers today which contain measurable amounts of polychlorinated biphenyls (PCBs). Concern for the potential health effects of PCBs led to bans on their production in many countries in the 1970s. Although continued use of electrical equipment containing PCBs generally was allowed, stringent regulation of their continued use, labeling, storage and disposal was introduced at the same time. A significant fraction of the mineral insulating oils in use in transformers today contain measurable amounts of PCBs, and it falls under these regulations. Regulations vary from country and from one subdivision to another within a country, and they also evolve with time. Users or testers of mineral insulating oils should inform themselves of the applicable requirements and follow them.

PCBs in low concentration have no effect on the functional characteristics of mineral insulating oils and were not added deliberately to transformer mineral oils. From the time they were introduced in the 1930s until they were banned, PCBs were a major constituent of the nonflammable askarel insulating liquids used in transformers manufactured for installation in enclosed spaces and other areas where fire risk is high, and in capacitors. Inadvertent mixing of these askarels with transformer mineral oils occurred prior to the introduction of regulations controlling their handling. Insulating oils typically have a useful life measured in decades and significant amounts of these contaminated oils remain in service. The amount of PCB-containing transformer oil in service continues to decrease as older transformers are replaced, and there are both public and private programs in place in many places to speed the replacement process.

4.4.3 Test Methods

4.4.3.1 Flash and Fire Point Temperatures—The Cleveland open-cup method (ASTM Test Method D 92) is the one most widely used in North America and is recommended for determining the flash and fire points of insulating oils. The methods involve passing a small flame over the gas space above a heated volume of oil and noting when the first flash of vapor ignition occurs, and when a flame is maintained for 5 s. The fire point is approximately 10°C higher than the open flash point. [A closed-cup method is also available for flash point (ASTM Test Method D 93)].

4.4.3.2 Auto-Ignition Temperature—The auto-ignition temperature can be evaluated according to ASTM Test Method E 659. A small (100 μL) sample of oil is inserted with a syringe into a uniformly heated 500-mL glass flask containing air at a predetermined temperature measured by a thermocouple. The contents of the flask are observed in a dark room. Auto-ignition is evidenced by the sudden appearance of a flame inside the oil flask at a temperature much higher than the flash-point and fire-point temperatures (Table 4.7).

4.4.3.3 Volatile Content—The boiling-range distribution and vapor pressure of oils can be measured according to ASTM Test Methods D 2551, D 86, D 2878, and D 2887 to evaluate the

TABLE 4.7—*Flammability properties of insulating oils as a function of viscosity* [7].

Viscosity, mm²/s, 40°C	3.5	16.5	100	460
Flash point, °C COC	65	145	204	237
Fire point, °C	75	155	214	247
Auto-ignition temperature, °C	230	250	345	410

presence of detrimental volatile compounds in oils or the maximum temperatures to apply during vacuum degassing and drying operations on a transformer in order to avoid eliminating the DBPC antioxidant (Fig. 4.20). In D 2551, a known volume of oil is introduced into an evacuated bulb. The rise in pressure resulting from partial oil vaporization in the vacuum is measured with a manometer and converted to vapor pressure. In D 86, a 100-mL sample of oil is placed in a distillation flask. The volumes of recovered condensates are recorded on an automatic apparatus as a function of vapor (boiling) temperatures, yielding a distillation curve from which the relative volatility of the sample can be evaluated. In D 2878, the weight loss of an oil sample is measured as a function of time under a circulation of air at a preselected temperature. The evaporation rate, compared to that of a standard compound, yields the apparent vapor pressure. In D 2887, the oil sample is introduced into a gas chromatographic column that separates hydrocarbons in boiling point order, determined

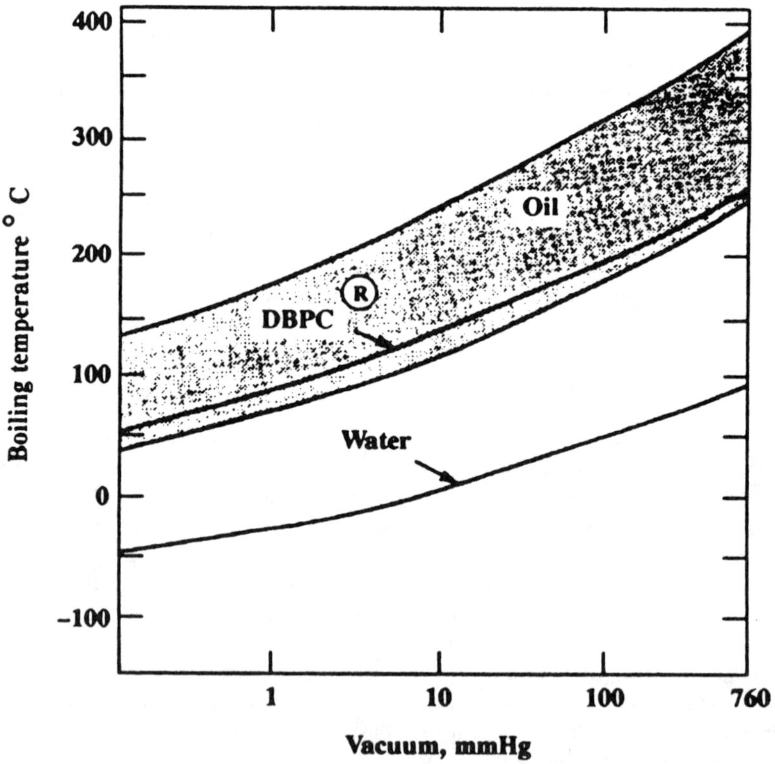

FIG. 4.20—*Boiling-temperature distribution of an oil versus the vacuum applied.*

by calibration with standard hydrocarbon mixtures. Low-molecular-weight oil decomposition material also has been detected by gel-permeation chromatographic techniques (see Fig. 4.6) [*28*].

4.4.3.4 Heat of Combustion—The net heat of combustion (net calorific value) can be determined according to ASTM Test Methods D 240 or D 2382. Results are expressed in MJ/kg.

A weighted amount of oil is burned in an oxygen bomb calorimeter under controlled conditions. The heat released by the combustion of a unit mass of oil at a constant pressure of 1 atm, computed from the temperature rise in the calorimeter and taking into account a number of experimental corrections, provides the net heat of combustion of the sample.

4.4.3.5 PCB Content—PCB contamination in oil can be measured according to ASTM Test Method D 4059 using gas chromatography. In this method, an oil sample is diluted with a suitable solvent (for example, heptane), treated with a polar absorbent or concentrated sulfuric acid to remove interfering substances, and injected in the gas chromatograph. The components are separated as they pass through the column with the carrier gas and measured by an electron-capture detector. Calibration is made by comparing with chromatograms obtained for standard Aroclors 1242, 1254, and 1260 under the same conditions. Alternative analytical techniques have also been proposed [*65*].

4.5 Insulation-Related Properties

4.5.1 In-Service Oxidation Resistance

Oxidation of mineral insulating oils results in the formation of organic acids, insoluble condensed products (sludges) and soluble polar intermediates. These products do not interfere with the operation of circuit breakers and tap changers where the arc-quenching characteristics of the insulating liquid are of primary concern. Acids and polar compounds contribute to dielectric loss in an insulating liquid and therefore oxidation can have a deleterious influence on the functioning of bushings and cables as well as current and voltage transformers [*127*]. However, because of their relatively simple construction, this equipment can be well sealed, and there is less opportunity for ingress of oxygen in service. The problem of oxidation control is greatest with transformers and reactors. The heat transfer characteristic of self and forced cooled liquid-filled transformers is the result of careful design to control the hot spot and winding temperatures in order to balance overall losses and to prolong the life of the thermally vulnerable solid cellulosic insulation. The formation of sludge on surfaces and in the liquid itself can interfere with the transfer of heat and cause the internal temperatures to rise to undesirable levels. The dielectric losses in the oil, on the other hand, contribute little to the total losses of a power transformer and, unless extraordinarily high, do not reduce dielectric breakdown strength or add to thermal runaway. Soluble oxidation products then have relatively little effect on the functioning of power transformers.

Design features of transformers can operate to reduce the oxidation of the insulating oil. Larger transformers are designed to reduce the access of oxygen (air) in one of several ways. Conservator units connected to the tank by means of a long oil tube not only compensate for variations in the oil volume with temperature but also reduce the ingress of air by reducing the surface of the oil-air interface and the convection of air-enriched oil back into the tank. Sealed, nitrogen-blanketed or continuously degassed [*87*] transformers further reduce, though never completely, the transfer of air into the oil.

The enamels widely used today as replacements for paper and cloth insulation and current magnetic steel coating reduce the contact of insulating oils with catalytic metals. Aluminum, a poor oxidation catalyst, has now largely replaced copper as the conductor in distribution transformers.

The various types of hydrocarbon insulating oils all oxidize according to the same general mechanism [*73*] (Table 4.8). The first stage is the attack by oxygen on hydrocarbon molecules to form peroxide or hydroperoxides, which then dissociate into very reactive free radicals (initiation)

TABLE 4.8—*Insulating-oil oxidation mechanism* [73].

Initiation	RH Initiator	$R^\cdot$
	$R^\cdot$ O_2	$RO_2^\cdot$
Propagation	$RO_2^\cdot + RH$	$RO_2H + R^\cdot$
	$R^\cdot + O_2$	$RO_2^\cdot$
Chain reaction	RO_2H	$RO^\cdot + OH^\cdot$
	$RO^\cdot + RH$ O_2	$ROH + RO_2^\cdot$
	$OH^\cdot + RH$ O_2	$H_2O + RO_2^\cdot$
Inhibition		
Chain-stopping inhibitors		
	$InH + RO_2^\cdot$	$In^\cdot + RO_2H$
Peroxide decomposers		
	RO_2H	inert products

RH–hydrocarbon
RO_2H–hydroperoxide
$R^\cdot$–alkylradical
$RO_2^\cdot$–peroxyradical
$RO^\cdot$–alkoxyradical
InH–inhibitor (phenol)
$In^\cdot$–inhibitor radical

which react with hydrocarbon molecules to form intermediate oxidation compounds such as aldehydes, alcohols and ketones (propagation) and new active radicals able to react with additional hydrocarbon molecules (chain reaction). Many complex reactions are involved at this stage, but the end products are organic acids or polymeric sludge materials (termination) or both.

Some compounds with a labile hydrogen or electron react more rapidly than the hydrocarbon molecules with the intermediate peroxide free radicals and are thus able to break the chain reaction and inhibit the overall oxidation reaction. Known as oxidation inhibitors, they can be present naturally in the oil or introduced deliberately as additives (Fig. 4.21(1)). Other compounds, called oxidation catalysts, accelerate the initiation step of reactive peroxide free-radical formation and the various other oxidation chain reaction steps (for example, copper).

The chemical composition of mineral oils strongly influences their oxidation resistance (see Fig. 4.2). Paraffinic compounds, if not protected by oxidation inhibitors, oxidize rapidly to ultimately form organic acids but, when a suitable inhibiting additive is present, they are more resistant to oxidation than aromatics. In uninhibited or inhibitor-depleted oils, some aromatics are natural inhibitors for the paraffinic compounds. Di- and tri-aromatics, able to form phenolic structures with good antioxidant properties during the oxidation process, are particularly favorable [*31*]. These aromatic compounds oxidize rapidly, however, with the possible formation of increasing quantities of oil-insoluble sludges, which are polymeric products resulting from aromatic-ring condensation.

An optimum compromise concentration of aromatics in insulating oils of between 5 and 15% C_A (IR) (see Fig. 4.2), therefore, is desirable to offer some natural oxidation protection to inhibitor-depleted paraffins (as well as gas absorption properties under electrical discharges, see Section 4.5.2.2) together with a still acceptable oxidation of the aromatics.

Some types of sulfur-containing hydrocarbons naturally present in the oil also act as very efficient and important natural inhibitors (see Fig. 4.21(1)). The precise chemical structure of such compounds in oil is not known but probable candidates are benzyl sulfides, sulfoxides and sulfones. These are all well-known hydroperoxide decomposers used extensively as synthetic additives in polyethylene and polyolefins, which have a paraffinic structure closely related to that of oils but a much higher molecular weight [*72*]. They are not likely to be present in H_2-treated oils, however. Other types of sulfur-containing hydrocarbons, act as oxidation catalysts; (Fig. 4.21(2)) mono- and di-substituted benzothiophenes, and cyclic sulfides and thiols may be in this category. Linear

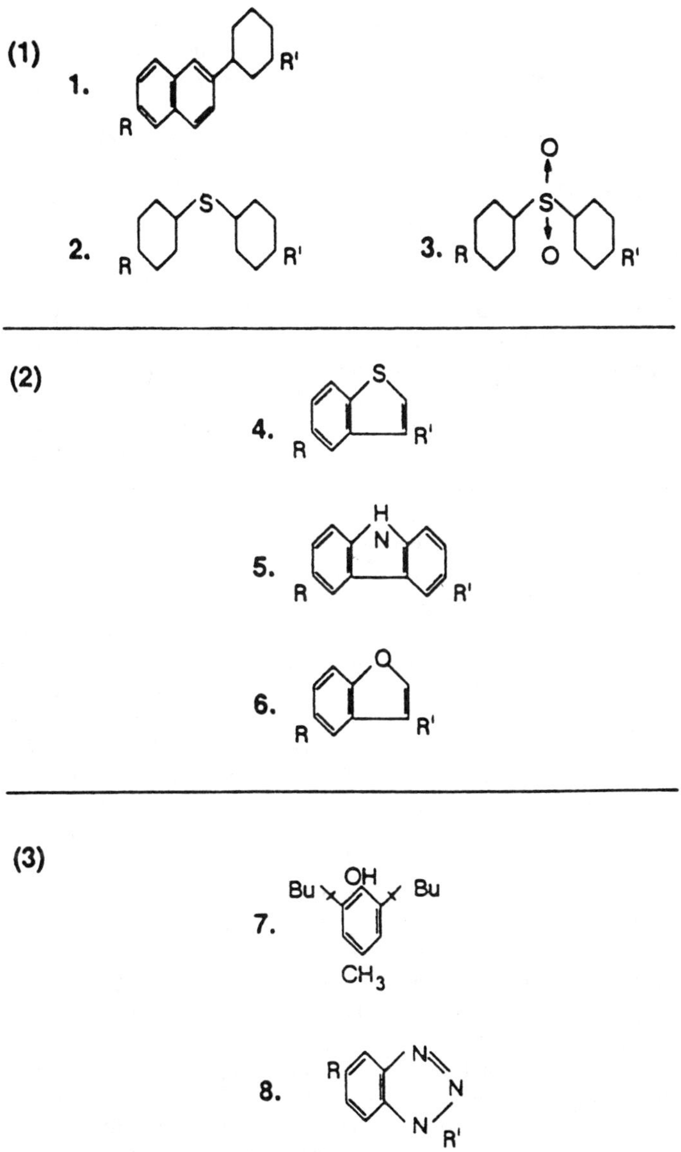

I. Natural oxidation inhibitors
1. Diaromatics
2. Benzylsulfides
3. Benzylsulfoxides

II. Natural oxidation catalysts
4. Benzothiophenes
5. Carbazoles
6. Benzofurans

III. Synthetic additives
7. Phenolic antioxidant (DBPC)®
8. Metal passivator (Reomet)

R, R^1 = alkyl groups

FIG. 4.21—*Oxidation inhibitors and catalysts.*

sulfides and mercaptans, on the other hand, have acidic properties which can cause metallic corrosion, but these usually are removed completely from the oil at the refining stage.

Hydrocarbons with nitrogen- or oxygen-containing rings or cycles, such as alkylated and carboxylated quinolines, indoles, carbazoles, porphyrins, benzofurans, benzoxazoles, and other heterocyclic compounds, may also act as oxidation catalysts (see Fig. 4.21(2)). Very little is known or reported about the specific influence of each of these compounds on the oxidation of insulating oils. When their presence is suspected, they usually are evaluated only on the basis of the total sulfur or nitrogen content of the oil. The content of these natural catalysts in some modern insulating oils, particularly thiophenes (see Fig. 4.21(2) and Table 4.5) is increasing as some oil refiners lose control over their supplies of crude oil.

Oil oxidation compounds, such as aldehydes, ketones, alcohols, and carboxylic acids, are themselves oxidation catalysts, so that the oxidation of hydrocarbon oils, once initiated, is said to be autocatalytic. The major initiator of oil oxidation, however, is the presence of dissolved metals, especially copper and to a lesser extent iron, which, even at very low contents (in the parts per billion to parts per million range), are responsible for accelerated peroxide free-radical formation. In electrical equipment, these metals originate from the windings and magnetic parts.

When the natural oxidation inhibitors cannot provide the required protection, it is often convenient to incorporate synthetic antioxidant or metal deactivator additives in the oil or both (see Fig. 4.21(3)). A widely used antioxidant additive in North America is DBPC® (2, 6 di-tert-butyl p-cresol) and, to a lesser extent, DBP (2, 6 di-tert-butyl phenol). Antioxidants do not prevent copper dissolution in oil (Fig. 4.22) but as long as they are present (induction period) they prevent the formation of free-radical peroxides, especially from the saturated hydrocarbon fraction, and the subsequent chain oxidation reactions. During this period they themselves are slowly oxidized and deactivated [*73*] and, when they have been almost completely consumed (end of the induction period), the above protection is no longer available. Accelerated formation of peroxides is then observed, followed by acids and polar oxidation compounds and, finally, by oil-soluble and -insoluble sludges. The relative saturation in the final stage of oxidation results from the buildup of inhibiting phenolic forms of the aromatics. The electrical properties ($\tan\delta$ and conductivity) are closely related to the observed oxidation steps and the mechanism applies to paraffinic as well as naphthenic oils [*68*].

Even though it has been demonstrated clearly that copper and similar metals dissolved in oil strongly accelerate the rate of oxidation of mineral oils, the copper dissolution mechanism and chemical nature of dissolved copper are more obscure. Some authors have suggested a reversible shift between the cuprous and cupric states in reactions involving hydroperoxides and peroxide radicals. These have not been identified experimentally however, and the copper usually is evaluated only as total copper content, with no attempt to determine whether it is ionic in form.

Recognizing the detrimental catalytic effects in insulating oils, authors proposed the use of metal-deactivator (or passivator) additives as far back as 1958 [*141,142*]. These have since been applied widely in Australia, but in North America economical and reliable metal deactivator additives did not become readily available until the recent development of commercial, oil-soluble derivatives of benzotriazole. They are now being reconsidered seriously in Canada. By absorbing on the surface of the copper metal, these compounds actually reduce the copper dissolution or its activity in oil or both (as well as the formation of peroxide free-radicals), preferably in combination with oxidation inhibitors, and therefore strongly reduce the oil oxidation rate. An additional advantage of importance may be the improvement in the dielectric properties, especially $\tan\delta$, which is strongly dependent on the dissolved-copper content [*127*] (Fig. 4.23).

Accelerated tests are used to compare the oxidation resistance of various types of oil and of additive-oil combinations, because testing under normal service conditions would take too long. In order to accelerate the oxidation process, the temperature is increased (oxidation rate doubles approximately per 7 to 10°C rise), as well as the amount of oxygen and copper. The relevance of

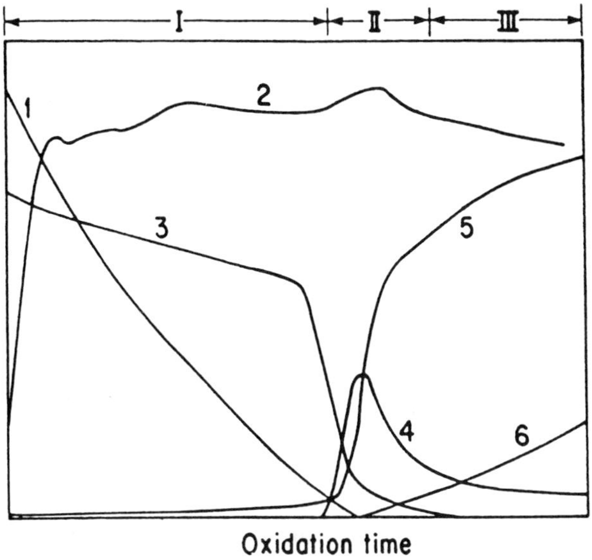

I. Induction period
II. Acceleration period
III. Saturation period
1. Antioxidant content
2. Copper content
3. Interfacial tension
4. Peroxide content
5. Acidity
6. Sludge content
Arbitrary relative ordinate units.

FIG. 4.22—*Oxidation of insulating oil versus time* [68,73].

such tests, however, is queried regularly, especially in Europe, on the grounds that chemical reactions occurring under accelerated conditions are not the same as in service, especially when using oxidation inhibitor additives. The latter would thus appear more efficient under the conditions of high oxygen availability used in accelerated tests than when performing under the low-oxygen conditions encountered in service. Inhibited oils might even start oxidizing while appreciable amounts of inhibitor are still present, which has raised some controversy [*93*] about the real advantage of using inhibitor additives, except for equipment operating under the most severe conditions. In any case, care always should be exercised when comparing very different oils. Highly refined oils with a low aromatic content will have a better inhibitor response during the induction period but will oxidize very rapidly thereafter, whereas oils with a higher aromatic content will start oxidizing during the induction period but will not oxidize so rapidly afterwards. A base oil with an optimum aromatic content is usually necessary as a compromise between these two conflicting behaviors (Fig. 4.24). Bearing these limitations in mind, accelerated tests are useful to assure the continuity of quality of relatively similar types of oils.

The most important physical and chemical properties to monitor during the induction period are the interfacial tension, antioxidant (metal passivator), and copper content, and the amount of polar-oxidation compounds. Then come the peroxide content, neutralization number, carbonyl infrared

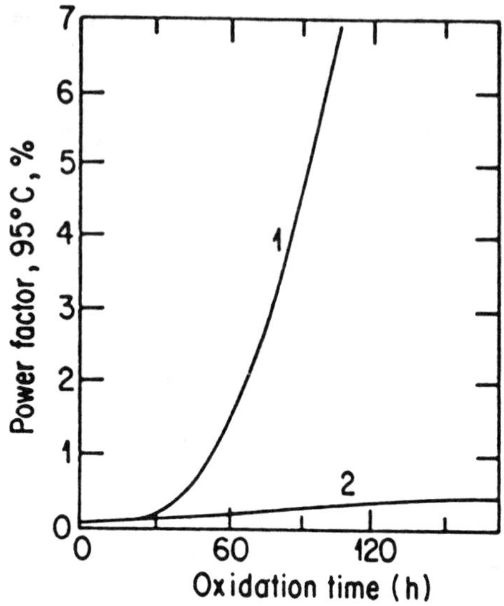

1. Original oil
2. Oil with metal deactivator additive

FIG. 4.23—*Influence of metal deactivator additive on the dielectric dissipation factor of an oxidizing oil* [127].

absorption and the sludge content during the acceleration and saturation periods (Fig. 4.22). Acceptable limits proposed for the interfacial tension and neutralization number for used transformer oils in service are indicated in Table 4.3. Oxidized oils can be reclaimed, when economically justifiable, to eliminate acids, sludges, and other oxidation compounds. However, it has been shown recently [*69*] that oils oxidized up to the saturation stage (Fig. 4.25) cannot be restored to

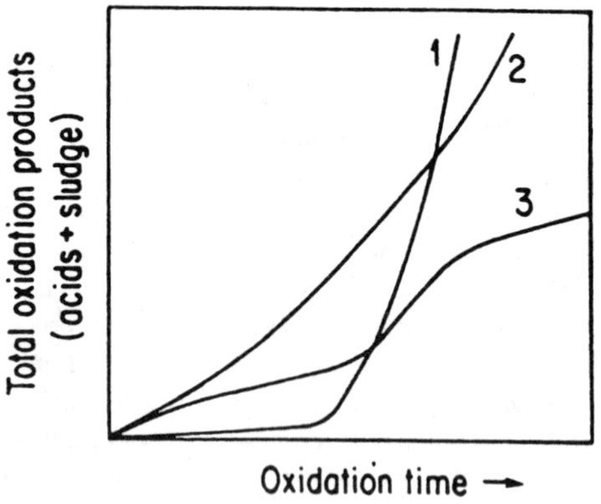

1. Low aromatic content (<5% C_A, IR), inhibited
2. High aromatic content (>15% C_A, IR)
3. Optimum aromatic content (≈5 to 10% C_A, IR), inhibited

Arbitrary relative units.

FIG. 4.24—*Typical oxidation behavior of insulating oils.*

heir original properties by reclamation. To be fully efficient, reclamation should be done at the end of the induction period, before the natural inhibitors of the oils have been consumed [*67,69*] (for the lefinition of the various oxidation periods, refer to p. 344 and Fig. 4.22).

Transformers contain relatively large amounts of cellulosic insulation, a number of other solid nsulating materials, and many materials of construction. All degrade in service. These materials or heir degradation products may intervene in the chain of oxidative reactions of the oil in actual service in transformers in ways which are not reflected in the laboratory studies or accelerated tests, liscussed previously. The oxidation of transformer oils in actual transformers undergoing accelerated aging was investigated in an EPRI sponsored study [*82–84*]. Commercially manufactured 25-kA distribution transformers were filled with several transformer oils, both with and without added DPBC, and tested according to Standard IEEE 345-1972 accelerated life test protocol for listribution transformers. Top oil temperatures were maintained at 135°C, and the units were tested at five to six times the standard ANSI life expectancy by maintaining the temperature of hot spots in he winding at 180 to 185°C for five of every seven days over a forty-week period.

Air was admitted to half of the transformers during the two day cooling period; the remaining units were kept sealed. The oils in the sealed units showed little degradation after the first week. Oxidation of the oils in the units with ingress of oxygen began at the start of test and continued hrough the testing. The dissipation factors, acidity, color, and visual sludge all increased. The rates were essentially the same for the several oils and were independent whether DBPC was present or not. In addition to examination of samples of oil taken from the units, several electrical measurements were made weekly on the transformers themselves. Transformers were tested to failure under full wave impulse voltage at the end of the forty weeks. The power factors (HV winding or LV winding to ground) of the units in which oxidation was occurring increased throughout the test; otherwise, the electrical characteristics were unaffected.

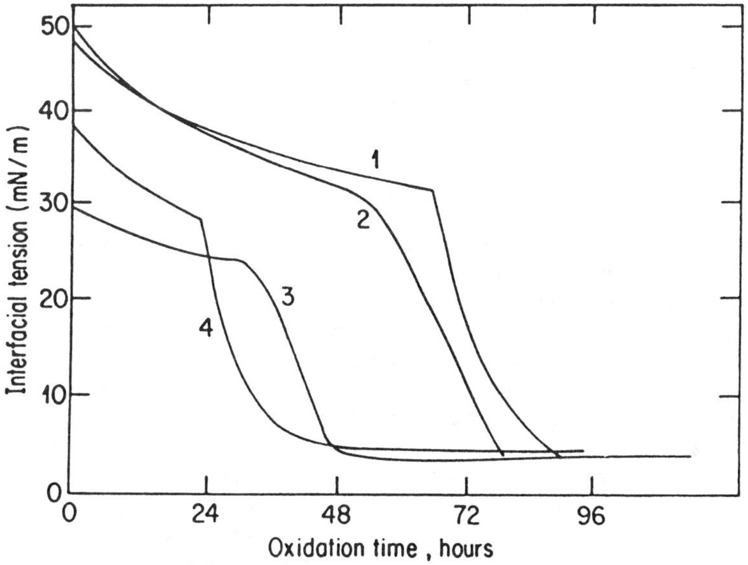

1. New, unaged oil, inhibited
2. Oil aged up to the end of the induction period, reclaimed and reinhibited
3. Oil aged up to the end of the induction period, not reclaimed but reinhibited
4. Oil aged up to the saturation period, reclaimed and reinhibited.

FIG. 4.25—*Oxidation resistance of reclaimed oils* [69].

Concurrently, the kinetics of the oxidation of the same oils were examined using the bom apparatus designed for the D 2112 test (see Section 4.5.4.1). Measurements were made at tw temperatures in order to determine the zero-order initiation step rate constant over the period wh little oxygen is being consumed and the first-order oxygen consumption rate constant as the oxyg partial pressure falls to zero. The activation energies for the two processes are determined from t temperature dependence of the rates. The rate constants for the initiation step (the inverse of t inhibition time) differed from oil to oil indicating that the natural inhibitor content of the o differed. The addition of DBPC both decreased the rate constants (lengthened the inhibition time and increased the activation energy, suggesting that the mechanism of inhibition was changed the DBPC. The ratios of acid and sludge in the product mix resulting from the oxygen upta appear to vary from oil to oil, but the rate constants and activation energies for oxygen uptake a surprisingly similar for long-accepted oils, new oils offered for approval, and service-aged oils th have been cleaned for one reason or another.

Taken together, these results suggest that the inhibition processes characterized in laborato studies play little or no part in the oxidation process actually occurring in transformers, an therefore, that neither the inhibitors present naturally in oils nor added DPBC have much effect in-service oxidation. In-service oxidation is controlled by the propagation process and its rate c be limited only by control of the availability of oxygen. The primary differences in oxidati behavior between oils meeting the other chemical, electrical, and physical requirements for use transformers are in the relative formation of acids and other soluble species, and sludges, presu ably resulting from rather gross differences in the relative quantities of aromatic, naphthenic, a paraffinic hydrocarbons making up the bulk of the oils.

Care should be taken in extrapolating these results. The transformers tested here were represen tive of one combination of materials; for example, coil conductors were made of enameled alun num, copper was available only from the brass in bushing studs, and paper insulation thermal upgraded with dicyandiamide was used. However, this combination is not atypical of those used modern transformers. These results suggest that further understanding of the mechanism of oxid tion occurring in actual transformers undergoing service aging is needed in order to predict a optimize the functional behavior of oils intended for this application.

4.5.2 Electrical and Thermal Decomposition Processes

Insulating oils in electrical equipment may be subjected to various thermal and electrical stress in service. Eddy currents and stray losses may induce very high temperatures (700 to 1000°C) transformer cores or loose metallic parts. Abnormally high dielectric losses in the paper- insulation may also lead to thermal runaway at lower temperatures (200 to 300°C). High-ener arcing in the high-voltage windings and towards the transformer tank, core, and low-volta windings, with arc temperatures above 6000°C, are observed as well as lower-energy arcing li sparks in oil or tracking along the paper insulation. Partial discharges of only a few picocoulom can occur in gas occlusions.

4.5.2.1 Gas Formation—Oil molecules subjected to electrical breakdown or high temperatu undergo various decomposition reactions, which lead to the formation of mixtures of acetyle ethylene, methane, ethane, and hydrogen, together with varying quantities of C_3 hydrocarbo These gases may be found in solution in the oil, at the transformer gas relay, and in the gas space nitrogen-blanketed oil preservation systems. Thermodynamic [*197,207,209*] and empirical obser tions indicate that higher reaction temperatures favor more dehydrogenation and the formation unsaturated gases, as these have higher bond energies and are more stable at high temperatur Acetylene (230 kcal/mol) is thus mainly associated with arcing (where temperature can rea thousands of degrees), ethylene (167 kcal/mol) with high temperature (300 to 800°C) hot spots a hydrogen (104 kcal/mol) with low-temperature gas-phase discharges. Further distinctions wit ch of these fault categories may be given by methane and ethane. Carbon monoxide and carbon oxide result from the decomposition of cellulose insulation, while carbon dioxide in conservator-pe equipment also results from atmospheric air ingress in the oil. The relative proportions of these ses formed under different conditions have been studied extensively [*163–214*] in order to agnose faults in electrical equipment. Various codes and interpretation schemes have been oposed, each related to a particular electrical network or piece of apparatus. Standardization has d to the internationally accepted interpretations table shown in Table 4.9 [*195*], although there is ll much diversity of opinion about this method. The ability of the guidelines in this table to present adequately all cases observed in service has been questioned recently, and it is currently der revision by the IEEE, IEC, and CIGRE. Other methods have been developed to overcome esent limitations, for example the triangular representation of Fig. 4.26 [*203*]. Identification of the tual cause of a fault can be further complicated by the fact that several faults can be occurring nultaneously. Distinction between the gases formed initially and as a later consequence of a fault ay also be a problem in some cases (for instance, a hot spot followed by short-circuit arcing).

Discovery of fault gases in low concentrations can serve as a sensitive early warning of uipment failure. The limits placed on the maximum concentrations of fault generated gases found ssolved in oil and gas formation rates judged to be acceptable in continued service are very portant factors for determining whether these incipient faults are significant and potentially rmful to equipment. Large variations can be observed in practice. High-voltage current trans-rmers, for example, have had severe failures in service with concentrations as low as 37 ppm H_6 and 21 ppm CH_4 [*127*], while other types containing over 100 000 ppm H_2 have been erating satisfactorily for months or years. Considerable experience is necessary to make the oper decisions, based on the type of equipment used, its loading, operating and maintenance story, the number of years in service, the nature of the fault involved, and gas formation trends. dgments based on so many issues depend on the skill and intuition of experienced and expert

TABLE 4.9—*IEC and IEEE codes for the interpretation of dissolved-gas analysis* [195].

$\frac{CH_4}{H_2}$	$\frac{C_2H_4}{CH_4}$	$\frac{C_2H_4}{C_2H_6}$	$\frac{C_2H_2}{C_2H_4}$	Suggested Diagnosis
).1 .0	<1.0	<1.0	<0.5	normal
.1	<1.0	<1.0	<0.5	partial discharge, corona
).1	<1.0	<1.0	≥0.5 <3.0 or ≥3.0	partial discharge, corona with tracking
).1 .0	<1.0	≥3.0	≥3.0	continuous discharge
.1 .0	<1.0	≥1.0 <3.0 or ≥3.0	≥0.5 <3.0 or ≥3.0	arc, with power follow-through
.1 .0	<1.0	<1.0	≥0.5 <3.0	arc, with no power follow-through
.0 .0 or ≥3.0	<1.0	<1.0	<0.5	slight overheating to 150°C
.0 .0 or ≥3.0	≥1.0	<1.0	<0.5	overheating 150 to 200°C
.1 .0	≥1.0	<1.0	<0.5	overheating 200 to 300°C
.1 .0	<1.0	≥1.0 <3.0	<0.5	general conductor overheating
.1 .0	<1.0	≥1.0 <3.0	<0.5	circulating currents in windings
.0 .0	<1.0	≥3.0	<0.5	circulating currents in core and tank; overloaded joints

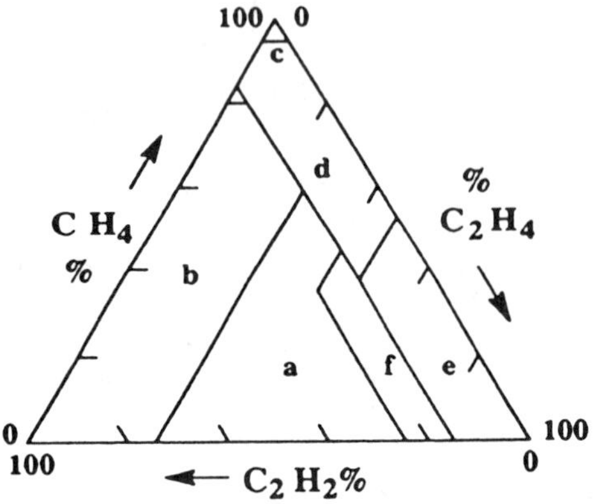

a. High-energy arcing (I>20 I_n)
b. Low-energy arcing, tracking
c. Corona discharges
d. Hot spots, T < 200°C
e. Hot spots, 200 < T < 400°C
f. Hot spots, T > 400°C

Triangle coordinates:

$\% C_2H_2 = \frac{100x}{x+y+z}$; $\% C_2H_4 = \frac{100y}{x+y+z}$; $\% CH_4 = \frac{10C}{x+y}$

with x= $[C_2H_2]$; y= $[C_2H_4]$; z= $[CH_4]$ in pp

FIG. 4.26—*Duval's triangle* [203], *revised 1980.*

engineers who must decide on proper withdrawal timing: if equipment is withdrawn from servi too early, the fault may be difficult to find by inspection and the resulting costs will be hig Catastrophic failure will be even more expensive if the fault is allowed to develop for too lor Valuable help may be found in the use of probability norms: the 90% norm, used for example IEEE, considers that the lower of the 90% values of gas concentrations in a given transform population are normal. This may not be the case, however, nor does it necessarily mean that t upper 10% are all abnormal. A more sophisticated approach was therefore developed at Hydı Québec [*167,171–173*] based on the actual probability of failure in service, which can be deduc from DGA results in the utility's computer data bank, and precise characterization of the state of transformer at the time of oil sampling. It can be seen in Figs. 4.27 to 4.29, for example, that t probability of a failure increases sharply above a given level of dissolved gas concentration. Tl level depends on the gas considered, on the type of fault involved (arcing or hot spot), on the ty of equipment (power or instrument transformer, with or without a tap-changer communicating w the main tank), on the age of the equipment, and on whether or not cellulose is involved in the fau Tables 4.10 to 4.12 give examples of acceptable and dangerous levels that can be deduced fro such curves, while more detailed information about the influence of the various parameters me tioned above can be found in Refs *171–173*. General trends indicate that safe levels are lower arcing than for thermal faults, for instrument transformers than for power transformers (especia those with a communicating tap changer), when cellulose is involved (in the case of C_2H_4), and the early and late years of the equipment (following a Weibull type of behavior). There are number of exceptions, however, and expert systems are available today to provide nonspecializ personnel with such complex information in a user-friendly form [*170,177,178*].

Unacceptable levels in service may be reached in the interval between two regularly schedul oil samplings. Permanently installed gas detectors, designed to detect the accelerated gas formati which usually immediately precedes a major fault, have been developed. Hydrogen detectors the most advanced so far [*215–219*], but detectors specific to other gases (acetylene and ethyle are under development. These devices are based on gas diffusion from the oil, through a semip meable membrane, towards sensitive elements (gas chromatographic or fuel-cell type) (Figs. 4.3 and *b*). In Fig. 4.30*b*, the hydrogen concentration is monitored continuously by the curr produced by the electrochemical reaction of H_2 with air drawn from the atmosphere.

In addition to gases, other products can result when oil is subjected to electrical or therr

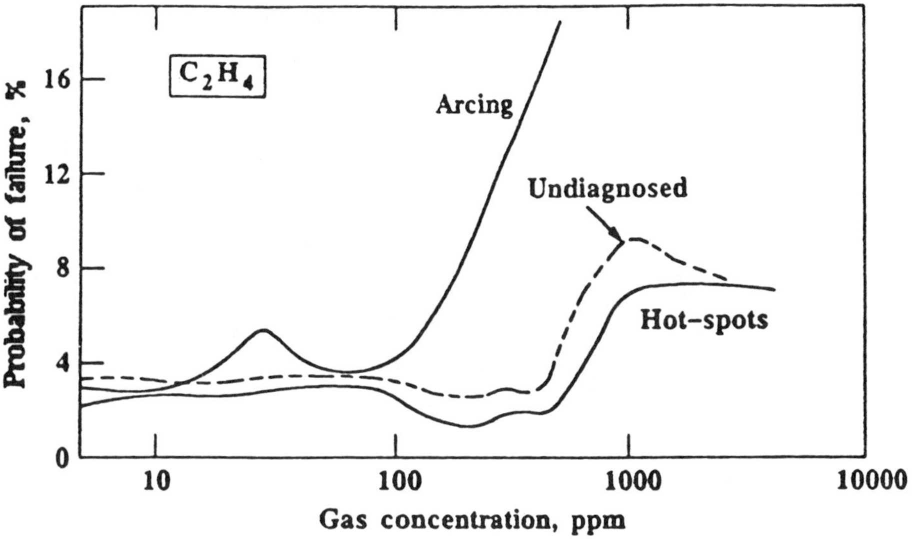

FIG. 4.27—*Probability of transformer failure in service versus dissolved C_2H_4 concentration. Influence of fault type* [173].

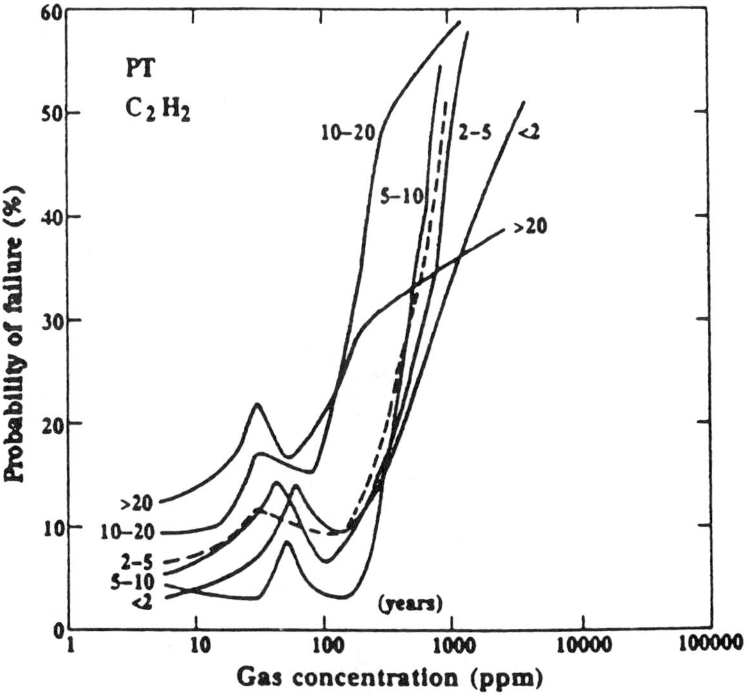

FIG. 4.28—*Probability of transformer failure in service versus dissolved C_2H_2 concentration. Influence of equipment age. PT = power transformer* [172].

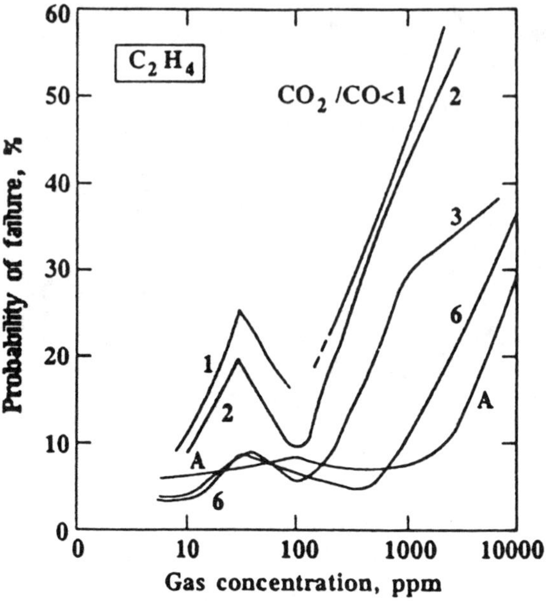

FIG. 4.29—*Probability of transformer failure in service versus dissolved C_2H_4 concentration. Increasing contribution of paper to the fault when the CO_2/CO ratio decreases* [171].

TABLE 4.10—*Dangerous levels of dissolved gases versus type of fault (power transformers PT) in ppm* [173].

Fault	C_2H_2	C_2H_4	H_2
Undiagnosed	200	500	200
Hot spot	600	600	400
Arcing	100	100	100

TABLE 4.11—*Acceptable levels of dissolved gases versus equipment age (instrument transformers IT) in ppm* [172].

Years	C_2H_2	C_2H_4	H_2
<2	20	10	200
2 to 5	40	40	5000
5 to 10	40	100	500
10 to 20	10	40	200
>20	10	...	40

degradation. For example, carbon particles are formed as a result of arcing and discharges in the oil phase; these are particularly evident in circuit breakers and tap-changers, where oil has an arc-quenching function. High-molecular-weight compounds are formed when a gas phase containing oil vapors is subjected to ionization and corona discharges (see Section 4.5.2.2). This is observed primarily in cables, bushings or high-voltage instrument transformers where small gas bubbles may

TABLE 4.12—*Acceptable levels of dissolved gases versus equipment type, in. ppm. PT = power transformer; IT = instrument transformer; TC = tap changer communicating with the main tank* [172].

Type	C_2H_2	C_2H_4	H_2
PT	20	500	200
PT-TC	200	500	250
IT	40	40	500

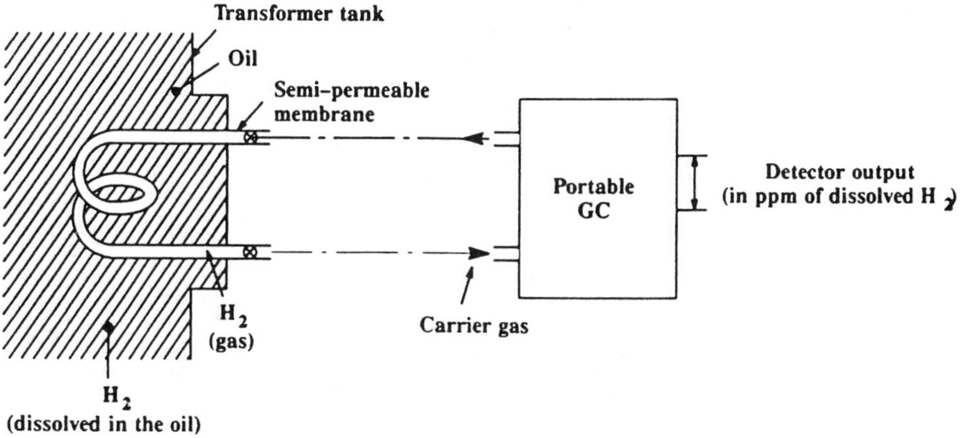

FIG. 4.30*a*—*Principle of hydrogen detector (semi-continuous).*

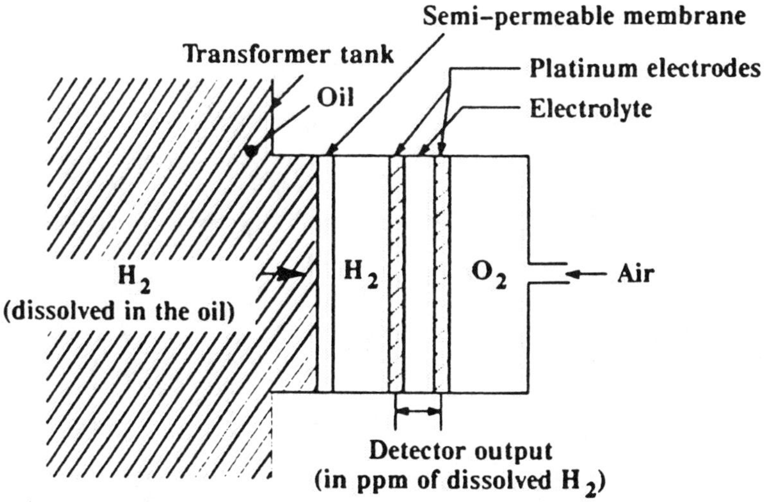

FIG. 4.30*b*—*Principle of hydrogen detector (continuous).*

be trapped in the thick layers of oil-impregnated paper insulation. These products (see Fig. 4.6), often referred to as X-wax, are polymers containing UV absorbing unsaturated double bonds [*28*]. Low-molecular-weight liquid compounds may also be formed as a result of thermal cracking or arcing [*28*] and may lower the flash point (see Section 4.4).

Recently, the presence of furfuraldehyde (2-furanal) and related compounds in oil (mostly, 5-hydroxymethyl-2-furanal, 2-acetylfuran, 5-methyl-2-furanal, 2-hydroxymethylfuran, and furoic acids, Fig. 4.31) has been associated specifically with decomposition of the cellulose insulation [*23–161,163*]. This could prove a powerful tool to complement the diagnosis given by dissolved-gas analysis when remaining interpretation problems, such as the distribution of these compounds between oil and paper, the effect of thermally upgrading additives to paper and the relative instability of some of these compounds, are solved.

The solubility coefficients of fault gases in insulating oils must be known to utilize some of the methods used for dissolved-gas analysis (see Section 4.5.4.9) and to understand the supersaturation phenomena (bubble formation) that may occur with temperature changes and pump cavitation (Table 4.13 and Fig. 4.32). The values in Table 4.13 are valid for oils with a density of 0.855 g/cm^3 at 15.5°C. Oswald coefficients for mineral oils of different densities can be calculated as follows: K

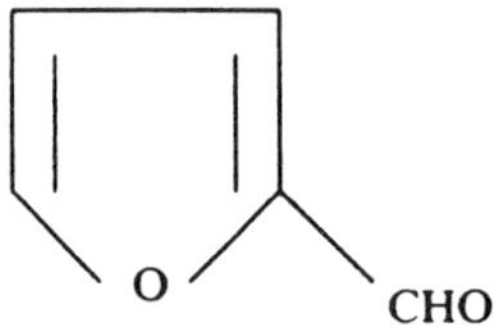

2-Furanal

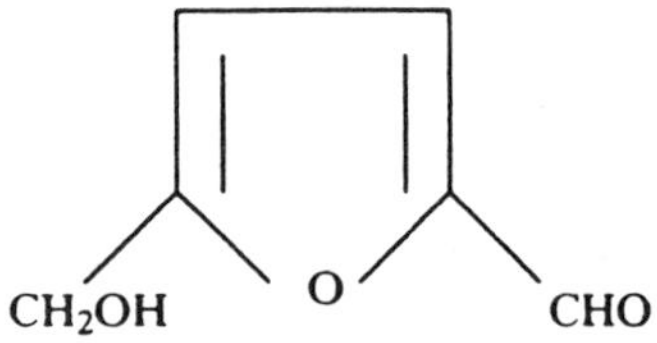

5-Hydroxymethyl-2-Furanal

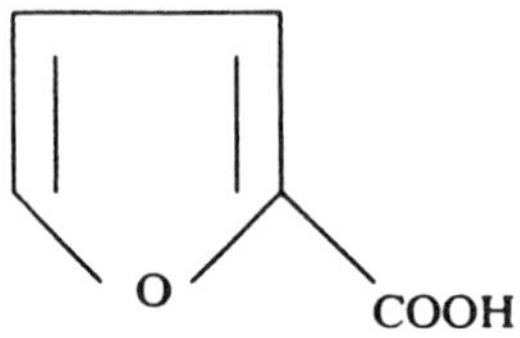

2-Furoic Acid

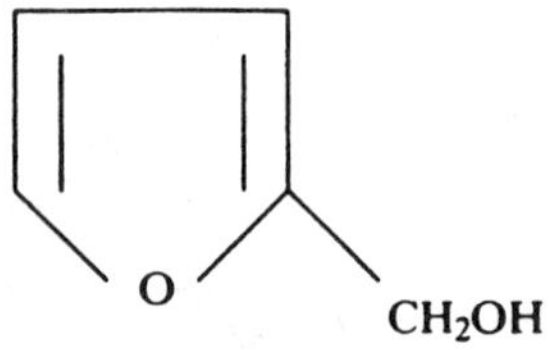

2-Hydroxymethylfuran

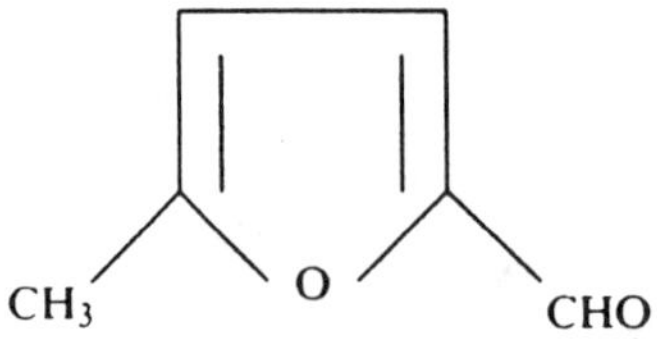

5-Methyl-2-Furanal

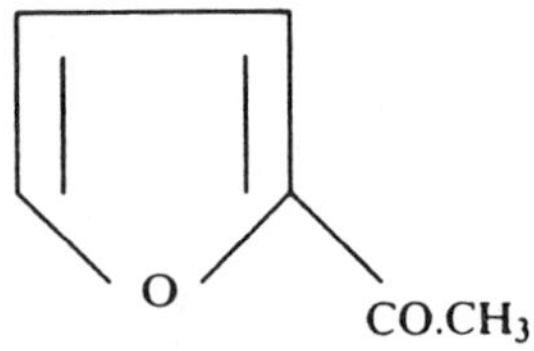

2-Acetyl furan

FIG. 4.31—*Furanic compounds in insulating oil.*

(corrected) = 0.980 density/0.1255 where the density is measured in grams per cubic centimeter at 15.5°C. While the values of Table 4.13 generally are accepted in North America, slightly different values may be used in other parts of the world. Solubility coefficients are also dependent on temperature. Methods described in Section 4.5.4.12 can be used to clear up such differences. Experimentally determined values are reported in Ref *84*.

Extension of dissolved-gas analysis to factory heat-run acceptance tests is being considered for transformers [*183,190,192*]. The reliability of the analytical methods at the low gas concentrations [*169*] measured during these tests (0.1 to 2 ppm) has yet to be demonstrated however (see Section 4.5.4).

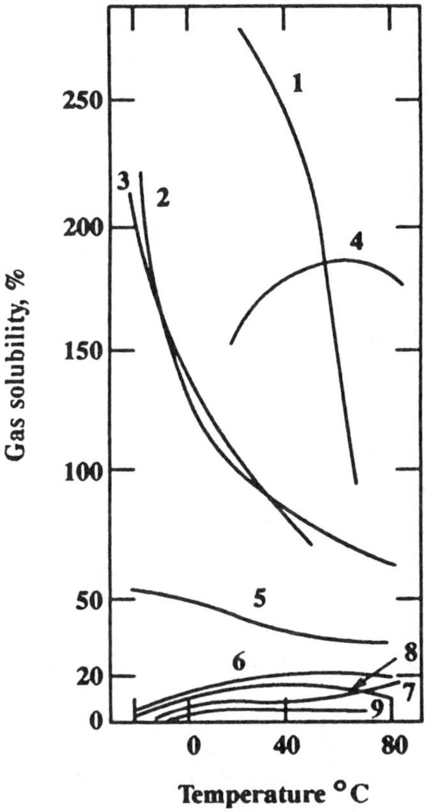

1. Ethane
2. Carbon dioxide
3. Acetylene
4. Ethylene
5. Methane
6. Oxygen
7. Nitrogen
8. Carbon monoxide
9. Hydrogen

FIG. 4.32—*Variations of gas solubility in oil versus temperature* [4,209,168].

TABLE 4.13—*Solubility coefficients of gases in insulating oil (ASTM D 3612).*

Gas	Ostwald Solubility Coefficient K at 25°C
H_2	0.0558
N_2	0.0968
CO	0.133
O_2	0.179
CH_4	0.438
CO_2	1.17
C_2H_2	1.22
C_2H_4	1.76
C_2H_6	2.59
C_3H_8	11.0

4.5.2.2 Gas Absorbing and Evolving Behavior—Gas bubbles may form in oil as a result of the thermal and electrical processes discussed previously. Poor drying or oil impregnation of the paper insulation or both may also lead to bubbles of air or water vapor, especially in overlapping areas of thick-layer insulation. Bubbles of air or nitrogen may form as a result of oversaturation in oil due to temperature or pressure changes and bubbles of water vapor or carbon monoxide as a result of hot spots in the paper insulation. These gases have a much lower dielectric strength than does oil and a gas bubble formed in or passing through an area of high electric stress.

Nascent hydrogen is also formed, either by the decomposition of oil vapors, or through reactions at the oil-gas interface, and may recombine to form hydrogen gas. A gas bubble will then increase in volume to a point where dielectric breakdown and arcing may occur; the oil is said to be "gas-evolving." Alternatively the nascent hydrogen can react with the oil molecules, and bubble formation is reduced or eliminated; the oil is said then to be "gas-absorbing." The gas-absorbing character of the oil increases with increasing aromatic, particularly polyaromatic, content. This may conflict with oxidation stability requirements (see Section 4.5.1) and in practice a compromise of around 10% C_A (IR) is generally used for free-breathing transformers (see Fig. 4.2). For cables and equipment where gas bubbles may be trapped in thick layers of insulation (current transformers, for example), gas-absorbing oils with higher aromatic contents are preferable and commonly used. Actually, the nature of the aromatics involved is more important than their total content. It has been shown [*146*] that gas-absorbing properties are affected by polyaromatics far more than by monoaromatics and, therefore, are related more closely to the polyaromatic content than to the total aromatic content of the oil.

Rates of gas formation generally increase with increasing temperature or applied voltage, although there may be slight variations in this behavior. Rates also depends on the relative volume of the gas phase as compared to the oil volume and on the nature of the gas phase. Thus it has been observed that even oils with high aromatic contents are gas-evolving when the gas phase is nitrogen and are gas-absorbing with hydrogen. With air there is an initial gas-absorbing stage corresponding to the consumption of oxygen through oil oxidation reactions followed by a gas-evolution stage as light hydrocarbons are formed. The protection given by aromatic oils, therefore, is limited to the case of bubbles formed by corona discharges, where hydrogen is the main gas formed.

All the preceding factors affect the rate of gas formation, so that an oil may become gas-absorbing or gas-evolving when the test conditions are changed (Fig. 4.33) [*7*] without necessarily changing the ranking of oils, as seen in Table 4.14 [*143*]. Correlation with the actual gassing tendency in service, however, is still subject to controversy. The IEC even recommends different

TABLE 4.14—*Gassing of mineral oils as a function of chemical composition and test methods used* [143].
IEC 628. Method A. Large gas-phase volume.
IEC 628. Method B. Small gas-phase volume.

Oil Type	Chemical Composition of Oil %			Gassing	
	C_A	C_N	C_P	Method A, mm^3/min	Method B, cm^3
	5.8	52.8	41.4	+22.1	+7
	13.4	44.6	42	+9.3	+7
Naphthenic	16	43	41	+24.6	−3
	17.2	41.6	41.2	−7.9	−0.3
	18	40.7	41.3	−6.8	−1.4
	7.9	30	62.1	−1.9	+5
Paraffinic	9.6	25.3	65.1	−35.7	−2.9
	13.8	25.8	60.4	−22.9	−0.5
	17	25	57	−34.3	−2.7

test methods for different equipment applications (see Section 4.5.4.11). Further there does not seem to be any relation between the gassing tendency measured by tests and breakdown initiation in oil.

4.5.2.3 Impulse Strength—The impulse strength, that is, the ability of an oil to withstand transient voltage stresses in highly divergent fields, is another electrical property which, like the gassing tendency, depends very much on the oil's chemical composition (especially in the negative-

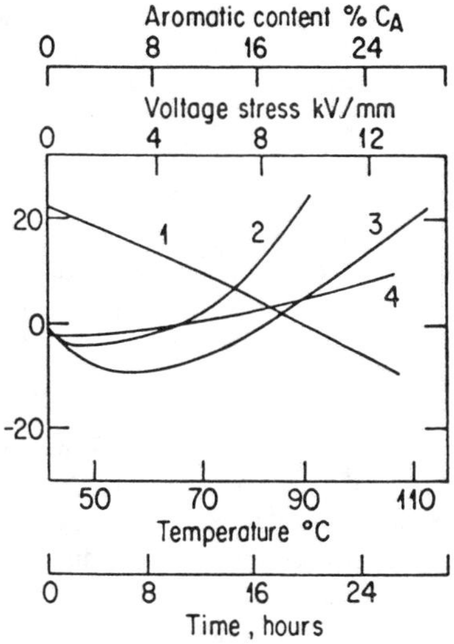

1. Aromatic content
2. Voltage stress
3. Time
4. Temperature

Ordinates in mm^3/min.

FIG. 4.33—*Effect of test conditions on the gassing characteristics of insulating oils* [7].

polarity mode) [*150*]. The amount and type of aromatics involved (mono- or poly-aromatics) also play a major role, but not necessarily in the same way as for the gassing tendency [*3,146*].

4.5.3 Contaminants

Unlike the gassing tendency and impulse strength, the dissipation factor (a-c loss) and the dielectric strength of a well refined and purified oil are very little, if at all, affected by the oil's basic chemical composition. The aromatic molecules act as weak dipoles which contribute to the low values of the dissipation factor found at lower temperatures [*234*]. The dissipation factor at higher temperatures and the dielectric strength are primarily affected by the contaminants that may have been introduced during handling and use of the oil, such as water, particles, and solubilized matter. Low values of the dielectric strength of the oil or high dielectric losses, therefore, do not necessarily mean that the insulation system is compromised but rather that the oil is dirty. Cleaning up the contaminated oil and drying the solid insulation (when necessary) are often enough to restore both the oil and the solid insulation to their original quality. Only in the case of thick solid insulation layers and prolonged contact with badly contaminated oil will cleaning of the solid insulation also be necessary.

During manufacture of electrical equipment, considerable attention is given to meeting levels of dryness appropriate to the type and voltage class of the particular equipment. The residual water in the insulation system redistributes itself between cellulosic insulation and oil as the temperature changes in operation. The saturation value for water in paper is decreased as the temperature increases, while the saturation value in an oil is increased. Water then moves from paper to oil as the temperature increases and the system reaches toward a new equilibrium in which the relative saturation of the two phases is the same. When the insulation system temperature falls, water is sucked back into the paper. Water from two sources can contaminate equipment during service. Water is the primary products of the degradation of cellulosic paper and press board insulation during aging. Water can also be absorbed from moist air through leaks or in free-breathing equipment. Regardless of its source, water will cycle between paper and oil as the temperature cycles.

The saturation limits of water in transformer oils increase with increasing temperatures, from a typical value of 50 ppm at 20°C to 800 to 1000 ppm at 100°C. Higher limits may be found in oxidized oil due to the greater attraction of the more polar products. Water is more harmful to dielectric strength in combination with fibers and other particulates which become more conductive when wet. Water alone affects the dielectric strength as saturation is approached.

The number of particles and their size distribution affect the dielectric strength of oil, although reliable correlations are difficult to establish [*220–228,232*]. According to Krasucki's theory [*233*], breakdown might be associated with the formation and growth in the liquid of a gas bubble created by the presence of a particle [*229*], but other theories have also been proposed. Particles, especially conductive ones may also induce partial discharges [*227,230*]. The nature of the particles (metal, dirt, fiber, sludge) certainly plays a major role, as well as the nature of the metal (iron, aluminum, copper, etc.) in the metallic particles, and several major failures in service have been related to accidental contamination by metallic particles. Recently the application of ferrography has been proposed to specifically detect magnetic particles [*222*]. ASTM Method D 1816, which measures the dielectric breakdown voltage with spherically capped electrodes of Verband Deutscher Electrotechniker Specification 0370 contour, is particularly sensitive to particles and dissolved water. Removal of these two types of contaminant by filtering and vacuum dehydration is one of the most widely used equipment maintenance operations.

Ionic and polar contaminants soluble in oils usually will be detected first by measuring the interfacial tension, resistivity, and dielectric dissipation factor. These are often influenced markedly by the first traces of soluble contaminants such as soluble copper or oxidation precursors [*68,127*], then change only slowly when higher levels are reached (see Figs. 4.22 and 4.25). Small-scale nonuniform field breakdown tests using either a point-to-sphere or point-to-point electrode configuration have also been proposed [*231*].

Apart from the products resulting from oxidation and stress decomposition of oil and paper in service (Section 4.5.1), soluble contaminants may result from refining treatments, which may leave traces of sulfates, chlorides, phenols, etc. Metals such as lead, zinc, and cadmium are not very active catalysts but can form metal salts and soaps in the presence of moisture and acids, while ferrous metals, and fluxes used for brazing, welding, and soldering can dissolve and affect the liquid properties. The metal content of the oil, therefore, should be checked in such cases. Equipment components such as transformer windings are often coated with a varnish which can partially dissolve in oil if not properly cured. Some rubbers used for gaskets (butyl, nitrile, and neoprene) and some plastic materials (PVC) are attacked by oils. Compatibility tests are available for closer checks of problems that may be encountered when using a more aggressive oil in conventional equipment.

Increases in the dielectric losses of oil, combined with higher operating temperatures of the equipment, may largely exceed losses in the paper insulation (Fig. 4.34) and lead to uncontrolled thermal runaway in thick layers of oil-impregnated paper [*127*].

Reclamation on clay or Fuller's earth may be necessary to remove these soluble contaminants when they exceed acceptable levels (see Table 4.3). Fuller's earth treatment consists of a selective

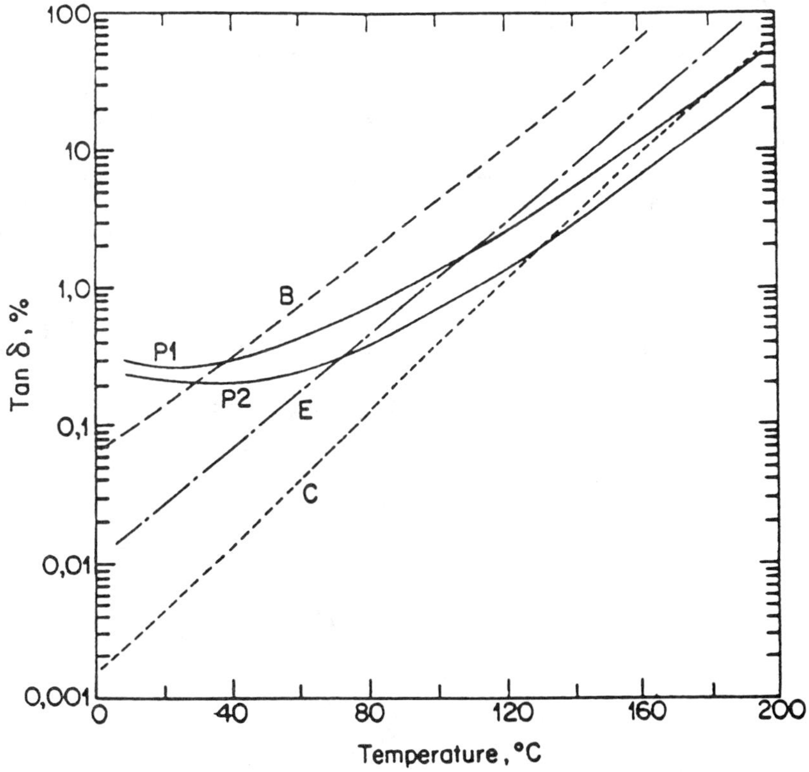

FIG. 4.34—*Dielectric losses versus temperature: (C) new, good-quality oil; (B) contaminated or unstabilized oil (C); (E) intermediate oil; P_1, P_2 paper insulation with 0.3% and 0.1% moisture content, respectively* [127].

and very efficient adsorption of the highest polar contaminants in oil (acids, esters, soluble coppe etc.), to the Si—O and Si—OH groups present at the surface of Fuller's earth, a microporous forr of magnesium aluminum silicate with a very high surface area. However the less polar contami nants in oil, such as oxidation precursors containing aldekyde or ketone groups, are not adsorbe [*69*] (see Section 4.5.1).

4.5.4 Test Methods

4.5.4.1 Oxidation Resistance—ASTM Test Method D 2440 and IEC Publication 74 are use widely for assessing the oxidation resistance of inhibited and uninhibited insulating oils. The tes tubes (25 mm in diameter, 210 mm long) (Fig. 4.35), are placed in a thermostat bath. A copper-wir spiral catalyst is introduced with 25 g of oil in the tubes and air or oxygen is bubbled through th oil. The test conditions are indicated in Table 4.15. The IEC test provides for low-oxyge conditions, which may be closer to actual service conditions (see Section 4.5.1). The oil i evaluated at the end of each aging period by measuring the amount of sludge and acid formed. Th sample is diluted with n-heptane and the solution filtered to remove the sludge which is then drie and weighed. The sludge-free solution is titrated at room temperature to determine the acidity. Th volatile acidity can also be collected and measured by installing a second test tube filled with wate in series with the oil oxidation tube.

The duration of the tests (72 h, 164 h, or multiples of 164 h) is rather arbitrary and, as mentione in Section 4.5.1, may not properly reflect the oxidation behavior of very different oils, especiall inhibited oils. When possible, determination of the entire oxidation profile (see Fig. 4.22) i preferable. For that purpose, several oil test tubes must be prepared and oxidized simultaneousl and removed at regular intervals to build the whole profile [*68*].

A much shorter-duration oxidation test used for the evaluation of inhibited oil, is available i ASTM Method D 2112. Operating at 140°C with a copper catalyst under 6.2 bar oxygen pressure i a rotating bomb, it determines the time to a specified pressure drop. Sophisticated interpretations o the results obtained already have been mentioned (Section 4.5.1). Whether the chemical reaction involved in these even more stringent conditions of temperature and oxygen supply are really th same as in actual service life may be even more debatable than for the other accelerated methods o Table 4.15 however.

Another oxidation test available in North America is the PFVO and sludge-free life test, devel oped by Doble Engineering, which measures the power factor and the incipient sludge point as function of oxidation time at 95°C under a prescribed flow rate of air or oxygen (see Fig. 4.23 Although not widely used in the industry, this test has been found quite useful to detect abnorma increases in the dielectric losses of new types of oils in their early life in the equipment, particularl

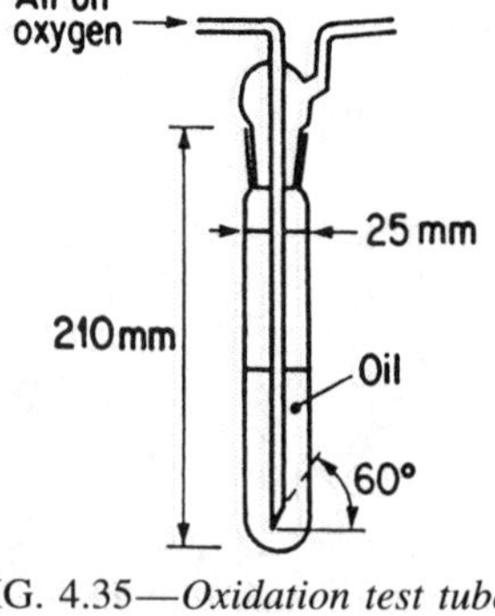

FIG. 4.35—*Oxidation test tube.*

TABLE 4.15—*Accelerated oxidation test conditions.*

Test	Temperature, °C	Gas	Gas Flow, L/h	Copper Catalyst Surface Area, cm^2/g oil	Duration
ASTM D 2440	110	oxygen	1	6.3	72 and 164
EC 74 (uninhibited oils)	100	oxygen	1	0.39	164
EC 474 (inhibited oils)	120	oxygen	1	1.14	induction period
EC 474 (low oxygen)					
uninhibited oils	120	air	0.15	1.14	164
inhibited oils	120	air	0.15	1.14	multiples of 164
ASTM D 943 (inhibited)	95	oxygen	3	0.57	time to 2.0 mg KOH/g
ASTM D 2112 (inhibited)	140	oxygen	...	3.4	time to specified oxygen pressure drop

nstrument transformers [*127*]. These losses were related to the special types of oxidation precur-ors formed, long before any measurable amounts of acids were produced.

4.5.4.2 Neutralization Number—The acidic or basic constituents present in insulating oils as a esult of oxidation or contamination or as additives can be determined by titrating with acids or ases, according to ASTM Test Method D 664, using potentiometric titration, or D 974, using olor-indicator titration (methyl orange or p-naphthol-benzein). Method D 664 is more appropriate or excessively dark-colored oils. In this method, a sample of oil is dissolved in a mixture of toluene nd isopropyl alcohol containing a small amount of water and titrated potentiometrically with lcoholic potassium hydroxide or hydrochloric acid solution, using a glass-indicating electrode and calomel reference electrode. The neutralization number is expressed in milligrams of KOH per ram of oil. When the amount of oil to be analyzed is too small to allow accurate analysis, the emimicro color indicator titration of ASTM Test Method D 3339 may be used. Sample sizes range rom 0.1 to 5 g oil, for acidities of 3 to 0.01 mg KOH/g, respectively.

The acidity can also be deduced from the infrared absorption of oil in the 1700 to 1780 cm^{-1} egion of carbonyl stretching vibrations (C = 0), according to ASTM Test Method D 2144 (see Section 4.2.3.6).

4.5.4.3 Peroxide Number—The amount of peroxides formed during oil oxidation can be mea-ured according to ASTM Test Method D 1563, where a sample of oil, dissolved in carbon etrachloride, is brought into contact with aqueous potassium iodine solution which reduces the eroxides present. The iodine liberated is titrated with sodium thiosulfate solution. The results are alculated to parts per million of oxygen. Although this test is no longer listed in ASTM D 117 for esting of insulating oils, it is still useful for research and special investigations, as peroxides are xidation precursors which are formed just before the massive buildup of acids and sludges in oils. They may thus provide an early warning of impending accelerated oxidation, or may be indicative of unusual behavior in a batch of oil. Values of up to 12 ppm have been measured during ccelerated oxidation tests [*68*].

4.5.4.4 Sediment and Soluble Sludge Content—Sludge formation starts at the acceleration stage of oxidation (see Fig. 4.22). Sludge is composed of high-molecular-weight molecules resulting rom oil oxidation, which are no longer soluble in oil and have precipitated. It can be determined by

ASTM Test Method D 1698, which involves the separation of insoluble sludge (sediment) by centrifugation and the precipitation of soluble sludge by dilution with n-pentane. "Soluble sludge" has not yet reached a molecular weight high enough to become insoluble in oil alone and is a precursor to insoluble sludge. Details concerning the membrane filtration of the pentane insolubles are given in ASTM Test Method D 4055. The inorganic portion of the sediment is measured after ignition at 500°C in a Gooch crucible. ASTM Test Method D 4310 is used specifically to obtain the sludge values of "inhibited" mineral oils.

4.5.4.5 Antioxidant Content—The most widely used oxidation inhibitors in insulating oils are 2,6 ditertiary-butyl para-cresol (DBPC)® and 2,6 ditertiary-butyl phenol (DBP). The DBPC content can be determined according to ASTM Test Methods D 2668, and D 4768. The first method uses the infrared absorption band of DBPC and DBP at 3680 cm^{-1}; bands at 860 and 750 cm^{-1} identify DBPC and DBP, respectively. Interference by oil oxidation products may be difficult to eliminate with this method, however. In the second method, after removing interfering substances on an extraction column with heptane, the inhibitors are recovered in methanol and separated by gas chromatography on a nonpolar, silicone-bonded packed column with FID detector.

HPLC (on LiChrosorb RP-18 columns with methanol as the mobile phase and UV detection at 283 nm) and cyclic voltametry techniques have also been used successfully to determine the antioxidant content [*33,34*].

4.5.4.6 Polar Oxidation Products—Besides acids and sludges, a large number of polar oxidation products are formed during oil oxidation. Those containing —OH groups (such as phenols and alcohols) are generally more polar than those containing —C= 0 groups only (such as aldehydes and ketones). The latter are oxidation precursors, which often precede the formation of acids and sludge and have a marked influence on the interfacial tension (Fig. 4.22) and the dielectric properties of oil. In addition, this latter group is not removed completely by reclaiming processes so that its content in oil should be closely monitored. This can be done by infrared spectrometry in the 1700 to 1780 cm^{-1} region of carbonyl vibration (see Section 4.5.4.2) or, more accurately, by HPLC techniques in the reverse-phase mode (C_{18} columns with a water-methanol mixture as the mobile phase) [*24*].

4.5.4.7 Copper Content—Dissolved copper compounds are active oxidation catalysts (see Section 4.5.1) and can also strongly influence oil properties such as interfacial tension and dielectric losses (see Section 4.5.3). Determining the presence of copper in oil, therefore, may be useful to explain changes in oil properties and general behavior. Copper dissolved in oil can be determined according to ASTM Test Methods D 2608 or D 3635. In the first method, copper species are extracted from the oil by a mixture of hydrochloric and acetic acid and determined colorimetrically as the neocuproine complex at 450 mm. Atomic absorption spectrophotometry is used in the second method after diluting oil in a solvent such as methylisobutyl ketone. Copper concentrations down to 0.1 ppm, thus can be determined. Neutron activation analysis is an alternative, even more sensitive, method [*74*] but it requires nuclear facilities.

Other metals (iron, nickel, aluminum, etc.) have effects similar to those of copper, although to a much lesser extent, and their content in oil can be determined in a similar fashion, using atomic absorption, emission spectroscopy, or neutron activation analysis.

4.5.4.8 Color—Color is a rapid and easily performed quality-control technique, which in some cases may indicate contamination of the oil and can reflect the extent of the oxidation very qualitatively. Color is determined by ASTM Test Method D 1500, by comparing the liquid sample with colored glass disks ranging in value from 0.5 to 8.0. Visual examination of oil, according to ASTM Test Method D 1524, is also useful to determine whether a sample should be sent to a central laboratory for full evaluation. Cloudiness, particles or other undesirable suspended material may be detected.

4.5.4.9 Dissolved-Gas Content—The determination of individual gases dissolved in insulating oil can be carried out according to ASTM Test Method D 3612 or IEC Publication 567. The gas extraction procedure common to both publications is partial degassing, which consists in introducing the oil sample in a preevacuated known volume, compressing the evolved gases to atmospheric pressure with mercury, and measuring the total gas volume in a buret, as shown in Fig. 4.36*a*. Correction for the amount of gases remaining in the oil must be made using the solubility coefficients of the component gases at the reduced partial pressure of the extraction flask.

Two alternative gas extraction procedures are proposed in IEC Publication 567: the Topler pump method and the stripping method. The former is a vacuum extraction method using a basic experimental setup similar to that depicted in Fig. 4.36*a* but modified to allow further extraction of the gases remaining in the oil sample. This is done by lowering the mercury level and compressing the additional gas extracted into the buret, combining it with the gas already extracted (using the valve No. 8 in Fig. 4.36*a*) and repeating this same procedure several times. Extraction of almost 100% of the gases contained in the oil is thus theoretically possible, given an appropriate number of strokes and provided no additional error is introduced by the numerous manipulations involved. The stripping method (see Fig. 4.36*b*) is the simpler of the two, consisting of bubbling the carrier gas of the gas chromatograph through the oil, directly into the chromatographic column. (In the vacuum method, transfer to the gas chromatograph is done by a gas syringe or switching valve from the buret.) The columns are usually Porapak N and molecular sieve 5A, the carrier gas helium (or argon for a greater sensitivity for hydrogen determination); a flame ionization detector is used for

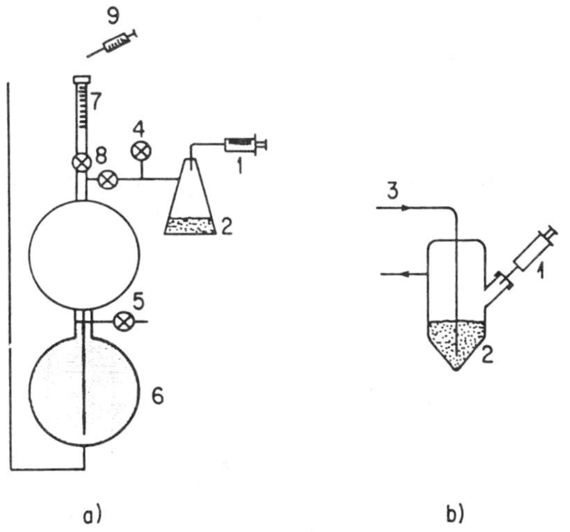

a. Vacuum methods
b. Stripping method
1. Oil syringe
2. Oil sample
3. Carrier gas (from and to GC)
4. Vacuum
5. Air pressure
6. Mercury
7. Buret
8. Valves
9. Gas syringe to GC

FIG. 4.36—*Dissolved-gas extraction methods.*

the hydrocarbons and a thermal conductivity detector for the other gases. A methanator is sometimes used for greater sensitivity for carbon monoxide and carbon dioxide.

Recent round-robin tests organized by the IEC have shown that, when used by normally skilled operators, the three gas-extraction methods gave adequate results and comparable precision (10 to 15% on average) for gas concentrations in the 15 to 500 ppm range. The precision obtained at the very low (1 to 5 ppm) levels observed during factory tests is much lower (40 to 50% for the best laboratories). These data concerning the precision and accuracy of analyses (Fig. 4.37) was obtained using special oil samples, containing standard amounts of dissolved gases and prepared according to a method described in Ref *185*.

Another stripping procedure, using a steel column filled with glass beads or steel turnings rather than a gas bubbler [*179*], has been developed successfully into a commercial instrument (TOGAS analyzer) [*166*], and its use may be incorporated in revised standard ASTM D 3612. Direct headspace injection in the chromatograph [*168*] is also being investigated [*165*]. In this method, the oil sample (10 ml) is introduced in a 20-ml closed vial and kept in contact with the remaining 10 ml of gas atmosphere in the vial, consisting of carrier gas from the chromatograph, at typically 80°C, until near equilibrium has occurred (*ca.* 4 h). A sample of the gas phase is then introduced directly in the chromatograph. This method is automated easily and allows a large number of samples to be analyzed, but it necessitates calibration with standard oil samples or with samples analyzed by another gas-in-oil technique. No reliable comparative data on precision and accuracy, based on international round robin tests, are available yet for these last two methods.

ASTM Test Methods D 831, D 2945, and D 1827 are also available to determine the total gas content of oils by means of a vacuum or by displacement with carbon dioxide, but they are seldom used, and only as a factory control for a quick check of low total gas content, with relatively low precision. In D 831 and D 2945, oil is allowed to flow into an evacuated chamber, allowing free volatilization of the gaseous components. The system is brought back to atmospheric pressure and the evolved gases are measured. In D 1827 the gases dissolved in oil are purged (stripped) with pure carbon dioxide gas into a graduated buret filled with a potassium hydroxide solution. The carbon dioxide is absorbed completely by the potassium hydroxide and the volume of the other gases are measured.

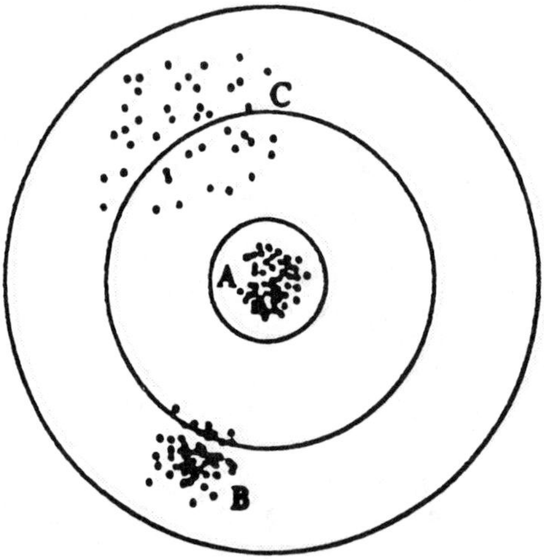

FIG. 4.37—*Precision and accuracy:* (a) *precise and accurate;* (b) *precise but not accurate;* (c) *not precise and not accurate.*

4.5.4.10 Solubility Coefficients—The solubility coefficients needed to calibrate the partial degassing method described previously and determine the concentration of gas in oil or the gas blanket when the concentration of one is known can be obtained according to two methods. With ASTM Test Method D 2779, the Oswald coefficients are deduced from those at 0°C for petroleum liquids with d = 0.85 using the following equations:

$$L_c = 7.70\,\mathrm{L}\,(0.980\,\mathrm{d})$$
$$\mathrm{L} = 0.3 \exp[(0.639\,(700\text{-}T)/T) \ln 3.333\, L_o]$$

with

L_c = Oswald coefficient at T for oil of density d
L = Oswald coefficient at T for oil of d = 0.85
L_o = Oswald coefficient at 0°C for oil of d = 0.85
(values of L_o are given in D 2779)

Best practice requires that the solubility of each of the gases of interest be measured directly, rather than calculated empirically. Method D 2780 is based on saturating a sample of oil with a fixed gas, extracting this gas quantitatively in a vacuum extractor similar to that described in Fig. 4.36*a*, transferring it to a gas buret where its volume is determined, and calculating its solubility coefficient from the result.

4.5.4.11 Gassing—The gassing tendency of insulating oils is measured commonly according to ASTM Test Method D 2300 (Method B) and IEC Publication 628 (Method A), both of which are based on a method originally developed by the Pirelli company, in a test cell like that in Fig. 4.38*a*). The oil-gas interface saturated with a gas (usually hydrogen), is then subjected to a radial electrical stress of 3 to 4 kV/mm (the nominal test calls for an applied stress voltage of 10 kV over a 3-mm oil gap). The gas phase above the oil is ionized and the oil-gas interface subjected to ion bombardment. The amount of gas evolved or absorbed in the oil is measured in the buret and expressed in the units of volume per unit of time, at a given temperature. The IEC publication suggests that this test is more applicable to the case of partial discharges in cables and capacitors, using hydrogen as the gas

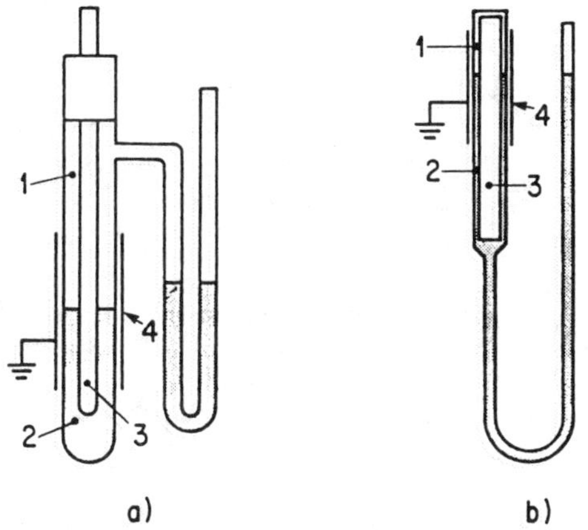

1. Gas phase
2. Oil sample
3. High-voltage electrode
4. Outer electrode

FIG. 4.38—*IEC gassing-test cells, types a and b.*

phase. It recommends another test cell (Fig. 438*b*)) with a smaller gas/oil ratio and nitrogen or air as the gas phase as more appropriate to simulate transformer conditions, where nitrogen or air bubbles are formed as a result of mechanical stresses or temperature changes in gas-saturated oil. The corresponding ASTM Test Method is D 2300 (Method A).

Some types of cable or current transformers may develop reduced pressures when subjected to load cycling, and ASTM Test Method D 2298, although no longer recommended for routine testing in ASTM D 117, can be used to simulate such a situation. In this method, the gas phase above the oil is evacuated and subjected to an electric discharge at a stress of 2 to 3 kV/cm for 1000 min. The pressure increase is related to the amount of gas evolved.

4.5.4.12 X-Wax Content—Unsaturated polymers (X-wax) are formed under gassing conditions such as those described above (gas-phase corona discharges). Relatively little is known of the physical and chemical structure of X-waxes. They appear to be amorphous, formed with all types of oils and chemically unsaturated, that is, containing several carbon-carbon double bonds which cause them to have fluorescent properties. In some cases they can be detected in paper insulation by their fluorescence under a UV lamp whereas their presence in oil can be analyzed more precisely using HPLC techniques [*28*] (see Fig. 4.6).

4.5.4.13 Furfural Content—Furfural and furanic compounds are produced by the thermal decomposition of the cellulosic paper insulation used in electrical equipment. Figure 4.31 shows the main furanic compounds that are at least partially soluble in oil and that have been detected in oil from operating equipment. These compounds give an early indication of paper decomposition and, at the low level that is usually detected (less than 5 ppm, in rare cases above 10 ppm), do not appear to affect the oil properties more markedly than do oil oxidation compounds.

Methods such as those described in Ref *23* and related papers have been developed to determine the contents of furfural and furanic compounds dissolved in oil. The oil sample is placed on a silica cartridge and the nonpolar oil compounds separated with pentane. The furanic compounds are then extracted with methanol and analyzed by HPLC on C_8- or C_{18}- bonded silica columns with water/methanol (60/40) as the mobile phase at 1 mL/min.

4.5.4.14 Water Content—The water content in insulating oils is determined commonly using the Karl Fischer reaction of iodine with water in SO_2, pyridine, methanol mixtures, or buffer solutions, according to ASTM Test Method D 1533. Basically, iodine (I_2) reacts with the water (H_2O) contained in the oil sample, to form iodine (I^-). The end point of the titration is reached when free iodine is sensed electrochemically. The amount of iodine reacted is thus directly proportional to the amount of water in the oil sample. A number of automatic titrators are now commercially available, offering sharp and stable end points and minimal manipulations. Such instruments are calibrated to deliver a known current which generates iodine to react with the unknown amount of water present, the result being given directly in weight (ng or μg) of water. Commercially available pyridine-free reagents also are widely used.

Other available methods involve the bubbling of a dry gas through the oil sample and determination of the moisture content of the humidified gas, either by dew-point measurement or by absorption on phosphorous pentoxide-coated electrodes calibrated to deliver a known current to electrolyse the unknown amount of absorbed water.

4.5.4.15 Particle Content—ASTM Test Method F 661-80 can be used for particle-content determination using an optical particle counter. Particles are classified in six classes, according to size (1 to 2 μm, 2 to 5 μm, 5 to 10 μm, 10 to 15 μm, 15 to 25 μm, and above 25 μm). Several commercial instruments are available, varying in sophistication with respect to computerized size counting and particle-shape evaluation.

In a particle counter, the oil sample is allowed to flow through an optical cell consisting of a glass capillary crossed by a narrow beam of light at a perpendicular angle. The reduction of intensity of the light beam brought about by the presence of a particle in the oil is detected by an optical sensor and related to the size of the particle, through proper calibration with particle standards. Electronic instrumentation records and relates the number of such events to the number of particles of this size. Optical sensors based on the light scattering principle detect the dark zones caused by the particles, as well as the gray zones of diffraction around them, allowing a better determination of the sizes of the nonspherical particles (such as fibers) which do not have their axis perpendicular to the light beam. The interpretation of results and calibration are complex, however. Optical sensors based on the light blocking principle, where only the dark zones are detected and empirical correction factors are used to compensate for variations in particle position, are increasingly favored in laboratory practice.

Further distinctions between metallic and nonmetallic particles involve filtering the oil sample on a fine-mesh cellulose acetate or teflon filter, and analyzing the filter for metallic content by neutron activation analyses, atomic absorption, or emission spectroscopy (see Section 4.5.4.7). Individual particles on the filter may be examined by scanning electron microscope for metal identification and content. Distinctions between magnetic and nonmagnetic particles also can be obtained by ferrographic analysis [*220*], a method used originally for lubricating oils. Oil is circulated on a glass plate above a magnet. Particles are immobilized more or less rapidly on the plate according to their magnetic character and size, the smaller, nonmagnetic ones being washed away by the oil flow. After drying and fixing the particles on the plate, these can be examined by microscopy for structure (ferrous, metallic, crystalline, etc.), shape (fibers) and size. Analysis can be extended to nonmagnetic particles by the use of special magnetizing liquids.

4.5.4.16 Dielectric Breakdown Voltage—There are two ASTM standard test methods for determining the dielectric breakdown voltage of electrical insulating oils at commercial power frequency (60 Hz). In Method D 877, the oil is tested between 25 mm diameter, square-edged disk electrodes spaced 2.5 mm apart, and in Method D 1816 between spherically capped (VDE) electrodes spaced either 1 or 2 mm apart. The voltage at which the current produced by breakdown of the oil reaches the range of 2 to 20 mA, tripping a circuit breaker, is considered the dielectric breakdown voltage.

The more uniform electric field of method D 1816 is much more sensitive to contaminating agents such as water or conducting particles. Adequate stirring of the liquid is necessary, to ensure that particles are dispersed uniformly in the oil. Ultrasonic treatment has been proposed to stabilize dispersions [*220*]. The scatter of data obtained with these methods remains high, however, with standard deviations of readings of up to 25% in the 45 kV range during IEC round-robin tests.

4.5.4.17 Inorganic Chlorides and Sulfates—As mentioned in Section 4.5.3, inorganic sulfates and chlorides may be left after refining and be a cause of increase of the oil conductivity and dielectric losses. Their presence may also point out some quality control problems in the oil refining process.

Inorganic chlorides and sulfates can be detected according to ASTM Test Method D 878, following extraction from the oil sample with water, by reaction with silver nitrate and barium chloride solution (appearance of white precipitates).

4.5.4.18 Compatibility—The compatibility of an insulating oil with the construction materials used for the electrical equipment concerned, such as varnishes, coatings, core steel, gaskets, and wire enamels, can be evaluated according to ASTM Test Method D 3455 or IEC Publication 588 by placing the material in a sample of the oil at 100°C for 164 or 168 h. Changes in color, interfacial tension, acidity, dielectric dissipation factor, and dielectric strength of the oil are indications of the solubility of a material in the oil and of reactions with or in the oil. More sensitive and extensive accelerated life testing procedures for examining the interactions of the major insulation materials have also been proposed [*82–84*]. 25-kvA distribution transformers were used for that purpose, modified to allow higher hot-spot temperatures of 185°C without exceeding the top-oil temperature of 140°C, and thermally cycled at overload for up to 4000 h. The units could be blanketed with nitrogen or air could be introduced periodically. Short circuit and a-c voltage withstand tests, and winding-winding and winding-ground power factor measurements were used to monitor the condition of the aging transformers. Impulse breakdown tests were used as end point tests. Accelerated testing of transformers is expensive and time consuming and can be justified only when a major change in type of insulating material is contemplated. Simpler tube aging test protocols involving paper-oil combinations, and model coils made from appropriate conductor, oil and paper, or thin board, were developed concurrently. The results of tests using these protocols paralleled those found in the accelerated transformer aging tests cited.

4.5.4.19 Interfacial Tension—The interfacial tension of oil is the surface tension of the interface between oil and water. When water is placed in contact with air or with a clean, new oil, the water molecules at the surface of the water tend to be attracted mostly towards the bulk of the water creating a force at the surface referred to as the surface tension (interfacial tension, if oil). The presence of hydrophilic compounds in oil will slightly attract the water molecules to the oil and thus reduce the overall interfacial tension. The extent of the reduction will depend on the amount and hydrophilic character (polarity) of these compounds.

Interfacial tension is thus highly sensitive to the presence of ionic or polar contaminants in oil including the polar oxidation products formed before the acids (see Section 4.5.4.6 [*24*], the acids themselves and the sludges formed by oxidation. The interfacial tension of insulating oils can be measured according to ASTM Test Methods D 971 or D 2285. The former involves measuring the force necessary to detach a planar ring of platinum wire from the water-oil interface. Two vertical wires welded to the ring are used to lift the ring upwards through the oil. Measurements are made under standard conditions (1 min after formation of the interface) and corrected by empirically determined factors that depend on the force applied, the oil and water densities, and the dimensions of the ring. Method D 2285 is a rapid procedure particularly applicable for field use, where interfacial tension is determined by measuring the volume of a drop of water that the oil will support. The larger the drop of water, the higher the interfacial tension. A piston containing water forces the formation of the drop of water in the oil sample, until the drop falls from the piston tip to the bottom of the oil. The volume of the water drop can be related to interfacial tension through calibration.

Interfacial tension is expressed in dynes per centimeter or millinewtons per meter.

4.5.4.20 AC Loss Characteristics and Resistivity—ASTM D 924, using a three-terminal or guarded-electrode test cell maintained at the desired test temperature and an appropriate bridge circuit, is recommended for measuring the power factor (or dielectric losses) of oil at a commercial power frequency of 60 Hz. The power factor is a dimensionless quantity usually expressed as a percentage. It increases with temperature (Fig. 4.34) and is often measured at 25 and 100°C, with temperature correction factors available [*3*].

Similar cells are used for the measurement of resistivity, according to ASTM D 1169, applying a d-c voltage so that the electric stress in the liquid is between 200 and 1200 V/mm, and measuring the current between the electrodes, which is related to the presence of free ions or ion-forming particles. For more details on the measurement, Chapters 1 and 2 should be consulted by the reader.

Acknowledgments

Special thanks are due to L. Régnier and J. P. Kinney for reviewing the manuscript.

References

Books and Technical Papers

General

[*1*] Johnson, D. L. in *Electrical Insulating Oils, ASTM STP 998,* American Society for Testing and Materials, Philadelphia, 1988, pp. 37–46.

[*2*] Pearce, H. A. in *Electrical Insulating Oils, ASTM STP 998,* American Society for Testing and Materials, Philadelphia, 1988, pp. 47–55.

[*3*] Griffin, P. J., *Doble Engineering Conference,* 1987, pp. 10–701/715.

[4] Vuarchex, P. J. in *Techniques de l'ingénieur,* Paris, 1986, pp. D230a-1/D233-6.
[5] Bradwell, E. in *Electrical Insulation,* Peter Peregrinus Publication, Stevenage, UK, 1983.
[6] Myers, S. D., Kelly, J. J., and Parrish, R. H. in *A Guide to Transformer Maintenance,* TMI Publication, Akron, Ohio, 1981.
[7] Wilson, A. C. M. in *Insulating Liquids: Their Uses, Manufacture, and Properties,* Peter Peregrinus Publication, Stevenage, UK, 1980.
[8] Allinson, J. P. in *Criteria for Quality of Petroleum Products,* Wiley, New York, 1973.
[9] Sillars, R. W. in *Electrical Insulating Materials and Their Applications,* Peter Peregrinus, Stevenage, UK, 1973.
[10] Hobson, C. D. and Pohl, W. in *Modern Petroleum Technology,* Wiley, New York, 1973.
[11] Lipshtein, R. A. and Shakhnovich, M. I. in *Transformer Oil,* Energiya Publication, Moscow, 1968 (Translation from Israel Program for Scientific Translation Jerusalem, 1970).
[12] Clark, F. M. in *Insulating Materials for Design and Engineering Practice,* Wiley, New York, 1962.

Oil Refining

[13] Lipscomb, T. G. in *Electrical Insulating Oils, ASTM STP 998,* American Society for Testing and Materials, Philadelphia, 1988, pp. 5–24.

Paraffin Hydrocarbons

[14] Redelius, P., *Thermochimica Acta,* Vol. 85, 1985, pp. 327–330.
[15] Crine, J.-P., Duval, M., and St-Onge, H., *Transactions on Electrical Insulation,* Institute of Electrical and Electronics Engineers, Vol. EI-20, No. 2, 1985, pp. 419–422.
[16] Yasufuku, S., *Journal of the Japanese Petroleum Institute,* Vol. 27, No. 6, 1984, pp. 525–532.
[17] Blaine, R. L., *American Laboratory,* Vol. 6, No. 1, 1974, pp. 18–22.
[18] Noel, F., *Journal of the Institute of Petroleum,* Vol. 57, No. 558, 1971, pp. 354–358.
[19] Fritz, B., *European Chemical News,* 1966, pp. 29–31.
[20] Asinger, F. in *Paraffins Chemistry and Technology,* Pergamon Press, Oxford, UK, 1965.
[21] Schiessler, R. W. and Flitter, D., *Journal of the American Chemical Society,* Vol. 74, 1952, p. 1720.
[22] Smith, A. E., *Journal of Chemical Physics,* Vol. 18, 1950, pp. 150–151.

Aromatic Hydrocarbons and Polar Compounds

[23] Burton, P. J., Carballeira, M., Duval, M., Fuller, C. W., Graham, J., De Pablo, A., Samat, J., and Spicar, E., CIGRE Report 15-08, 1988.
[24] Duval, M., Giguère, Y., and Lamarre, C., *Journal of Chromatography,* Vol. 284, 1984, pp. 273–280.
[25] Murray, D. W., Clarke, C. T., MacAlpine, G. A., and Wright, P. G., Technical Paper Series No. 821236, Society of Automotive Engineers, 1982.
[26] Dark, W., *Journal of Liquid Chromatography,* Vol. 5, No. 9, 1982, pp. 1645–1652.
[27] Sodhi, J. S., Chadha, B. R., and Chamdra, D., *Proceedings,* Institution of Electrical Engineers, Vol. 128A, No. 2, 1981, pp. 128–131.
[28] Duval, M. and Lamarre, C., *Transactions on Electrical Insulation,* Institute of Electrical Engineers, Vol. EI-12, No. 5, 1977, pp. 340–348.
[29] Suatoni, J. C. and Swab, R. E., *Journal of Chromatographic Science,* Vol. 13, 1975, pp. 361–368.
[30] Melchiore, J. J. and Mills, I. W., *Industrial Engineering Chemistry,* Product Research Development, Vol. 6, 1967, pp. 40–42.
[31] Jezl, J. L., Stuart, A. P., and Ross, E. S., *Industrial Engineering Chemistry,* Vol. 50, No. 947, 1968, pp. 715–721.

Heat Transfer

[32] Duval, M., Lamothe, S., Lamarre, C., and Giguère, Y., *Journal of Chromatography,* Vol. 244, No. 1, 1982, pp. 169–173.
[33] Lamarre, C., Duval, M., and Gauthier, J., *Journal of Chromatography,* Vol. 213, 1981, pp. 481–490.
[34] Bélanger, G., Lamarre, C., and Crine, J.-P., *Minutes of the Doble Conference,* 48A1C81, 1981, pp. 10-601–10-610.

Cooling Hot Spots

[35] Long, L. W., *Electra,* Vol. 86, 1982, pp. 34–51.

[36] Yamaguchi, M., Kumasaka, T., Inui, Y., and Ono, S., *Transactions,* Institute of Electrical and Electronics Engineers, PAS-100 (3), 1981, pp. 956–963.
[37] Grubb, R. L., Hudis, M., and Traut, A. R., PES SM, Institute of Electrical and Electronics Engineers Minneapolis, MN, Paper 80 SM 511-6, 1980.
[38] Heinrichs, F. W., *Transactions,* Institute of Electrical and Electronics Engineers, PAS-98(5), 1979, pp 1576–1582.
[39] Kelly, J. J., Parrish, R. H., and Myers, S. D., *Electrical/Electronics Insulation Conference,* 1979, pp 233–238.
[40] ANSI Guide for Loading Mineral Oil Immersed Overhead-Type Distribution Transformers with 55°C or 65°C Average Winding Rise, C 57-91, 1969.

Hot Spots Detectors

[41] Lampe, W., Petersson, L., Ovren, C., and Wahlstrom, B., CIGRE Report 12-02, 1984.
[42] McNutt, W. J., McIver, J. C., Leibinger, G. E., Fallon, D. J., and Wickersheim, K. A., *Transactions* Institute of Electrical and Electronic Engineers, PAS-103(6), 1984, pp. 1155–1162.
[43] Buchan, P. G., Report No. 157 T 351, Canadian Electrical Association, 1984.
[44] Duval, M., Aubin, J., Giguère, Y., Paré, G., and Langhame, Y., *Transactions,* Institute of Electrical and Electronics Engineers, Vol. EI-17, No. 5, 1982, pp. 414–422.
[45] Yannucci, D. A. and Thompson, J. H., *Electric Light Power,* 1979, pp. 60–61.
[46] Bell, R. and Bertin, M., EPRI Report EL-573, Electric Power Research Institute, Palo Alto, CA, 1977
[47] Hampton, B. F. and Medhurst, D. R., CEGB Technical Disclosure, Bulletin No. 266, 1976.

Low-Temperature Behavior

[48] Lampe, W., *Transactions,* Institute of Electrical and Electronics Engineers, PWRD-1(1), 1986, pp 217–224.
[49] Langhame, Y., Castonguay, J., Crine, J.-P., Duval, M., and St-Onge, H., *Transactions,* Institute of Electrical and Electronics Engineers, Vol. EI-20, No. 3, 1985, pp. 629–638.
[50] St-Onge, H., Langhame, Y., Duval, M., Crine, J.-P., and Castonguay, J., International Conference on Properties and Applications of Dielectric Materials, Xian, China, 1985.
[51] Duval, M., Cauchon, D., Lamothe, S., and Giguère, Y., *Transactions,* Institute of Electrical and Electronics Engineers, Vol. EI-18, No. 6, 1983, pp. 586–590.
[52] Duval, M., Cauchon, D., Lamothe, S., and Giguère, Y., *Conference,* Electrical Insulation and Dielectric Phenomena, Amherst, PA, 1982, pp. 451–457.
[53] Perdomo, T., *Future Heavy Crude Tar,* Sands Institute Conference, Second, 1982, McGraw Hill, New York, 1984, pp. 81–87.
[54] Clark, C. A., *Proceedings,* Institute of Petroleum, London, Vol. 1, 1981, pp. 213–230.
[55] Marin, P. D. and Idas, A., *ETG Fachteberichte,* 1981, pp. 12–16.
[56] Urabe, K., Masuda, T., and Kurahashi, A., *Conference,* Institute of Electrical Engineers, Japan, 1977.
[57] Bartz, W. J., *Tribology International,* 1976, pp. 13–19.
[58] Knepper, J. I. and Hutton, R. P., *Hydrocarbon Processing,* 1975, pp. 129–136.
[59] Price, R. C., *Journal of the Institute of Petroleum,* Vol. 57, No. 554, 1971, pp. 106–109.
[60] Mulhall, V. R. and Thompson, H. E., *Electrical Insulation Conference,* Institute of Electrical and Electronics Engineers, Chicago Paper No. 32C79-56, 1967.
[61] Holder, G. A. and Winkler, J., *Journal of the Institute of Petroleum,* Vol. 51, No. 499, 1965, pp 228–234.
[62] Lorensen, L. E., *Symposium on Polymer Lubricating Oils,* American Chemistry Society Meeting Atlantic City, 1962, pp. B61–B69.

Flammability and Environmental Effects

[63] Orbeck, T., Buch, R., and Tewarson, A., *Electrical Insulation Magazine,* Institute of Electrical and Electronics Engineers, Vol. 6, No. 3, 1990, pp. 8–23.
[64] Kuchta, J. M. and Cato, R. J., Society of Automotive Engineering, Airport Transportation Meeting, New York, 1986.
[65] *Proceedings,* Montech' 86 Conference on PCBs and Replacement Fluids, Institute of Electrical and Electronics Engineers, Montréal, 1986.
[66] Régis, J., *Doble Engineering Conference,* 1972, pp. 10–301.

Oxidation

[*67*] Lamarre, C., Duval, M., and Crine, J.-P., *Doble Engineering Conference,* 1988, pp. 10-6.1/10-6.15.
[*68*] Lamarre, C., Crine, J.-P., and Duval, M., *Transactions,* Institute of Electrical and Electronics Engineers, Vol. EI-22, No. 1, 1987, pp. 57–62.
[*69*] Lamarre, C., Duval, M., and Crine, J.-P., Ninth International Conference on Dielectric Liquids, Salford, UK, 1987.
[*70*] Lamarre, C., Crine, J.-P., and Duval, M., International Symposium on Electrical Insulation, Institute of Electrical and Electronics Engineers, Washington, DC, 1986.
[*71*] Al-Sammerai, D. A. and Barbooti, M. M., *Industrial Engineering Chemistry,* Product Research Development, Vol. 24, 1985, pp. 171–175.
[*72*] Crine, J.-P. and Duval, M., *Proceedings,* International Symposium on Electrical Insulations, Institute of Electrical and Electronics Engineers, Montréal, 1984, pp. 68–71.
[*73*] Murray, D. W., McDonald, J. M., White, A. M., and Wright, P. G., *Proceedings,* World Petroleum Congress, Vol. 11, No. 4, 1983, pp. 447–457.
[*74*] Crine, J.-P., Duval, M., and Lamarre, C., International Symposium on Electrical Insulation, Institute of Electrical and Electronics Engineers, Philadelphia, PA, 1982, pp. 35–37.
[*75*] Al-Wahaib, I. H., Al-Samarai, D. A., Majeed, Q. S., Yahia, W. M., Selman, Q. J., and Al-Azawi, Z., *Journal of Petroleum Research,* Vol. 1, No. 1, 1982, pp. 39–46.
[*76*] Fallou, B., Despiney, Ph., Perret, J., and Vuarchex, P., CIGRE Report 15-11, 1982.
[*77*] Studmiarz, S. A., Mullen, G. A., and Dakin, T. W., *Transactions,* Institute of Electrical and Electronics Engineers, PAS-99(3), 1980, pp. 1130–1136.
[*78*] Edilashvili, I. L., *Nefteckhimiya,* Vol. 20, No. 6, 1980, pp. 852–857.
[*79*] Fournié, R., Le Gall, Y., Perret, J., and Recoupé, P., *Revue Générale de l'Électricité,* Vol. 1, 1979, pp. 31–45.
[*80*] Frey, R. M., Gedemer, T., and Maller, G., PES, Institute of Electrical and Electronics Engineers, New York, A79-083-7, 1979.
[*81*] Dudukovic, P. and Pejovic, V., CH 1287, Institute of Electrical and Electronics Engineers, 1978.
[*82*] Fessler, W. A., Nichols, F. S., and Rouse, T. O., *Proceedings,* International Symposium on Electrical Insulation, Institute of Electrical and Electronics Engineers, 1978, Paper H9, pp. 265–274.
[*83*] Fessler, W. A., Kaufmann, G. H., and Rouse, T. O., IEEE CH 1287, 1978, *Proceedings,* International Symposium on Electrical Insulation, Institute of Electrical and Electronics Engineers, Paper H 11, 1978, pp. 279–290.
[*84*] Rouse, T. O., International Symposium on Electrical Insulation, Institute of Electrical and Electronic Engineers, Paper H 7, 1978, pp. 250–260.
[*85*] Gedemer, T. J. and Maller, G., PES SM, Institute of Electrical and Electronics Engineers, Los Angeles, Paper A 78561-3, 1978.
[*86*] Baranowski, L. B. and Kelly, J. J., *Minutes of the Doble Engineering Conference,* 44 A1C 77, 1977, pp. 10-301–10-304.
[*87*] Lampe, W. and Spicar, E., CIGRE Report 12-05, 1976.
[*88*] Studt, P., *Gemeinschaftstagung,* OGEW/DGMK, Salzburg, 1976.
[*89*] Frey, R. M. and Gedemer, T. J., PES, Institute of Electrical and Electronics Engineers, Paper A 76435-8, 1976.
[*90*] Gedemer, T. J. and Frey, R. M., PES, Institute of Electrical and Electronics Engineers, Paper A 76-030-7, 1976.
[*91*] Frey, R. M. and Gedener, T. J., PES, Institute of Electrical and Electronics Engineers, Paper SM, San Francisco, Paper F 75 434-1, 1975.
[*92*] Narasimban, V. R., Viswanathan, P. N., and Krishnaswamy, K. R., *Chemical Age of India,* Vol. 26, No. 12, 1975, pp. 1026–1034.
[*93*] Waldon, P. L. and Hopkinson, P. J., PES, Institute of Electrical and Electronics Engineers, C 75 138-3, New York, 1975.
[*94*] Noël, R. and Cranton, G. E., *Analytical Calorimetry,* Vol. 3, 1974, pp. 305–320.
[*95*] Dovgopolyi, E. E., Eminov, E. A., Zimina, K. I., Siryuk, A. G., and Kozlova, E. K., *Khimiia Tekhnologiia Topliv Masel,* Vol. 7, No. 2, 1972, pp. 22–25.
[*96*] Kovgopolyi, E. E., Eminov, E. A., Lipshtein, R. A., and Kozlova, E. K., *Khimiia Tekhnologii Topliv Masel,* Vol. 17, No. 2, 1972, pp. 18–22.
[*97*] Pass, F., *Energie and Machinen,* Vol. 7, 1971, pp. 290–296.
[*98*] Dotterer, G. O., *Review Scientific Instruments,* Vol. 42, No. 4, 1971, pp. 471–474.
[*99*] Schober, J., *Bulletin de l'Association Suisse de l'électricité,* Vol. 62, No. 25, 1971, pp. 1210–1215.
[*100*] Reich, L. and Stivala, S. S., Marcel Dekker, New York, 1969.

[101] Dimeler, G. R., Mills, I. W., and Melchiore, J. J., *Transactions,* Institute of Electrical and Electronic Engineers, Vol. EI-4, No. 1, 1969, pp. 7–12.
[102] Nanda, J. R. and Krishnaswamy, K. R., *Electrical Times,* 1968, pp. 858–860.
[103] Emanuel, N. M., Denisov, E. T., and Maizus, Z. K., Plenum Press, New York, 1967.
[104] Shroff, D. H. and Wilson, A. C. M., *Proceedings,* Institution of Electrical Engineers, Vol. 114, No. 6 1967, pp. 817–823.
[105] *The Oxidation of Hydrocarbons in the Liquid Phase,* N. M. Emanuel, Ed., Pergamon Press, Oxford, UK 1965.
[106] Wilson, A. C. M., *Proceedings,* Institution of Electrical Engineers, Vol. 112, No. 3, 1965, pp. 617–63
[107] Massey, L. and Wilson, A. C. M., *Journal of the Institute of Petroleum,* Vol. 44, 1958, p. 336.
[108] Morton, F. and Bell, R. T., *Journal of the Institute of Petroleum,* Vol. 44, 1958, pp. 260–272.
[109] Chernozhukov, N. I. and Krein, S. E. in *Oxidation Properties of Mineral Oils,* Moscow, 1955.
[110] Rumpf, K. K. and Jahn, E., *World Petroleum Congress,* Paris, Sect. V/A, Paper 6, 1955, pp. 119–128
[111] Dietsch, M. C., *Bulletin de la Societe Francaise des Electriciens,* Vol. 46, 1954, pp. 601–606.
[112] Salomon, M. T., *Revue de la Societe Francaise Electriciens,* Seventh Series, Vol. IV, No. 46, 1954, pp 570–600.
[113] Liander, H. and Ericson, G., *Evaluation of Insulating Oils—European Development, ASTM STP 172* American Society for Testing and Materials, Philadelphia, 1954.
[114] Ham, A. J. and Thompson, C. N., *Journal of the Institute of Petroleum,* Vol. 36, 1950, pp. 673–678.
[115] Keulen, D. J. W. and Kreulen Van Selms, F. G., *Journal of the Institute of Petroleum,* Vol. 35, 1949, pp 88–96.
[116] Denison, G. H., *Industrial Engineering Chemistry,* 1944, pp. 477–482.
[117] Balsbaugh, J. C. and Assaf, A. G., *Transactions,* American Institute of Electrical Engineers, Vol. 62 1943, pp. 311–322.
[118] Balsbaugh, J. C., Assaf, A. G., and Pendleton, W. N., *Industrial Engineering Chemistry,* Vol. 33, 1941 pp. 1321–1330.
[119] Evans, R. N. and Davenport, J. E., *Industrial Engineering Chemistry,* Vol. 9, 1937, pp. 321–323.
[120] Whitehead, J. B. and Maurite, F. E., *Transactions,* American Institute of Electrical Engineers, 1937, pp 465–474.
[121] Rodman, C. J., *Journal of the Electrochemical Society,* Vol. 40, 1921, pp. 99–108.

Role of Copper and Metal Deactivators

[122] Lamarre, C., Duval, M., and Crine, J.-P., *Proceedings,* Tenth International Conference on Dielectric Liquids, Grenoble, France, 1990.
[123] El-Sulaiman, A. A., Ahmed, A. S., Qureshi, M. I., and Hassan, M. M. A., *Arab Gulf Journal of Scientific Research,* Vol. 4, No. 1, 1986, pp. 385–400.
[124] El-Sulaiman, A. A., Ahmed, A. S., Qureshi, M. I., and Alrashood, K. A., *Journal of Engineering Science* King Saudia University, Vol. 12, No. 2, 1986, pp. 185–195.
[125] Alexandrov, V. S. and Ivanov, S. K., *Oxidation Communications,* Vol. 8, Nos. 1–2, 1985, pp. 77–86.
[126] Clark, B., Klaus, E. E., and Hsu, S. M., *Journal of the American Society of Lubricating Engineers,* 1985 pp. 280–287.
[127] Duval, M. and Crine, J.-P., *International Conference on Conductance Breakage of Dielectric Liquids* Institute of Electrical and Electronic Engineers, Pavia, Italy, 1984, pp. 421–425.
[128] Alexandrov, V. S. and Ivanov, S. K., *Ropa a Uhle,* Vol. 26, 1984, pp. 671–675.
[129] Wright, P., MacDonald, J. M., and MacAlpine, G., U.S. Patent No. 4267152, 1983.
[130] El-Sulaiman, A. A., Ahmed, A. S., and Qureshi, M. I., PES WM, Institute of Electrical and Electronic Engineers, New York, Paper 82 WM 190-7, 1982.
[131] Bronger, W., Kranz. H. G., and Moller, K., *Journal of Electrostatics,* Vol. 12, 1982, pp. 573–579.
[132] Ishii, T. and Ueda, M., *Conference on Electrical Dielectric Phenomenon,* 1978, pp. 175–184.
[133] Yasufuku, S., Umemura, T., and Tanii, T., *Transactions,* Institute of Electrical and Electronics Engineers Vol. EI-14, No. 1, 1979, pp. 28–35.
[134] Christy, R. A., *Minutes of Doble Conference,* 45A1C78, 1978, pp. 10-201/10-217.
[135] Melchiore, J. J. and Mills, I. W., *Transactions,* Institute of Electrical and Electronics Engineers, Vol EI-2, No. 3, 1967, pp. 150–155.
[136] Melchiore, J. J. and Mills, I. W., *Journal of the Electrochemical Society,* Vol. 112, 1965, pp. 390–395
[137] King, R. W., Ceratp, C. C., and Melchiore, J. J., American Petroleum Institute Meeting, St. Louis, Paper No. 25-64, 1964.
[138] Hughes, F. and Haydock, P. T., *Journal of the Institute of Petroleum,* Vol. 50, No. 490, 1964, pp 239–254.

[139] Gemant, A., *Ions in Hydrocarbons,* Wiley, New York, 1962.
[140] Loveland, J. W., Dimeler, G. R., Bostwick, L. G., and Cali, L. J., *Journal of the Institute of Petroleum,* Vol. 40, No. 3, 1960, pp. 410–415.
[141] Thompson, C. N., *Journal of the Institute of Petroleum,* Vol. 44, No. 417, 1958, pp. 295–317.
[142] Wood-Mallock, J. C., Steiner, H., and Wood, L. G., *Journal of the Institute of Petroleum,* Vol. 44, 1958, pp. 320–332.

Gassing

[143] Fallou, B., Samat, J., Perret, J., and Vuarchex, P. J. CIGRE Report 15-10, 1986.
[144] Saunders, B. L. and Gibbons, J. A. M., Conference Publication No. 177, Institution of Electrical Engineers, 1979.
[145] Harrold, R. T., PES SM, Institute of Electrical and Electronics Engineers, Vancouver, Paper 79-1B5-CORNA-PI, 1979.
[146] Rouse, T. O. and Fessler, W. A., *Conference on Electrical Insulation and Dielectric Phenomena,* 1978, pp. 120–128.
[147] Povey, E. H., *Doble Clients Conference,* 43 AIC 76, 1976, pp. 10-0401–10-0404.
[148] Zaky, A. A., Megahed, I. Y., and El-Sharawy, M. E., CIGRE Report 25, 1976.
[149] Mathes, K. N., *Transactions on Electrical Insulation,* Institute of Electrical and Electronics Engineers, Vol. T2, EI-11 (4), 1976.
[150] Mathes, K. N. and Rouse, T. O., *Conference on Electrical Insulation and Dielectric Phenomena,* 1975.
[151] Zaky, A. A. and Hawley, R. in *Conduction and Breakdown in Mineral Oil,* Peter Peregrinus, UK, 1973, pp. 50–65.
[152] Reynolds, E. H. and Black, R. M., *Proceedings,* Institution of Electrical Engineers, Vol. 11, No. 94, 1972, pp. 497–504.
[153] Wilputte, R., CIGRE Report 24, 1970.
[154] Pedersen, B., *Revue Brown Boveri,* Vol. 55, No. 415, 1968, pp. 222–228.
[155] Kalantar, N. G., Varshavskii, D. S., and Glazunov, V. I., Klimiia Teknologii Toplev I Masel, Vol. 12, No. 5, 1967, p. 54.
[156] Sloat, T. K., Johnson, J. L., and Sommerman, G. M. L., *Transactions,* Institute of Electrical and Electronics Engineers, PAS-86(3), 1967, pp. 374–384.
[157] Black, R. M. and Reynolds, E. H., *Proceedings,* Institution of Electrical Engineers, Vol. 112, No. 6, 1965, pp. 1226–1236.
[158] Basseches, H. and Mc Lean, D. A., *Industrial Engineering Chemistry,* 1955, pp. 1782–1794.
[159] Salomon, T., *Revue de la Societe Francaise des Electriciens,* 1954, pp. 557–569.
[160] Sticher, J. and Piper, J. D., *Industrial Engineering Chemistry,* Vol. 33, No. 12, 1941, pp. 1567–1574.

Dissolved Gases

[161] Carballeira, M., *Electra,* Vol. 133, 1990, pp. 44–51.
[162] Unsworth, J. and Mitchell, F., *Transactions,* Institute of Electrical and Electronics Engineers, Vol. EI-25, No. 4, 1990, pp. 737–746.
[163] Carballeira, M., Harmand, D., Maujean, J. M., Sapet, J., Samat, J., and Vuarchex, P., *Revue Generale de l'Électricité,* Vol. 8, 1990, pp. 18–24.
[164] Duval, M., *Electra,* Vol. 133, 1990, pp. 38–44.
[165] Leblanc, Y., Hubert, J., Gilbert, R., and Duval, M., Chemistry Institute of Canada Conference, Halifax, 1990.
[166] Ferrito, S., *Transactions on Power Delivery,* Institute of Electrical and Electronics Engineers, Vol. 5, No. 1, 1990, pp. 220–225.
[167] Duval, M., *Electrical Insulation Magazine,* Institute of Electrical and Electronics Engineers, Vol. 5, No. 6, 1989, pp. 22–27.
[168] Hinshaw, J. V. and Seferovic, W., Pittsburgh Conference Paper 279, American Chemical Society, 1989.
[169] Azizian, H., Report 260T507, Canadian Electrical Association, Montréal, 1989.
[170] Marin, M. A. and Jasmin, J. L., Conference on Expert Systems, Applications for the Electric Power Industry, EPRI, Orlando, 1989.
[171] Duval, M., Langdeau, F., Gervais, P., and Bélanger, G., *Conference on Electric Insulation and Dielectric Phenomenon,* Annual Report, 1989, pp. 358–362.
[172] Duval, M., Langdeau, F., Gervais, P., and Bélanger, G., *Doble Engineering Conference,* 1989, pp. 10-3.1/10-3.9.

[*173*] Duval, M., Langdeau, F., Gervais, P., and Bélanger, G., *Doble Engineering Conference,* 1988, pp 10-7.1/10-7.10.
[*174*] Savio, L. J. in *Electrical Insulating Oils, ASTM STP 998,* American Society for Testing and Materials Philadelphia, 1988, pp. 83–88.
[*175*] Griffin, P. J. in *Electrical Insulating Oils, ASTM STP 998,* American Society for Testing and Materials Philadelphia, 1988, pp. 89–107.
[*176*] Haupert, T. J. and Jakob, F. in *Electrical Insulating Oils, ASTM STP 998,* American Society for Testing and Materials, Philadelphia, 1988, pp. 108–115.
[*177*] Recrosio, N., Jegou, Y., and Carballeira, M., Société des Électriciens et Électroniciens, Club 11, Études Générales, Diagnostic et Maintenance par Système Expert, Paris, 8 Oct. 1987.
[*178*] Riese, C. E. and Stuart, J. D. in *Artificial Intelligence Applications in Chemistry,* American Chemical Society, 1986, pp. 25–30.
[*179*] Fisher, D. J., U.S. Patent No. 4,587,834, 1986.
[*180*] Wilputte, R. and Randoux, R., CIGRE Report 12-07, 1986.
[*181*] Kawamura, T., Yamaoka, M., Kawada, H., Ando, K., Maeda, T., and Takatsu, T., CIGRE Report 12-05 1986.
[*182*] Sobral Viera, C. L. C., *Doble Clients Conference,* 1986.
[*183*] Douglas, D. H., Lawrence, C. O., and Templeton, J. B., *Transactions,* Institute of Electrical and Electronics Engineers, PAS-104(9), 1985, pp. 2492–2500.
[*184*] Bartl, J., Fidler, A., Gregor, P., and Vavra, Z., *Acta Physica Slov.,* Vol. 35, No. 3, 1985, pp. 174–178
[*185*] Duval, M. and Giguère, Y., *Doble Clients Conference,* 51AIC84, 1984, pp. 10C-01/10C-07.
[*186*] Burton, P. J., Graham, J., Hall, A. C., Laver, J. A., and Oliver, A. J., CIGRE Report 12-09, 1984.
[*187*] Burton, P., Carballeira, M., Crespo, M., Foschum, H., Gandillon, M., Knab, H. J., Markestein, T Praehauser, T. H., Rindfleisch, H. J., Samat, J., Schliesing, H., Serena, E., Soldner, K., Vuarchex, P Whitelock, K., and Wolf, P. G., CIGRE Report 15-11, 1984.
[*188*] Musil, R. J. and Foschum, H., *Doble Clients Conference,* 50AIC83, 1983, pp. 10-701A/10-705A.
[*189*] Oommen, T. V., *Transactions,* Institute of Electrical and Electronics Engineers, PAS-101(6), 1982, pp 1716–1722.
[*190*] Oommen, T. V., Moore, R. H., and Luke, L. E., *Transactions,* Institute of Electrical and Electronic Engineers, PAS-101(5), 1982, pp. 1048–1052.
[*191*] Kelly, J. J., *Transactions,* Institute of Electrical and Electronics Engineers, IA-16 (6), 1980, pp. 777–78
[*192*] Musil, J. and Foschum, H., *Doble Clients Conference,* 47AIC80, 1980, pp. 6-801/6-804.
[*193*] MacDonald, J. D. and Vitols, A. P., *Doble Clients Conference,* 47AIC80, 1980, pp. 6-901/6-904.
[*194*] Rickley, A. L., Baker, A. E., and Armstrong, G. W., *Doble Clients Conference,* 45AIC78, 1978, pp 10-401/10-407.
[*195*] Rogers, R. R., *Transactions,* Institute of Electrical and Electronics Engineers, Vol. EI-13, No. 5, 197 pp. 349–354.
[*196*] Shirai, M., Ishii, T., and Makino, Y., *Transactions,* Institute of Electrical and Electronics Engineers, Vol EI-12, No. 4, 1977, pp. 266–271.
[*197*] Shirai, M., Shimoji, S., and Ishii, T., *Transactions,* Institute of Electrical and Electronics Engineers, Vol EI-12, No. 4, 1977, pp. 272–280.
[*198*] Rabaud, J., Thibault, M., Verdon, J., and Viale, F., CIGRE Report 12-06, 1976.
[*199*] Viale, F., CIGRE Report 12-06, 1976.
[*200*] Serena, E. et al., CIGRE Report 12-09, 1976.
[*201*] Dind, J. E. and Regis, J., *Pulp and Paper,* Canada, Vol. 76, No. 9, 1975, pp. 61–64.
[*202*] Rogers, R. R., *Doble Client Conference,* 42AIC75, 1975, pp. 10-201/10-210B.
[*203*] Duval, M., PES SM, Institute of Electrical and Electronics Engineers, Anaheim, CA, Paper C74 476-8 1974.
[*204*] Pugh, D. R., *Doble Clients Conference,* 41AIC74, 1974, pp. 10-1201/10-1208.
[*205*] Wadington, F. B., *Science Technology,* Vol. 41, 1974, pp. 89–93.
[*206*] Yamadka, M., Obara, S., Hoshikawa, H., Sagara, H., and Machida, S., *Fiji Electric Review,* Vol. 19, No 2, 1973, pp. 83–86.
[*207*] Halstead, W. D., *Journal of the Institute of Petroleum,* Vol. 59, No. 569, 1973, pp. 239–241.
[*208*] Galand, J., Thibault, M., Viale, F., Samat, J., and Vuarchex, P., *Revue Générale de l'Electricite,* Vol. 81 No. 1, 1972, pp. 727–739.
[*209*] Thibault, M. and Galand, J., *Revue Générale de l'Electricite,* Vol. 80, No. 1, 1971, pp. 45–53.
[*210*] Dind, J. E., Daoust, R., Regis, J., and Morgan, J., *Doble Clients Conference,* 1971.
[*211*] Ishii, T. and Namba, S., *Electrical Engineering in Japan,* Vol. 90, No. 3, 1970, pp. 53–61.
[*212*] Fallou, B., Davies, I., Rogers, R. R., Reynolds, E. H., Viale, F., Devaux, A., Fournié, R., Galand, J Vuarchex, P., and Dornenburg, E., CIGRE Report 15-07, 1970.

[213] Dornenburg, E. and Gerber, O. E., *Brown Boveri Review,* Vol. 54, No. 2/3, 1967, pp. 104–111.
[214] Wilputte, R., CIGRE Report 125, 1966.

Dissolved Gas Detectors

[215] Tsukioka, H., Sugawara, K., Mori, E., and Yamaguchi, H., *Transactions,* Institute of Electrical and Electronics Engineers, Vol. EI-21 No. 2, 1986, pp. 221–229.
[216] Graham, J., *Electrical Review,* Vol. 211, No. 8, 1982, pp. 29–30.
[217] Bélanger, G. and Duval, M., *Transactions,* Institute of Electrical and Electronics Engineers, Vol. EI-12, No. 5, 1977, pp. 334–340.
[218] Morgan, J. E., U.S. Patent No. 4112737, 1977.
[219] Pearce, H. A., U.S. Patent No. 3866460, 1975.

Contaminants

[220] Crine, J.-P. in *Electrical Insulating Oils, ASTM STP 998,* American Society for Testing and Materials, Philadelphia, 1988, pp. 59–79.
[221] Vincent, C. and Crine, J.-P., *Conference on Records,* International Symposium on Electrical Insulation, 1986, pp. 310–314.
[222] Olivier, R. G., Vincent, C., and Crine, J.-P., *Doble Engineering Conference,* 1986, pp. 10-401.
[223] Kako, Y. et al., PES-SM, New York, Institute of Electrical and Electronics Engineers, 1982.
[224] Oommen, T. V. and Petrie, E. M., PES-SM, New York, Institute of Electrical and Electronics Engineers, 1982.
[225] Miners, K., *Transactions,* Institute of Electrical and Electronics Engineers, PAS, Vol. 101, No. 3, 1982.
[226] Cooke, C. M., Conference on Electrical Insulation and Dielectric Phenomenon, 1982.
[227] Trinh, N. G. et al., CH 1496-9, Institute of Electrical and Electronics Engineers, 1980.
[228] Skog, J. E., de Giorgio, J. B., Jakob, F., and Haupert, T. J., *Doble Engineering Conference,* 1980.
[229] Saunders, B. L. and Gibbons, J. A. M., *Conference Publication 177,* Institution of Electrical Engineers, 1979, pp. 6–9.
[230] Mathes, K. N. and Atkins, J. M., International Symposium on Electrical Insulation, Institute of Electrical and Electronics Engineers, 1978.
[231] Ast, P. F., *Conference on Electrical Insulation and Dielectric Phenomenon,* 1978, pp. 203–212.
[232] Flanagan, P. E. et al., *Doble Engineering Conference,* 1976, pp. 10-801.
[233] Krazucki, Z., International Colloquium on Conditional Proceedings in Dielectric Liquids, Grenoble, 1968.
[234] Bartnikas, R., *Transactions,* Institute of Electrical and Electronics Engineers, Vol. EI-2, No. 1, 1967, pp. 33–54.

STANDARD TEST METHODS AND SPECIFICATIONS

ASTM standards recommended by ASTM Committee D 27 on electrical insulating oils, as part of Vol. 10.03 of the Annual Book of ASTM Standards are underlined once when referred to only in D 117 (test descriptions) and underlined twice when also specified in D 3487 (property requirements).

General

ASTM D 3487-88,	Specification for Mineral Insulating Oils Used in Electrical Apparatus.
ASTM D 117-83	Testing and Specifications for Electrical Insulating Oils of Petroleum Origin.
IEC Publ. 296 (1982)	Specification for Unused Mineral Insulating Oils for Transformers and Switchgear.
IEEE 64 (1977)	Guide for Acceptance and Maintenance of Insulating Oil in Equipment.

ASTM D 923-91	Sampling Electrical Insulating Liquids.
IEC 422 (1973)	Maintenance and Supervision Guide for Insulating Oils in Service.

Composition

ASTM D 1218-92	Refractive Index and Refractive Dispersion of Hydrocarbon Liquids.
ASTM D 1807-84 (1989)	Refractive Index and Specific Optical Dispersion of Electrical Insulating Liquids.
ASTM D 2224-78 (1983)	Mean Molecular Weight of Mineral Insulating Oils by the Cryoscopic Method.
ASTM D 2502-92	Method for Estimation of Molecular Weight of Petroleum Oils from Viscosity Measurements.
ASTM D 2503-92	Molecular Weight of Mineral Insulating Oils by the Cryoscopic Method.
ASTM D 611-82 (1987)	Aniline Point and Mixed Aniline Point of Petroleum Products and Hydrocarbon Solvents.
ASTM D 2140-91	Carbon-Type Composition of Insulating Oils of Petroleum Origin.
ASTM D 2786-91	Hydrocarbon-Type Analysis of Gas-Oil Saturate Fraction.
ASTM D 3238-90	Calculation of Carbon Distribution and Structural Group Analysis of Petroleum Oils by the n-d-M Method.
ASTM D 2159-93	Naphthenes in Saturates Fractions by Refractivity Intercept.
ASTM D 2007-91	Characteristic Groups in Rubber Extender and Processing Oils by the Clay-Gel Adsorption Chromatographic Method.
ASTM D 1319-93	Hydrocarbon Types in Liquid Petroleum by Fluorescent Indicator Adsorption.
ASTM D 2549-91	Separation of Representative Aromatics and Nonaromatic Fractions of High Boiling Oils by Elution Chromatography.
ASTM D 2144-90	Examination of Electrical Insulating Oils by Infrared Absorption.
ASTM D 2008-91	Ultraviolet Absorbance and Absorptivity of Petroleum Products.
IEC Publ. 590 (1977)	Determination of the Aromatic Hydrocarbon Content of New Mineral Insulating Oils.
ASTM D 3239-91	Aromatic Type Analysis of Gas-Oil Aromatic Fractions by High Ionizing Voltage Mass Spectrometry.
ASTM D 2786-91	Method for Hydrocarbon Types Analysis of Gas-Oil Saturate Fractions by High Ionizing Voltage Mass Spectrometry.
ASTM D 3228-92	Total Nitrogen in Lubricating Oils by Modified Kjeldahl Method.
ASTM D 3431-87	Trace Nitrogen in Liquid Petroleum Hydrocarbons (Microcoulometric Method).
ASTM D 1275-86 (1991)	Sulfur Corrosive in Electrical Insulating Oils.
ASTM D 2622-92	Sulfur in Petroleum Products (X-ray Spectrographic Method).
ASTM D 3120-92	Trace Quantities of Sulfur in Light Liquid Petroleum Hydrocarbons by Oxidative Microcoulometry.
ASTM D 129-91	Sulfur in Petroleum Products (General Bomb Method).

Heat Transfer

ASTM D 2717-90	Thermal Conductivity of Liquids.
ASTM D 2766-91	Specific Heat of Liquids and Solids.
ASTM D 1903-92	Coefficient of Thermal Expansion of Electrical Insulating Liquids of Petroleum Origin, and Askarels.

ASTM D 1298-85 (1990)	Density, Relative Density (specific gravity), or API Gravity of Crude Petroleum and Liquid Petroleum Products by Hydrometer Method.
ASTM D 287-92	API Gravity of Crude Petroleum and Petroleum Products (Hydrometer Method).
ASTM D 1481-91	Density and Relative (specific gravity) of Viscous Materials by Lipkins Bicapillary Pycnometer.
ASTM D 1217-93	Density and Relative Density (Specific Gravity) of Liquids by Bingham Picnometer.
ASTM D 445-88	Viscosity, Kinematic, or Transparent and Opaque Liquids (and the calculation of dynamic viscosity).
ASTM D 88-81 (1987)	Viscosity, Saybolt.
ASTM D 2161-87	Viscosity, Conversion of Kinematic to Saybolt Universal or to Saybolt Furol.
ASTM D 2532-87	Viscosity and Viscosity Change after Standing at Low Temperature of Aircraft Turbine Lubricants.
ASTM D 2983-87	Low-Temperature Viscosity of Automotive Fluid Lubricants Measured by Brookfield Viscometer.
ASTM D 2602-86 (1993)	Apparent Viscosity of Engine Oils at Low Temperature Using the Cold-Cranking Simulator.
ASTM D 2422-86 (1993)	Recommended Practice for Viscosity System for Industrial Fluid Lubricants.
ASTM D 2162-93	Method of Basic Calibration of Master Viscometer and Viscosity Oil Standards.
ASTM D 446-83a	Specification and Operating Instructions for Glass Capillary Kinematic Viscometers.
IEC Document 10A-98	Behaviour of Oils Containing Pour Point Depressant Additives and Fulfilling the Requirements of Publication 296.
ASTM D 97-87	Pour Point of Petroleum Oils.
ASTM D 2500-91	Cloud Point of Petroleum Oils.
ASTM D 3117-87 (1993)	Wax Appearance Point of Distillate Fuels.

Flammability and Environmental Effects

ASTM D 92-90	Flash and Fire Points by Cleveland Open Cup.
ASTM E 659-78 (1983)	Auto Ignition Temperature of Liquid Chemicals.
ASTM D 56-87	Flash Point by Tag Closed Tester.
ASTM D 93-90	Flash Point by Pensky-Martens Closed Tester.
ASTM D 2887-89	Boiling Range Distribution of Petroleum Fractions by Gas Chromatography.
ASTM D 2551-80	Vapor Pressure of Petroleum Products (Micro Method).
ASTM D 2878-87	Method for Estimating Apparent Vapor Pressures and Molecular Weights of Lubricating Oils.
ASTM D 240-92	Heat of Combustion of Liquid Hydrocarbon Fuels by Bomb Calorimeter.
ASTM D 2382-88	Heat of Combustion of Hydrocarbon Fuels by Bomb Calorimeter (High-Precision Method).

Factory Mutual Association—Specification for Less Flammable Transformer Fluids—FM 545/14-85 (1979).

ASTM D 4059-91	Polychlorinated Biphenyls in Mineral Oils by Gas Chromatography.

Oxidation

ASTM D 2440-93	Oxidation Stability of Mineral Insulating Oils.
ASTM D 2112-93	Oxidation Stability of Inhibited Mineral Insulating Oil by Rotating Bomb.
ASTM D 943-81 (1991)	Oxidation Characteristics of Inhibited Mineral Oils.
IEC Publ. 474 (1974)	Test Method for Oxidation Stability of Inhibited Mineral Insulating Oils.
IEC Publ. 813	Test Method for Evaluating the Oxidation Stability of Hydrocarbon Insulating Liquids (in preparation).
ASTM D 974-92	Neutralization Number by Color-Indicator Titration.
ASTM D 664-89	Neutralization Number by Potentiometric Titration.
ASTM D 3339-92	Total Acid Number of Petroleum Products by Semi-Micro Color Indicator Titration.
ASTM D 1563-84	Peroxide Number of Mineral Insulating Oils (withdrawn).
ASTM D 1698-84 (1990)	Sediment and Soluble Sludge in Service-Aged Insulating Oils.
ASTM D 4310-83	Determination of Sludging Tendencies of Inhibited Mineral Oils.
ASTM D 4055-81	Pentane-Insolubles by Membrane Filtration.
ASTM D 2668-92	2,6 Ditertiary-Butyl Para-Cresol and 2,6 Ditertiary Butyl Phenol in Electrical Insulating Oil by Infrared Absorption.
ASTM D 4768-88	Analysis of 2,6 Ditertiary-Butyl Para-Cresol and 2,6 Ditertiary-Butyl Phenol in Insulating Fluids by Gas Chromatography.
IEC Publ. 666 (1979)	Detection and Determination of Specified Antioxidant Additives in Insulating Oils.
ASTM D 2608-84	Copper in Electrical Insulating Oils by Photometric Analysis.
ASTM D 3635-90	Copper in Electrical Insulating Oils by Atomic Absorption Spectrophotometry.
ASTM D 1500-91	Color, ASTM, of Petroleum Products (ASTM Color Scale).
ASTM D 1524-84 (1990)	Examination, Visual, of Used Electrical Insulating Oils of Petroleum Origin in the Field.

Gassing

ASTM D 2300-85 (1991)	Gassing of Insulating Oils Under Electrical Stress and Ionization (Modified Pirelli Method).
ASTM D 2298-81	Stability of Insulating Oils Under Electrical Stress (Merrel Test) (withdrawn).
IEC Publ. 628 (1978)	Gassing of Cable and Capacitor Insulating Oils Under Electrical Stress and Ionization.
IEC Publ. 815	Gassing of Insulating Liquids Under Electrical Stress and Ionization (in preparation).
ASTM D 3300-85	Dielectric Breakdown Voltage of Insulating Oils of Petroleum Origin Under Impulse Conditions.

Dissolved Gases

ASTM D 3612-93	Analysis of Gases Dissolved in Electrical Insulating Oils by Gas Chromatography.
ASTM D 2945-90	Gas Content of Insulating Oils.

ASTM D 831-84 (1989)	Gas Content of Cable and Capacitor Oils.
ASTM D 1827-92	Gas Content (nonacidic) of Insulating Liquids by Displacement with Carbon Dioxide.
ASTM D 1827-64 (1979)	Gas Content (nonacidic) of Insulating Liquids by Displacement with Carbon Dioxide.
ASTM D 2780-92	Solubility of Fixed Gases in Liquids.
ASTM D 3613-92	Sampling Electrical Insulating Oils for Gas Analysis and Determination of Water Content.
ASTM D 2779-92	Estimation of Solubility of Gases in Petroleum Liquids.
IEC Publ. 599 (1978)	Interpretation of the Analysis of Gases in Transformer and Other Oil-Filled Electrical Equipment in Service.
IEC Publ. 567 (1977)	Guide for the Sampling of Gases and Oil from Oil-Filled Electrical Equipment and for the Analysis of Free and Dissolved Gases.
ANSI-IEEE C57.104 (1978)	American National Standard Guide for the Detection and Determination of Generated Gases in Oil-Immersed Transformers and their Relation to the Serviceability of the Equipment.

Contamination

ASTM D 971-91	Interfacial Tension of Oil Against Water by the Ring Method.
ASTM D 2285-85 (1990)	Interfacial Tension of Electrical Insulating Oils of Petroleum Origin Against Water by the Drop-Weight Method.
ASTM D 1533-88	Water in Insulating Liquids (Karl Fischer Reaction Method).
ASTM F 661-92	Practice for Particle Count and Size Distribution Measurement in Batch Samples Using an Optical Particle Counter.
ASTM D 4056-82	Estimation of Solubility of Water in Hydrocarbon and Aliphatic Ester Lubricants.
IEC Publ. 733 (1982)	Determination of Water in Insulating Oils, and in Oil-Impregnated Paper and Pressboard.
IEC Publ. 814	Determination of Water in Insulating Liquids by Automatic Coulometry Karl Fischer Titration.
ASTM D 3455-83	Compatibility of Construction Material with Electrical Insulating Oil of Petroleum Origin.
ASTM D 878-91	Inorganic Chlorides and Sulfates in Insulating Oils.
ASTM D 877-87	Dielectric Breakdown Voltage of Insulating Liquids Using Disk Electrodes.
ASTM D 1816-84a (1990)	Dielectric Breakdown Voltage of Insulating Oils of Petroleum Origin Using VDE Electrodes.
ASTM D 924-92	A-C Loss Characteristics and Relative Permittivity (Dielectric Constant) of Electrical Insulating Liquids.
ASTM D 1169-89	Specific Resistance (Resistivity) of Electrical Insulating Liquids.

G. A. Vincent[1]

Chapter 5

Molecular Structure and Composition of Liquid Insulating Materials

5.1 Atomic and Molecular Bonding

It is necessary to know something about the chemical bonding holding atoms together to form molecules and the interactions of molecules in order to have some understanding of matter. Why is a liquid a liquid and not a gas or a solid? What is the basis of the physical and chemical properties of liquids?

5.1.1 Atomic Structure

In order to discuss the structure of a molecule one must have some understanding of the chemical bonds involved in forming that molecule. Molecules are a collection of chemically bound atoms. An atom of an element has a nucleus composed of positively charged protons and uncharged neutrons. There are negatively charged electrons orbiting the nucleus. The number of electrons and protons in an atom are equal. The lower the electron energy, the closer it is to the nucleus. Electrons orbiting about the nucleus of an atom do not occupy narrow, exactly defined orbits like planets about the sun. Instead they exist somewhere in a volume of space located near the nucleus of an atom. Electrons having different energies occupy different volumes, also called orbitals, having different shapes. Even within this orbital, the electron spends more time in some portions than in others. Thus, the probability of finding the electron varies even within the volume to which it is restricted 99 + % of the time. This orbital may be viewed as a distribution of probabilities of finding an electron having the energy defined by that volume. Electrons associated with an atom may possess only discrete levels of energy. An electron can absorb energy while in one orbital and jump to another orbital. However, it can not absorb less or more energy than that required to make the transition from one orbital to another.

In the simplest of elements, hydrogen, one electron orbits a single proton. The probability distribution for the electron in a hydrogen atom takes the form of a sphere with the nucleus at the center. The electron is said to occupy an orbital. The lowest energy atomic orbital is designated 1S. The next highest orbital is also spherical; however, it has a larger radius than that of the 1S orbital and is designated the 2S atomic orbital. The S orbitals may contain zero, one, or two electrons. Lithium is the first element on the periodic chart of the elements to have an electron in the 25 orbital.

The 2*P* orbitals are the next highest available atomic orbitals. Boron is the first element on the periodic chart to have an electron in a 2*P* orbital. There are three 2*P* orbitals, and each may contain

[1]Dow Corning Corporation, Midland, MI 48686.

zero, one, or two electrons. These orbitals are shaped approximately like a figure eight with the center of the eight at the atomic nucleus. If the nucleus of an atom is placed at the intersection of *x*, *y*, and *z* coordinates, then the three *P* orbitals extend out from the origin along each of the axes in both a positive and a negative direction. The three 2*P* orbitals are then designated as 2P*x*, 2P*y*, and 2P*z*. The 2P*x* orbital is illustrated in Fig. 5.1.

There are several higher energy orbitals above the 2P; however, we need not go further with the discussion in order to cover the subject of this chapter.

5.1.2 Covalent Bonds

Electrons are involved in the formation of bonds between atoms to form molecules. If two atoms of hydrogen are moved towards each other from a distance there will initially be a weak attraction. Eventually there will be a repulsion due to the interaction of two like charges. However, if the two atoms are energetic enough to overcome repulsion, their electron clouds will overlap. When this happens, the atomic orbitals interact to form two molecular orbitals. One of these molecular orbitals is a bonding orbital and the other is an antibonding orbital, which is higher in energy than the bonding orbital. The two electrons, one from each hydrogen, occupy the bonding orbital. Electrons occupy the antibonding orbitals in a molecule only when enough energy has been supplied to the molecule to cause an electron to jump to the antibonding orbital. As with electrons in an atom,

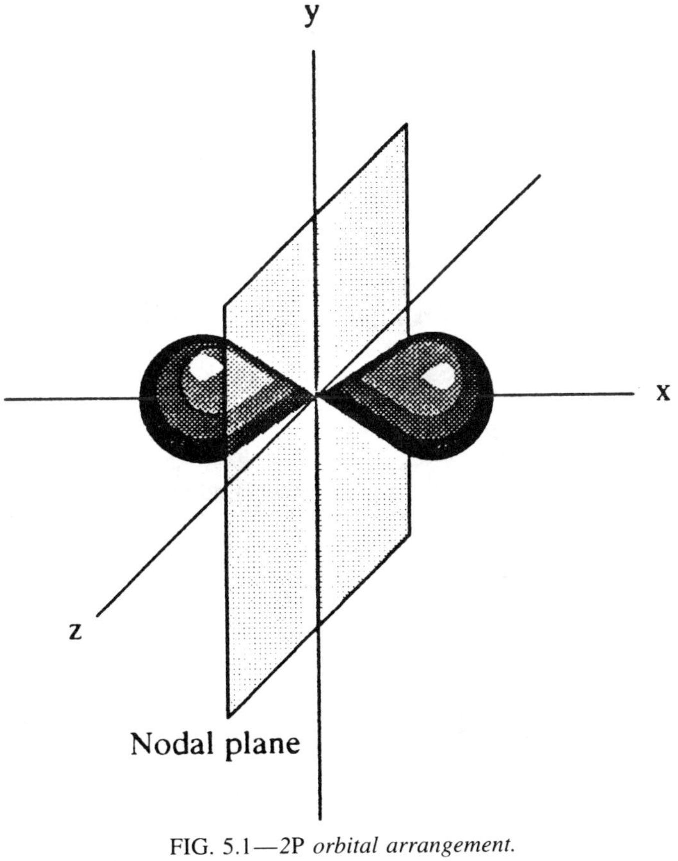

FIG. 5.1—2P *orbital arrangement.*

electrons involved in bonding are limited to discrete transitions. A bonding electron must occupy either a bonding or a nonbonding molecular orbital. It can not be somewhere in between these two alternatives. Although antibonding orbitals will not be mentioned again in this discussion, it should be remembered that when atomic orbitals combine to form molecular orbitals, an equal number of bonding and antibonding orbitals are formed. The electrons normally occupy the bonding orbitals. As the name implies, antibonding orbitals destabilize a molecular bond. If an antibonding orbital is occupied by an electron, the bond between atoms is weaker than it would be with the same number of electrons in a bonding orbital. The bonding molecular orbital formed when two hydrogen atoms combine surrounds the two atoms and is symmetrical through their centers. The bond energy holding two hydrogen atoms together is 104 kcal/mol [*1*]. This means that by combining two hydrogen atoms to form one hydrogen molecule the energy of the system is lowered by 104 kcal; and, that much energy would have to be supplied to hydrogen molecules to again separate them into hydrogen atoms. The molecular bonding orbital is formed by the constructive overlapping of the 1S atomic orbitals of two hydrogen atoms, as schematically depicted in Fig. 5.2.

Not all molecular orbitals result from so simple a process, however. Carbon is the basic building block for organic chemicals. Many insulating liquids, such as mineral oils, polybutenes, esters, polychlorinated biphenyls, etc., are based upon carbon. The carbon atom contains six electrons. Two of them occupy the 1S orbital, two of them occupy the 2*S* orbital and the remaining two are in two of the three 2*P* orbitals (Electrons will always singly occupy orbitals of equal energy before pairing with another electron. This is known as Hund's rule.) When carbon forms chemical bonds it does not do so simply by combining a 2*P* atomic orbital with an atomic orbital from another atom. Instead the 2*S* and 2*P* atomic orbitals undergo a hybridization, or mixing, process. When carbon is bonded to four other atoms the hybridization process produces four energetically equivalent orbitals. Each of these orbitals contains one electron.

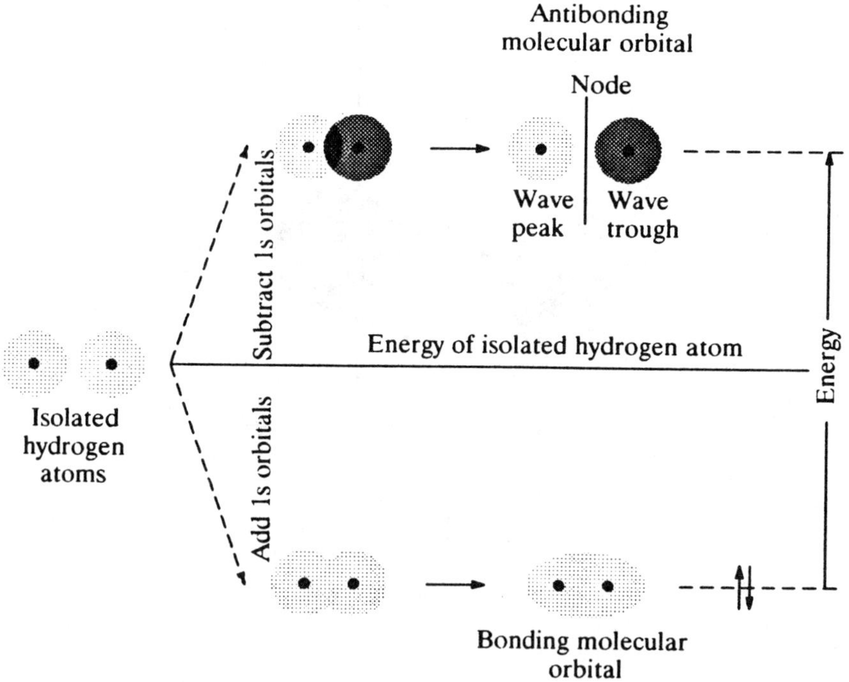

FIG. 5.2—*Molecular bonding orbital of hydrogen.*

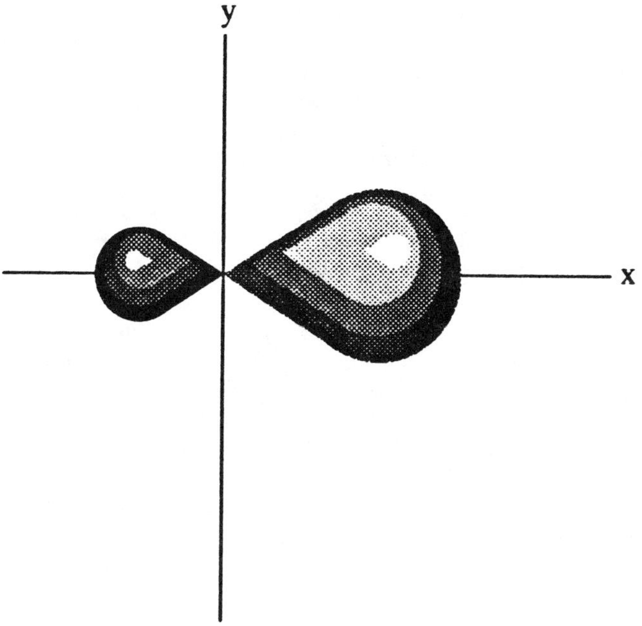

FIG. 5.3—*An* SP^3 *orbital arrangement.*

These four new orbitals are highly directional and form an angle of 109.5° with each other. They are one part $2S$ and three parts $2P$ in character. For this reason they are called SP-three (SP^3) orbitals. One such orbital would have the shape shown in Fig. 5.3.

These four orbitals are oriented in space in such a fashion that they are directed towards the four corners of a regular tetrahedron. As a result of this orientation of the hybrid carbon orbitals, methane has the geometric form of a regular tetrahedron with the carbon atom at its center and a hydrogen atom at each of the four apexes. Higher alkanes can be represented as a series of tetrahedra linked together through SP^3-SP^3 bonding between carbon atoms. The carbon-hydrogen bonds result from overlapping of the $1S$ orbital of hydrogen with an SP^3 hybrid orbital of carbon.

The bonds between carbon and carbon and carbon and hydrogen in ethane are covalent, σ, bonds. "Covalent" means that the two electron bond results from combining of atomic orbitals, each containing one electron, from each of the two atoms. The carbon-carbon bonds are nonpolar, and the carbon-hydrogen bonds are relatively nonpolar. This means that the bonding electrons used in those bonds spend approximately an equal amount of time near each of the atoms involved in the bond. Not all bonds are nonpolar, however.

5.1.3 Polar Covalent Bonds

Not all atoms have the same affinity for electrons. The atoms of some elements have a strong attraction for electrons and easily form an anion. Elements such as oxygen, fluorine, chlorine, etc., tend to easily acquire extra electrons, while atoms of elements such as sodium, lithium, magnesium, etc., easily lose electrons to become cations. The tendency to acquire extra electrons increases across the periodic chart of the elements and decreases down that same chart. Thus, fluorine has the strongest affinity for extra electrons while lithium has the lowest preference for a negative charge. This affinity for electrons is known as electronegativity. When a bond is formed between two different elements, the electrons forming that bond are more likely to be found near the more electronegative atom, although they spend some time near the other element too. This behavior results in a concentration of charge in one end of the bond. For example, in methyl chloride (CH_3Cl) the electrons forming the carbon-chlorine bond are attracted to the more electronegative chlorine atom. In spite of the fact that the bond between carbon and chlorine is longer in length and more polar than the carbon hydrogen bond, the molecule is still tetrahedral in shape albeit slightly distorted. The time averaged uneven distribution of electrons produces a permanent dipole in the methyl chloride molecule. It is measured by a quantity called the dipole moment.

The dipole moment is expressed in terms of debyes. The debye is a product of the distance, in angstroms, separating two charges and the absolute value of those two charges expressed in electrostatic units. The dipole moment is a vector sum which means that it has both magnitude and direction components. The fact that dipole moment is a vector quantity means that not all molecules containing polar bonds are themselves polar, that is, have a dipole moment. Thus, methyl chloride has a polar bond between carbon and chlorine and a dipole moment associated with it. The methyl chloride molecule also has a dipole moment and is said to be a polar molecule. Its bond dipole moment is shown schematically in Fig. 5.4.

The carbon-oxygen bonds in carbon dioxide also are polar. However the carbon dioxide molecule lacks a dipole moment. The reason for this is that there are two carbon-oxygen bonds, one on each side of the carbon atom. The polarity of the two bonds is the same but geometrically orientated in opposite directions as shown in Fig. 5.5. As a result, the vector sum of the two polar bonds is zero, leaving the carbon dioxide molecule without a permanent dipole moment. Because the molecule lacks a permanent dipole moment, it is said to be nonpolar.

5.1.4 The Carbon-Carbon Double Bond

Compounds containing carbon-carbon double bonds are known as alkenes or olefins. The bonding and reactivity of olefins are quite different from the singly bonded, saturated organic compounds known as alkanes. Hybridization of the atomic orbitals in carbon to form ethylene involves only two of the $2P$ orbitals and the $2S^2$ orbital. The three hybrid orbitals which are formed are labeled as SP^2 orbitals. Ethylene is a planar molecule. The bond angle between carbon and hydrogen is 121°, while the bond angle between hydrogen atoms on the same carbon is 118° [2]. There are four σ bonds connecting the hydrogen atoms to the two carbon atoms and one σ bond between the two carbon atoms. This leaves the unhybridized $2P$ orbitals of each carbon atom apparently unused. Because the ethylene molecule is planar these two orbitals overlap above and below the molecular plane (Fig. 5.6).

This overlapping of the P orbitals is the basis of a new bond. The new bond is called a π bond. It is of higher energy than the σ bond. The carbon-carbon σ bond has a bond strength of approximately 109 kcal/mol, while the π bond has a bond strength of only about 63 kcal/mol. The total energy bonding the two carbon atoms together is the sum of the energies of the two bonds. Thus the homolytic scission of a carbon-carbon double bond requires the input of 172 kcal/mol

$$H_2C{=}CH_2 + 172\,\text{kcal} \longrightarrow 2\,H_2C \tag{5.1}$$

$$H_3\overrightarrow{C{-}Cl}$$

FIG. 5.4—*Direction of dipole moment in the methyl chloride molecule.*

$$\overleftarrow{O{=}}C\overrightarrow{{=}O}$$

FIG. 5.5—*Carbon-oxygen dipole directions in a CO_2 molecule.*

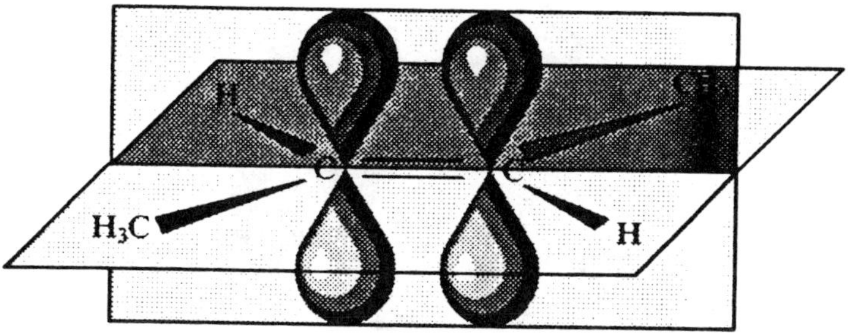

FIG. 5.6—*Orbitals in an ethylene molecule.*

In addition to the difference in the amount of energy bonding the two carbon atoms using SP^2 hybrid orbitals, there is a difference in the carbon-carbon bond length compared to that obtained with SP^3 hybridization. The bond length is 1.54 Å in ethane but only 1.34 Å in ethylene [*2*]. Alkenes can be reduced to the corresponding alkanes by reacting them with hydrogen in the presence of a number of metal catalysts.

$$H_2C{=}CH_2 \xrightarrow[Pt]{H_2} H_3C{-}CH_3 \tag{5.2}$$

This exothermic reaction proceeds with the evolution of approximately 30 kcal/mol of heat [*1*], which is a measure of the higher energy state of the carbon-carbon double bond compared to the carbon-carbon single bond. Alkenes also react with halogens and hydrogen halides

$$H_2C{=}CH_2 + Br_2 \longrightarrow BrH_2C{-}CH_2Br \tag{5.3}$$

$$H_2C{=}CH_2 + HCl \longrightarrow H_3C{-}CH_2Cl \tag{5.4}$$

In addition, alkenes are more susceptible to oxidation than are alkanes. Oxidation leads to the formation of polar materials such as organic acids.

5.1.5 The Carbon-Carbon Triple Bond

Compounds containing carbon-carbon triple bonds are known as alkynes. Alkynes result from the *SP* hybridization of the 2*S* and 2*P* atomic orbitals of carbon. This hybridization produces two bonding orbitals which are at an angle of 180°. Acetylene, HC≡CH, the simplest of the alkynes, is a linear molecule. The two hybrid bonds on each carbon are used [*3*] to form a bond between carbon and hydrogen and another a bond between the carbon atoms. This leaves two unhybridized *P* orbitals apparently uninvolved in bonding. Because acetylene is a linear molecule, the two *P* orbitals unused in σ bonding are able to overlap and form bonds. The *P* orbitals in an atom are perpendicular to one another; so, there is an overlap of *P* orbitals above and below and to either side of the molecule. There is sufficient interaction between all of the bonding π orbitals so that the electron density lies in a cylinder about the axis of the acetylene molecule (Fig. 5.7).

The carbon-carbon bond length in a triple bond is 1.20 Å. The decreasing bond length in going from alkane to alkene to alkyne is the result of the increasing *S* character of the hybrid orbitals. An *S* orbital is shorter than a *P* orbital [*1*].

Alkynes readily add two moles of hydrogen with the evolution of heat

$$HC{\equiv}CH + 2H{-}H \longrightarrow H_3C{-}CH_3 \tag{5.5}$$

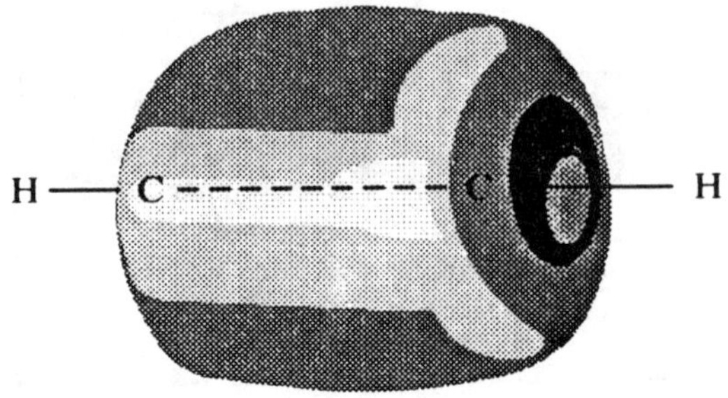

FIG. 5.7—*Electron density of an acetylene molecule.*

They also add two moles of halogen or hydrogen halide

$$HC{\equiv}CH + 2Br{-}Br \longrightarrow Br_2HC{-}CHBr_2 \quad (5.6)$$

$$HC{\equiv}CH + 2HCl \longrightarrow H_3C{-}CHCl_2 \quad (5.7)$$

Alkynes, like alkenes, are more susceptible to the oxidation than are alkanes. Oxidation leads to polar products, like organic acids. As with alkenes, alkynes can be extracted from mineral oils by treatment with sulfuric acid. Both react with sulfuric acid to form acid soluble addition products.

5.1.6 Aromatic Compounds

There are a number of organic compounds which, although unsaturated, behave totally differently from alkenes and alkynes. These materials are known as aromatics. This term originated with early investigators who noted that many of them were fragrant.

5.1.6.1 Benzene—Benzene, the parent compound of a series of aromatics, was discovered by Michael Faraday in 1825. Benzene was shown to have the molecular formula of C_6H_6. This formula indicates that benzene is highly unsaturated. Yet, it does not behave at all like other unsaturated materials. It does not react with cold or dilute permanganate solutions. It does not add halogen or hydrogen halides except under specific conditions. When it does react with a halogen, it is by means of substitution not addition

$$C_6H_6 + Br{-}Br \longrightarrow C_6H_5Br + HBr \quad (5.8)$$

Further, it reacts with hydrogen only at elevated temperature and pressure.

In 1865, August Kekulé proposed a six membered ring with alternating single and double bonds (Fig. 5.8).

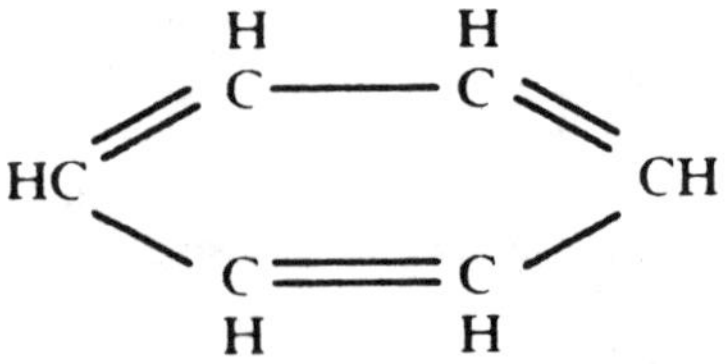

FIG. 5.8—*Molecular structure of benzene.*

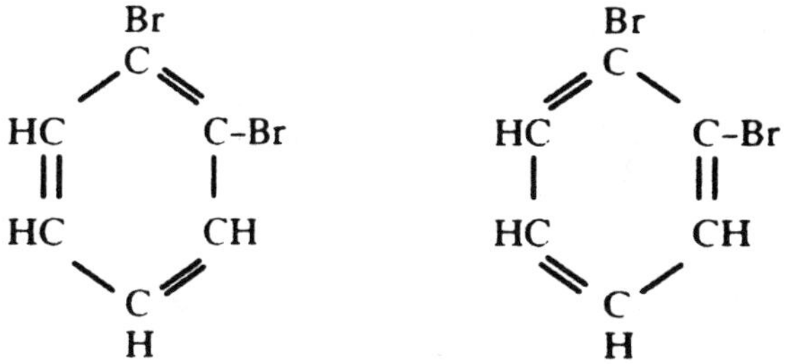

FIG. 5.9—*Two structures of 1,2-dibromobenzene.*

One of the problems with the structure was that it required the existence of two 1,2-dibromobenzenes, while only one was known (Fig. 5.9).

Kekulé then proposed that benzene was actually two compounds in rapid equilibrium

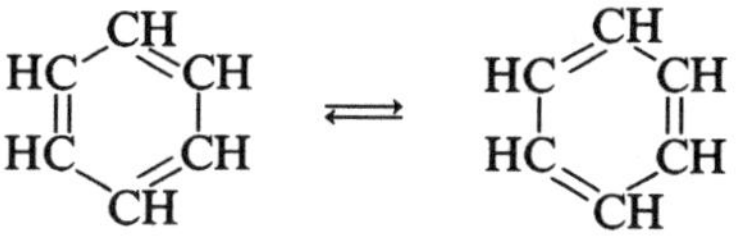

(5.9)

The uniqueness of benzene is evidenced by its heat of hydrogenation. If cyclohexene is reacted with hydrogen to form cyclohexane, the reaction proceeds with the release of 28.6 kcal/mol of heat [2]

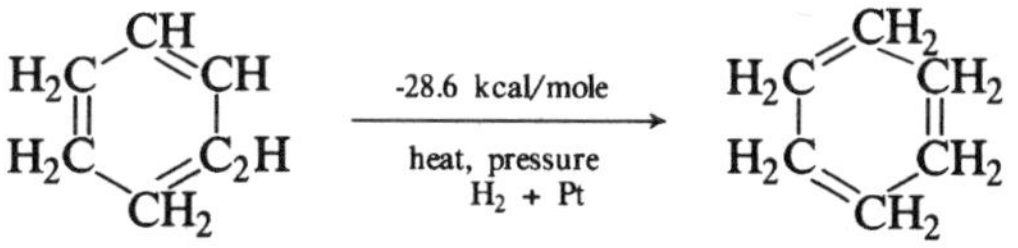

(5.10)

If benzene were simply cyclohexatriene the amount of heat released by reaction of 3 mol of hydrogen should be approximately 3 × 28.6 kcal/mol = 85.8 kcal/mol. This value contrasts sharply with the experimentally determined value of only 49.8 kcal/mol. The difference, 85.8 − 49.8 = 36.0 kcal/mol, between the theoretical and experimental values is called the resonance energy of the compound [2]. In the case of benzene, this difference means that benzene is 36 kcal/mol less energetic than cyclohexatriene would be.

With modern molecular orbital theory, it is not difficult to offer an explanation for the chemistry and bond energy of benzene. The benzene molecule is known to be planar. The carbon-carbon bond angles are 120°, as are the carbon-hydrogen bonds. The carbon-carbon bond length is 1.39 Å, which is between that for a carbon-carbon double bond (1.33 Å) and a carbon-carbon single bond (1.47 Å) [2]. The known facts concerning the benzene molecule suggest that the σ bonding carbon orbitals are SP^2 hybridized. If a six membered planar ring is drawn with the known bond lengths a configuration results, which permits the overlapping of the unhybridized *P* orbital of each carbon (Fig. 5.10).

The overlapping *P* orbitals combine to form a continuous ring. Because the *P* orbitals have lobes above and below the plane of the molecule, there are two rings of delocalized electron clouds. It is

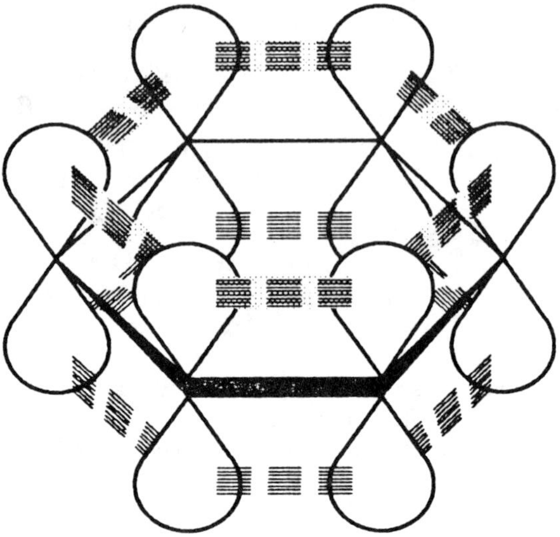

FIG. 5.10—*Overlapping* P *orbitals in benzene.*

the freedom of the six *P* electrons to move about all of the carbon atoms in the molecule which produces the observed resonance energy stabilization in benzene.

Thus, there is but one form of benzene. It can be represented by the two Kekulé structures, however, the two structures must be viewed as two possible representations of the same molecule, not as two isomers in rapid equilibrium. Another way to represent an aromatic structure is to place a circle in the center of the ring of atoms comprising the aromatic compound. A single conjugated double bond drawing is still used commonly to represent benzene, and other aromatic compounds; however, it is done with the understanding that the drawing is a simple representation of a ring structure with electron clouds above and below the plane of the molecule.

5.1.6.2 Other Aromatic Compounds—Materials other than benzene and its derivatives are known which have large resonance energy stabilization. In order for a molecule to contain aromatic resonance stabilization it must conform to all of the following criteria: (*i*) it must have a cyclic arrangement of *P* orbitals; (*ii*) there must be a *P* orbital available from every atom in the ring; (*iii*) the molecule must be planar so that the *P* orbitals on each atom can overlap; (*iv*) the cyclic arrangement of *P* orbitals must contain $4n + 2\pi$ electrons. These criteria were first recognized in 1931 by Erich Hückel. The criteria are known as the Hückel $4n + 2$ rule [*1*]. If n is zero the number of π electrons is two. The cyclopropenyl cation (Fig. 5.11) fulfills the electron requirement and is known to be aromatic. If n is one, the required number of π electrons is six. Besides benzene and its derivatives, the cyclopentadienyl anion (Fig. 5.12) and the cycloheptatriene cation (Fig. 5.13) also

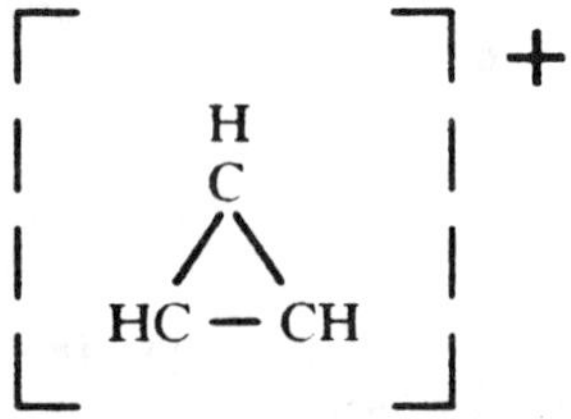

FIG. 5.11—*Molecular structure of the cyclopropenyl cation.*

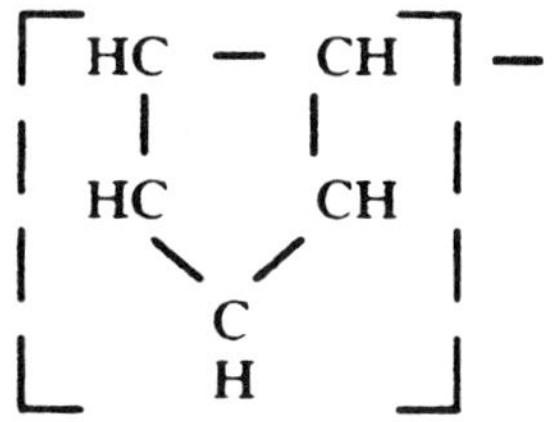

FIG. 5.12—*Molecular structure of the cyclopropenyl cation.*

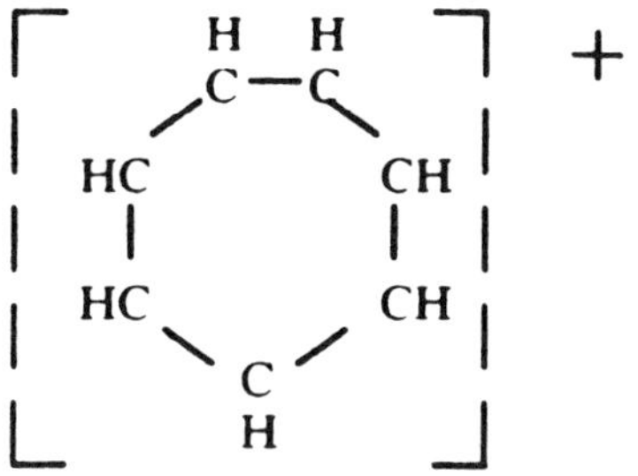

FIG. 5.13—*Molecular structure of the cycloheptatriene cation.*

have six π electrons and are aromatic. Examples of the 10 electron aromatic system are naphthalene (Fig. 5.14) and cyclooctatetraenyl dianion (Fig. 5.15) [*2*].

5.1.6.3 Heterocyclic Aromatic Compounds—There are many cyclic molecules which contain elements in addition to hydrogen and carbon and are aromatic. Examples of these types of compounds are thiophene, furan, and pyrrole (Fig. 5.16).

The aromaticity of these compounds is derived from two carbon-carbon double bonds contributing a total of four π electrons and from the hetero atom which contributes two electrons from a *P* orbital not otherwise used in bonding. These materials undergo the same sort of substitution reactions as seen with benzene. The resonance stabilization energy is on the order of 22 to 26 kcal/mol. This is significantly less than the 36 kcal/mol for benzene but much more than the 3.0 kcal/mol of stabilization found in a simple conjugated diene system: —CH═CH—CH═CH— [*1*].

5.1.7 Molecular Geometry

Molecular geometry is important because the way a molecule reacts chemically is closely linked to its structure. The geometry of a covalent molecule is defined by the bond length and bond angle. The bond length is the distance between the centers of nuclei of bound atoms. Bond lengths are

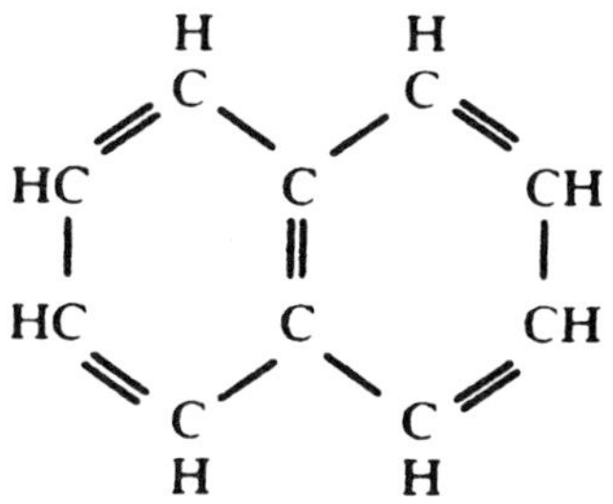

FIG. 5.14—*Molecular structure of naphthalene.*

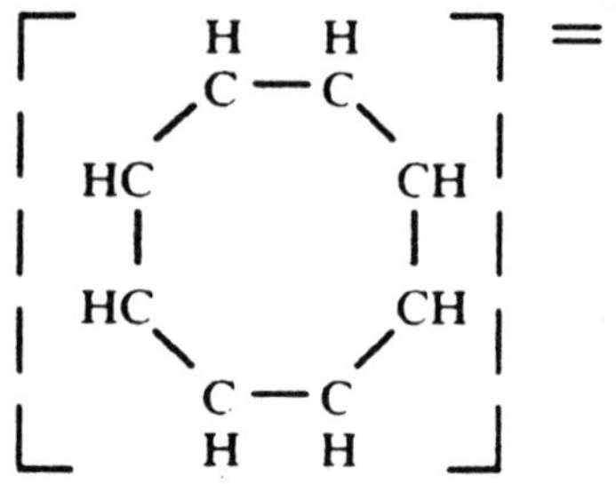

FIG. 5.15—*Molecular structure of the cyclooctatetraenyl cation.*

expressed generally in terms of angstroms (1 Å = 1 × 10^{-10} m). The bond length between atoms of the same element decrease with an increase in the multiplicity of the bond between them. Because bond length is the distance between nuclear centers, the lengths increase with the size of the bonded atom. Bonds become shorter as one of the bonding atoms is exchanged for an atom of a more electronegative element.

Bond angle is the angle between two bonds to the same atom. Bond angles dictate the shape of a molecule. Groups bound to a central atom seek to be as far apart from one another as possible. This determines whether a molecule will be tetrahedral, planar, linear, or some other geometry. It is this mutual aversion of substituents on carbon which generates the tetrahedral geometry for tetra substituted carbon, planar geometry for doubly bonded carbon, and the linearity of the triply bonded carbon [*3*].

5.2 Intermolecular Forces

In addition to the atomic forces involved in bonding of elements there are molecular forces which influence the nature of matter. These forces are dipole-dipole interactions and van der Waals forces. They serve to hold neutral molecules to each other and appear to be electrostatic in nature.

5.2.1 Dipole-Dipole Interactions

Dipole-dipole interaction is the attraction of the positive end of one polar molecule for the negative end of another polar molecule. If a molecule is composed of a number of different elements, a dipole can result from the difference in electronegativity of two different atoms sharing a chemical bond. In a chemical bond between atoms of elements having different electronegativities the electrons forming the bond will spend more time near the more electronegative element's atom. This has the effect of imparting a partial negative charge near that atom and a partial positive

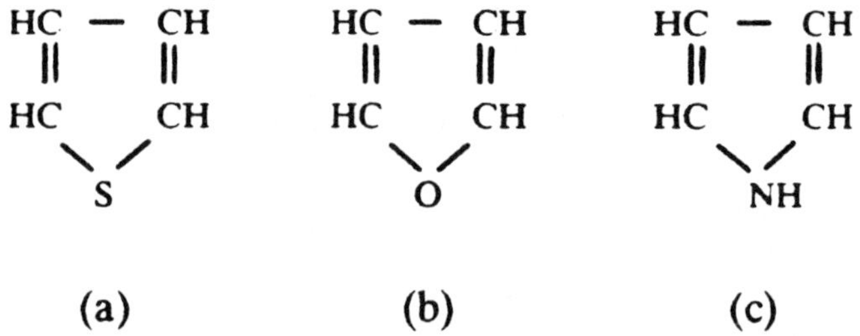

FIG. 5.16—*Molecular structures of:* (a) *triophene,* (b) *furan, and* (c) *pyrrole.*

$$\ldots\ \overset{\delta+\ \delta-}{H_3C\text{–}Cl}\ \ldots\ \overset{\delta+\ \delta-}{H_3C\text{–}Cl}\ \ldots\ \overset{\delta+\ \delta-}{H_3C\text{–}Cl}\ \ldots\ \overset{\delta+\ \delta-}{H_3C\text{–}Cl}\ \ldots$$

FIG. 5.17—*Dipole-dipole interaction in methyl chloride.*

charge near the atom of the less electronegative element. The effect of this can be seen in the dipole-dipole interaction for methyl chloride (Fig. 5.17).

The δ is read "somewhat," so that the carbon atom is "somewhat positive" and the chlorine is "somewhat negative." The energy involved in this association (shown by the dotted lines) is much less than 1 kcal/mol [*1*]. None the less, polar molecules are held together more strongly than nonpolar molecules because of this interaction. Another type of dipole-dipole interaction is the hydrogen bond. It is an especially strong and important interaction and deserves to be discussed in detail.

5.2.2 Hydrogen Bonding

In a hydrogen bond a hydrogen atom serves as a bridge between two highly electronegative atoms. It is bonded covalently to one of these atoms and associated only through electrostatic forces with the second atom. When hydrogen is bonded to nitrogen, oxygen, or fluorine, the electron cloud bonding the two elements is distorted greatly toward the electronegative atom. This exposes the positively charged hydrogen nucleus. The hydrogen nucleus is attracted strongly to the negative charge of the electronegative atom of a second molecule.

Compounds which display hydrogen bonding include organic acids, alcohols, phenols, amines, amides, hydrogen fluoride, and water. The hydrogen bond has a dissociation energy of about 5 kcal/mol [*1*], which is much greater than other dipole-dipole interactions. Hydrogen bonding strongly affects the boiling point, solubility properties and even the shape of large molecules. Thus, ethyl alcohol, which can form hydrogen bonds in the fashion depicted in Fig. 5.18, has a higher boiling point (78°C) than dimethyl ether (−25°C), which has the same molecular weight but can form only dipole-dipole attractions (Fig. 5.19). Hydrogen bonding is sufficiently strong so that it persists even in the vapor phase for formic acid [*1*].

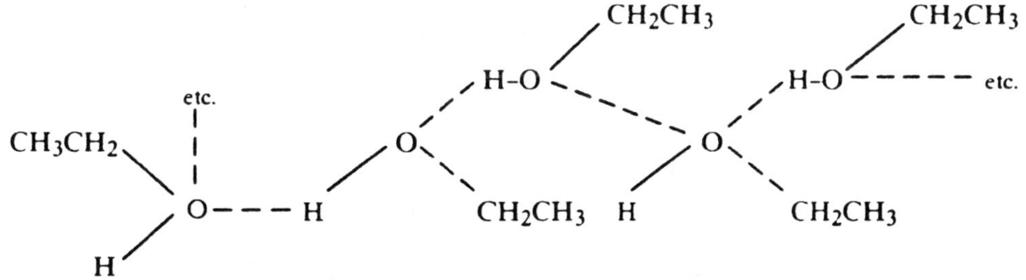

FIG. 5.18—*Hydrogen bonding in ethyl alcohol.*

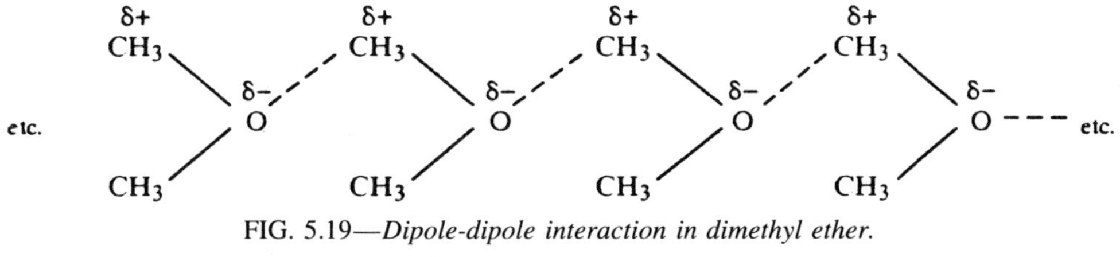

FIG. 5.19—*Dipole-dipole interaction in dimethyl ether.*

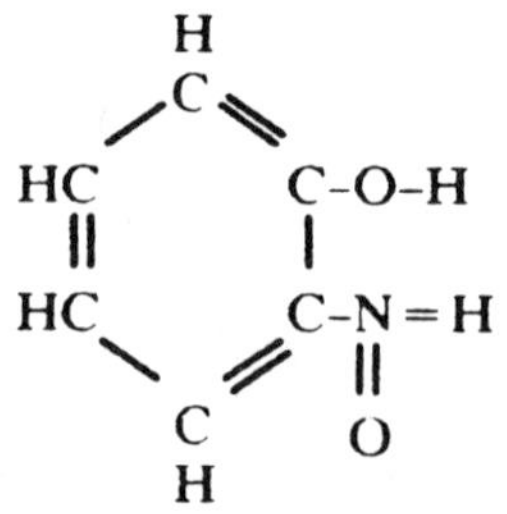

FIG. 5.20—*Intramolecular bonding in o-nitrophenol.*

If an OH or NH moiety is contained in a molecule possessing another nitrogen or oxygen atom, intramolecular hydrogen bonding can occur. Examples are o-nitrophenols (Fig. 5.20) and the enol form of acetylacetone (Fig. 5.21). Such intramolecular interactions lead to higher melting points and lower boiling points than for equivalent molecular weight molecules where such interaction is not possible [*1*].

5.2.3 van der Waals Forces

A comprehensive explanation of van der Waals forces requires the use of quantum mechanics, which is beyond the scope of this chapter. The nature and origin of these forces, however, can be described qualitatively. van der Waals forces are attractive intermolecular forces. They result from the polarizing effect molecules have on one another. The average distribution of charge in a nonpolar molecule is uniform. At any given instant, however, the electrons may not be distributed uniformly. Because of the momentary concentration of electrons in one portion of a molecule, a small dipole is created temporarily. The existence of this dipole induces a dipole of opposite charge in an adjacent molecule. Because the dipoles have opposite charge, they attract each other. These dipoles rapidly disappear only to be replaced by others. This interaction takes place at the surface of molecules; so, the more surface area a molecule has the more it will interact with its neighbor through transient dipole formation. This molecular interaction is reflected in the boiling points of liquids. The boiling points of linear alkanes increases as their molecular weight (surface area) increases. The boiling points of branched chain liquids are lower than that of the linear chain liquids. This results from the lower surface area presented by the branched chain molecule. This effect is particularly true for small molecules which can approach a spherical shape. A sphere has the smallest possible surface area for a given volume [*1,3*].

5.3 Properties of Insulating Liquids

Knowledge concerning composition and molecular structure of the constituents of an insulating liquid is of great importance. This results from the fact that much of the performance potential of an insulant is determined by its composition. The properties, which are of importance, vary depending

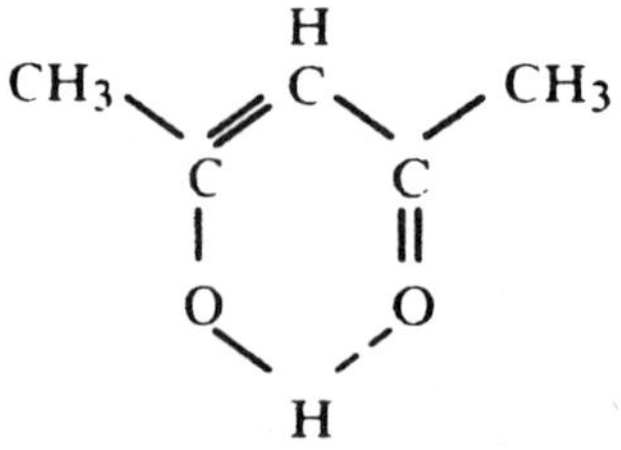

FIG. 5.21—*Intramolecular bonding in the enol form of acetylacetone.*

upon the application. The characteristics, which one would seek in a fluid for use in a high stress capacitor, are not the same as those needed for a transformer coolant. Thus, an understanding of both the application requirements and the properties of candidate liquids is necessary in order to insure selection of an appropriate material. Important performance characteristics which are influenced strongly by compositions include: electric strength, impulse strength, resistivity, permittivity, dissipation factor, flow and heat properties, gas absorbing characteristic, and flammability properties.

5.3.1 Electric and Impulse Strength

With tests such as ASTM D 877 and IEC 156, the electrical strength is affected mainly by contaminants such as water and solid particles [*4–6*]. However, in carefully purified liquids the dielectric strength of a homologous series has been shown to increase with increasing molecular weight. This has been demonstrated for normal alkanes [*7*], polydimethylsiloxanes [*8*], branched alkanes [*9*], and alkyl benzenes. Branched chain hydrocarbons have lower dielectric strengths than do the linear molecules of equal chain length. Benzene has a higher dielectric strength than hexane and alkyl substituted benzenes have higher dielectric strengths than alkanes containing the same number of carbon atoms [*10*]. The influence of molecular weight is seen only in short chain length hydrocarbons. Variation in dielectric strength is not seen in multicomponent insulating oils differing only in viscosities [*11*].

In a uniform field, the impulse strength also is affected mainly by contaminating water and particles. As with uniform field dielectric strength testing, a direct correlation between molecular weight and impulse strength has been shown for a homologous series [*12*]. Both impulse and alternating voltage breakdown have been shown to be affected strongly by composition when conducted under nonuniform field conditions. If the electric field is divergent, the presence of even very small amounts of aromatic compounds produces a lower impulse withstand and dielectric breakdown values than are obtained under the same test conditions in the absence of the aromatic materials [*13,14*]. The difference in impulse breakdown values between aliphatic and aromatic hydrocarbons is due largely to their intrinsic electrical conductivity. Aromatic molecules have a higher conductivity than do aliphatic hydrocarbons; thus, they have lower breakdown values.

The effects of composition can be further demonstrated in the case of point to plane impulse breakdown voltages. Many materials show a significantly higher breakdown voltage when the point is negative than when it is positive. However, if the test liquid contains an electronegative substituent or contains a species which includes an electronegative substituent, the point to plane impulse breakdown voltage will be nearly the same for both polarities. The negative point to positive point breakdown value ratios for toluene and n-octane are approximately 2.0 and 2.7, while the same ratios for chlorobenzene and carbon tetrachloride are 1.07 and 1.04. Similar results have been obtained with perfluoroalkanes [*15*]. It has been suggested that this effect is due to the injected electrons attaching themselves to the electronegative groups to form negative ions. Because breakdown voltage is inversely proportional to the mobility of the charge carriers generated at the point, and negative and positive ions have similar mobilities which are less than that of electrons, the negative point breakdown voltage of an insulating fluid containing electronegative groups is similar to the positive point value [*16*]. For additional readings on electrical breakdown processes in liquids, the reader is referred to Chapter 3 of this volume.

5.3.2 Resistivity

Under low stress conditions, <500 V/mm, the resistivity of a fluid is dependent greatly upon ionic or easily disassociated compounds. Contaminants such as water, acids, and extraneous polar materials are the usual causes of low resistivity (see Chapters 1 and 2 in this volume). Saturated hydrocarbon oils are dependent upon polar impurities or thermalized electrons to supply the charge carriers required for conduction [*17*]. Aromatic compounds lower the resistivity of an oil, probably due to the intermolecular transfer of π electrons associated with the aromatic ring. This intermolecular electron pumping process is a conduction mechanism available to all unsaturated molecules, including olefins [*18*]. Conduction resulting from intermolecular electron pumping does not change the composition of the fluid as is the case when conduction is due to ionic impurities. It has been shown [*19*] that application of a constant direct voltage to a sample of clean aromatic fluid will produce a constant current over a long period of time. Application of a constant direct voltage to a saturated hydrocarbon liquid produces a decreasing current as the ionic charge carriers are removed from the oil.

Resistivity is associated with the freedom of movement, or mobility, of charge carriers. Electron mobility decreases in normal alkanes as the molecular weight increases up to a chain length of about six carbons, where it levels off at about $5 \times 10^{-2} cm^{2, -1} s^{-1}$. Branching results in increased electron mobility, and the greatest electron mobility is found with compounds containing a quaternary atom, such as neopentane or tetramethylsilane. The increased electron mobility results from the electron spending less time associated with a liquid molecule and more time as a free species. Cycloalkanes have a higher electron mobility than the corresponding alkanes. Introduction of a double bond can either increase or decrease mobility. Electron mobility in benzene is greater than that in cyclohexane but less than that in cyclohexene [*20*]. Table 5.1 [*20*] clearly shows how structure affects free electron mobility. Higher electron mobility has been shown [*10,21*] to be associated with lower dielectric strength. These effects can be seen only in very pure liquids. Those interested in an extensive treatment of the subject of conduction in liquids are referred to Chapter 2 of this volume.

5.3.3 Permittivity

The dielectric constant or permittivity of a material results from polarization of electrons, nuclei, and molecules or portions of molecules. Exposure of a material to an electric field produces three types of polarization within a liquid: electronic polarization, atomic polarization, and orientation polarization. A fourth type of polarization, interfacial polarization, takes place at interfaces between materials or different phases of the same material. It is not considered in this discussion.

Electronic polarization is the displacement of an electron cloud relative to the nuclei when a molecule is acted upon by an electric field [*22*]. Electron response to an electric field occurs within 10^{14} to 10^{-16}/s. The nuclei also are affected by the electric field and are displaced relative to one another. Nuclei response to a field occurs in 10^{-12} to 10^{-13}/s. The contribution of this atomic polarization to the permittivity of a molecule is relatively small, being on the order of 5 to 10% of the value of electronic polarization. The sum of electron and atomic polarization is known as the deformation polarization [*23*] (see Chapter 1 of this volume).

In the case of polar molecules, in addition to the deformation polarization, there is another type of polarizability caused by the orientation of the permanent dipoles in the electric field. This type of polarizability is known as orientation polarizability. It responds in a 10^{-2} to 10^{-6}/s time frame. This polarizability results from the presence of a permanent dipole in a molecule. A molecule having an unsymmetrical distribution of electrons (charge) has a permanent dipole. Thus, methane has a symmetrical distribution of electrons and has no permanent dipole. Methyl chloride, however, has an unsymmetrical distribution of charge in the molecule. There is a higher electron density around the chlorine atom than around carbon. This net separation of charge results in a permanent dipole present even in the absence of an orientating electrical field. In bulk, the random distribution of molecules produces a solution with no net dipole. However, if an electric field, sufficient to overcome thermal energy causing random distribution, is applied, some of the molecules will align themselves in the field to produce a net dipole moment for the material within the field. The number of dipoles which manage to arrange themselves parallel to the field is small under normal measure-

TABLE 5.1—*Excess electron mobilities at room temperature (after Schmidt* [20]*).*

Liquid	μ ($cm^2V^{-1}s^{-1}$)
N-ALKANES	
n-butane	0.34
n-pentane	0.15
n-hexane	0.09
n-heptane	0.046
n-octane	0.040
n-nonane	0.047
n-decane	0.038
n-dodecane	0.030
C4 ISOMERS	
n-butane	0.34
2-methylpropane	70
C3 ISOMERS	
n-pentane	0.15
2,2-dimethylpropane	70
C6 ISOMERS	
n-hexane	0.09
2-methylpentane	0.29
3-methylpentane	0.22
2,3-dimethylbutane	1.1
2,2-dimethylbutane	12
CYCLOALKANES	
Cyclopentane	1.1
Cyclohexane	0.35
Cycloheptane	0.44
Cyclooctane	0.17
DOUBLE BONDS	
2,2-dimethylbutane	12
3,3-dimethyl-1-butene	1.9
2,3-dimethylbutane	1.1
2,3-dimethyl-2-butene	5.
n-butane	0.34
l-butene	0.064
cis-2-butene	2.2
trans-2-butene	0.029
Cyclopentane	1.1
Cyclopentene	1.3
Cyclohexane	0.35
Benzene	0.6
Cyclohexene	1.0
Cycloheptane	0.44
Cycloheptene	0.35
SPHERICAL SHAPE	
Tetramethylsilane	90
Tetramethylgermanium	90
Tetramethyltin	70
2,2-dimethylpropane	70
2,2-dimethylbutane	12
3,3-dimethylpentane	2
3,3-diethylpentane	0.76
2,2,4-trimethylpentane	7
2,2,4,4-tetramethylpentane	24
2,2,3,3-tetramethylpentane	5.2
2,2,5,5-tetramethylhexane	12

ment conditions, amounting to only about one in 100 000 at 300 V/cm [*22*]. The difficulty in achieving alignment is derived from thermal energy which strives for randomness of orientation. The contribution of orientation polarization to the dielectric constant is given by $\alpha = \mu/3kT$ [*24*]. Because of the sensitivity of orientation polarization to temperature, the permittivity of a polar material is much more sensitive to temperature changes than is that of a nonpolar material [*24–26*] (see Chapter 1 of this volume).

5.3.4 Dissipation Factor

The dissipation factor is related to the dielectric constant. This relationship is shown by the equation $\epsilon^* = \epsilon' - j\epsilon''$, where ϵ^* is the complex permittivity, or dielectric constant, ϵ' is the measured dielectric constant, or real value of the permittivity and ϵ'' is the loss or dissipation factor of the fluid. The loss factor results from the unsuccessful attempt of molecules, or portions of molecules, to orient themselves within an alternating electrical field. Under the influence of an alternating field, polar molecules rotate towards an equilibrium distribution. There is a field frequency above which the molecules cannot reach this equilibrium distribution between cycles. The orientation polarization then acquires a component which is out of phase with the field, and the displacement current acquires a conductance component in phase with the field resulting in thermal dissipation of energy. This resistive loss is referred to as tanδ and is equal to ϵ''/ϵ' [*27*]. Tanδ refers to the tangent of the angle constructed by the vector sum of the capacitive and resistive currents and the purely capacitive current, which is 90° out of phase with the voltage. Dissipation factor and power factor, which is the sine of the loss angle, are essentially equal up to a value of approximately 0.05.

Molecular size, chemical composition, and the specific orientation and electronic configuration of the elements will affect both the dielectric constant and dissipation factor. At low frequencies, permittivity is affected by molecular size, substitution, chemical functionality, molecular configuration, and chemical composition. In a homologous series, the permittivity increases with increasing molecular weight [*28*].

The dissipation factor at power frequencies is not affected greatly by structural and compositional related molecular absorption. Losses at these frequencies are analogous to resistivity and are controlled by the same factors (contaminants) (see Chapters 1 and 2 of this volume). Tanδ increases as frequency changes and a part of an atom or molecule fails to respond with that change in frequency. Except for high polymers, the onset of an increase in tanδ will start at a frequency of about 10^{10} Hz as the response time of small molecules is exceeded. Tanδ will increase with frequency through a peak and then decrease again. The number of such dispersion curves one notices with a material depends upon its structure. Changes in temperature can lead to phase changes and other phenomena which are reflected in the tanδ measurements also. The subject of dielectric loss is treated in greater detail in Chapter 1.

5.3.5 Gas Absorbing Characteristic

The gas absorbing characteristic of a liquid is of particular importance in closed, high stress devices, such as capacitors. This is true because high stresses often result in the formation of gasses, particularly hydrogen, resulting from ionization of the insulating liquid. Because the dielectric strength of a gas is less than that of a liquid, the presence of a gas bubble can result in electrical failure. The gas absorbency of a liquid is derived from chemical unsaturation. This can be either aromatic or aliphatic in nature. The gas absorbing capacity of a fluid is proportional to the amount of unsaturation. If a liquid lacks unsaturation, hydrogen gas will be evolved in the presence of electrical discharges (see Chapter 4).

5.3.6 Flow and Heat Properties

In some applications, such as transformers, one of the major functions of an insulating liquid is to help in dissipation of heat generated by resistive losses in an electrical circuit. This ability to transport heat is determined by the amount of heat the liquid can absorb and the rapidity with which it can transport that heat to an external sink. The thermal conductivity and specific heat capacity are important thermal properties of a liquid. Viscosity and the variation of viscosity with temperature are important in determining how well a fluid will transport heat from one place to another. All of these characteristics are influenced by the composition of a liquid.

The viscosity of liquids increases with decreasing temperature; and the rate of that change is dependent upon the constitution of the oil [*29,30*]. Aromatic hydrocarbons exhibit the greatest viscosity change with temperature among the mineral oils. Cycloaliphatics (naphthenics) are less affected by temperature changes; and, paraffinics are the least affected in this series. Synthetic polybutenes are less affected by temperature changes than are natural paraffinic oils. The temperature sensitivity of polychlorinated biphenyls is dependent upon the level of chlorination. The higher the level of chlorination, the steeper the slope of the viscosity versus temperature data. Aliphatic esters can have a smaller viscosity response to temperature changes than even polybutylene. As with hydrocarbons, the presence of aromatic molecules produces a greater change in viscosity with temperature. Polydimethyl siloxanes have the least change in viscosity with temperature of all insulating liquids. This is believed to result from the large free volume of silicones derived from the free rotation about the Si—O—Si bonds. A general equation according to ASTM D 341-87 which describes the viscosity behavior of any Newtonian liquid with changes in temperature, is given by

$$\log \log (\nu + 0.7) = A + B \log T \tag{5.11}$$

where ν is the kinematic viscosity in mm^2/s or cSt, T is the absolute temperature in K, and A and B are constants. The constants A and B may be evaluated for a fluid from two data points and ν can be calculated readily for other points of temperature. Comparative viscosity versus temperature data for several insulating liquids can be seen in Fig. 5.22.

Thermal conductivity, specific heat and the coefficient of expansion are thermal characteristics of liquids. They are dependent upon the density, d, of insulating liquids. Thermal conductivity is proportional to T/d while specific heat is proportional to $T/d^{1/2}$. The thermal conductivity of dielectric liquids may be determined by using ASTM Method D 2717 which measures the temperature gradient, ΔT in °C across the liquid specimen as a result of a known amount of energy being introduced into a cell by electrically heating a platinum element. The thermal conductivity of the liquid, K_L, in cal/s cm °C, is given by

$$K_L = A/[(4.19\,\Delta T/I^2R) - B] \tag{5.12}$$

where I^2R is the power dissipated in the platinum element of resistance R due to the current I; A and B are constants, given by $A = [\ln (r_i/r_f)]/2\pi L$ in cm^{-1} and $B = [\ln (r_o/r_i)]/2\pi L\ K_G$ in s °C/cal. Here r_f is the filament radius in cm, and r_i and r_o are the internal and external radii, respectively of the glass test cell in cm. K_G represents the thermal conductivity of the glass tube in cal/s cm °C.

ASTM Method D 2766 may be used to determine the heat capacity or specific heat of a dielectric liquid. Since the specific heat is a thermodynamic property that indicates the amount of energy required to produce a given temperature change within a unit quantity of the liquid, the technique involves the use of a calorimeter. The net enthalpy change per gram of liquid, ΔH_s, thus, can be expressed as an analytical function of the temperature differential, $T' = T_f - T_c$; where T_c is the initial temperature of the calorimeter in °C and T_c is the hot zone temperature. That is

$$\Delta H_s = BT' + CT'_2 \tag{5.13}$$

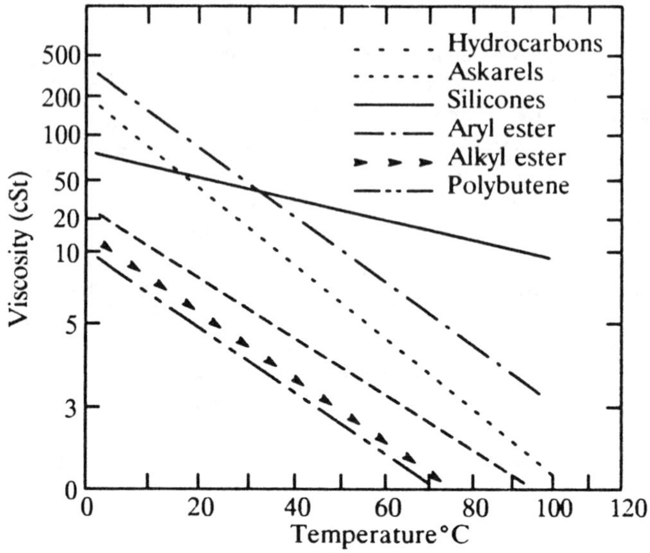

FIG. 5.22—*Temperature-viscosity relationship of some insulating liquids.*

where B and C are constants that can be evaluated, using different values of ΔH_s with corresponding values of T'. By definition, the specific heat of the liquid is thus equal to the differential $d(\Delta H_s)/dT'$—that is, it is derived by differentiating Eq 5.13 and substituting specific values of T' for given values of the "hot zone" temperature, T'_f at which the specific heat or heat capacity must be known.

Comparative thermal conductivity and specific heat data for several insulating oils are shown in Figs. 5.23 and 5.24 [*31*]. It is of interest to note as concerns the interpretation of the foregoing

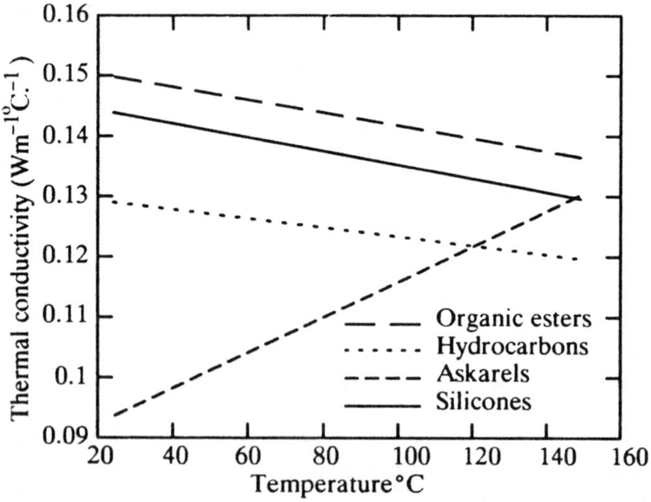

FIG. 5.23—*Thermal conductivity of a number of representative insulating liquids (after Wilson [31]).*

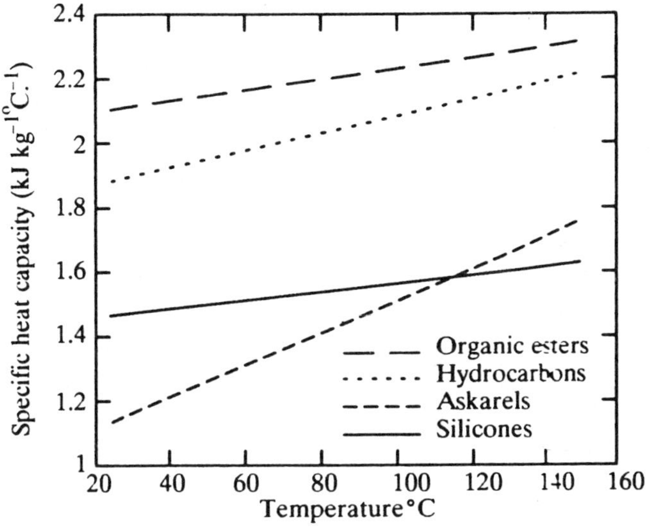

FIG. 5.24—*Specific heat capacity of a number of representative insulating liquids (after Wilson [31]).*

figures that the density of aromatic hydrocarbons is greater than that of cycloaliphatics (naphthenics). Paraffinic hydrocarbons have the lowest density of the various mineral oil types (see Table 1.1 in Chapter 1 of this volume).

The coefficient of thermal expansion of a liquid is defined as the change in volume per unit volume per degree increase in temperature and may be expressed as (ASTM D 1903)

$$\text{Coefficient of thermal expansion} = (S - S_1)/S(T_1 - T) \quad (5.14)$$

where S represent the relative density or specific gravity at the lower temperature T, and S_1 is the specific gravity at the higher temperature T_1. If the temperature is measured in °C, its units are °C^{-1}; for temperatures in °F, the units become °F^{-1}. Since the coefficient of thermal expansion is used commonly as an average value over a given temperature range, ASTM Method D 1903 specifies that the specific gravities be measured below 90°C not less than 14° apart in accordance with D 1810 (using a glass hydrometer for which the expansion of the glass must be taken into account). The coefficient of thermal expansion is of particular importance in determining the size of a container to permit accommodation of a given volume of insulating liquid over the full range of temperatures over which the electrical apparatus is to be subjected. For mineral oils in the range from −17.7°C (0°F) to 65.5°C (150°F), the coefficient of thermal expansion is of the order of 4×10^{-4} °F^{-1}.

5.3.7 Flammability Properties

Prior to the banning of polychlorinated biphenyls (PCBs) there were, essentially, only two types of insulating liquids: mineral oils and askarels (solutions of PCBs in chlorinated benzenes). The issue of flammability was simple. Mineral oils could support combustion while askarels could not. Thus, if fire safety was of concern, and all other factors were equal, an askarel would be used instead of a mineral oil. As a result of the restrictions placed on the manufacture and use of PCBs, a large number of liquids have been developed as askarel replacements. Nearly all of these materials are capable of supporting combustion in a normal atmosphere. Thus, the issue of fire safety has become much more complex.

The fire properties of liquids are determined by both their chemical and structural composition. Flash and fire points are influenced mainly by the volatile, lower boiling constituents of the oil. In some cases, such as the higher chlorinated PCBs, components of a fluid may be volatile but unable to support combustion in the atmosphere. This behavior results from the presence of fire retarding elements, such as chlorine.

Because combustion takes place in the gas phase, the flammability of a liquid can be reduced by using only higher molecular weight components. In that case, enough energy must be put into the system to cause volatilization of the large molecules composing the oil, or chemical bonds must be ruptured to permit the evolution of volatile species. There are two major types of liquids presently used as insulating materials which derive their fire properties, at least in part, from the fact that they are composed of high molecular weight components. These two materials are polyolefins and silicones. Both types of materials have flash points on the order of 300°C and fire points typically in excess of 300°C. As a result of this higher molecular weight, these liquids have much higher viscosities than typical mineral oils or askarels. In order to obtain desired fire properties, the required molecular weight of polyolefins results in a liquid of about 300 cSt. Polydimethylsiloxanes achieve similar flash and fire points with fluids having a viscosity of 50 cSt. In addition, the combustion of silicones produces a crust of SiO_2 held together with crosslinked silicone liquid. This crust will, at a minimum, retard the rate at which the silicone is consumed in a fire; and, in some cases, it may be sufficient to extinguish the flames entirely.

In the following pages, a number of insulating liquids will be examined in detail. Special attention is given to composition, structure, manufacturing process, oxidation characteristics, and electrical properties.

5.4 Liquid Electrical Insulating Materials

5.4.1 Mineral Oils

Mineral oils have been used in electrical equipment since the late nineteenth century. They are obtained by refining specific fractions of petroleum. These fractions are isolated by fractional distillation under reduced pressure. Petroleum is extracted from the earth in most parts of the world; however, the quality and compositions vary considerably [*32,33*]. The complexity of the composition is illustrated by the identification of 234 different hydrocarbons representing only 45% of the total, in one particular crude [*34*]. In attempt to simplify at least the nomenclature, the U.S. Bureau of Mines has established nine descriptive categories which make use of the words paraffinic, naphthenic (cycloaliphatic), and intermediate (Table 5.2) [*35–37*].

The categories were devised by Smith [*35*] and Lane and Garton [*36*], based on their work carried out between 1927 and 1935; the categories were derived from the American Petroleum Institute (API) gravity values of two key distillate fractions. In their method the fractions were collected at temperature steps of 25°C in boiling point at key fraction No. 1 from 250 to 275°C at atmospheric pressure in the kerosene range and at key fraction No. 2 from 275 to 300°C at 40 Torr in the lubricating oil range [*37*]. The API gravity scale given in Table 5.2 has been adopted by the API in 1921, because it permitted a linear scale calibration for the constant mass-variable displacement hydrometers of uniform stem section [*37*]. The API gravity scale is based on the Baumé scale and is defined as [ASTM D 1298],

$$\text{API gravity} = \frac{141.5}{\text{specific gravity } 60/60°\text{F}} - 131.5 \tag{5.15}$$

TABLE 5.2—*U.S. Bureau of Mines classification of crude oils, based on the API gravity scale of two key fractions (after Gruse of Stevens* [37]*).*

Crude Type	Key Fraction No. 1 API Gravity, deg	Key Fraction No. 2 API Gravity, deg
Paraffinic	40° or lighter	30° or lighter
Paraffinic—intermediate	40° or lighter	20° to 30°
Intermediate—paraffinic	33° to 40°	30° or lighter
Intermediate	33° to 40°	20° to 30°
Intermediate—naphthenic	33° to 40°	20° or heavier
Naphthenic—intermediate	33° or heavier	20° to 30°
Naphthenic	33° or heavier	20° or heavier
Paraffinic—naphthenic	40° or lighter	20° or heavier
Naphthenic—paraffinic	33° or heavier	30° or lighter

with the units given in "degrees." The specific gravity (relative density) in the denominator of Eq 5.15 is by definition the ratio of the mass of a given volume of specimen liquid at 15°C (60°F) to the mass of an equal volume of pure water at the same temperature; this is the basis for the notation specific gravity 60/60°F. For comparison purposes, it is helpful to note that for crude oils whose specific gravities usually extend from 0.5 to 1.06, the corresponding values on the API gravity scale range from 87 to 2°, respectively.

In reference to Table 5.2, it is apparent that the terms do not recognize the existence of aromatic constituents in the crude oil; however, aromatic fractions do exist. These aromatic materials have a significant impact on mineral oils suitable for use in electrical apparatus. Mineral oils are composed principally of carbon and hydrogen; however, relatively low levels of sulfur, nitrogen, and oxygen may also be present. The hydrocarbons in petroleum can be split into three major groups: paraffins, naphthenic (cycloparaffins), and aromatics [*38–40*]. The paraffinic hydrocarbons are composed of linear $CH_3—(CH_2)n—CH_3$ and branched $CH_3—\underset{}{\overset{\overset{\displaystyle CH_3}{|}}{CH}}—(CH_2)n—CH_3$ species.

The naphthenic hydrocarbons have a saturated cyclic structure. These materials vary in complexity from cyclohexane to multiple fused ring structures containing alkyl substitution (Figs. 5.25, 5.26, 5.27, and 5.28).

The aromatics assume forms very similar to those common in the naphthenic series. Single structure representations of the resonance stabilized aromatic molecules are depicted in Figs. 5.29, 5.30, 5.31, 5.32, and 5.33.

The oxygen, sulfur, and nitrogen content of mineral oils is derived generally from heterocyclic aromatic such as those depicted in Figs. 5.34, 5.35, and 5.36. Typical amounts of sulfur are 0.05 to 5% mass/mass, of nitrogen 0.001 to 0.2% mass/mass, and a trace of oxygen [*31*].

Because of cracking, reforming, and hydrofining processes, refineries are able to custom make

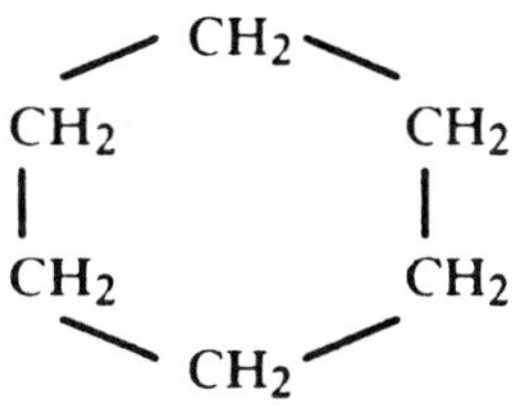

FIG. 5.25—*Single ring naphthenic hydrocarbon.*

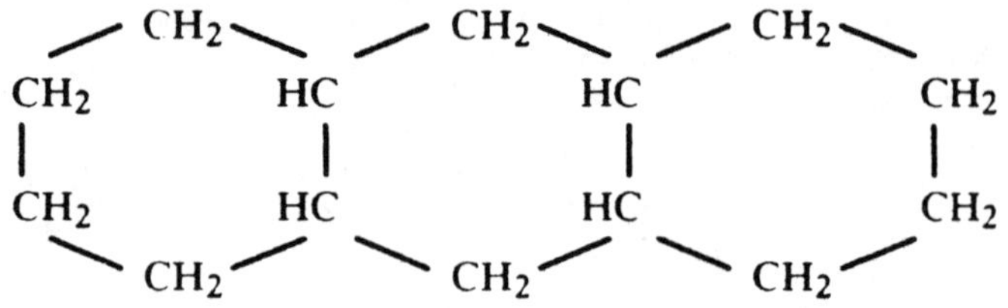

FIG. 5.26—*Linear condensed ring naphthenic hydrocarbon.*

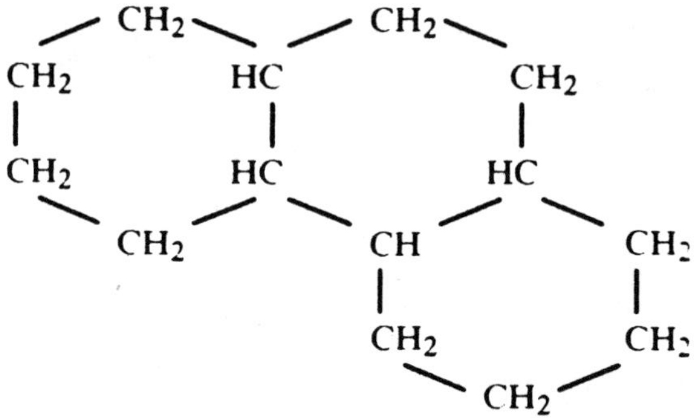

FIG. 5.27—*Angular condensed ring naphthenic hydrocarbon.*

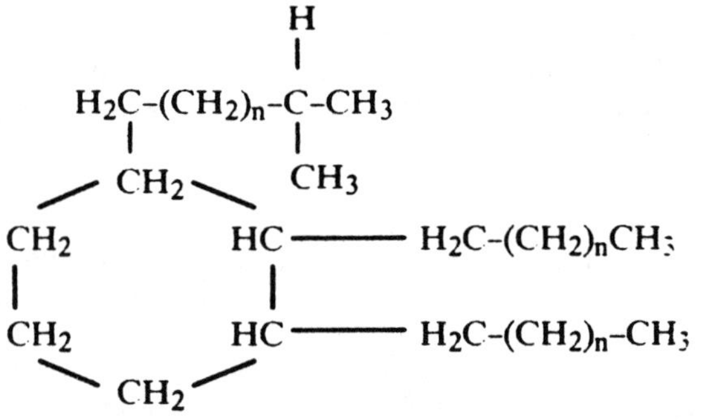

FIG. 5.28—*Alkyl substituted condensed ring naphthenic hydrocarbon.*

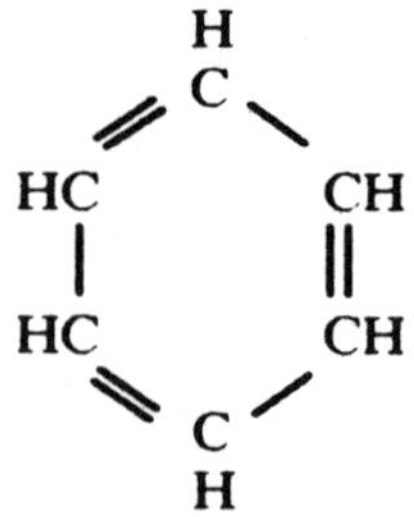

FIG. 5.29—*Single ring aromatic hydrocarbon.*

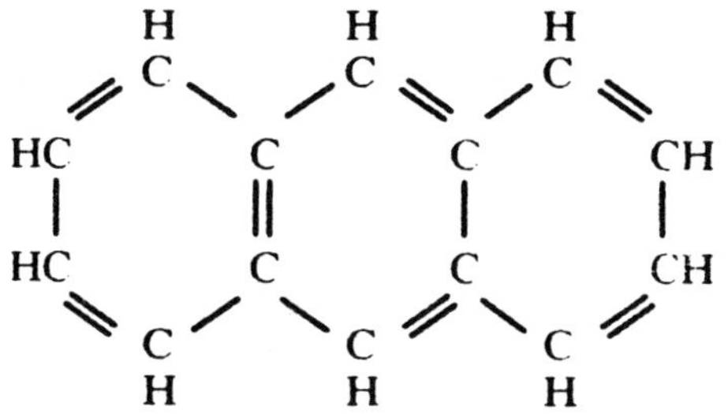

FIG. 5.30—*Linear condensed aromatic hydrocarbon.*

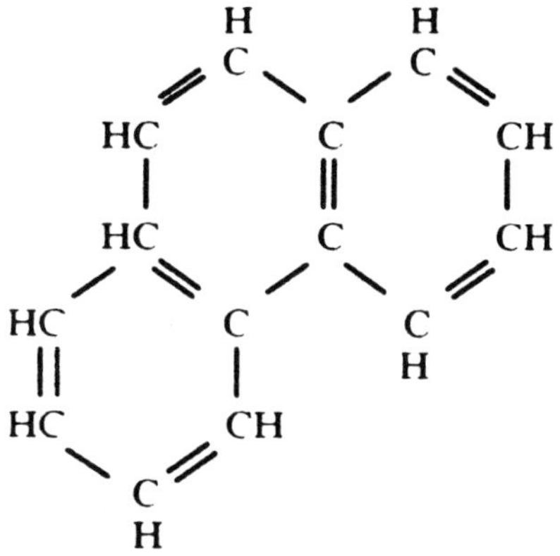

FIG. 5.31—*Angular condensed aromatic hydrocarbon.*

insulating oils from a considerable variety of crude oils. The portion of crude oil used for insulating purposes is obtained by vacuum distillation. The distillate is then subjected to one or more of a number of processes such as: solvent extraction, sulfuric acid extraction, earth filtration, hydrogenation, redistillation, filtration, and dehydration. The intent of these operations is to remove waxes, sulfur, nitrogen, and oxygen containing compounds and aromatic compounds. Sufficient aromaticity must remain, however, to maintain satisfactory gas absorbing capability and oxidation stability.

Solvent refining relies upon the selective solubility of waxes, sulfur, and nitrogen containing compounds and aromatic hydrocarbons in carefully formulated solvent systems. Treatment with sulfuric acid results in the removal of additional amounts of sulfur and nitrogen containing materials and aromatics through the reaction of the materials with the sulfuric acid. Earth filtration removes polar compounds. Hydrogenation is utilized to reduce the sulfur and nitrogen containing compounds and aromatics; the following reactions may be envisaged [*41*]

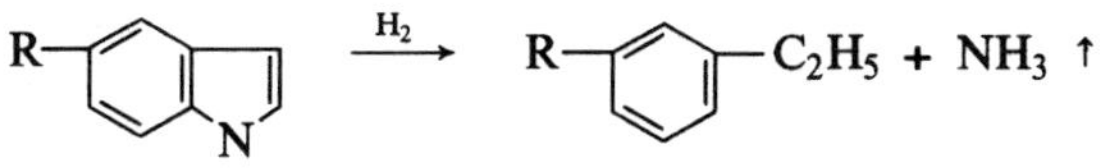

(5.16)

and

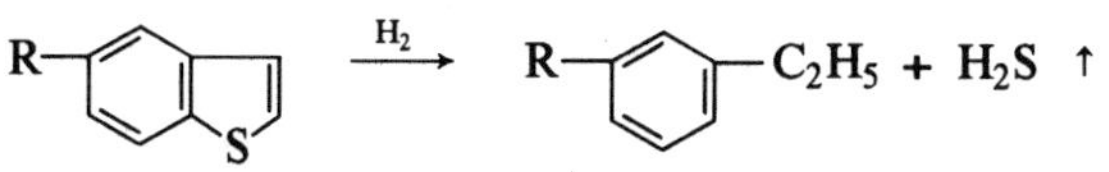

(5.16*a*)

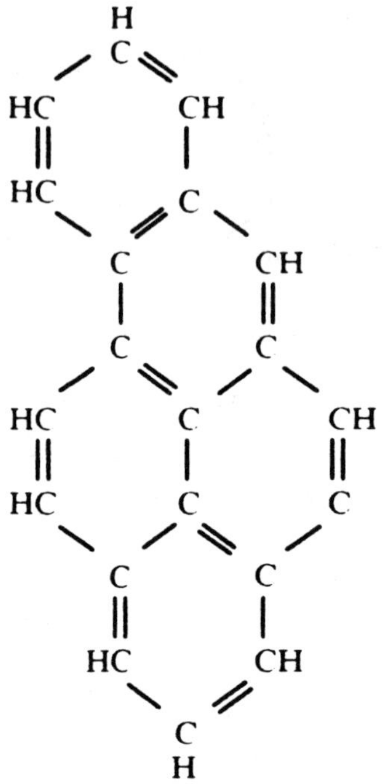

FIG. 5.32—*Peri condensed aromatic hydrocarbon.*

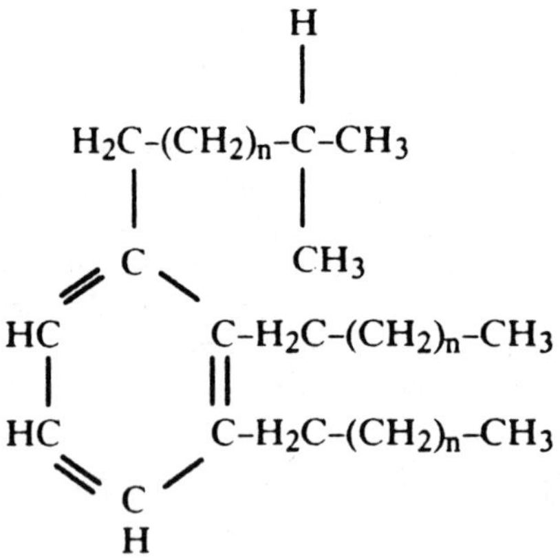

FIG. 5.33—*Alkyl substituted aromatic hydrocarbon.*

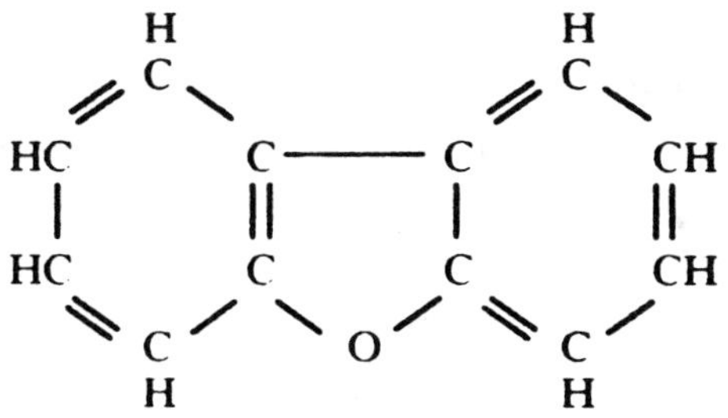

FIG. 5.34—*Oxygen containing hetero-cyclic aromatic.*

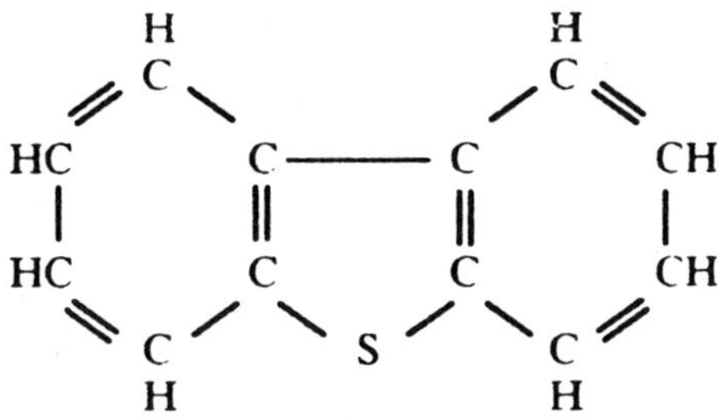

FIG. 5.35—*Sulfur containing hetero-cyclic aromatic.*

The nitrogen compounds are seen to be converted irreversibly to ammonia and the sulfur compounds to hydrogen sulfide. The hydrogenation process may be employed also to convert some of the aromatics to cycloaliphatics (naphthenics), as, for example [41]

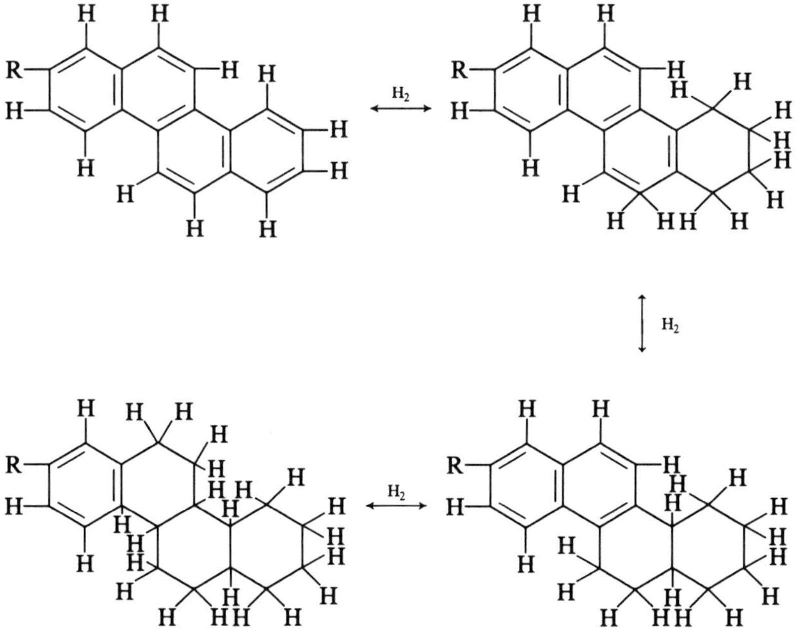

(5.16*b*)

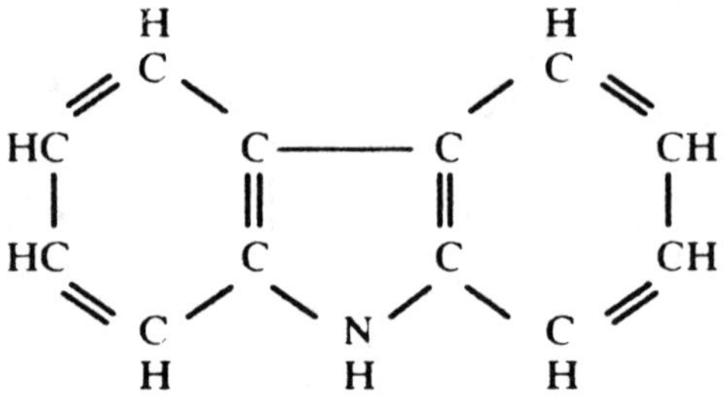

FIG. 5.36—*Nitrogen containing hetero-cyclic aromatic.*

If the refining severity is further increased, the reaction reverses and dehydrogenation occurs as some of the naphthenics are reconverted to aromatics.

The finished mineral oil used in electrical applications generally has a molecular weight of 200 to 400 and is composed of a large number of components. The species present in the fluid are a combination of those originally present in the distillate and those generated by the refining process.

In addition to components originally present or generated by the refining process, some types of oils have oxidation inhibitors added. The most common of these additives are 2,6-ditertiary-butyl paracresol and 2,6-ditertiary-butyl phenol. They are present at a level of 0.3% mass/mass, or less [*42*].

Oxidation of a hydrocarbon may proceed as follows

$$RCH_2CH_3 + O_2 \longrightarrow R\dot{C}(H)CH_3 + .OOH \tag{5.17}$$

$$\longleftarrow HOO. + RCH_2CH_3 \longrightarrow R\dot{C}(H)—CH_3 + HOOH \tag{5.17a}$$

$$R\dot{C}(H)—CH_3 + O_2 \longrightarrow RC(H)(O—O)—CH_3 \tag{5.18}$$

$$RC(H)(O—O.)—CH_3 \longrightarrow RC(H)(H)—CH_3 \longrightarrow R—C(H)(O—OH)—CH_3 + R—\dot{C}(H)—CH_3 \tag{5.19}$$

$$RC(H)(O—OH)—CH_3 \longrightarrow RC(O)—CH_3 + H_2O \tag{5.20}$$

$$R-\overset{\overset{\displaystyle O}{\|}}{C}-CH_3 + O_2 \longrightarrow R\overset{\overset{\displaystyle O}{\|}}{C}-OH + H_2\overset{\overset{\displaystyle O}{\|}}{C} \tag{5.21}$$

The presence of an oxidation inhibitor retards the oxidative process by terminating the propagation step through the donation of a hydrogen atom. The resultant aromatic free radical is stabilized by resonance, thus preventing it from propagating the oxidative process. The inhibitor can finally form a dimer structure [*42,43*]

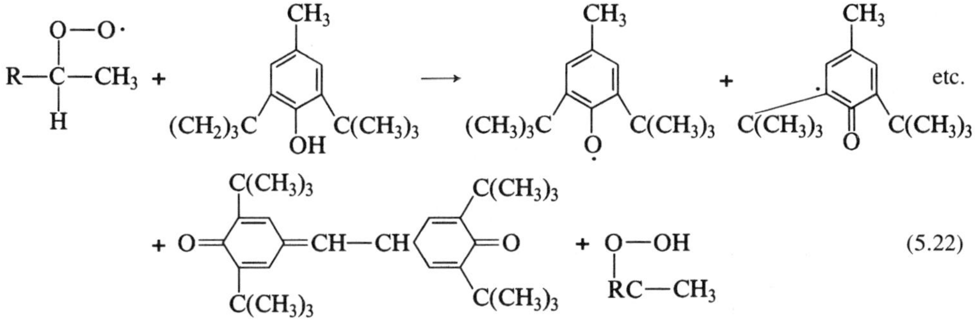

(5.22)

Composition can affect physical, electrical, and chemical properties of an oil. The specifications placed on mineral oils for use as insulants by such standards as ASTM D 3487 and IEC 296 serve to impose compositional constraints upon the oils. The aniline point is intended to characterize an oil by giving an indication of the composition. Aromatic hydrocarbons exhibit the lowest and paraffin oils the highest values. By specifying the aniline point the composition is fixed, at least to a degree. Fire properties set a limit in the amount of low molecular weight material that can be present while the pour point and viscosity requirements control the allowable amount and type of high molecular weight species.

The composition of mineral oils has not been shown to be a significant factor in the normal breakdown voltage at commercial power frequencies. Contaminants, most importantly particulate matter and water, or other polar materials, exert the major influence on this electrical property [*28*]. The presence of certain aromatic materials has been shown, however, to affect the divergent field alternating and impulse breakdown voltages [*13,14*].

The resistivity and dissipation factor of mineral oils is controlled mainly by contaminants. The presence of aromatic components produces an oil with a lower resistivity than one lacking such materials. The dielectric constant of a typical mineral oil is about 2.2. This parameter too varies somewhat depending upon the composition of the fluid. Because most mineral oils used as insulants in electrical applications have relatively low viscosities, on the order of 20 cSt at 25°C, they are quite efficient in removing heat. This low viscosity (molecular weight) and lack of fire retarding substituents, however, results in mineral oils being among the most flammable of liquid insulating materials. As was mentioned in the introduction, the gas absorbing properties of a liquid depends upon its level of unsaturation. Aromatic containing mineral oils display gas absorbancy capability proportional to the amount of aromatics present.

5.4.2 Askarels

Chlorinated aromatic hydrocarbons were first introduced as dielectric fluids in the early 1930s.[2] Their use continued and grew until the early 1970s when they were shown to be persistent in the

[2]For example see White, A. H. and Morgan, S. O., "The Dielectric Properties of Chlorinated Diphenyl," *Journal of the Franklin Institute,* Vol. 216, 1933, p. 635.

environment, nonbiodegradable, and subject to bioconcentration. Accusations of carcenogenity were also made. Although the manufacture of polychlorinated biphenyls, a major constituent of these fluids, has been discontinued in the United States and many other countries, devices containing these materials continue in operation today; however, they are in many cases subject to governmental regulation. The polychlorinated aromatics most commonly used as electrical insulating liquids are composed of chlorinated benzenes and biphenyls. They are known generically as askarels.

Benzene is obtained, principally, by the catalytic reforming of the heavy gasoline fractions of petroleum. A mixed catalyst, such as platinum and rhenium, is used in the reforming process. Several low molecular weight materials are obtained through this process. The benzene is isolated by fractional distillation, solvent extraction, crystallization, or similar methods.

Biphenyl is produced by passing benzene vapor over hot metal, such as lead, at elevated temperatures. Small amounts of higher molecular weight materials are also formed.

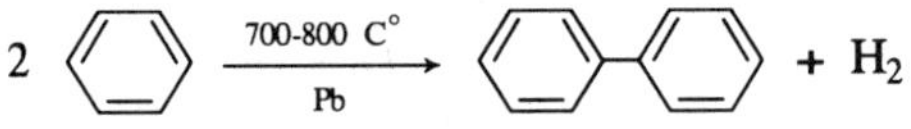

(5.23)

Chlorine, normally produced by the electrolysis of sodium chloride, is reacted with benzene and biphenyl in the presence of a Lewis acid, such as aluminum chloride, to form chlorinated products. The electrophylic substitution is controlled by chlorine concentration, temperature, and reaction time. Various levels of substitution are possible. The equation for the reaction is

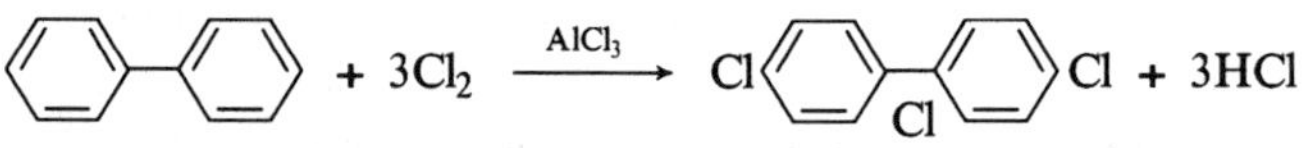

(5.24)

Separation of the various isomers (only one of 209 possible isomers is shown) resulting from chlorination of biphenyl is possible by means of fractional distillation and several other methods. This is difficult to do and generally is not necessary. The possible isomers are shown in Table 5.3 [*31*].

The two major divisions of use for askerals are in capacitors and in transformers. The capacitor compositions contain more di- and tri-chlorophenyl because of the higher dielectric constant. The transformer askarels contain a higher level of tetra-, penta-, and hexa-chlorobiphenyl because of the decreased combustibility of these materials. A list of typical properties for the two types of askarels is given in Table 5.4 [*44*].

Commercial products composed of combinations of chlorinated benzenes and chlorinated biphenyl are known most commonly as "askarels" and have been available under such trade names as Arochlor, Dykanol, Pyralene, Clophen, and Nepolin [*34*]. In addition to having acceptable electrical properties, most askarel compositions will not support combustion.

Because of the inherent oxidative resistance of the aromatic systems, the inclusion of oxidation inhibitors or pacifiers is not necessary with polychlorinated aromatics. Askarels absorb hydrogen gas under discharge conditions. However, if the discharges become intense enough hydrogen chloride (HCl) can be released. It is necessary, therefore, particularly in capacitor applications, to add something to neutralize any HCl which may be generated in the system. Additives include tetraphenyl tin and various organic epoxides. Under d-c stress at elevated temperatures, metal electrodes can catalyze the degradation of askarels. Stabilizers such as anthraquinone and its chlorinated compounds are used to minimize such degradation.

The askarels' electrical properties are affected by their dipolar activity which results in an increasing and then decreasing permittivity with increasing chlorine content. Their resistivity and dissipation factor are also affected. Both quantities are worse than those for mineral oils. Dielectric strength is comparable to that for mineral oils and is affected by the same factors. The impulse strength is lower than that for mineral oil.

TABLE 5.3—*Isomers of chlorinated biphenyl (after Wilson* [31]*).*

		Chlorine			Chlorine			Chlorine			Chlorine
Product	Total Isomers	1st Ring	2nd Ring	Isomers	1st Ring	2nd Ring	Isomers	1st Ring	2nd Ring	Isomers	% Mass, Mass^{-1}
Mono-chloro biphenyl	3	1	0	3							18.6
Di-chloro biphenyl	12	1	1	6	2	0	5				31.4
Tri-chloro biphenyl	24	2	1	18	3	0	6				41.0
Tetra-chloro biphenyl	42	2	2	21	3	1	18	4	0	3	48.3
Penta-chloro biphenyl	46	3	2	36	4	1	9	5	0	1	54.0
Hexa-chloro biphenyl	42	3	3	21	4	2	18	5	1	3	58.7
Hepta-chloro biphenyl	24	4	3	18	5	2	6				62.5
Octa-chloro biphenyl	12	4	4	6	5	3	6				65.7
Nona-chloro biphenyl	3	5	4	3							68.5
Deca-chloro	1	5	5	1							70.9

Because of the high viscosity of the more completely chlorinated biphenyls, askarels, particularly those used in transformers, have thermal properties inferior to those of mineral oil at low temperatures. A steep viscosity versus temperature relationship results in askarels having thermal properties at higher temperatures similar to mineral oils.

5.4.3 Chlorinated Diphenyl Oxides

Diphenyl oxide has the molecular structure shown in Fig. 5.37.

Upon chlorination, the same number of chloroisomers can be obtained as is possible upon chlorination of biphenyl. Because of the oxygen in the ether linkage, the chlorinated diphenyl

TABLE 5.4—*Typical properties and uses of the nonflammable askeral insulating liquids (after Clark* [44]*).*

Use	Capacitors	Capacitors	Transformers	Transformers
Condition, 25°C	clear	clear	clear	clear
Color	yellow tint	light yellow	yellow tint	yellow tint
Acid value, mg KOH/g	<0.01	<0.01	<0.01	<0.01
Free chloride, ion	nil	nil	nil	nil
Specific gravity, 25°C	1.40 to 1.45	1.54 to 1.55	1.54 to 1.55	1.55 to 1.57
Fire point, °C	none	none	none	none
Pour point, °C	0 to −7	6 to 12	−40 to −50	−35 to −40
Viscosity, 37.8°C	200	300	42	54
SSU, 98.9°C	40	46	...	33
Coefficient of expansion, 25 to 65°C, cc/cc/°C	68×10^{-5}	65×10^{-5}	67×10^{-5}	67×10^{-5}
Heat conductivity, 40°C, kcal/m/h/°C	0.091	0.087	0.087	0.087
Refractive index, 20°C (D)	1.623	1.638	1.607	1.614
Dielectric strength, 20 to 90°C				
kV/cm (VDE)	200	200	200	200
VPM (ASTM D824)	750	750	750	750
Dielectric constant	5.3 (20°C)	5.0 (20°C)	4.5 (25°C)	4.2 (25°C)
	4.4 (90°C)	4.2 (90°C)	3.9 (100°C)	3.6 (100°C)
Specific resistance, 90°C, ohm·cm	1.5×10^{12}	1.5×10^{12}	1.5×10^{12}	1.5×10^{12}
Power factor, 90°C	0.015	0.015	0.01–0.05	0.01–0.05

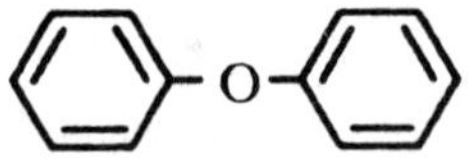

FIG. 5.37—*Molecular structure of diphenyl oxide.*

oxides have lower viscosities than the biphenyl containing the same number of chlorine atoms. A list of typical properties for chlorinated diphenyl oxides is compiled in Table 5.5 [*23*].

5.5 Esters

5.5.1 Organic Esters

Organic esters are obtained by the reaction of an organic acid with an alcohol

$$RC(=O){-}OH + R'CH_2OH \longrightarrow RC(=O){-}OCH_2R' + H_2O \tag{5.25}$$

It is necessary to remove the water as it is formed in order to force the reaction to completion. Typically, the reaction is run in the presence of an acid catalyst, such as sulfuric or phosphoric, and a solvent which azeotropes with water. The reaction is run at reflux in order to remove water from the reaction vessel. It is collected in the bottom of a receiver as the solvent is returned to the reactor at the top of the receiver. Reaction temperatures are generally less than 190°C. Following completion of the reaction, the product is neutralized. The solvent and excess alcohol are removed by distillation. The ester often is distilled and treated with carbon or fuller's earth or both prior to its use.

Esters can be aliphatic, such as, butyl sebacate, $CH_3(CH_2)_3O{-}C(=O)(CH_2)_3C(=O){-}O(CH_2)_3CH_3$, or

TABLE 5.5—*Properties of chlorinated diphenyl oxides* [23].[a]

	Mono	Di	Tri	Tetra	Penta
Refractive index, 25°C	1.5845	1.5950	1.6030	1.6100	1.6220
Fire point, °C	146	200	200+	200+	none
Specific gravity, 25/15.5°C	1.180	1.294	1.390	1.450	1.535
Viscosity SSU, 37.8°C	36	41	52	76	390
Pour point, °C	−55	−40	−24	−12	0
Distilling range at 15 mm, °C	170 to 180	190 to 200	210 to 220	220 to 235	230 to 250
Dielectric strength, kV, 2.5 mm gap, 25°C	35[a]	35	35	35	35
Resistivity, ohm·cm × 10^9					
100°C	10	25	50	75	100
30°C	150	250	400	500	1000
Dielectric constant, 60 cycles					
100°C	3.8	3.9	4.2	4.2	4.3
30°C	4.5	4.3	4.7	4.7	5.0

[a]The dielectric strength in all cases averages irregularly higher than the value stated.

aromatic, such as, dioctyl phthalate, 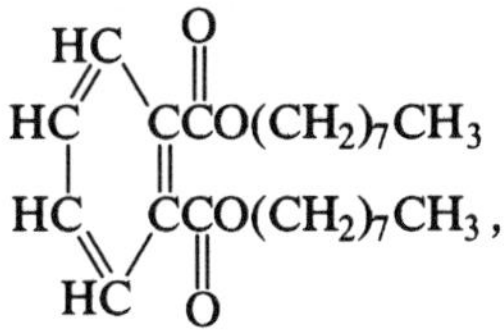 , or benzyl benzoate

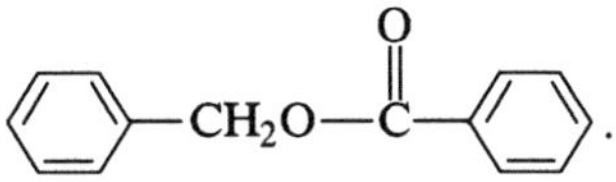

Complex esters are obtained from dibasic acids and polyhydric alcohols. An example of such an ester is given by Wilson [*31*]

$$CH_3(CH_2)_7O\overset{O}{\overset{\|}{C}}—(CH_2)_8\overset{O}{\overset{\|}{C}}[(CH_2CH_2O)_n—\overset{O}{\overset{\|}{C}}(CH_2)_8\overset{O}{\overset{\|}{C}}O—]_x(CH_2)_7\overset{O}{\overset{\|}{C}}H_3 \qquad (5.26)$$

The total of *n* and *x* are such as to give an ester molecular weight of about 1000. Monobasic acids can be used also as an endblock for these short chain polymeric materials instead of the monohydric alcohol.

Another type of ester is derived from neopentyl glycols and monobasic acids. The neopentyl glycol penta-erythritol is used commonly in the preparation of this type of ester

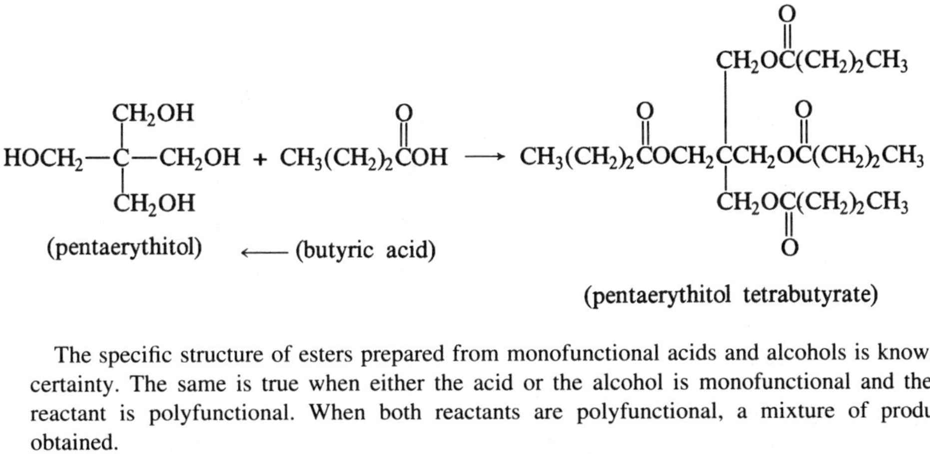

(5.27)

(pentaerythitol) ⟵ (butyric acid)

(pentaerythitol tetrabutyrate)

The specific structure of esters prepared from monofunctional acids and alcohols is known with certainty. The same is true when either the acid or the alcohol is monofunctional and the other reactant is polyfunctional. When both reactants are polyfunctional, a mixture of products is obtained.

The number of possible esters is nearly limitless. However, the more commonly used ones are listed in Table 5.6 [*31,44*]. An ester having the desired properties can be obtained by the selection of the appropriate alcohol and acid combination. In general, organic esters offer low viscosity for their boiling points; low viscosity change with change in temperatures; low pour points; high flash, fire, and auto ignition temperatures; good oxidation response, low toxicity; and adequate biodegradability. Gas absorbancy is dependent upon the use of aromatic reactants. Highest thermal stability is achieved by the selection of a hindered alcohol lacking hydrogen atoms on the carbon adjacent to the hydroxyl substituted carbon. This prevents the following reaction [*31*]

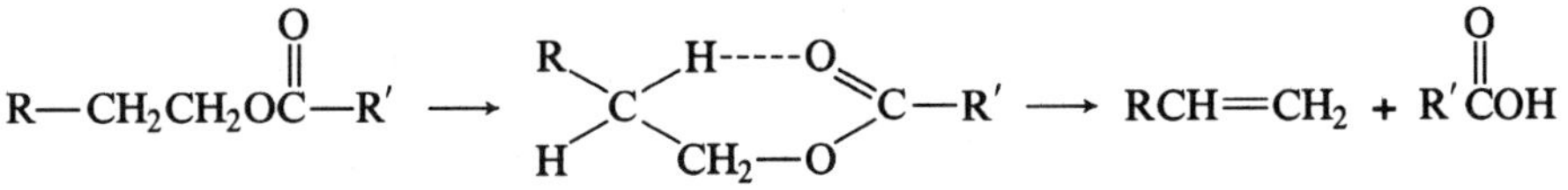

(5.28)

TABLE 5.6—*Characteristics of organic ester insulating liquids* [44].

Name	Structure	Pour, Pt °C	Flash, Pt °C	Viscosity SSU, 37.8°C	Dielectric Constant, 25°C, 60 Hz	tanδ × 10^{-3}, 25°C, 60 Hz	Resistivity ohm·cm × 10^{12}, 25°C
amyl oleate	$CHC_3(CH_2)_4O\overset{O}{\overset{\|}{C}}(CH_2)_7HC{=}CH(CH_2)_7CH_3$	−12	175	47	2.9	2.3	
butyl oleate	$CHC_3(CH_2)_3O\overset{O}{\overset{\|}{C}}(CH_2)_7HC{=}CH(CH_2)_7CH_3$	−8	179	49	2.9	8.0	1.0
tetra-hydro furfuryl oleate	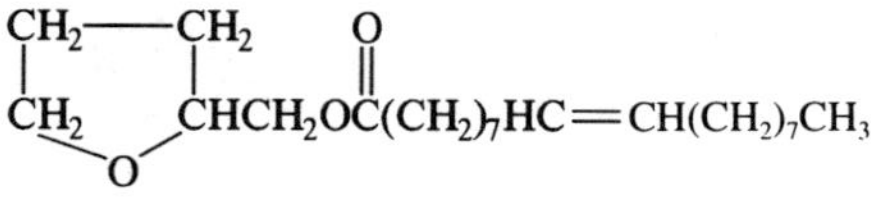	−18	204	61	4.3	19	<0.1
benzyl benzoate	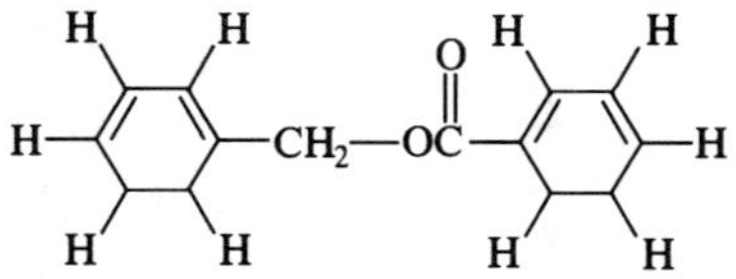				4.5	41	1.0
iso-amyl benzoate	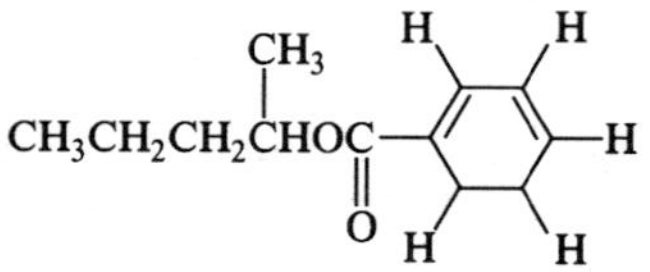				(5.04 @ 1 kHz)	>500	
iso-propyl benzoate	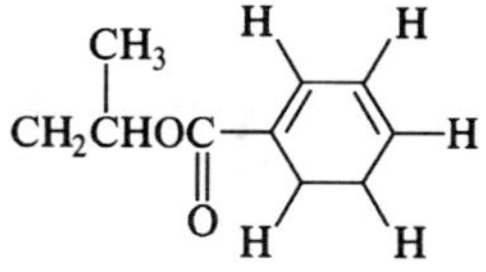				5.0	47	

methyl benzoate	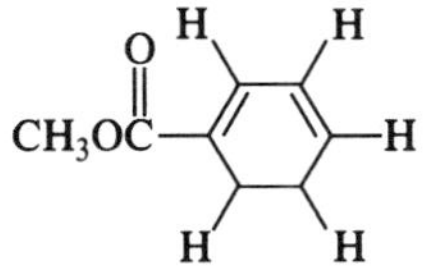					>500	
butyl stearate	$CH_3(CH_2)_3OC(=O)(CH_2)_{16}CH_3$	21	167	48	3.3	10	5.0
methyl stearate	$CH_3OC(=O)(CH_2)_{16}CH_3$	26	156	44	2.9	>500	
butyl ricinoleate	$CHC_3(CH_2)_3OC(=O)(CH_2)_7HC{=}CHCH_2CH(CH_2)_5CH_3$				4.0	150	
cyclohexyl ricinoleate	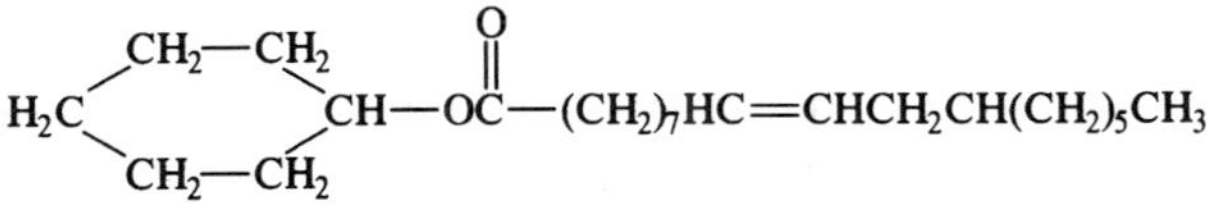				4.3	>500	
methyl ricinoleate	$CH_3OC(=O){-}(CH_2)_7HC{=}CHCH_2CH(CH_2)_5CH_3$				4.3	510	
tri-cresyl phosphate	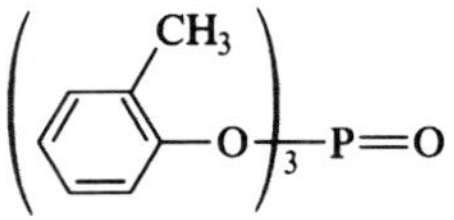	−27	250		7.17	7.15	
tri-xylenyl phosphate	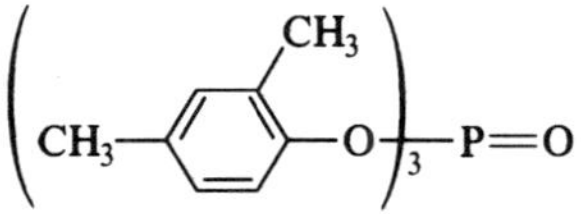	−18	260				

TABLE 5.6—*Continued.*

Name	Structure	Pour, Pt °C	Flash, Pt °C	Viscosity SSU, 37.8°C	Dielectric Constant, 25°C, 60 Hz	tanδ × 10^{-3}, 25°C, 60 Hz	Resistivity ohm·cm × 10^{12}, 25°C
butyl citrate	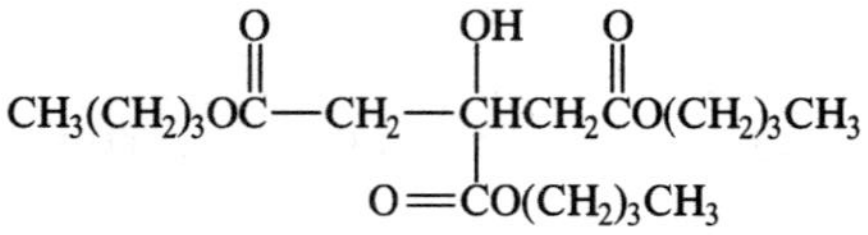					>500	<0.1
butyl carbonate	$CH_3(CH_2)_3OC(=O)O(CH_2)CH_3$				2.3	4.0	<0.1
butyl laurate	$CH_3(CH_2)_3OC(=O)(CH_2)_{10}CH_3$	−9	140	39	3.0	26	
butyl borate	$[CH_3(CH_2)_3O]_3B$				2.1	3.7	
butyl naphthenate	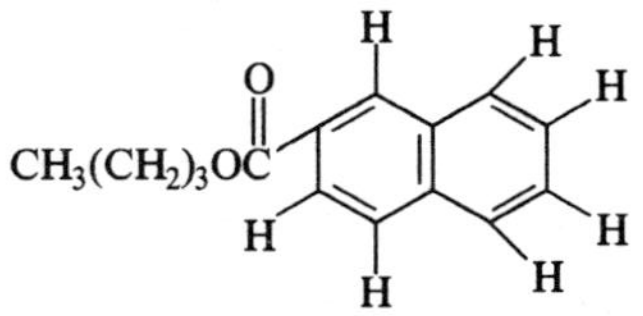				3.1	16	1.0
diethyl hexyl adipate	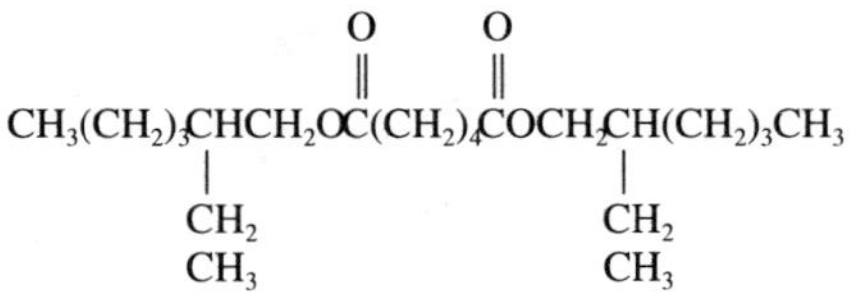	−66	202	52	4.1	26	1.0

diethyl hexyl phthalate	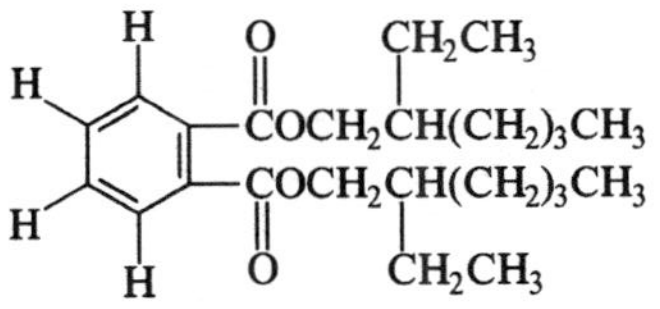	−30	200			
dibutyl silicate	$CH_3(CH_2)_3O\overset{O}{\overset{\|}{C}}(CH_2)_8\overset{O}{\overset{\|}{C}}O(CH_2)_3CH_3$	−15	186	46	4.4	10
diethyl hexyl sebecate	$CH_3(CH_2)_3\underset{\underset{CH_2CH_3}{\|}}{C}HCH_2O{-}\overset{O}{\overset{\|}{C}}(CH_2)_8\overset{O}{\overset{\|}{C}}OCH_2{-}\underset{\underset{CH_2CH_3}{\|}}{C}H(CH_2)_3CH_3$			67	3.8	20.7
tetrahydro furfuryl oxalate	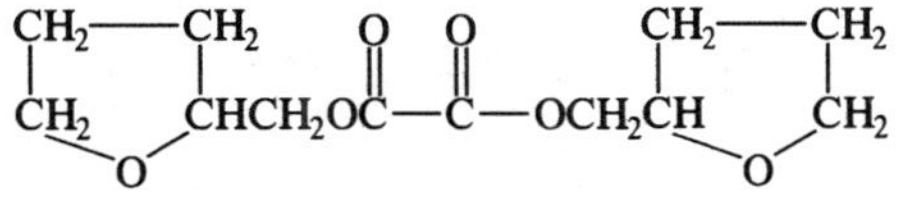	−54		97	12.8	(4.6 @ 1 kHz)

Esters require addition of antioxidants to impart an acceptable level of oxidation resistance. These antioxidants can also improve the hydrolytic stability. Hydrolytic stability can be maximized by selection of dibasic acids having substitution adjacent to the carboxy (COOH) group. Esters have water saturation levels of 200 to 400 ppm. The existence of this much water in any insulating liquid can affect adversely its performance. These high water levels coupled with hydrolysis products (acid and alcohol) can reduce the electrical properties of the liquid to an unacceptable level.

Organic esters have been used in cables and high frequency capacitors [*45*]; in the latter application, they have been used because of their relatively low losses at the more elevated frequencies. Some esters, are characterized by unusually high dielectric constants; for instance, ethylene glycol possesses a dielectric constant of *ca.* 40, making it especially suitable for to capacitors. The physical and electrical properties of a few representative esters are listed in Table 5.7 [*45*].

5.5.2 Phosphate Esters

Triaryl phosphate esters also have been suggested as insulating liquids. Alkyl esters find no use as insulating liquids. The phosphate esters are prepared by reaction of phosphorous oxychloride with phenols, cresols, and xylenols

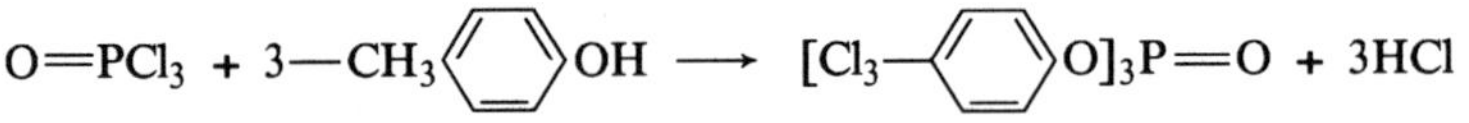

(5.29)

The reaction is forced to the right by using excess phenol, in the presence of a metal chloride, and removing the HCl gas as it is formed. The crude product is washed with alkali to remove acidic materials. It is then distilled, filtered, and dried. There are three cresol isomers. Thus, there are nine possible tricresyl phosphates. There are six possible dimethyl phenols (xylenols) which would produce 36 different phosphate esters. In addition, commercial phenols often contain small amounts of other phenolic materials. These change the ester compositions and can affect properties. One possible isomeric structure for tri-xylenyl phosphate is portrayed in Fig. 5.38 [*31*].

These esters are more fire resistant than silicones, but less so than askarels. Their electrical properties tend to be poor as are their low temperature flow characteristics and hydrolytic stability. They have a permittivity similar to that of askarels. Their oxidation resistance and hydrolytic stability is superior to the uninhibited organic esters.

TABLE 5.7—*Physical and electrical properties of organic esters (after Saums and Pendleton* [45]*).*

Property	Castor Oil	Dibutyl Sebacate	Butyl Stearate	Silicate Ester
Pour point, °C	−23	−10	21	−60
Flash point, °C	291	175	167	188
Viscosity, 37.8°C, SUS	...	46	48	12.2
Dielectric strength, 100°C, kV	...	38	...	27
Dielectric constant, 100°C, 60 Hz	3.74	4.4	3.3	2.65
Dissipation factor, 100°C				
at 60 Hz	0.06	0.01	0.01	...
at 1 MHz	...[a]	0.00015	0.00013	0.042
Resistivity, ohm·cm	3×10^{10}	5×10^{12}	5×10^{12}	9×10^{10}

[a]The high frequency value is not available at 100°C; but at 25°C, $\tan\delta \simeq 1.0$ at 1 MHz, according to Ref *46* which is due to the appreciable polarity of the castor oil molecule.

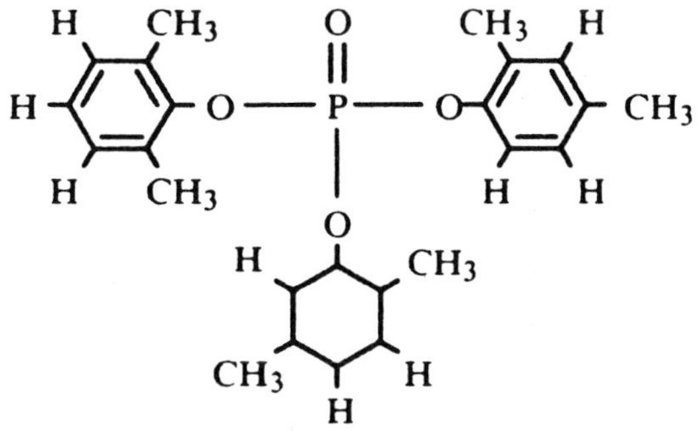

FIG. 5.38—*A possible isomeric structure for trixylenyl phosphate* [31].

5.6 Synthetic Hydrocarbons

5.6.1 Polybutenes

Two types of hydrocarbons synthesized from petroleum feedstocks and used as insulating liquids are polyolefins and alkylaromatic. These materials are predominantly mixtures of isomers and homologs of one type of hydrocarbon. Their composition and characteristics, however, vary due to the many sources of feedstocks and different methods of manufacturing.

Polybutenes are the simplest of the synthetic hydrocarbons to find use as electrical insulants [*47*]. They are formed by the acid catalyzed polymerization of butenes. Butenes are isolated during the distillation of crude oil. They are also formed as a by-product in the thermal and catalytic cracking, hydrocracking, and reforming of distillates. Butene exists as two isomers: $CH_3CH_2CH{=}CH_2$ (n-butylene) and $(CH_3)_2C{=}CH_3$ (isobutylene). One synthesis scheme used to prepare polybutenes is represented by the equation [*30*]

$$(CH_3)_2C{=}CH_2 \xrightarrow[25\text{-}50^\circ C]{AlCl_3} CH_3{-}C(CH_3)_2{-}\left(C(CH_3)_2{-}CH_2\right)_x{-}CH_2C(CH_3){=}CH_2 \qquad (5.30)$$

The molecular weight is determined by the reaction temperature. Because the reaction is exothermic, cooling is required to obtain the desired product. The product is purified by the removal of the catalyst, neutralization of residual acid, distillation, drying, and filtration.

It is isobutylene which is used as the monomer to prepare the oils referred to as polybutenes. The polybutenes are prepared in the molecular weight range of 200 to 3000. These products vary from low viscosity fluids to sticky, semisolids. The molecular weight and viscosity have an effect on some of the electrical and physical properties, as can be perceived from Table 5.8 [*48*], which provides data on a number commercially available electrical grade polybutene oils.

Because of the terminal unsaturation, polybutenes are gas absorbing under electrical stress or partial discharge. That same unsaturation leads to ready oxidation to form organic acids. Therefore, the use of oxidation inhibitors and metal passivators is necessary to achieve acceptable oxidation stability.

TABLE 5.8—*Physical and electrical properties of polybutene oils with different viscosities* [48].

Property	Oil Type			
	0SH[a]	06SH[a]	015SH[a]	30SH[a]
Average molecular weight	340	450	570	1350
Viscosity, SUS				
100°F	110	575	3440	115000
210°F	41	63	158	3070
Flash point, °F	270	310	320	450
Pour point, °F	−60	−40	−10	+40
Acidity, mg KOH/g	0.01	0.01	0.01	0.01
Water content, ppm	15	15	15	15
Dissipation factor, 60 Hz, 20 V/mil, 100°C	0.0003	0.0003	0.0002	0.0002
Dielectric constant, 1 MHz, 25°C	2.14	2.16	2.17	2.24
Dielectric strength, kV, 0.1 in gap, 80°C	>35	>35	>35	>35
Volume resistivity, ohm·cm, 100°C	8×10^{14}	1×10^{15}	1.2×10^{15}	1.5×10^{15}

[a]Registered trade names of Gosden Oil and Chemical Company.

Polybutenes are also subject to depolymerization at elevated temperatures. The temperature required, on the order of 200°C, is in excess of any temperatures likely to be seen in electrical equipment, however.

The dielectric strength of polybutenes is essentially the same as that of paraffinic oils. The impulse strength could be lower due to the terminal unsaturation. The dissipation factor and volume resistivity of polybutenes are superior to that of mineral oils while the permittivity is similar, 2.1 to 2.3 (see Table 5.8 [*48*]). The permittivity increases with increasing molecular weight and the dissipation factor tends to decrease [*48*].

Polybutenes have found limited use in pressurized pipe type alternating power cable. Their attempted use in low pressure and solid type cables has failed. It is probable that this limitation results from the ''cracking'' of the polybutene molecule in the highly stressed alternating fields producing gaseous decomposition products which lowered the ionization voltage. Such gas evolution would not be noticed in the high pressure hollow core cables. Polybutene performs well, however, in direct voltage systems.

5.6.2 Polyalpha-Olefins

Higher molecular weight alpha-olefins also can be polymerized to form insulating oils. These materials are obtained in their purest form through the controlled polymerization of polyethylene

$$nH_2C{=}CH_2 \xrightarrow[175°C\ 195\ bar]{AlR_3} CH_3(CH_2CH_2)_{n-1}HCHCH{=}CH_2 \qquad (5.31)$$

Polymerization of an alpha-olefin proceeds similarly to the polymerization of butene

$$nRCH{=}CH_2 \longrightarrow (R{-}\overset{\overset{H}{|}}{\underset{\underset{H}{|}}{C}}{-}\overset{\overset{H}{|}}{\underset{\underset{H}{|}}{C}}{-}\overset{\overset{H}{|}}{\underset{\underset{H}{|}}{C}}{-}\overset{\overset{R}{|}}{\underset{\underset{H}{|}}{C}}{-})_{n-2} \longrightarrow \overset{\overset{R}{|}}{\underset{\underset{H}{|}}{C}}{=}CH_2 \qquad (5.32)$$

These materials have many of the same characteristics as polybutenes, that is, low dissipation factor, high volume resistivity, and a permittivity similar to mineral oil. They also require oxidation inhibitors and metal passivators to permit extended oxidation free operation. In recent years at least one manufacturer's product has found use as a PCB replacement fluid in transformers. Its use in this application is based principally upon the high flash and fire points of such materials. Representative properties of a commercial polyalpha-olefin are given in Table 5.9 [*42,49*].

5.6.3 Alkyl Substituted Aromatic

The alkyl-aromatic compounds used as dielectric liquids are composed of compounds having alkyl chains of up to 25 carbon atoms attached to aromatic ring structures which include benzene, tetralin, naphthelene, biphenyl, and diaryl alkanes (Fig. 5.39).

Alkyl aromatics are manufactured from an aromatic and an olefin or halogen substituted alkyl hydrocarbon by use of Friedel Crafts condensation reactions [*50*]

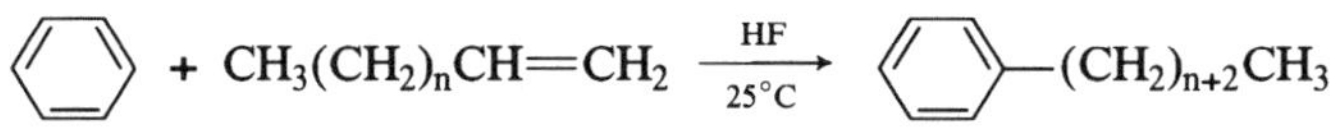

(5.33)

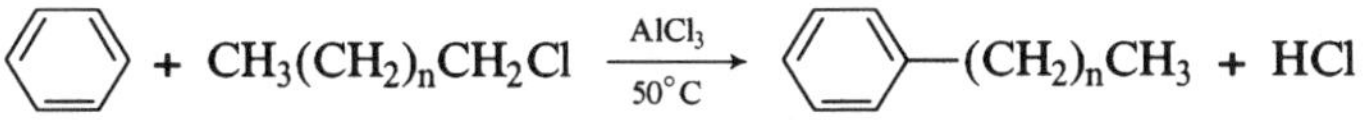

(5.34)

TABLE 5.9—*Properties of polyalpha olefins* [42,49].

Specific gravity @ 25°C	0.87
Viscosity, cSt, ASTM D 445	
@ 40°C	140
@ 100°C	14
Flash point, °C, ASTM D 92	284
Fire point, °C, ASTM D 92	312
Pour point, °C, ASTM D 97	−24
Dielectric strength, 25°C ASTM D 1816	56
Relative permittivity, 25°C ASTM D 824	2.2
Dissipation factor, 100°C ASTM D 0924	0.0016

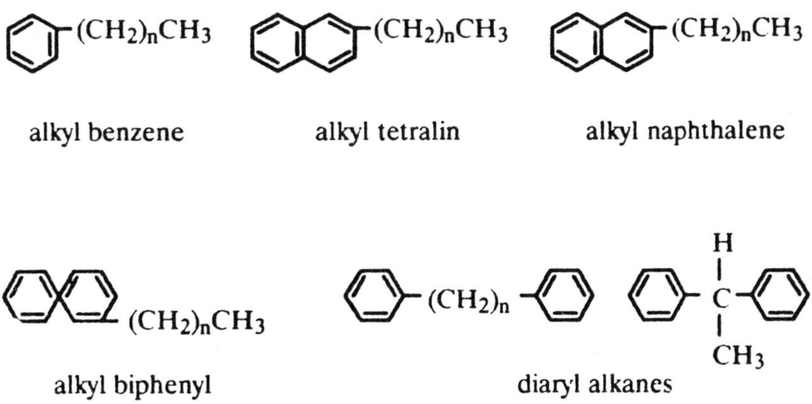

FIG. 5.39—*Molecular structures of some alkyl-aromatic compounds.*

Products are purified by catalyst removal, acid neutralization, distillation, earth filtration, and drying. Chlorine containing materials can be persistent contamination in liquids prepared via alkyl chlorides. Such contaminants can adversely affect the properties of the finished product.

These materials have excellent gas absorbing characteristics. Their thermal properties are very similar to those of mineral oil. They have low pour points and high viscosity indexes. Flash and fire points are controlled by the length of the alkyl chain substitution. The oxidation stability of alkyl aromatics in the presence of limited oxygen is poor compared to mineral oil. They do respond well, however, to the addition of antioxidants. Due to their solvent and detergent nature, they are appreciably more susceptible to contamination than mineral oils.

Dielectric strength for alkyl aromatics, as with other oils, is controlled by particle and water contamination not by the chemical composition of the oil. The dissipation factor for alkyl aromatics is very comparable to that for mineral oils, while their permittivity is only slightly higher than for petroleum and polyolefin hydrocarbons. The physical and chemical properties of several commercial electrical grade alkyl benzenes are provided in Table 5.10 [*48*].

5.6.4 Halogenated Aliphatic Hydrocarbons

Stable halogenated hydrocarbons are obtained when all hydrogens have been replaced by halogens. The presence of residual hydrogen can lead to the elimination of a hydrogen halide acid. Halogenated alkenes require the presence of conjugated, —C=C—C=C—, unsaturation for max-

TABLE 5.10—*Properties of alkyl benzenes* [48].[a]

Property	Chevron Alkylate[b] 21	Dichevrol DF100[c]	Dodecyl Benzene
Average molecular weight	246	320	249
Specific gravity, 15.6°C	0.87	0.87	0.85
Viscosity (SUS)			
37.8°C		100	
98.9°C		38	
Viscosity (cS)			
37.8°C	6.0	20.6	4.30 to 7.37
98.9°C		3.5	
Flash point	127°C	350°F	130°C
Pour point, C	−50	−50	−50
Water content, ppm	24	25	≤30
Acidity, mg KOH/g	0.01	0.02	0.02
Dissipation factor			
60 Hz, 210°F		0.0002	0.001
60 Hz, 25°C	0.0004		
Dielectric constant			
60 Hz, 210°F	2.1	2.1	2.1
60 Hz, 25°C			2.2
Dielectric strength, 77°F, kV, 0.254 cm gap	>35	50	>35
Volume resistivity			
ohm·cm, 210°F,		1×10^{15}	
ohm·cm, 25°C	3×10^{14}		2×10^{13}
Gas evolution under stress, μL/min	−30	(−46 to −18)	−30

[a]Data taken on dichevrol dielectric fluids provided by the Chevron Chemical Co. and Shrieve Chemical Products Inc.

[b]An alkyl benzene for use as additive in a hollow core type mineral oil.

[c]An alkyl benzene for use on self-contained hollow core cables and forced cooled pipe cables.

imum stability. The moste stable of halo alkanes are the fluorocarbons [*45,51*]. Fluoromethanes and ethanes first found use as refrigerants. These materials contain both fluorine and chlorine, and sometimes, bromine, in the same molecule. Because the reaction of fluorine with a hydrocarbon is too energetic to control, these materials are made generally from the totally chlorinated alkane; that is, carbon tetrachloride and perchloroethane. The reaction uses hydrogen fluoride and a catalyst

$$CCl_4 + HF \xrightarrow{SbF_5} CFCl_3 + HCl \xrightarrow[SbF_5]{HF} CCl_2F_2 + HCl \text{ etc.} \tag{5.35}$$

$$Cl_3C{-}CCl_3 \xrightarrow[SbF_5]{HF} Cl_2FC{-}CCl_3 \xrightarrow[SbF_5]{HF} Cl_2FC_2{-}CFCl_2 + HCl \text{ etc.} \tag{5.36}$$

Some of these materials have found commercial application in closed transformer systems which make use of the material in its vapor state for cooling [*51*].

Chlorotrifluoroethylene also is used to prepare liquids of interest. This monomer is obtained by the partial fluorination of perchloroethane followed by selective dehalogenation.

$$CCl_3{-}CCl_3 \xrightarrow[SbF_5]{3HF} F_2ClC{-}CCl_2F + 3HCl \tag{5.37}$$

$$F_2ClC{-}CCl_2F \xrightarrow[\text{alcohol}]{Zn} F_2C{=}CFCl + ZnCl_2 \tag{5.38}$$

or

$$F_2ClC{-}CCl_2F + H_2 \longrightarrow F_2C{=}CFCl + 2HCl \tag{5.39}$$

The monomer is polymerized via a free radical process to produce short chain polymeric materials.

$$F_2ClC{=}CCl_2F \xrightarrow[CCl_4,\text{ heat}]{\text{peroxide}} Cl_3C(F_2C{=}CClF)_nCl \tag{5.40}$$

Increased stability is obtained by additional fluorination

$$Cl_3C(F_2C{-}CClF)_nCl \xrightarrow{CoF_3} Cl(CF_2{-}CClF)_{n+1}Cl \tag{5.41}$$

Liquid products have n ranging from 2 to 5.

As has been pointed out by Wilson [*31*], the unsaturated chlorinated hydrocarbon hexachlorobutadiene also has found limited use as an insulant. It has the structure shown in Fig. 5.40.

These listed halocarbons are characterized by high-densities, low-surface tension, viscosity, boiling point, and, except for hexachlorobutadiene, pour point. Their thermal properties are superior to mineral oil. They are highly resistant to fire and tend to selfextinguish if they are ignited [*45,51*]. They are stable to 250°C at which point the rupture of carbon-carbon bonds may begin with the less

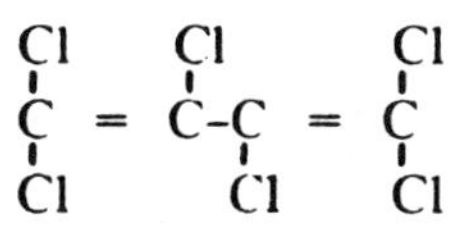

FIG. 5.40—*Molecular structure of hexachlorobutadiene* [31].

stable species. Hexachlorobutadiene is less stable than the halo alkanes. The electrical properties are similar to those obtained with mineral oils [*31*].

The term "fluorocarbons" used in the context of this section may be misleading in the sense that a number of common fluorocarbons used for electrical applications have oxygen and nitrogen in their molecular structures. These particular fluorocarbon liquids have been used as dielectric test bath media and as convective and evaporative dielectric insulants. The physical and electrical properties of some of these types of fluorocarbon fluids are listed in Table 5.11 [*51*].

5.6.5 Silicones

The term "silicone" has been applied to any material containing the element silicone. In fact, the term silicone is used correctly only with organic substituted silicone atoms alternating with oxygen in a polymeric fashion (Fig. 5.41) where *R* is any hydrocarbon.

Commercial silicones used for insulants are substituted with either methyl or phenyl radicals. The major product contains only methyl substitution. Table 5.12 gives the physical and electrical properties of a recently developed polydimethylsiloxane transformer fluid.

Organic compounds of silicone were first prepared in the late nineteenth century. The synthesis and understanding of silicone chemistry was further developed by Kipping in the early part of this century. In 1941 E. G. Rochow was issued U.S. Patent 2380995, which described the reaction of silicon metal with alkyl halides. This patent is the basis for most commercial processes used to prepare alkyl, in particular methyl, substituted silicon, which is then converted to silicones. The reaction needs high temperature, about 300°C, and a copper or silver catalyst. Many products are formed.

TABLE 5.11—*Physical and electrical properties of several common fluorocarbons* [51].

Property	$(C_9F_9)_3N$	$F(CFCF_2O)_nCHFCF_3$ (with CF_3 branch on CF)	$C_8F_{16}O$
Viscosity, cSt			
25°C	2.59		0.82
37.8°C	1.80	2.7	0.64
99 to 100°C	0.48	0.88	0.30
Pour point, °C	−50.0	−154 to −84	−100.0
Specific gravity			
15.6°C	1.89	1.57 to 1.82	1.79
25°C	1.88	1.54 to 1.79	1.77
Coefficient of expansion, °C	0.0012	0.00084 to 0.00036	0.0016
Thermal conductivity, BTU/h, ft², °F, ft	0.049	0.070 to 0.049	0.081
Boiling point at 760 mm, °C	177.8	40.8 to 224.2	102.2
Dielectric strength, kV/0.1 in., 0.254 cm	>35	28 to 50	>35
Dielectric constant			
60 Hz	1.90	3.02 to 2.45	1.86
10^3 Hz	1.89	3.02 to 2.45	1.84
10^6 Hz	1.90		1.87
Dissipation factor			
60 Hz	<0.0005[a]	<0.00006	<0.0005[a]
10^3 Hz	<0.0005	<0.00006	<0.0005
10^6 Hz	<0.0005		<0.0005
Volume resistivity, ohm·cm	3×10^{14}	4×10^{14}	6×10^{14}

[a]At 100 Hz.

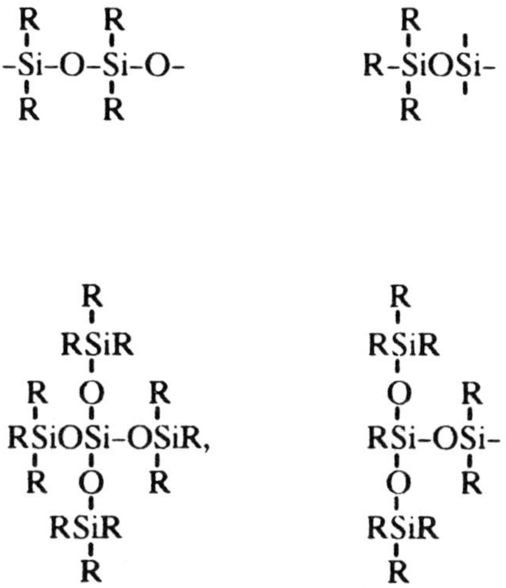

FIG. 5.41—*Molecular structure of silicones.*

$$CH_3Cl + Si \xrightarrow[300°C]{Cu} (CH_3)_2SiCl_2 + HSiCl_3 + CH_3SiCl_3 + (CH_3)_2SiHCl + (CH_3)_3SiCl + CH_3SiHCl_2 + (CH_3)_4Si + \text{very low levels of ethyl analogs of these compounds} + SiCl_4 + Cl(CH_3)_2Si\text{—}[(CH_3)_2SI]_n\text{—}Cl + SiCH_2Si \text{ compounds} + \text{numerous hydrocarbons} \quad (5.42)$$

The reaction mixture is separated into its components by fractional distillation. The purified dimethydichloro- and trimethylchloro-silane can then be used to prepare silicone fluids [*52*].

$$(CH_3)_3SiCl + (CH_3)_2SiCl_2 + H_2O \longrightarrow (CH_3)_2SiOH + (CH_3)_2Si(OH)_2 + HCl \quad (5.43)$$

$$2(CH_3)_3SiOH + n(CH_2)_2SiOH \longrightarrow (CH_3)_3Si\text{—}[OSi(CH_3)_2]_n\text{—}CH_3 + H_2O \quad (5.44)$$

Because of the very careful purification of the chlorosilanes, there is an extremely low level of branching due to the presence of tri- or tetra-functional species. In addition, the need to achieve the

TABLE 5.12—*Physical and electrical.*

Properties of a New Polydimethylsiloxane Transformer Fluid	
Specific gravity, 25°C/15.6°C	0.957 to 0.963
Refractive index, 25°C	1.4010 to 1.4030
Viscosity, cSt @ 25°C	47.5 to 52.5
Pour point, °C	<-50
Flash point, °C	>300
Fire point, °C	>343
Dielectric strength, V/mil	>350
Dissipation factor, 60 Hz, 25°C	<0.0001
Volume resistivity, ohm·cm, 25°C	$>1 \times 10^{14}$
Dielectric constant, 60 Hz, 25°C	2.50 to 2.60
Water content, ppm	<50
Volatile content	$<0.5\%$

highest possible flash and fire points results in all of the cyclic polysiloxanes being removed from polydimethyl siloxanes used in transformers. The fluid used in this application is 50 cSt and has an average molecular weight of about 3000 (n = 40). The fluid is composed of linear molecules having molecular weights varying over a range determined by the specific manufacturing process and conditions. The major use of silicone fluids involving electrical equipment is as a substitute for askarels in transformers. This application results directly from the high flash and fire points and low environmental concerns of silicones.

Compared to any other insulating liquid, polydimethyl siloxane fluids have the highest viscosity temperature index. Relatively speaking, their viscosity is little changed by changes in temperature. This viscosity stability results from the ease of rotation about the Si—O—Si bond, which prevents close packing of the molecules and reaction between the weak intermolecular forces or dipoles.

The dielectric strength is comparable to that of other insulating liquids. The negative point to where sphere impulse withstands capability is nearly 2.5 times that for mineral oil and askarels; however, the positive polarity value is essentially the same as that for mineral oils and askarel. The dissipation factor and resistivity of dimethyl silicones are better than the values for the other insulating liquids. Permittivity of 2.7 is somewhat higher than that of mineral oil.

Polydimethyl silicones have a high thermal stability and low dielectric losses to temperature as high as 150°C or more. At 200°C oxidation can be significant. Oxidation produces formaldehyde, water, carbon dioxide, carbon monoxide, hydrogen, formic acid, and crosslinked polymer. Oxidation includes the following reactions [*53*]

$$\equiv SiCH_3 + O_2 \longrightarrow \equiv SiCH_2 + OH \quad (5.45)$$

$$\equiv SiCH_2 + O_2H \longrightarrow \equiv SiCH_2 + 2OH \quad (5.46)$$

$$\equiv SiCH_3 + OH \longrightarrow \equiv SiCH_2 + H_2O \quad (5.47)$$

$$\equiv SiCH_2 + O_2 \longrightarrow \equiv SiCH_2O_2 \quad (5.48)$$

$$\equiv SiCH_2O_2 \longrightarrow \equiv SiO + CH_2O \quad (5.49)$$

Although polydimethyl siloxanes have found the greatest use, polyphenylmethyl siloxanes also have been used where even higher oxidative and radiation resistance are desired. Because of their low flammability and high radiation resistance, 10 rad, these fluids have found acceptance in nuclear power plants.

The production of phenylchloro silanes can involve the use of a Grignard reaction

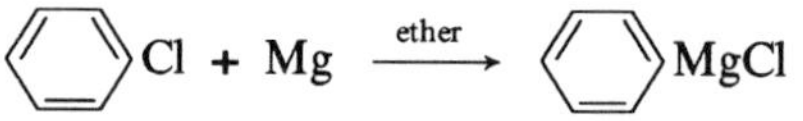

(5.50)

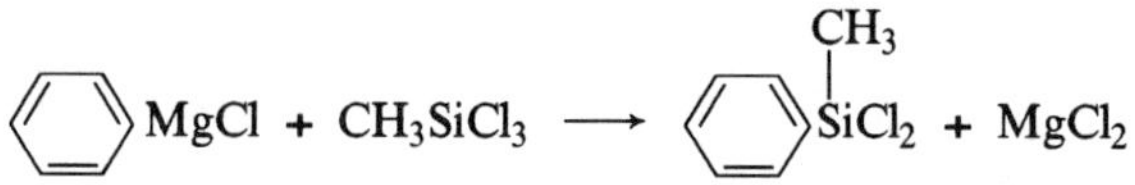

(5.51)

The phenylmethyl dichlorosilane is isolated from other reaction products, mainly silicon with a higher phenyl content, mixed with trimethylchloro silane and hydrolyzed. The product is shown in Fig. 5.42.

Copolymers also can be obtained by including, for example, dimethyldichloro silanes in the hydrolysis reaction. Then the product is as shown in Fig. 5.43.

Phenyl containing fluids have somewhat higher permittivity and dissipation factor than the dimethyl analogs. In addition the flat viscosity temperature relationship of the completely methylated series is lost. The viscosity of these fluids changes with temperatures in a fashion similar to

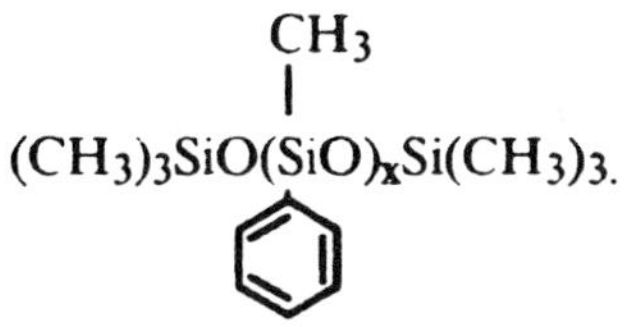

FIG. 5.42—*A trimethylchlorosilane derived silicone copolymer.*

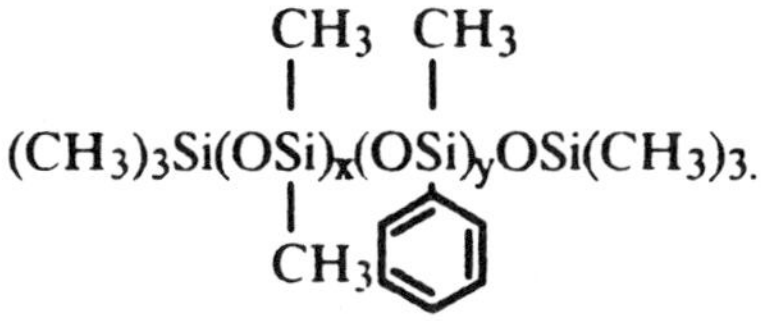

FIG. 5.43—*A dimethyldichlorosilane derived silicone copolymer.*

that for hydrocarbons. The presence of the phenyl moiety produces a gas absorbing fluid, which the dimethyl liquids are not.

5.7 Liquids for Use in Capacitors

Since the banning of PCBs, a substantial amount of effort has gone into developing new liquids which can be used in capacitors [*54*]. In transformers, the major action of the insulating liquid is the removal of heat. In capacitors, the major concern is prevention or extinction of partial discharges. Many liquids, including paraffinic mineral oils, polydimethyl siloxanes, and askarels, have high discharge inception voltages (DIV). This is the voltage stress at which electron injection into the liquid, with gas formation, begins. Both aliphatic and aromatic liquids have relatively high DIVs. Only unsaturated insulating liquids have the ability to extinguish discharges near the inception voltage. It is necessary, in many cases, to lower the voltage only a small percentage in order to reach the discharge extinction voltage (DEV) with aromatic liquids. However, when discharges have been initiated in saturated liquids, the voltage stress has to be lowered from perhaps 2000 V/mil, or more, to only 400 to 500 V/mil before the discharges cease. Because of this characteristic, new capacitor liquids are generally low molecular weight derivatives of benzene. Included in these materials are:

5.7.1 Isopropylbiphenyl (Fig. 5.44)

This material has a boiling point of 270°C, a specific gravity of 1.0 [*54*], a viscosity of nine cSt at 20°C, a density of 0.988, and a relative permittivity of 2.83 [*23*].

5.7.2 Propylbiphenyl (Fig. 5.45)

This material has a boiling point of 280°C.

5.7.3 Methylated Diphenylethane (Fig. 5.46)

5.7.4 Phenylxylylethanes (Fig. 5.47)

5.7.5 Benzyl Neocaprate (Fig. 5.48)

This material has a density of 0.957, a viscosity of 6 cSt at 20°C, and a relative permittivity of 3.75.

5.7.6 Cumylphenylethane (Fig. 5.49)

Cumylphenylethane has a density of 0.959, and viscosity of 9 cSt at 20°C. It has a relative permittivity of 2.60 [55].

5.7.7 Ditolylether (Fig. 5.50)

Ditolylether has a density of 1.035 and a viscosity of 6 cSt at 20°C. It has a relative permittivity of 3.50 [55].

5.7.8 Dioctylphthalate (Fig. 5.51)

This material has a density of 0.98, a viscosity of 80 cSt at 20°C, and a relative permittivity of 5.25 [55].

5.7.9 Phosphate Esters (Fig. 5.52)

At 30°C, the kinematic viscosity of phosphate esters falls between 15 and 30 cSt, while the pour points range from −32.5 to −45°C [56]. The breakdown voltage for a 2.5 mm gap at 25°C is found to be greater than 60 kV. At 80°C the permittivity and dissipation factor values for tri-cresyl, tri-ethylphenyl, and tri-isopropyl phenyl phosphate esters are as follows: 3.7 and 0.005, 4.3 and 0.006, and 4.2 and 0.006, respectively.

5.7.10 Mono/Dibenzyltoluene (Fig. 5.53)

The mixture of mono- and di-benzyltoluene has a viscosity of 6.5 cSt at 20°C and a permittivity of 2.66 at 25° [57]. At 100°C, its resistivity is 9×10^{14} ohm cm and the $\tan\delta$ value is <0.004 at a frequency of 50 Hz. Its breakdown voltage is in excess of 70 kV for a 2 mm gap with the VDE electrode system. The liquid has a gas absorption rate of 2.3 cm^3/h. Its other physical properties are as follows: specific gravity, 1.0 g/cm^3 at 20°C; pour point, <−50°C; flash point, 144°C. It appears to possess good partial discharge extinction characteristics.

In addition, chlorinated benzenes have been considered for use in capacitors. The foregoing examples of the liquid insulants described are intended for use in high stress applications and illustrate the type of structures which commonly are used. Because of the continuing development of new materials, a comprehensive listing is not possible.

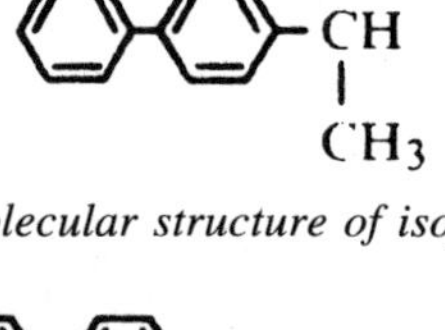

FIG. 5.44—*Molecular structure of isopropylbiphenyl.*

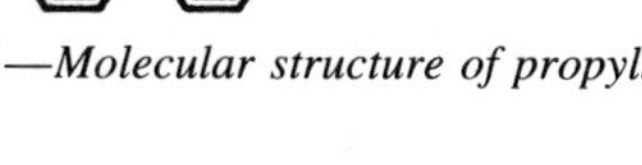

FIG. 5.45—*Molecular structure of propylbiphenyl.*

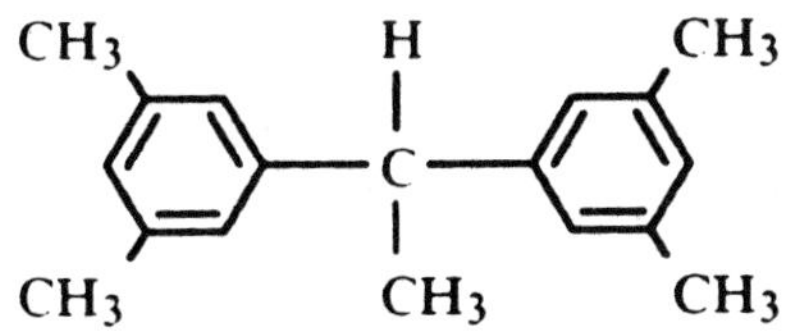

FIG. 5.46—*Molecular structure of methylated diphenylethane.*

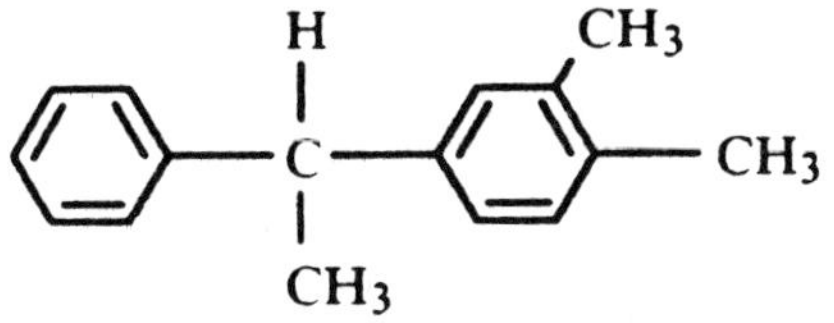

FIG. 5.47—*Molecular structure of phenylxylyethane.*

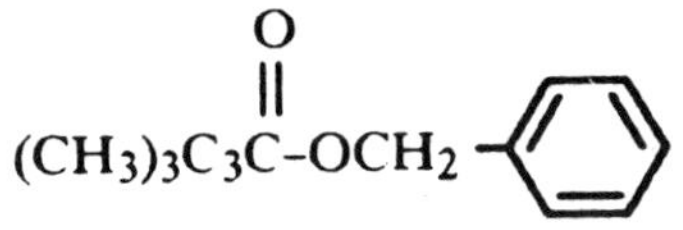

FIG. 5.48—*Molecular structure of benzyl neocaprate.*

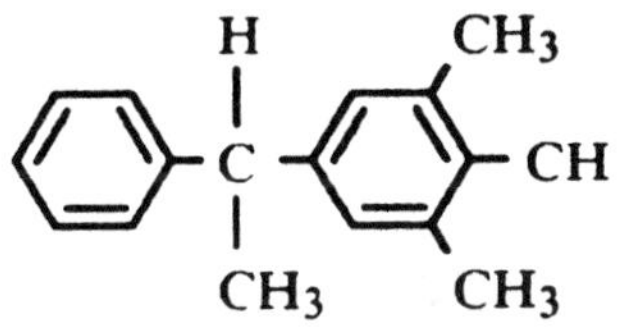

FIG. 5.49—*Molecular structure of cumylphenylethane.*

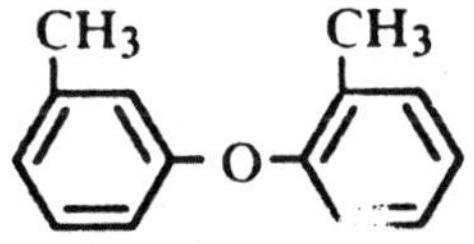

FIG. 5.50—*Molecular structure of ditolylether.*

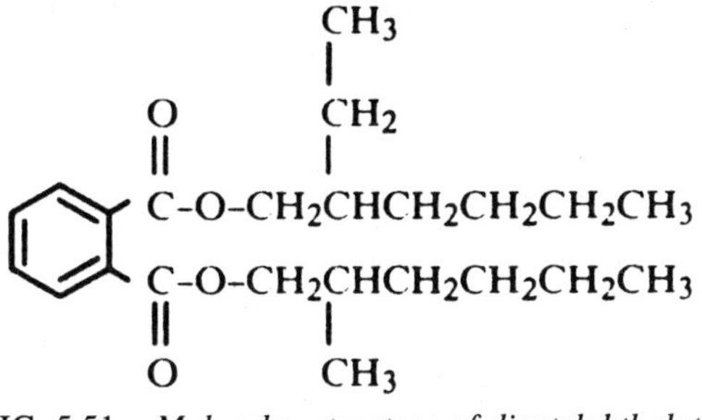

FIG. 5.51—*Molecular structure of dioctylphthalate.*

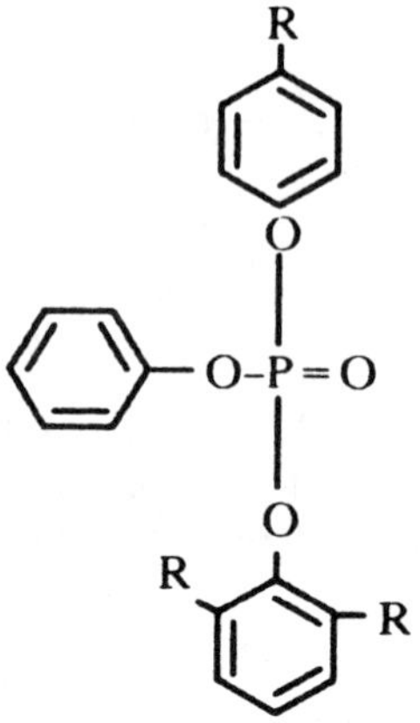

FIG. 5.52—*Molecular structure of phosphate ester.*

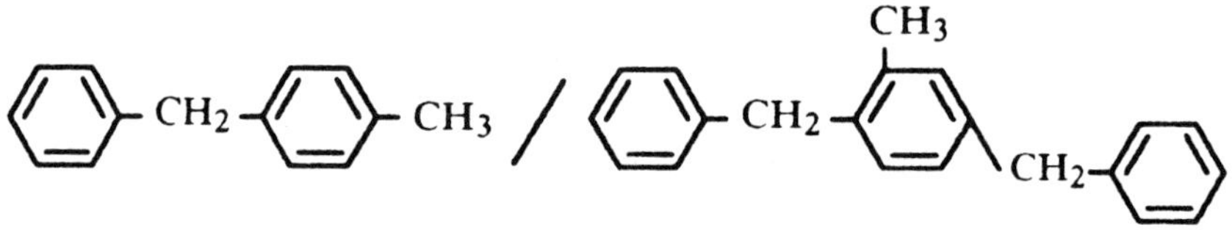

FIG. 5.53—*Molecular structure of mono-dibenzyltoluene (after Burger and Jay* [57]*).*

References

[*1*] Loudon, G. M., *Organic Chemistry,* Addison-Wesley, Reading, MA, 1984.

[2] Salomons, T. W. G., *Organic Chemistry,* Wiley, New York, 1984.

[*3*] Morrison, R. T. and Boyd, R. N., *Organic Chemistry,* Allyn and Bacon, Boston, 1983.

[*4*] Palmer, S. and Sharpley, W. R., "Electrical Strength of Transformer Insulation," *Proceedings,* Institution of Electrical Engineers, Vol. 116, 1969, pp. 2029–2037.

[*5*] Abgrall, F. and Gordon, J. M., "Influence of Solid Impurities on the Electric Strength of Transformer Oils," *V^th^ International Conference on Conditions and Breakdown in Dielectric Liquids,* Noordwijkerhout, Netherlands, Delft University Press, 28–31 July 1975, pp. 79–83.

[*6*] Giao Trinh, N., Olivier, R., Vincent, C., and Régis, J., "Effect of Impurity Particles in Transformer Oil Under Normal Operating Conditions," *1980 International Symposium on Electrical Insulation,* Institute of Electrical and Electronics Engineers Conference Record No. 80CH 1496-9EI, 9–11 June 1980, Boston, pp. 225–228.

[*7*] Salvage, B., "The Electric Breakdown of Some Simple Organic Fluids," *Proceedings,* Institution of Electrical Engineers, Vol. 98, 1951, pp. 15–22.

[*8*] Lewis, T. J., "The Electric Breakdown of Some Simple Organic Fluids," *British Journal of Applied Physics,* Vol. 9, 1958, pp. 30–33.

[*9*] Goodwin, P. W. and MacFadyen, K. A., "Electrical Conduction and Breakdown in Liquid Dielectrics," *Proceedings Physics Society,* Vol. B-66, 1953, pp. 85–97.

[*10*] Sharbaugh, A. H., Crowe, R. W., and Cox, E. B., "Influence of Molecular Structures Upon the Electric Strengths of Hydrocarbons," *Journal of Applied Physics,* Vol. 27, 1956, pp. 806–808.

[*11*] Devins, J. C., Rzad, S. J., and Schwaze, R. J., "The Role of Electronic Processes in the Electrical Breakdown of Liquids," *1978 Annual Report,* Conference on Electrical Insulation Dielectric Phenomena, National Academy of Sciences, Washington, DC, 1978, pp. 182–192.

[*12*] Crowe, R. W., "Formative Time Lags in the Electric Breakdown of Liquid Hydrocarbons," *Journal of Applied Physics,* Vol. 27, 1956, pp. 156–160.

[*13*] Mathes, K. N. and Rouse, T. O., "Influence of Aromatic Compounds in Oil on Pirelli Gassing and

Impulse Surge Breakdown,'' *1975 Annual Report,* Conference on Electrical Insulation and Dielectric Phenomena, National Academy of Sciences, Washington, DC, 1978, pp. 129–140.

[*14*] Hosticka, C., ''Dependence of Uniform/Nonuniform Field Transformer Oil Breakdown on Oil Composition,'' *Transactions on Electrical Insulation,* Institute of Electrical and Electronics Engineers, Vol. EI-14, Feb. 1979, pp. 43–49.

[*15*] Yoshino, K., ''Electrical Conduction and Dielectric Breakdown in Liquid Dielectrics,'' *Transactions on Electrical Insulation,* Institute of Electrical and Electronics Engineers, Vol. EI-21, 1986, pp. 847–853.

[*16*] Schmidt, W. F. and Pugh, D. R., ''On the Polarity Effect of the Breakdown Voltage for Dielectric Liquid in an Inhomogeneous Field,'' *Journal of Physics, D: Applied Physics,* Vol. 10, 1977, pp. 1139–1141.

[*17*] Maruska, H. P., Forster, E. O., and Enard, J. H., ''Electrical Transport Processes in Heavy Hydrocarbon Fluids,'' *Transactions on Electrical Insulation,* Institute of Electrical and Electronics Engineers, Vol. EI-20, 1985, pp. 947–955.

[*18*] Forster, E. O., ''Electronic Conduction in Liquid Hydrocarbons I. Benzene,'' *Journal of Chemistry and Physics,* Vol. 37, 1962, pp. 1021–1028.

[*19*] Forster, E. O., ''Electric Conductance in Liquid Hydrocarbons,'' *Transactions on Electrical Insulation,* Institute of Electrical and Electronics Engineers, Vol. EI-2, 1967, pp. 10–18.

[*20*] Schmidt, W. F., ''Electron Mobility in Nonpolar Liquids: The Effect of Molecular Structure, Temperature and Electric Field,'' *Canadian Journal of Chemistry,* Vol. 55, 1977, pp. 2197–2210.

[*21*] Thomas, W. R. L., ''The Influence of Molecular Structure on the Dielectric Strength of Insulating Liquids,'' *Conference on Dielectric Materials, Measurements, and Applications,* 21–25 July, 1975, Cambridge, UK, pp. 311–314.

[*22*] Smyth, C. P., *Dielectric Behavior and Structure,* McGraw Hill, New York, 1955.

[*23*] Bradwell, A., Ed., *Electrical Insulation,* Peter Peregrinus Ltd., London, 1983.

[*24*] Minker, V. J., Osipov, O. A., and Zhdanov, Y. A., *Dipole Moments in Organic Chemistry,* Plenum Press, New York, 1970.

[*25*] Azaroff, V. L. and Brophy, J., *Journal of Electronic Process in Materials,* McGraw-Hill, New York, 1963.

[*26*] Hill, N., Vaughan, E., Worth, E., Price, A. H., and Davies, M., *Dielectric Properties and Molecular Behavior,* Van Nostrand Reinhold, London, 1969.

[*27*] Harrop, P. J., *Dielectrics,* John Wiley, New York, 1972.

[*28*] Sugden, T. M., Ed., *Dielectric Properties and Molecular Behaviors,* Van Norstrand Reinhold, London, 1969.

[*29*] Kurtz, S. S., King, R. W., Stout, W. J., Partikian, D. G., and Skrabek, E. A., ''Relationship between Carbon-type Composition, Viscosity-Gravity Constant and Refractivity Intercept of Viscous Fractions of Petroleum,'' *Analytical Chemistry,* Vol. 28, 1956, pp. 1928–1936.

[*30*] Barry, A. J., ''Viscometric Investigation of Dimethylsiloxane Polymers,'' *Journal of Applied Physics,* Vol. 17, 1946, pp. 1020–1024.

[*31*] Wilson, A. C. M., *Insulating Liquids: Their Uses, Manufacture, and Properties,* Peter Peregrinus, Stevenage, UK, and New York, 1980.

[*32*] Mair, B. J. and Rossini, F. D., *Science of Petroleum,* Vol. 5, Part I, B. T. Brooks and A. E. Dunstan, Eds., Oxford University Press, London, 1950.

[*33*] Van Nes, K. and Van Western, H. A., *Aspects of the Constitution of Mineral Oils,* Elsevier, New York, 1951.

[*34*] Hobson, G. D. and Pohl, W., *Modern Petroleum Technology,* Applied Science, Publishers, London, 1973.

[*35*] Smith, N. A. C., *U.S. Bureau of Mines Report,* 2806, Washington, DC, 1927.

[*36*] Lane and Garton, *U.S. Bureau of Mines Report,* 3279, Washington, DC, 1935.

[*37*] Gruse, W. A. and Stevens, D. R., *Chemical Technology of Petroleum,* McGraw-Hill, New York, 1960.

[*38*] Kurtz, S. S. and Martin, C. C., ''The Hydrocarbon Composition of Rubber Processing Oils,'' *India Rubber World,* Vol. 126, 1952, pp. 495–499.

[*39*] Lipkin, M. R., Sankin, A., and Martin, C. C., ''Double Bond Index. A Correlation Useful for Classification of Aromatics and Olefins in Petroleum Fractions,'' *Analytical Chemistry,* Vol. 20, 1948, pp. 598–606.

[*40*] Martin, C. C. and Sankin, A., ''Determination of Aromatic and Naphthene Rings in Aromatics from Petroleum Fractions,'' *Analytical Chemistry,* Vol. 25, 1953, pp. 206–214.

[*41*] Dimeler, G. R., Mills, I. W., and Melchiore, J. J., ''The Scope of Hydrogeneration as a Refining Tool for the Manufacture of Transformer Oils,'' *Transactions on Electrical Insulation,* Institute of Electrical and Electronics Engineers, Vol. EI-4, 1969, pp. 7–12.

[*42*] ASTM D 3487, ''Standard Specification Mineral Insulating Oil Used in Electrical Apparatus,'' American Society for Testing and Materials, Philadelphia.

[*43*] Elias, H.-G., *Macromolecules. 2. Synthetics and Materials,* Plenum Press, New York, 1977, pp. 844–845.

[*44*] Clark, F. M., *Insulating Materials for Design and Engineering Practice,* John Wiley, New York, 1962.

[*45*] Saums, H. L. and Pendleton, W. W., *Materials for Electrical Insulating and Dielectric Functions,* Hayden, Rochelle Park, NJ, 1973.

[46] Ramu, T. S., "On the High Frequency Dielectric Behavior of Castor Oil," *Transactions on Electrical Insulation,* Institute of Electrical and Electronics Engineers, Vol. EI-14, 1979, pp. 136–144.
[47] Sillars, R. W., *Electrical Insulating Materials and Their Applications,* Peter Peregrinus Ltd., Stevenge, UK, 1969.
[48] Bartnikas, R. and Srivastava, K. D., Eds., *Power Cable Engineering,* Sandford Educational Press, Waterloo, Ont., Canada, 1987.
[49] RTE Corporation, Product Literature for RTEmp Fluids.
[50] Surrey, A. R., *Name Reactions in Organic Chemistry,* Academic Press, New York, 1961.
[51] *Insulation/Circuits,* Encyclopedia Issue, June/July 1972.
[52] Eaborn, L., *Organosilicon Compounds,* Butterworth, London, 1960.
[53] Gunderson, R. C. and Hart, A. W., Eds., *Synthetic Lubricants,* Reinhold, New York, 1962.
[54] *Toxicity Profiles of PCB Substitutes,* EPRI EA 3567, Project 2378-8, Electric Power Research Institute, Palo Alto, CA, July 1984.
[55] Kron, R., "New Non-Chlorinated Insulating Liquids as Impregnants for Capacitors," *Eighth International Conference on Conductance Breakdown Dielectrics Liquids,* Molinari, G. and Viviani, A., Eds., Conference Record No. 84CH2055-2, Institute of Electrical and Electronics Engineers, 24–27 July 1984, Pavia, pp. 426–432.
[56] Kamata, Y., "New Liquid Insulating Materials," *Transactions on Electrical Insulation,* Institute of Electrical and Electronics Engineers, Vol. EI-21, 1986, pp. 929–931.
[57] Burger, N. and Jay, P., "A New Impregnant for HV Power Capacitors," *Transactions on Electrical Insulation,* Institute of Electrical and Electronics Engineers, Vol. EI-21, 1986, pp. 59–63.

Appendix I

Units and Equivalents

Area

1 square metre (m^2)	$= 10^4$ cm^2 $= 10^6$ mm^2
1 square centimetre (cm^2)	$= 0.1550$ in.^{2} $= 1.076 \times 10^{-3}$ ft^2
1 circular mil	$= 5.067 \times 10^{-6}$ cm^2 $= 7.854 \times 10^{-7}$ in.^{2}

Length

1 metre (m)	$= 39.370$ in. $= 3.9370 \times 10^4$ mil
	$= 10^{10}$ Angstrom unit ($\mathring{A}$)
1 millimetre (mm)	$= 40$ mil $= 10^3$ micrometer (μm) $= 10^7$ $\mathring{A}$
1 mil	$= 25.4$ μm
1 inch (in.)	$= 1000$ mil $= 2.54$ cm
1 centimetre (cm)	$= 10^8$ $\mathring{A}$
1 μm	$= 10^4$ $\mathring{A}$

Density

1 kilogram per cubic metre (kg/m^3)	$= 10^{-3}$ g/cm^3 $= 0.0624$ lb/ft^3
1 g/cm^3	$= 10^3$ kg/m^3 $= 62.43$ lb/ft^3 $= 0.03613$ lb/in.^{3}

Dipole Moment

1 debye (D)	$= 10^{-18}$ electrostatic unit (ESU) $= 3.33 \times 10^{-30}$ C m

Force

1 newton (N)	$= 10^5$ dyne $= 1$ kg m/s^2 $= 7.233$ poundal

Volume

1 cubic metre (m^3)	$= 10^6$ cm^3 $= 10^9$ mm^3
1 cubic centimetre (cm^3)	$= 6.1024 \times 10^{-2}$ in.^{3}
1 litre	$= 1000$ cm^3 $= 0.2642$ gal $= 1000$ mL $= 10^{-3}$ m^3
1 gallon	$= 231$ in.^{3}

Mass

1 kilogram (kg)	$= 1000$ gram (g) $= 2.2046$ lb
1 gram (g)	$= 2.2046 \times 10^{-3}$ lb $= 3.5273 \times 10^{-2}$ oz

Pressure

1 pascal (Pa)	$= 1\ N/m^2 = 1 \times 10^{-5}$ bar $= 9.869 \times 10^{-6}$ atm
	$= 7.501 \times 10^{-3}$ torr $= 1.450 \times 10^{-4}$ psi $= 10\ dyne/cm^2$
1 atmosphere (atm)	$= 760$ mm Hg (0°) $= 14.7$ psi $= 1.033\ g/cm^3$

Energy

1 joule (J)	$= 1$ N·m $= 10^7$ erg $= 0.239$ cal $= 0.1020$ kg·m
1 electron volt (eV)	$= 1.6 \times 10^{-19}$ J $= 1.6 \times 10^{-12}$ erg
	$= 23.06$ kcal/mol
1 calorie (cal)	$= 3.9683 \times 10^{-3}$ Btu
1 British thermal unit (Btu)	$= 1.055 \times 10^3$ J $= 778$ ft·lb
1 foot-pound (ft·lb)	$= 1.355$ J
1 watt-hour	$= 3.6 \times 10^3$ J
1 watt-second (Ws)	$= 1$ J
1 million electron volts (MeV)	$= 1.602177 \times 10^{-13}$ J

Absolute Viscosity

1 poise (P)	$= 1\ dyne\ s/cm^2 = 0.1$ Pa s
1 centipoise (cP)	$= 0.01$ P

Kinematic Viscosity

1 stoke (St)	$= 1\ P\ cm^3/g = 1\ cm^2/s$
1 centistoke (cSt)	$= 0.01$ St

Power

1 horsepower	$= 746$ W $= 746$ J/s
1 W	$= 0.001$ kW $= 0.738$ ft·lb/s $= 0.239$ cal/s

Capacitance

1 farad (F)	$= 9 \times 10^{11}$ ESU (statfarad)
	$= 1 \times 10^6$ pF $= 1 \times 10^9$ nF

Charge

1 coulomb (C)	$= 3 \times 10^9$ ESU (statcoulomb)
	$= 1 \times 10^6$ pC $= 1 \times 10^9$ nC

Current

1 ampere (A)	$= 3 \times 10^9$ ESU (statampere)
	$= 10^3$ mA $= 10^6\ \mu A$

Resistance

1 ohm (Ω)	$= 0.11 \times 10^{-11}$ ESU (statohm)
	$= 10^6\ \mu\Omega = 10^{-6}\ M\Omega$

Voltage

1 volt (V)	= 0.0033 ESU (statvolt)
	= 10^3 mV = 10^6 μV

Electric Stress

1 megavolt per metre (MV/m)	= 10 kV/cm = 1 kV/mm = 25.4 V/mil

Appendix II

General Physical Constants

Atomic mass unit (amu)		$= 1.6605402 \times 10^{-24}$ g
Avogadro's number	N_A	$= (6.0228 \pm 0.0011) \times 10^{23}$/mol
		$= (6.0228 \pm 0.0011) \times 10^{26}$/kmol
Loschmidt's number	N_l	$= 2.687 \times 10^{25}$ m^{-3}
Gas Constant	R	$= (8.31436 \pm 0.00038) \times 10^{7}$ erg/deg mol
		$= (8.31436 \pm 0.00038) \times 10^{3}$ J/K mol
Boltzmann constant	k	$= R/N_A = (1.38047 \pm 0.00026) \times 10^{-16}$ erg/K
		$= (1.38047 \pm 0.00026) \times 10^{-23}$ J/K
Electronic charge	e	$= (4.8025 \pm 0.001) \times 10^{-10}$ ESU
		$= 1.60202 \times 10^{-19}$ C
Electron volt	eV	$= 1.60202 \times 10^{-19}$ J
Rydberg constant	R_∞	$= 1.0974 \times 10^{7}$ m^{-1}
		$= 1.0974 \times 10^{5}$ cm^{-1}
Madelung constant	α_M	$= 1.748$ (for NaCl)
		$= 4.80$ (for TiO_2)
Faraday constant	F	$= N_\alpha e = 96.501 \pm 10$ C/mol
Mass of electron	M_e	$= (9.1066 \pm 0.0032) \times 10^{-28}$ g
Hydrogen atom	M_H	$= (1.67339 \pm 0.0031) \times 10^{-24}$ g
Proton	M_p	$= (1.67248 \pm 0.0031) \times 10^{-24}$ g
α particle	M_α	$= (6.6442 \pm 0.0012) \times 10^{-24}$ g
Ratio of mass hydrogen atom to mass electron	M_H/M_e	$= (1.8375 \pm 0.5) \times 10^{3}$
Mechanical equivalent of heat	J_o	$= 4.1855 \pm 0.0004$ J/cal
Permittivity of free space	ϵ_o	$= 1$ ESU
		$= 8.854 \times 10^{-14}$ F/cm
		$= 8.854 \times 10^{-12}$ F/m
		$= 8.854 \times 10^{-12}$ C^2/C m^2
Planck's constant	h	$= (6.624 \pm 0.002) \times 10^{-27}$ erg s
		$= (6.624 \pm 0.002) \times 10^{-34}$ J s
		$= 4.1357 \times 10^{-15}$ eV s
Permeability of free space	μ_o	$= 4\pi \times 10^{-7}$ N/A^2
		$= 4\pi \times 10^{-7}$ T m/A
		$= 4\pi \times 10^{-7}$ H/m
Velocity of light	c	$= (2.99776 \pm 0.00004) \times 10^{10}$ cm/s
		$= (2.99776 \pm 0.00004) \times 10^{8}$ m/s

Appendix III

Greek Alphabet

Alpha	A, α	Nu	N, ν
Beta	B, β	Xi	Ξ, ξ
Gamma	Γ, γ	Omicron	O, o
Delta	Δ, δ	Pi	Π, π
Epsilon	E, ϵ	Rho	P, ρ
Zeta	Z, ζ	Sigma	Σ, σ
Eta	H, η	Tau	T, τ
Theta	Θ, θ	Upsilon	Υ, υ
Iota	I, ι	Phi	Φ, ϕ
Kappa	K, κ	Chi	X, χ
Lambda	Λ, λ	Psi	Ψ, ψ
Mu	M, μ	Omega	Ω, ω

Appendix IV

Prefixes of the SI System of Units

Multiplication Factor	Prefix	Symbol in SI System
1 000 000 000 000 = 10^{12}	tera	T
1 000 000 000 = 10^{9}	giga	G
1 000 000 = 10^{6}	mega	M
1 000 = 10^{3}	kilo	k
100 = 10^{2}	hecto	h
10 = 10^{1}	deka	da
0.1 = 10^{-1}	deci	d
0.01 = 10^{-2}	centi	c
0.001 = 10^{-3}	milli	m
0.000 001 = 10^{-6}	micro	μ
0.000 000 001 = 10^{-9}	nano	n
0.000 000 000 001 = 10^{-12}	pico	p
0.000 000 000 000 001 = 10^{-15}	femto	f
0.000 000 000 000 000 001 = 10^{-18}	atto	a

Author Index

C

I

J

K

L

Subject Index

B

C

K

L

M

T